Habitat Suitability and Distribution Models
With Applications in R

This book introduces the key stages of niche-based habitat suitability model building, evaluation and prediction required for understanding and predicting future patterns of species and biodiversity. Beginning with the main theory behind ecological niches and species distributions, the book proceeds through all major steps of model building, from conceptualization and model training to model evaluation and spatio-temporal predictions. Extensive examples using R support graduate students and researchers in quantifying ecological niches and predicting species distributions with their own data, and help to address key environmental and conservation problems. Reflecting this highly active field of research, the book incorporates the latest developments from informatics and statistics, as well as using data from remote sources such as satellite imagery. A website at www.unil.ch/hsdm contains the codes and supporting material required to run the examples and teach courses.

All three authors are recognized specialists of and have contributed substantially to the development of spatial prediction methods for species' habitat suitability and distribution modeling. They published a large number of papers, overall cumulating tens of thousands of citations, and are ISI Highly Cited Researchers.

ANTOINE GUISAN is Professor at the University of Lausanne, Switzerland, where he leads the ECOSPAT Spatial Ecology group. Besides being a specialist in habitat suitability and distribution models, his interests also include ecological niche dynamics in space and time, community and multitrophic modeling, very high resolution spatial modeling in mountain environments, and applications of models to environmental decision-making and transfer of scientific knowledge to society.

WILFRIED THUILLER is a senior scientist at the National Center for Scientific Research, Laboratory of Alpine Ecology in Grenoble, France. Besides being a specialist in habitat suitability and distribution models, his interests include macroecology, macroevolution, conservation, biodiversity modeling with both mechanistic and phenomenological models, community ecology, functional ecology, and ecosystem functioning in alpine environments.

NIKLAUS E. ZIMMERMANN is a senior scientist and directorate member of the Swiss Federal Research Institute WSL, and an adjunct professor at the Swiss Federal Institute of Technology ETH. Besides being a specialist in habitat suitability and distribution models, his interests include macroecology, macroevolution, biodiversity and community modeling using both empirical and mechanistic approaches, as well as conservation and applied biodiversity management support.

ECOLOGY, BIODIVERSITY AND CONSERVATION

Series Editors
Michael Usher   *University of Stirling, and formerly Scottish Natural Heritage*
Denis Saunders   *Formerly CSIRO Division of Sustainable Ecosystems, Canberra*
Robert Peet   *University of North Carolina, Chapel Hill*
Andrew Dobson   *Princeton University*

Editorial Board
Paul Adam   *University of New South Wales, Australia*
H. J. B. Birks   *University of Bergen, Norway*
Lena Gustafsson   *Swedish University of Agricultural Science*
Jeff McNeely   *International Union for the Conservation of Nature*
R. T. Paine   *University of Washington*
David Richardson *University of Stellenbosch*
Jeremy Wilson   *Royal Society for the Protection of Birds*

The world's biological diversity faces unprecedented threats. The urgent challenge facing the concerned biologist is to understand ecological processes well enough to maintain their functioning in the face of the pressures resulting from human population growth. Those concerned with the conservation of biodiversity and with restoration also need to be acquainted with the political, social, historical, economic and legal frameworks within which ecological and conservation practice must be developed. The new Ecology, Biodiversity, and Conservation series will present balanced, comprehensive, up-to-date, and critical reviews of selected topics within the sciences of ecology and conservation biology, both botanical and zoological, and both "pure" and "applied". It is aimed at advanced final-year undergraduates, graduate students, researchers, and university teachers, as well as ecologists and conservationists in industry, government and the voluntary sectors. The series encompasses a wide range of approaches and scales (spatial, temporal, and taxonomic), including quantitative, theoretical, population, community, ecosystem, landscape, historical, experimental, behavioral, and evolutionary studies. The emphasis is on science related to the real world of plants and animals rather than on purely theoretical abstractions and mathematical models. Books in this series will, wherever possible, consider issues from a broad perspective. Some books will challenge existing paradigms and present new ecological concepts, empirical or theoretical models, and testable hypotheses. Other books will explore new approaches and present syntheses on topics of ecological importance.

Ecology and Control of Introduced Plants
Judith H. Myers and Dawn Bazely

Invertebrate Conservation and Agricultural Ecosystems
T. R. New

Risks and Decisions for Conservation and Environmental Management
Mark Burgman

Ecology of Populations
Esa Ranta, Per Lundberg, and Veijo Kaitala

Habitat Suitability and Distribution Models

With Applications in R

ANTOINE GUISAN
University of Lausanne

WILFRIED THUILLER
CNRS, Université Grenoble Alpes

NIKLAUS E. ZIMMERMANN
Swiss Federal Research Institute WSL

With contributions from

VALERIA DI COLA
University of Lausanne

DAMIEN GEORGES
CNRS, Université Grenoble Alpes

ACHILLEAS PSOMAS
Swiss Federal Research Institute WSL

CAMBRIDGE
UNIVERSITY PRESS

CAMBRIDGE
UNIVERSITY PRESS

University Printing House, Cambridge CB2 8BS, United Kingdom

One Liberty Plaza, 20th Floor, New York, NY 10006, USA

477 Williamstown Road, Port Melbourne, VIC 3207, Australia

314-321, 3rd Floor, Plot 3, Splendor Forum, Jasola District Centre, New Delhi - 110025, India

79 Anson Road, #06-04/06, Singapore 079906

Cambridge University Press is part of the University of Cambridge.

It furthers the University's mission by disseminating knowledge in the pursuit of education, learning and research at the highest international levels of excellence.

www.cambridge.org
Information on this title: www.cambridge.org/9780521758369
DOI: 10.1017/9781139028271

First published 2017

A catalogue record for this publication is available from the British Library

ISBN 978-0-521-76513-8 Hardback
ISBN 978-0-521-75836-9 Paperback

Contents

Color plates can be found between pages 238 and 239

Foreword

As society confronts the impact of human population growth and the increasing threat of climate change on biodiversity and species extinction, ecology is becoming central to discussions of how society can respond.

Heavy demands will be placed on ecologists to help provide possible solutions to environmental problems. Greater collaboration, communication, and synthesis of ideas and goals will be required among ecologists, conservationists, land-use planners, and politicians if progress is to be made. There are similar needs for increased collaboration and synthesis of new ideas among ecologists, particularly those from sub-disciplines directly related to the current environmental crisis.

Habitat suitability modeling, the subject of this book, is one such area of research. This research area is central to understanding the variables determining species distribution and predicting the likely response of the world's biota to climate change and disturbance. The last 20 years has seen a remarkable expansion in understanding of the conceptual framework, methodology for, and prediction of species distribution. The three authors of this book are key contributors to these advances.

Today, progress is dependent on researchers having a wide range of knowledge, skills, and awareness in numerous areas in addition to basic biological and ecological knowledge. These include

- Conceptual framework of ecological theory
- Data measurement and sampling
- Statistical analysis
- Relational data bases
- Geographic information systems
- Computing skills

The importance of the digital environment today in combining these areas to address ecological questions is self-evident. However, there is no substitute for experience in using these approaches. The contents demonstrate the need to evaluate each step in the process of habitat suitability modeling in terms of the technical options available and the ecological assumptions each implies. Of equal importance is the provision of R code procedures making available to many the details of how to implement different methods and compare their performance. This book provides an invaluable learning experience for all ecologists interested in habitat suitability modeling based as it is on the combined experience of three of the pioneers of many of the ideas presented.

Mike Austin
CSIRO Land & Water Flagship, Canberra, Australia

Preface

Why This Book and How Should It Be Read?

This book explains how to fit habitat suitability models (HSMs) in ecology to quantify species' ecological niches and predict species distributions. It is a textbook with practical examples in R, intended for readers who want to familiarize themselves with this field. The book does not presume to review the whole field, as it is developing extremely rapidly, but rather to explain the main steps of model building, evaluation, and application in a simple and illustrated way. It should thus prove useful for master or doctoral courses, or specialized workshops, and help researchers unfamiliar with this field to analyze their own data. We advise that readers start their reading with Chapter 1, which explains in more detail what the book is about.

Acknowledgments

Our warmest thanks go to:Valeria Di Cola, Damien Georges, and Achilleas Psomas for their invaluable efforts in scripting much of the example material and other technical contributions to this book, Valeria Di Cola for her central role in coordinating the book progress, Mike Austin for his critical review of Part I and for writing the Foreword, Joana Vicente, Olivier Broennimann, Manuela D'Amen, and Rui Fernandes for additional comments on Part I, Brody Sandel for reviewing Part II, Jane Elith and Manuela D'Amen for reviewing Part III, Tom Edwards and Luigi Maiorano for reviewing Part IV, Laura Pollock for reviewing Part V, Daniel Scherrer for his careful review of several codes, and Joana again for the Glossary. We are grateful to Hamid Taleshi for carefully reading the whole manuscript, and to Lidong Mo for taking care of checking all the code and content of the book. We also thank our families and friends for their invaluable support, patience, and trust while we were writing this book, most often during our spare time. We would also like to thank our respective research groups and close colleagues in our institutions for their great support and enthusiasm, as such outcome would not be possible without the stimulating environment each of us had in his academic environment. AG, WT, and NEZ would like to add very special thanks to VD for her invaluable support and dedication during the last two years in pushing and allowing us to finish this book.

Authors' Contributions

AG, WT, and NEZ conceived and designed the book and wrote all the main texts. AG led the book development with large support by VD in the last four years. VD acted as the book manager to coordinate all contributions and scripts. AG led the writing of Parts I and IV, NEZ led Part II, WT led Parts III and VI, and all three led Part V. AP and NEZ developed all script examples in Part II, DG and WT in Parts III and VI, VD and AG in Part IV, and DG, VD, AP, NEZ, WT, and AG in Part V.

Introduction

Here, we present the main features of the book: its aims, structure, content, terminology, readership, supporting material, expected pre-requisites, and how it differs from other books already available. The structure of the book follows the main modeling steps, so we recommend reading the sections about the book's structure and content before reading the other parts.

1 · *General Content of the Book*

1.1 What Is This Book About?

This book aims to present a class of models used to quantify taxa–environment relationships (i.e. taxa's habitat suitability or ecological niche), and how these can be used to predict the geographical distribution of taxa. This book is also about understanding the drivers behind those distributions. This book mainly focuses on modeling the distribution of single species, yet the concept of a species itself is increasingly being called into question (especially in some groups). It is therefore important to note that these approaches can also be used to model other biological entities, for instance, genes, haplotypes, or clades, within or across species. Communities, ecosystems, or biomes can also be modeled using these approaches, directly as fixed entities or by assembling individual species predictions. Further examples of modeling these new entities are rapidly being added to the literature. This is why we prefer to use the term "habitat suitability modeling" (see below), as it does not solely refer to modeling species, but more generally modeling suitable conditions for any given biological entity. For the sake of simplicity, throughout this book we will remain at species level for all examples and explanations, but most of the techniques, metrics, and approaches presented can be directly applied to almost any other modeling entity. We also mainly discuss distributions in terms of simple occurrences or presence–absence of species, as these are the most commonly available data, but much of the reasoning can be extended to use abundances when available, provided the appropriate statistical methods are also used.

1.2 How Is the Book Structured?

The book is made of seven main parts, including this introduction. The first five each address one or several of the modeling steps described in Chapter 2 (see also Guisan and Zimmermann, 2000; Austin, 2002),

Part VI introduces the tools and datasets used in this book, with two developed examples, and Part VII provides some conclusions and perspectives. The seven main parts are as follows:

- Part I (Chapters 2–5) presents the theoretical framework on which these models are built, i.e. what drives species distributions and how to use it when building models.
- Part II (Chapters 6–8) presents the data necessary to build HSMs, i.e. what we would need theoretically, what is available or what should be sampled, on what scale, how to prepare or sample them, and what associated problems might occur.
- Part III (Chapters 9–14) presents a representative sample of the statistical techniques commonly used to model habitat suitability, how they work, and how to parameterize them in R.
- Part IV (Chapters 15–16) presents ways of evaluating the models fitted in Part III, presenting different approaches to keep some data "independent", and the metrics used to compare predictions with observations.
- Part V (Chapter 17) presents ways of deriving spatial predictions and projections in time and space from the fitted models, with the associated uncertainties and additional assumptions specific to model transferability.
- Part VI (Chapters 18–19) presents the data and tools used in this book, as well as two developed case studies, all freely available and explained. Chapter 18 presents an overview of all data and tools used throughout the book, but because many of the model-fitting examples are developed with the biomod2 R package. Chapter 19 develops two illustrative case studies specifically with biomod2, and in particular ensemble models using a variety of techniques.
- Part VII (Chapter 20) provides a short conclusion and some perspectives. HSM research has become a very active field, with fast internal dynamics, and there is still much to be done to make it a more mature field. We suggest some directions that we think are interesting and worth pursuing in this regard.

1.3 Why Write a Textbook with R Examples?

The book is conceived both as a textbook to teach the main modeling steps, but also as a practical guide to run these models in R.[1] The choice of R is deliberate here, as it is freeware available on all platforms, and is now by far

---

[1] www.r-project.org

the most commonly used statistical software for building HSMs, and more generally, for running ecological analyses of niches and distributions. The book is richly illustrated with examples and guidelines for building HSMs in R from sampling design and data construction to model building, testing, projections and interpretation. We have attempted to build all examples from publicly available information, which can be partly retrieved directly *from within* R (official packages), or at least from internet downloads.

1.4 What Is This Book Not About?

This book is not meant to be an exhaustive review of all the recent developments in the field of habitat suitability modeling *sensu lato* (see terminology issues below), which is best found in review papers or other work (e.g. Franklin, 2010a). It is not intended to provide detailed explanations of standard statistical methods (best found in statistical textbooks), or of standard geographic analyses (best found in geography and geographic information systems textbooks), nor does it provide a detailed review of fundamental knowledge in ecology and biogeography (best found in the relevant textbooks).

1.5 Why Was This Book Needed?

The twenty-first century is experiencing a major biodiversity crisis and our planet now faces the risk of a sixth – human-driven – major species extinction. Major threats currently include pollution, habitat destruction, modification and fragmentation, whereas future threats also include biological invasions and climate change. Such threats to species, biodiversity, and ecosystems are usually estimated using multi-scale assessments based on International Union for Conservation of Nature (IUCN) geographic criteria (e.g. geographic extent and/or area of occupancy). Therefore, sound approaches and methods are needed to forecast the future distribution of life on Earth. At the same time, science is progressing fast. The last century was paralleled by major technological advances, especially in bioinformatics and biomathematics. Not only have we improved our knowledge and understanding of the living world, and particularly what drives species distributions, but we have also improved our ability to model and predict it, with applications in evolutionary biology, biogeography and conservation biology. As a result, interest in predictive HSMs of organisms has grown dramatically during the twenty-first century. The reasons for this interest are twofold. First, the last two decades witnessed an exponential increase in computing power, with the advent of geographic information systems and remote-sensing technologies leading to the development of high-resolution environmental datasets, which together gave greater freedom to analyse, model, and predict a large number of taxa over various spatial scales (from the very local to global). Second, boosted by these numerical developments and with no other approaches applicable to large number of species in many different regions or over large geographic extents, HSMs have emerged as a vital tool in applied ecological and environmental sciences. They have proved particularly useful for evaluating the potential impact of global anthropogenic environmental changes on biodiversity and ecosystems (e.g. climate change, biological invasions, habitat destruction; (Johnson and Gillingham, 2005; Rodríguez et al., 2007; Franklin, 2010b; Schwartz, 2012; Guisan et al., 2013; Tulloch et al., 2016), as demonstrated by their key role in the first assessments by the Intergovernmental Panel on Biodiversity and Ecosystem Services (IPBES[2]). The strong societal need and political pressure to address these environmental problems have consolidated the role of HSMs in conservation sciences, but also require the formalization of the modeling methods used for predictions. Consequently, HSMs have also been increasingly commonly used in fundamental ecological and evolutionary sciences, to improve our understanding of species' ecological niches or to test biogeographic hypotheses (see more examples in Scott et al., 2002; Guisan and Thuiller, 2005; Franklin, 2010a; Peterson et al., 2011). This book can thus be seen as an additional contribution to establish modeling standards in basic and applied HSM research.

1.6 Who Is This Book For?

We have written this book for advanced students for use as textbook in university classes, at the level of third-year bachelor, masters, and doctoral studies, or for any scientist who wants to familiarize themselves with the principles and methods of habitat suitability modeling and species distribution predictions.

1.7 Where Can I Find Supporting Material?

To facilitate the learning process and transfer of technology to real case problems, supporting material (scripts, data, and manuals) is available on the companion website to the book.[3]

---

[2] www.ipbes.org
[3] www.unil.ch/hsdm

1.8 What Are Readers Assumed to Know Already?

We assume that readers have basic understanding of ecological theory, biogeography, and macroecology, in biostatistics – such as multivariate statistics, inference tests, bootstrapping, etc. – and in GIS and spatial analyses – such as geodata handling, spatial interpolation, and spatial autocorrelation. We also assume that readers have a grasp of the R environment.[4]

1.9 How Does This Book Differ From Previous Ones?

HSMs have become very popular and have undergone tremendous development in the past 10 to 15 years, yet there are still very few books on the subject. Our book is an ideal complement to the two most recent works published: Franklin (2010a) and Peterson et al. (2011). Franklin (2010a) is an excellent monograph on these models, but it does not include the codes explaining how to practically run the different analyses. Peterson et al. (2011) is a multi-authored book on species' ecological niches and geographic distributions that provides a series of chapters on specific topics, with some detailed examples, but no code for running the analyses. Other books are more miscellaneous, by having each chapter contributed by different authors, and a large number of authors overall (e.g. Corsi et al., 2000; Scott et al., 2002).

Our book has three unique features, namely:

1. It is a textbook which can be used to teach classes in universities, technical high schools or other courses. It follows a simple sequential structure, introducing the theoretical concepts, data preparation, model building, model evaluation, and spatial predictions, which thus follows the logical successive modeling steps.
2. It is unique in providing fully developed examples and practical case studies that can be run in the R language, using the most advanced tools available in R to model habitat suitability and niches, and to predict species distributions.
3. It comes with a companion website,[5] where additional resources can be found, regularly updated, and discussed in an online forum.

---

[4] www.r-project.org
[5] www.unil.ch/hsdm

It is therefore a more practical book, which – we hope! – will allow interested students and scientists to quickly get off to a good start in the field.

1.10 What Terminology Is Used in This Book?

Many different names and acronyms are used in the literature for the same class of models (see also Glossary): habitat suitability models (HSM), the term mainly used throughout this book, but also ecological niche models (ENM), species distribution models (SDMs), habitat distribution models (HDM), climate-envelope models (CEM), resource selection functions (RSF), and many other more minor variants. It appears that all the models in this class can be used to investigate both species niches (i.e. niche modeling, NM) and species distributions (i.e. distribution modeling, DM).

Although we recognize that all terms can be used depending on the study context, we decided to use the HSM terminology in the title and throughout the book because:

- it best reflects the basis of what all these models do: quantifying the species–environment (habitat) relationship;
- not all applications of these models need to predict the geographic distribution of the modeled entity;
- it is more generally applicable to entities other than species (e.g. genes or communities), to which the niche concept may not apply;
- it is still applicable to cases when a niche is only partially captured in a model, i.e. when an envelope of suitable habitats is modeled, but possibly not the full species niche, as typically observed in studies confined to geographic extents smaller than the full distribution range of the modeled species.

There are also issues of scale and hierarchy in the use of these different terms. As we will see in Chapters 2 and 3, the environmental niche is the envelope of all suitable habitats for a species across its whole range. All the conditions the species can withstand therefore needs to be encompassed to capture the whole species niche in a model.

The main terms used throughout the different parts are defined in the glossary at the end of the book.

PART I · Overview, Principles, Theory, and Assumptions Behind Habitat Suitability Modeling

In this first part of the book, we begin by briefly presenting the general procedure (i.e. the series of methodological steps) used to build and apply HSMs (Chapter 2). We next summarize our ecological and evolutionary understanding of the factors driving species distributions and related biogeographical theory (Chapter 3). It is by no means our intention to present an exhaustive review of all existing theories, which can best be found in textbooks (Lomolino et al., 2010; Smith and Smith, 2015), but rather to focus on the most useful concepts for HSMs. Readers familiar with the theory behind species' niches and geographic distributions may prefer to start directly with Chapter 4, where we explain the main principles of habitat suitability modeling, how predictions for individual species can be assembled to predict communities, and what the main applications of these models are. We finally present the main working assumptions that are made when fitting such models (Chapter 5; see also Part V for assumptions specifically related to projections in time and space).

2 · *Overview of the Habitat Suitability Modeling Procedure*

HSMs are empirical methods that relate species' field observations or museum-type species data to environmental predictor variables, based on a combination of statistically or theoretically derived response curves (Guisan and Zimmermann, 2000; Guisan and Thuiller, 2005; Chapter 4) that best reflect the ensemble of ecological requirements of the species (i.e. an approximation of the species' realized environmental niche described in Sections 3.4 and 3.5). The development of HSMs has resulted from successive improvements over the few last decades, marked by the emergence of, and increase in, computing tools (see Box 2.1 for a brief history of HSMs). A series of modeling stages need to be followed to build and use an HSM, each of which is important in its own right (Section 2.1).

2.1 The Different Methodological Steps of Habitat Suitability Modeling

Habitat suitability modeling ideally follows five steps (modified from Guisan and Zimmermann, 2000; Guisan and Thuiller, 2005, step 6 omitted). Each part of this book corresponds to one of these steps (Figure 2.1 and Table 2.1):

1. Conceptualization (Part I, Chapters 2–5)
2. Data preparation (Part II, Chapters 6–8)
3. Model calibration (fitting) (Part III, Chapters 9–14)
4. Model evaluation (Part IV, Chapters 15–16)
5. Spatial predictions (Part V, Chapters 17)

As in other sciences, HSMs must rely both on robust methodological principles, as well as sound biogeographical, ecological, and evolutionary theory to explain the patterns and causes of species distributions, and from these, of community assembly and distribution. In addition, it is important to identify the underlying working assumptions behind HSMs (this part, Chapters 2 to 5).

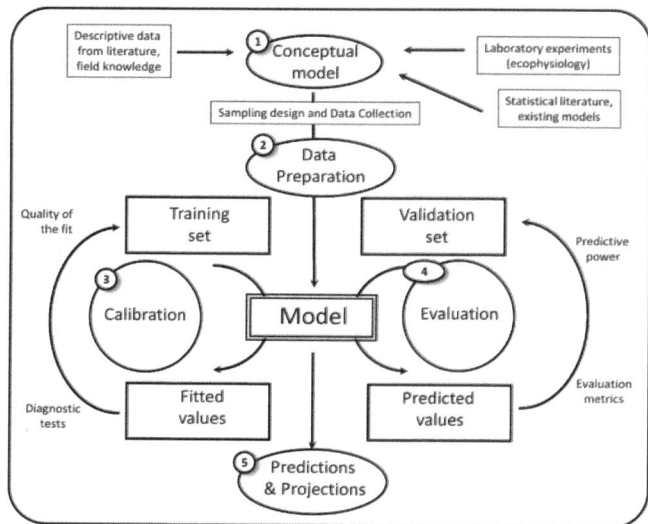

Figure 2.1 The five main modeling steps to be followed when building HSM. Simplified from Guisan and Zimmermann (2000), with permission. See also Table 2.1.

The success of HSMs is largely down to the increasing availability of spatially explicit biological (Chapter 6) and environmental data (Chapter 7) at different spatial and temporal scales (Chapter 8). The different chapters in Part II concern the preparation of data, be it biological or environmental, sampled in the field or derived from remote-sensing sources. This preparation is usually done in a GIS environment, yet recent advances in R mean it now has full GIS functionality. The latter greatly facilitates data preparation for modeling, designing stratified sampling or the scaling of data for modeling purposes (Chapters 6–8).

In the last two decades, an impressive array of modeling techniques has been made available, using very different but related statistical approaches (Chapters 9–14). Part III presents a selection of conventional and powerful approaches. Here, we will focus on a small number of techniques for three reasons. First, we are focusing on approaches available in the R statistical framework (see above) and associated packages. Second, we prioritize techniques based on a sound theoretical and ecological rationale, i.e. we tend to avoid "black-box" algorithms. Third, many "new" approaches proposed in the HSM literature are indeed methodological variations of standard techniques (e.g. bagging or boosting), and so understanding the main techniques is often sufficient to understand and explore the range of custom model implementations. It is also evident that newer or more

Table 2.1 *Important features to consider when building an HSM, working through the five steps described in this book, criteria for detecting potential problems, and some proposed solutions. Based on Table 2 in Guisan and Thuiller (2005). These steps include: (1) Conceptualization (see Part I): identifying the type of organisms being modeled, and accordingly deciding on which type of environmental predictors, model for the data (for instance to avoid unrealistically shaped response curves), sampling design, and accordingly what ecological theory needs to be included; (2) Data preparation (see Part II): identifying the available data for the species, environmental predictors, and the scale (grain/extent); (3) Model fitting (see Part III): identifying the type of statistical model for the data, checking for and addressing multicollinearity and spatial autocorrelation (SAC; i.e. non-independence of observations), identifying appropriate ways of selecting relevant predictors in the model, or preventing overfitting of the models (i.e. when the number of predictors is too large compared to the number of observations); (4) Model evaluation (see Part IV): identifying the type of evaluation data, design the evaluation framework, choose the appropriate metrics for evaluating the models and tracking uncertainty; (5) Spatial predictions and model applicability (see Part V): anticipate the domain of applicability of the models, identify potential problems in projecting models to future climates, assess the potential use of models for management, etc.*

Feature to consider	Possible problems	Detection criterion	Examples of proposed solution
1. Conceptualization (conceptual model)			
Type of organism	Mobile species in unsuitable habitats	Radio-tracking; continuous time field observations	Neighborhood focal functions; choice of grain size
	Sessile species (e.g. plants) in unsuitable habitats	Lack of fitness (e.g. no sexual reproduction)	Use fitness criteria to select the species observations to be used in model fitting
	Species not observed in suitable habitats	Knowledge of species life strategy (e.g. dispersal)	Interpret the various types of errors
	Low detectability species	Field knowledge, literature	Correct test for detectability
	Sibling species or ecotypes in the same species	Genetic analyses	Test niche-differentiation along environmental gradients

(continued)

Table 2.1 (*cont.*)

Feature to consider	Possible problems	Detection criterion	Examples of proposed solution
	Invasive species	Mostly commission errors	Fit models in the area of origin
Type of predictors	Direct or indirect predictors	Ecophysiological knowledge	Avoid indirect predictors
Type of model needed	Need to include environmental or biotic interactions, dispersal, autocorrelation, abundance, etc.	Ecological theory and available data	Consider an HSM technique which takes these features into account (e.g. tree based for abiotic interactions, spatial autologistic for autocorrelated data)
Designing the sampling	Selecting sampling strategy	Simulation tests with virtual species in a real landscape	Random- or random stratified (by the environment) sampling
Incorporating ecological theory	Linear or unimodal response of species to the predictors	Partial plots, smoothing curves	Generalized additive model (GAM) or quadratic terms in generalized linear model (GLM)
	Skewed unimodal response curves	Skewness test	HOF, beta functions in GLM, GAM, fuzzy envelope models
	Bimodal response curves	Smoothing curves, partial plots	GAMs, $\geq$ third-order polynomials or beta functions in GLMs

Table 2.1 (*cont.*)

Feature to consider	Possible problems	Detection criterion	Examples of proposed solution
2. Data preparation (data model)			
Species data	Bias in natural history collections (NHC)	Cartographic and statistical exploration	Various ways of controlling bias
	Heterogeneous location accuracy in NHC	Only detectable if recorded in the database	Selecting only observations of known accuracy below the threshold
	No absences	Type of database and source (metadata)	Generating pseudo-absences
Environmental predictors	Errors in environmental maps	Cartographic and field-proofing exploration	Incorporating error into the models
	Missing key mapped environmental predictors	Ecological theory or reduced variance explained	remote-sensing (RS) data as an alternative source
Scale (grain/ extent)	Different grain sizes for the various predictors	GIS exploration	Aggregating all GIS layers at the limiting grain
	Truncated gradients within the considered extent	Preliminary exploration of species response curves	Enlarging the extent of the study area to cover full gradients
3. Model fitting			
Type of data	No absences	Type of database and source (metadata)	Using profile methods
Multicollinearity	Correlations between the predictor variables	Variance Inflation Factor (VIF)	Removing correlated predictors; orthogonalization

(continued)

Table 2.1 (*cont.*)

Feature to consider	Possible problems	Detection criterion	Examples of proposed solution
Spatial autocorrelation (SAC)	Non-independence of the observations	SAC indices	Resampling strategies to avoid SAC; correcting inference tests and possibly incorporating SAC in models
Type of statistical model	Overdispersion	Residual degrees of freedom > residual deviance	Quasi-distribution in GLMs and GAMs; scaled deviance
Model selection	Which approaches and criteria?	–	AIC-based model averaging; cross-validation; shrinkage
4. Model evaluation			
Type of data	No absences; usual measures not applicable	Type of database and source (metadata)	New methods emerging for evaluating predictions of presence-only models (Boyce index, POC-plot, MPA, area-adjusted frequency index)
Evaluation framework	No independent set of observations	–	Resampling procedures
Association metrics	Choice of a threshold for evaluation	–	Threshold-independent measures (AUC, max-TSS, max-Kappa, etc.)
	Error costs	–	Error weighting (e.g. weighted Kappa)
Model uncertainty	Lack of confidence	Uncertainty map, residual map	Bayesian framework; spatial weighting

Table 2.1 (*cont.*)

Feature to consider	Possible problems	Detection criterion	Examples of proposed solution
Model selection uncertainty	Lack of confidence	Comparing selection algorithms and competing models	Model averaging, ensemble modeling
Spatial predictions			
Projection into the future and in new areas	Range of new predictors falls outside the calibration domain	Statistical summaries	Restrict area of projection; control the shape of response curves
5. Spatial predictions and model applicability			
Scope and applicability	Model not applicable to a distinct area	Unrealistic response curves	Avoid overfitting
Problem of future projections	HSMs not transposable to distinct environments	Strongly dependent on the scale considered	Spatially explicit models incorporating population dynamics, dispersal and habitat
Use in management	Difficult to implement in a management context	Software not available to managers and decision-makers	Free software
	Not interpretable	Black-box algorithms?	Choice of easy-to-read and easy-to-interpret methods (GLMs, GAM, CART)

refined versions of the models discussed here will be implemented in the coming years. However, it is very likely that the basic principles presented will still hold true for a reasonable amount of time.

The evaluation of models and their spatial predictions (Chapters 15–16) has also benefited from major advances in recent years, as being of crucial importance for many applications to conservation and management.

In Part IV, we discuss the most important aspects from the two angles of "which evaluation metric to use?" (Chapter 15) and "which evaluation data to use?" (Chapter 16). For metrics, we mainly consider those for presence–absence and presence-only data, on which most of the examples in this book are based, but we also briefly refer to metrics for other types of response variables (e.g. abundances). For the evaluation data, we discuss what we mean by independent data and present the different evaluation frameworks, including the most common resampling approaches such as traditional cross-validation, jackknife, repeated split sample, and bootstrap.

Among the different features of HSMs, the capacity to predict or project over large geographic extents or across long time periods (Chapter 17) is certainly part of what has made them so popular. In Part V, we define predictions as model application within the same area or time period, and projections as those in a different area or time period, and we describe ways of producing these and the different types of corresponding outputs. We also extensively discuss a number of additional assumptions that are made specifically when projecting models to other areas or time periods (Chapter 17).

2.2 The Initial Conceptual Step

In this first part, we will focus on the initial "conceptual" step, which takes place before model building. The four subsequent steps (see Table 2.1) are discussed in the next four parts of the book. This conceptual phase of the habitat suitability modeling procedure is important as it should serve to identify all the aspects requiring methodological decision to be taken, at the earliest possible stage of the process. These aspects can be divided into two main categories:

1. *Theory and data*: One of the first requirements is to define: (i) clear scientific question(s) and objectives for the modeling study (Austin, 2002, 2007; Huston, 2002); (ii) a good conceptual view of the model system used to answer the question(s), based on sound ecological knowledge (as presented in the next Chapter 3); (iii) the main underlying assumptions made when building the model (e.g. pseudo-equilibrium; Guisan and Zimmermann, 2000; see Chapter 5) and identifying the necessary proximal environmental predictors (see Chapters 3 and 4) for the focal species (e.g. Mod et al., 2016), including which of these are available or missing (data model; Austin, 2002). Furthermore, it requires identifying, if necessary, an appropriate sampling strategy for collecting species observations (Hirzel and

Guisan, 2002, Chapter 6), and choosing the appropriate spatio-temporal resolution and geographic extent for the study (Chapter 8). Species data can be simple presence (i.e. occurrences), presence–absence, or abundance observations, based on random and/or stratified field sampling, or based on observations obtained opportunistically, such as those from natural history collections (Graham et al., 2004a). The issues relating to species data are discussed in more detail in Chapter 7. There are also a number of implicit assumptions about the data and methods used that need to be checked early in the model building process (see Chapter 4). These theoretical aspects are mostly developed in this part of the book.

(2) *Modeling methods*: The second requirement is to identify: the most appropriate method(s) for modeling the response variable (e.g. ordinal generalized linear model (GLM) for semi-quantitative species' abundance, such as Braun–Blanquet's abundance–dominance plant cover classes in phytosociological surveys; Guisan and Harrell, 2000); the most optimal evaluation framework (e.g. resampling techniques vs. truly independent observations); the statistics needed to assess the predictive accuracy of the model (Pearce and Ferrier, 2000; Fielding, 2002); and the methods to be used to derive spatial and temporal predictions. These methodological issues are further developed in Parts III, IV and V.

Numerous other conceptual features – methodological, statistical, or theoretical – relating to the different steps of the HSM building process (see Guisan and Zimmermann, 2000; Guisan and Thuiller, 2005; Elith and Leathwick, 2009; Table 2.1) need to be assessed as early as possible, ideally during the conceptual phase. However, it is not always possible to make all the necessary decisions at the very beginning of a study. This might be due to a lack of knowledge of the target organisms (Part I), or of the study area and related data (Part II). For instance, the choice of the appropriate spatial resolution might depend on the size of a species' home range and the way this species uses resources in the landscape. The choice of the geographic extent might depend on prior knowledge of environmental gradients in the study area (to ensure that complete gradients are sampled; Austin, 2002; Van Horne, 2002). For animal species, males vs. females, or summer vs. winter habitats might require separate models (see Guisan and Thuiller, 2005). Answers to these questions usually require collecting preliminary field observations, running exploratory analyses on existing data, or conducting experiments (e.g. Kearney and Porter, 2004).

Box 2.1 A Short (Non-Exhaustive) History of HSMs (Adapted From Guisan and Thuiller, 2005).

Most modeling approaches developed for predicting plant or animal species' distributions have their roots in quantifying species–environment (habitat) relationships, i.e. fitting the species niche or part of it. Three phases seem to have marked the history of HSMs (therein called SDMs; Guisan and Thuiller, 2005): (i) the non-spatial statistical quantification of the species–environment relationship based on empirical data, (ii) the expert-based (non-statistical, non-empirical) spatial modeling of species distribution, and (iii) the spatially explicit statistical and empirical modeling of species distribution.

The earliest examples of correlative studies between species and climate distributions that we found are those produced by Johnston (1924), predicting the invasive spread of a cactus species in Australia, and Hintikka (1963) assessing the climatic determinants of the distribution of several European species (quoted in Pearson and Dawson, 2003). Early niche quantification studies then additionally measured niche breadth and overlap (e.g. Colwell and Futuyma, 1971; Green, 1971). The earliest developments in computer-based predictive modeling of species distribution seem to have originated in the early 1970s, stimulated by the extensive quantification of species–environment (e.g. multivariate ordinations) available at that time (Austin, 1971). Two examples of early attempts at computer-based species distribution modeling are the plant distribution studies of Jardine (1972) in UK, and the spatial predictions of crop species by Nix and co-workers in Australia (Nix et al., 1977). These were followed, in the early 1980s, by the pioneering simulations of species distribution by Simon Ferrier (1984) and others in Australia. At about the same time, the publication of two seminal works (Verner et al., 1986; Margules and Austin, 1991, resulting from a workshop in 1988) also contributed significantly to promoting this new approach, resulting in the increase in the number of species distributions models proposed in the literature. These advances were supported by the parallel developments in computer and statistical sciences, and by strong theoretical support to make predictive ecology "more rigorously scientific, more informative and more useful ecology" (Peters, 1991).

3 · *What Drives Species Distributions?*

The most obvious questions in biogeography, and the starting points for this book, are: what are we observing (i.e. which species, communities, or ecosystems?)? Where in space and time? Why are organisms distributed where they are? The quest to answer these questions is an age-old one (Guisan and Thuiller, 2005; Franklin, 2010a), but which really took off scientifically with the eighteenth and nineteenth centuries' first biogeographers (i.e. Von Humbolt, De Candolle, Darwin, and Wallace). Since the steady rise of ecological research during the twentieth century, explaining and understanding the distribution of biodiversity at various spatial and temporal scales has continued to be an important field of macroecological and biodiversity research (Lomolino et al., 2010; McGill, 2010). Numerous new or refined theories were proposed (metapopulation dynamics, e.g. Hanski and Gilpin, 1997; neutral theory, Hubbell, 2001; metabolic theory, Brown et al., 2004) in which geographic space was explicitly considered (Moloney and Jeltsch, 2008). This ultimately fostered the development of predictive models of species and biodiversity (Côté and Reynolds, 2002; Guisan et al., 2013).

This chapter does not attempt to review every single step in this long history, nor does it aim to provide an exhaustive review of all the theoretical aspects of species' niches and distributions. Instead, we aim to present the theories and findings most relevant to habitat suitability modeling. However, and although habitat suitability remains the major principle behind this type of model, we will begin with a more general biogeographical perspective. First, we will present the three specific key drivers of species distributions (3.1), then introduce each of them more detail in the following sections: speciation, dispersal, species pools and neutral theory (3.2), the abiotic environment: habitat and fundamental niche (3.3), and the biotic environment: species interactions, community assembly and the realized niche (3.4). Habitat and niche issues will then be further discussed in more detail in Chapters 4 and 5, from the angle of habitat suitability modeling. The aim here is to introduce the basic knowledge around these concepts that is useful for habitat suitability modeling rather than reviewing them exhaustively.

3.1 The Overall Context: Dispersal, Habitat, and Biotic Filtering

When building models to explain and predict the distribution of organisms we necessarily need to ask the same questions as the early biogeographers. It is now clear that **three main conditions** need to be met for a species to occupy a site and maintain populations (Figure 3.1, see Pulliam, 2000; Lortie et al., 2004; Soberón, 2007):

(i) the species has to reach the site, i.e. to access the region (Barve et al., 2011) and disperse there;

(ii) the abiotic environmental conditions must be ecophysiologically suitable for the species;

(iii) the biotic environment (interactions) must be suitable for the species.

The first condition is a matter of species **dispersal** capacity from those areas previously occupied by the species. It includes the biogeographic history of the species, and thus all factors limiting its distribution from the place where it first originated, such as barriers to migration, biotic and abiotic dispersal vectors, etc. (Section 3.2).

The second condition is the matter of abiotic **habitat suitability** for the target species, which implies that the combination of abiotic environmental variables at the site – often referred to as environmental suitability – is included in the environmental conditions that a species needs to grow and maintain viable populations, i.e. its environmental niche (sensu Hutchinson, 1957) and constitutes the basis of the habitat suitability modeling approach (Guisan and Zimmermann, 2000) presented in this book (Section 3.3 and Chapter 4).

The third condition concerns **biotic interactions**, i.e. interactions with other organisms, either positive (commensalism, mutualism) or negative (competition, predation), which themselves are dictated by the environment through their influence on all organisms in the local community (Section 4.4). This component can also include top-down environmental constraints on communities, if applicable, such as the idea that whole communities and ecosystems may also have their species composition limited by some form of environmental carrying capacity (Del Monte-Luna et al., 2004; but see Buckley et al., 2010; see Guisan and Rahbek, 2011). This may also involve some species engineering the

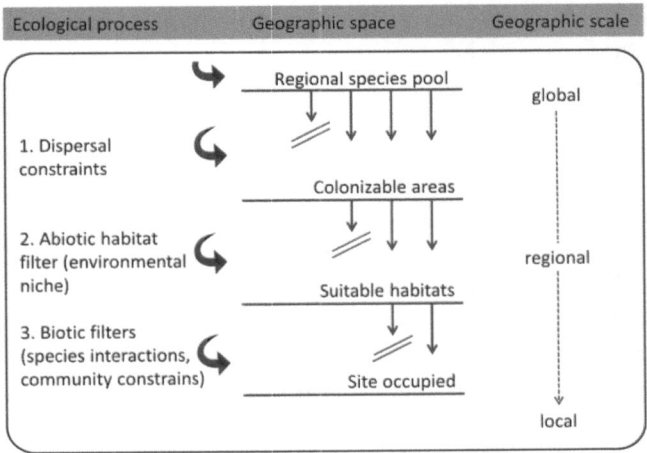

Ecological process	Geographic space	Geographic scale

Figure 3.1 Hierarchical view of the three main influences determining species occurrence at a given site: dispersal limitation, abiotic habitat filtering, and biotic filters, with corresponding geographic space at the successive scales, from global to regional to local; see also Pulliam (2000), Lortie et al. (2004), and Soberón (2007). Note that these processes may not act in the sequence and hierarchy suggested, and stochastic or neural processes might counteract some of these deterministic processes and add randomness to the assemblages (modified from Guisan and Rahbek, 2011, with permission). See Figure 3.2. for a more balanced view of the three drivers operating simultaneously.

environment for other species, such as trees providing favorable canopy cover conditions for shade-tolerant species (Nieto-Lugilde et al., 2015).

As we will see later, assessing habitat suitability constitutes the core of the whole HSM approach that is the focus of this book (Guisan and Zimmermann, 2000). However, it is also intimately related to, and depends on, the other two drivers: dispersal limitations and biotic interactions (i.e. through environmental engineering), which can be added as additional steps to constrain habitat suitability predictions. What is less clear are the relative roles these three drivers play in shaping species distributions and assemblages. Addressing this issue primarily requires gathering knowledge about these processes along geographic and environmental gradients. This, however, cannot be achieved with the sole use of *in situ* observations; it requires complementary approaches such as *in situ* measurements (e.g. ecophysiological) along key environmental gradients, reciprocal transplant experiments, removal experiments in natural or artificial communities, common garden experiments, or controlled *ex situ* experiments in the laboratory (e.g. ecotons). For instance, one can move high-elevation plants to low elevations to test the hypothesis that, despite the fact that they can grow and reproduce well at low elevations, they can systematically be excluded by more competitive plants, as well as determining to what extent phenotypic differences between local populations along such gradients have a genetic basis (see Clausen et al., 1948; Hautier et al., 2009). The data and results collected using these approaches constitute the basis of the knowledge used to untangle the three main drivers of species distributions and can be used to build ecologically more meaningful HSMs (Austin, 2002).

This triple influence – dispersal, niche, and biotic – shaping species distributions, and their interactions, can be viewed schematically as separate ensembles defined by specific boundary conditions (Soberón, 2007). Suitable conditions for a species according to all three factors are found at the intersection of the three ensembles (Figure 3.2, case 1). We discuss these three factors and their interactions in more detail in the next three sections, including their role in explaining local species occurrences and their implications for modeling species distributions.

Individual populations of a species may be present in suboptimal situations (cases 2–4 in Figure 3.2) outside the intersection of suitable conditions (i.e. case 1 in Figure 3.2), either in suitable environments in which the species would usually be excluded by biotic interactions (Figure 3.2, case 2), or in non-suitable sink environments colonized through high propagule pressure (source–sink dynamics, see Pulliam (2000); Figure 3.2, case 3). The latter is the most common case, defining what are known as "sink populations" where the intrinsic growth rate r is lower than zero and which can only persist as long as recruitment is maintained through the constant immigration of seeds (plants) or young adults (animals) from neighboring sites that harbor suitable environments and have positive growth rates (source populations; Pulliam, 2000). Finally, the species may also be observed in environments that are unsuitable due to both unsuitable abiotic conditions and negative biotic interactions (Figure 3.2, case 4). This latter case however, represents the least likely unsuitable condition under which a species can sometimes be observed.

A species may also be absent on a site for reasons other than the three main factors discussed so far, such as natural or human disturbances and/or intrinsic population stochasticity, leading to fluctuations of population size in time and space, with possible local and temporary extinctions in some places (Pulliam, 2000). Yet, once extinct locally, whether the species will be able to colonize this same location again once again depends on dispersal limitation (neutral process; Hubbell, 2001). Such fluctuations

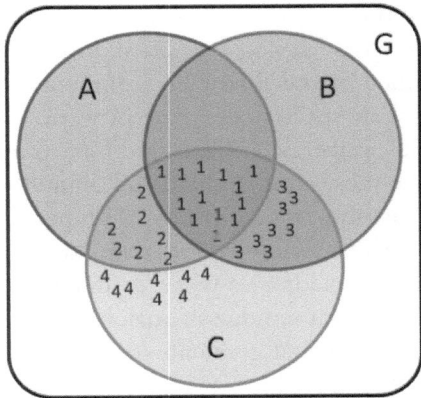

Figure 3.2 The three main factors that drive species ranges. G: studied geographic area; A: suitable abiotic environment (niche); B: suitable biotic environment; C: colonizable range. Observations in the field could result from four situations, from most to least likely: 1. Realized niche (suitable with regard to all three aspects). 2. Suitable abiotic environment with unsuitable biotic conditions, for instance due to strong competition. 3. Colonization outside the suitable environment, maybe due to facilitation (sink). 4. Sink in unsuitable biotic and abiotic conditions, maybe due to historical effect (e.g. trees persisting in unsuitable conditions). Modified from Soberón (2007), with permission.

caused by population dynamics that result in environmentally suitable sites being unoccupied or unsuitable sites being occupied complicate the quantification of species–environment relationships by increasing the percentage of unexplained variation. The question – still largely unresolved for most species – is how much of the variation (if any) in species distributions remains purely stochastic (unexplained variation), once dispersal (for species still undergoing expansion), habitat, and biotic influences have been accounted for? This is even more difficult to assess since population dynamics is itself modulated by these three influences. Dispersal effects (e.g. on the recruitment), the abiotic environment and biotic factors (e.g. the degree of shading for plants or the presence–absence of a mutualist species) can modify the performance of individuals in a population, as well as their birth and death rates.

Despite stochastic fluctuations, species are generally expected to be ruled by a significant local environmental determinism. The quantification of their environmental niches can therefore be expected to provide information for predicting their distribution in environmental, and thus also in geographic, space. The rather unlikely exception to this would be for species where the fluctuations in population dynamics and the strength of biotic interactions would be so high that these processes would override, and thus entirely hide, the species–environment relationship. On very large spatial scales, the distributions of nearly all species seem to be determined by the abiotic environment, and in particular the climatic conditions. Few species occurring in warm tropical habitats will be observed in cold temperate habitats and vice versa. This means that all species – even the most cosmopolitan and ubiquitous ones, which are found across large distributions in all possible types of habitats – usually exhibit quantifiable climatic preferences (e.g. Ba et al., 2010). The key questions that need to be addressed when defining the link between species distribution and habitat conditions mainly depend on the extent to which proximal environmental gradients (those having causal effects on species) and their variation are included in the analysis. Another key aspect of species–environment responses is that these can also change depending on the spatial resolution used. For instance, responses could change from being narrow unimodal and skewed at 90 m resolution to become piecewise linear and unbounded at 4 km resolution along a gradient of "precipitation in the wettest quarter" for a plant species in California (Franklin et al., 2013). Therefore, the spatial scale of observation and environmental descriptors can also influence our perception and explanation of ecological phenomena (Wiens, 1989; Willis and Whittaker, 2002). These scale considerations will be discussed more thoroughly in Part II, especially in Chapter 8, as the resolution of analysis is important to consider in any modeling exercise and is tightly related to the data used and their availability on various scales. In the following, we will assume that the right resolution and extent has been chosen for the modeled species.

In the following, we will define the three main drivers of species distributions individually – dispersal (3.2), habitat (3.3), and biotic interactions (3.4) – before focusing on the methodological aspects of fitting and evaluating habitat models for the rest of the book (Chapter 4 onwards). For aspects of population dynamics and their variations in space, we refer interested readers to the relevant literature (e.g. Jeltsch et al. (2008) for plants, Buckley (2008) and Morales et al. (2010) for animals; Nenzén et al. (2012) for a spatial population dynamics tool).

3.2 Speciation, Dispersal, Species Pools, and Neutral Theory

All species have emerged from evolutionary processes at some time and place on Earth. How and where are questions that can help explain patterns of biodiversity, in space on a local, regional, or global scale, but also over time (through e.g. studying bursts of diversification; Willis and McElwain, 2002). There are numerous causes of speciation, the two most commonly discussed being allopatric and sympatric speciation (Doebeli and Dieckmann, 2003; Wiens and Graham, 2005). Allopatric speciation arises when geographic barriers emerge that split the initial range of an ancestor species into two or more disconnected sub-ranges, leading to a disruption in gene flow between the separated populations, and ultimately to speciation into two or more distinct taxa (species or subspecies) on each side of the barrier (Hoskin et al., 2005; Grant and Grant, 2009). Sympatric speciation arises when divergence between populations take place within the initial range of the ancestor species, usually due to the ecological specialization of some populations to distinct environmental conditions (e.g. Filchak et al., 2000; Tautz, 2003). Intermediate or different speciation models also exist (parapatric speciation, quantum speciation, etc.).

For the purposes of introducing habitat suitability modeling, it is only necessary to know that such geographic or ecological speciation processes have occurred and are continuing to happen, though at a very slow pace. More important for habitat suitability modeling are the consequences such allopatric and sympatric speciation types may have in terms of shaping past, present, and future patterns of species distributions. For instance, do species that have resulted from sympatric speciation tend to be ecologically or behaviorally more specialized than species resulting from allopatric speciation? To what extent do species retain the ecological characteristics of their ancestor species after divergence (i.e. phylogenetic niche conservatism; Münkemüller et al., 2015) and what role does niche conservatism play in speciation processes (Wiens and Graham, 2005)? These are important issues to address in order to understand current distributions (Pearman et al., 2008a), with potentially important applied implications, such as the effect niche changes between native and invaded ranges can have on our capacity to predict and anticipate invasions (Guisan et al., 2014).

What we learn from evolutionary processes is that many species are today found in a given area (such as e.g. a subcontinent) primarily because they originated there or nearby (Cox, 2001). On such very large scale, biogeographic history and dispersal limitation predominate, with environmental suitability playing only a secondary role in explaining where on the globe the species is currently found. Worldwide biological invasions provide clear evidence of dispersal limitation for most floras and faunas. When brought to a new area, a certain proportion of the introduced species will establish and reproduce even in a new environment (alien or exotic species). In plants, we distinguish the biological invasions that began with the first expeditions across the oceans (around 1500; now referred to as neophytes) from the invasions that occurred even earlier (referred to as archeophytes). A small proportion of the established species spread successfully across the landscape (invasive species; Richardson et al., 2000), ultimately affecting native communities and ecosystems and causing damage to the human economy (e.g. in agriculture) or health (e.g. allergenic plants or pathogens). We learn from these processes that maladaptation to the environment is only one of several possible causes for the absence of a species in one specific location, dispersal limitation being another alternative.

The pool of species that are present in, or have dispersed to, a region is thus an important foundation for the types of communities and ecosystems that can be assembled from them (Zobel, 1997; Aarssen and Schamp, 2002; Graves and Rahbek, 2005), illustrating that local species diversity depends, at least partly, on regional species diversity (Caley and Schluter, 1997). Large-scale evolutionary processes that shape regional species pools therefore also have an impact on the expected local diversity, defining the source pool of species for a given site. The latter thus results from the balance between speciation, colonization, and local extinctions, and can be used to define the pool of species that can be modeled within a given area (Ricklefs, 1987, 2008; Guisan and Rahbek, 2011).

This idea that dispersal plays an important role in shaping the distribution of species and thus communities is not new. Early biogeographers had already noticed similarities between flora or fauna on separate continents. While some proposed strict long-distance dispersal to explain these patterns, others proposed "past continental bridges" to explain the same pattern, and this at a time when nothing was yet known about plate tectonics and continental drift (see Lomolino et al., 2010). There remains the question as to the extent to which dispersal accounts for observed patterns of species distributions (Svenning and Skov, 2007)? Taking an extreme view, one might wonder as to whether dispersal alone could account for all the observed variation in species composition, thus asking: "are communities mostly shaped by dispersal?" This is part of the view (together with speciation and extinction) taken to define the unified neutral theory of biodiversity and biogeography (Hubbell, 2001), which can be used as a null hypothesis to test the effect of the environment in shaping species distributions (i.e. neutral patterns assume neutral, equivalent environmental niches across species).

These dispersal-related aspects can be used as the basis to define the accessible area for a given species (Barve et al., 2011), at a given time, as the whole area within the colonizable reach of existing populations, i.e. where migration is not impeded by natural or human-made barriers to dispersal. For the future conditions, it encompasses those areas that can be naturally colonized in the future, e.g. if species change their distribution in response to climate change or following invasions. Historic factors (e.g. glaciations, climate shifts, etc.) are thus accounted for in the accessible area concept. The latter can reveal possible dispersal limitations and limited range filling, when the observed distribution range is smaller than the full accessible area (e.g. Svenning and Skov, 2004).

3.3 The Abiotic Environment: Habitats and Fundamental Niches

How does the abiotic environment influence the distribution of organisms? What are the different types of environmental influences on species distributions? How do multiple variables jointly determine a species' geographic distribution? These are the main questions that underpin this book, as they all relate to assessing species' "habitat suitability."

Early biogeographers (e.g. Von Humbolt or Darwin) observed that a same species could occur in sites with different environmental conditions, each single combination representing a distinct habitat (sensu Kearney, 2006; see Glossary), and thereby occupy a range of different habitats. Yet, most species have limited geographic ranges (Woodward and Kelly, 2003). From this, one can deduce that a species can colonize a range of conditions along environmental gradients, but that in most cases this range only occupies a proportion of all the possible habitat conditions available, resulting in the species occupying only a limited geographic and environmental range where these specific conditions are met. Why is this so? Why aren't all species found everywhere?

The answer lies in the physiology of organisms (Woodward and Kelly, 2003) and the specialized physiological adaptations that most species have undergone through evolution in order to survive and be competitive in specific habitats. These adaptations come at the cost of being maladapted (and thus unable to survive) or less competitive (and thus excluded) in other habitats, representing adaptation or functional trade-offs (Woodward, 1987; Kearney and Porter, 2009). The physiological specialization usually results in different shapes of responses being expected for different species along environmental gradients (Whittaker, 1967;

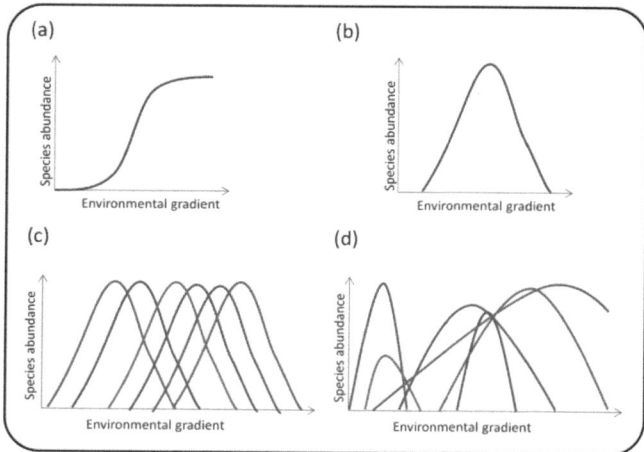

Figure 3.3 Fundamental response curves of species along a hypothetical environmental gradient. (a) Typical sigmoidal species response curve, usually expected at the end of gradients or resulting from the truncation of a unimodal curve. (b) Unimodal response curve. (c) Species packing hypothesis (unimodal responses for several species with regular placement of optima). (d) The response curves of different species vary in shape, amplitude and width (inspired from Austin and Gaywood 1994; Austin 2002, with permission). See Figure 3.4 and Section 3.4 for how fundamental curves can be modified by biotic interactions.

Austin, 1985; Austin and Gaywood, 1994). In most cases this is associated with a position along the gradient where the species performs best – the physiological optimum – and a gradual decrease in performance the further one moves away from this optimum, in either direction (Ellenberg, 1953, 1954; Hector et al., 2012). Such physiological response curves can therefore be represented as sigmoidal or unimodal shapes (Figure 3.3), with unimodal responses appearing to be dominant in nature (e.g. linear responses can be observed for species at the end of gradients or along very stressful gradients). The width of the curve also documents the physiological tolerance of species along the gradient. Different species have different optima and tolerance along a same gradient (Figure 3.3c), from narrow to wide. A species can have a wide tolerance along one variable but a narrow tolerance along another. A species that has a very broad tolerance along a specific gradient may be considered indifferent to variations in this variable, and thus be considered a generalist species (for that gradient).

The transition from optimal to poor performance can be smooth or abrupt, depending on the types of physiological mechanisms involved. An abrupt transition can result from limiting factors with threshold effects below or above which some metabolic pathways (e.g. photosynthesis) abruptly change. For instance, in conifer trees, the cambium activity allowing root growth tends to stop rather abruptly below some threshold value of soil temperature (around 7°C at -10 cm for Arolla pine *Pinus cembra* in the Alps; Körner, 2003; Körner and Paulsen, 2004). Smoother transitions occur when the gradient has a more or less linear effect on some metabolic rates (e.g. carbon sequestration, water use efficiency), progressively lowering individuals' fitness.

When looking at the effect of each environmental variable independently, one may fail to identify the combined effects (i.e. interactions) of some of these variables on the species' physiology, one possibly dampening or amplifying the effect of another. Therefore, all important variables should be considered jointly in an analysis, in order to define what is known as the environmental niche of species (Chase and Leibold, 2003), a term initially coined by Grinnell (1917). When considered jointly, the physiological responses of a given species to several environmental variables define a multidimensional volume called a species' fundamental environmental niche (see Figure 3.5a). The fundamental niche concept was quantitatively formalized by Hutchinson in (1957), based on earlier insights from Grinnell (1917) and others (see Chase and Leibold, 2003 for a review of the niche concept in ecology). Hutchinson defined it as an *n*-dimensional hypervolume (i.e. of possibly $n > 3$ dimensions) in a space defined using environmental variables hypothesized to have a direct influence on a species' physiology, and within which the population growth rate is positive (see Pulliam, 2000; Kearney, 2006). Hutchinson illustrated the concept with animal species, but the same concept applies to plants and other organisms (i.e. fungi, protista, bacteria), and Ellenberg had already proven this concept experimentally with graminoid plants along a water table gradient a few years before Hutchinson's conceptual summary (Ellenberg, 1953, 1954; Austin, 1990; Hector et al., 2012). For plants, the axes of the fundamental niche may typically be resource variables related to light, heat, water and nutrient availability, or regulators such as too high or too low temperatures or other extreme climatic conditions (Guisan and Zimmermann, 2000). For animals, they may be thermal limits, and/or water/food/habitat availability (Kearney and Porter, 2009).

If one were able to identify all the important environmental variables for a species, quantify its fundamental niche (i.e. physiological responses) along these gradients, and have these environmental variables available as global maps, one could predict all the geographic locations where the species could establish and maintain viable populations in the hypothetical absence of biotic interactions (Clark et al., 2007), and provided the species can access all locations (Barve et al., 2011). In practice, however, measuring the fundamental niche based on field observations in natural conditions is practically impossible, especially since most species require interactions with other species (e.g. pollinators, symbionts, etc.) to persist. As a result, it is often stated that only "mechanistic" models based on measured physiological or behavioral parameters (e.g. Kearney and Porter, 2009), or analyses based on *ex situ* data (e.g. a plant grown in botanical gardens outside its natural range; Vetaas, 2002), can approach the fundamental niche. There are two main reasons for our inability to capture the fundamental niche from field observations. First, the responses of species in nature are modified by their interactions with other species, within the same group (e.g. competition for light in plants or for food resources in animals) or across groups at different hierarchy or trophic levels (e.g. predation, parasitism, etc.), combined with the effects of migration limitation in response to changing environments (land use, stochastic disturbances). The biotic component, with the different types of possible interactions, will be discussed extensively in Section 3.4. These effects constrain the fundamental niche to what is considered to be the realized niche, the only one that can be observed from field observations (see Section 3.4 for an extended coverage of this aspect). Second, it is not always possible to spatially map and measure the variables that have a significant effect on a species' physiology (e.g. minimum absolute temperature during the relevant part of the year for a plant or animal, exact soil moisture for a plant). This is often due to material limitations (i.e. no system capable of measuring the target variable in the wild, or too costly to monitor numerous individuals in numerous populations), or due to our inability to map such important variables precisely, and to link species observations to these physiologically important predictors.

Furthermore, knowing which variables have a direct physiological effect (i.e. should be used to define the fundamental niche) requires, for each species, prior experimental laboratory measurements (e.g. measuring metabolic rates and individual fitness while varying environmental variables), which is not feasible for all species. This becomes even more difficult if such responses are measured over a large number of populations to account for genetic differences among populations. It is therefore often easier to use measurements of surrogate environmental variables that are hypothesized to best correlate with the physiologically meaningful variables (e.g. altitude for temperature, or minimum of mean monthly temperature for absolute minimum temperature), but with the consequence to reduce both the predictive power (e.g. for projections in space or time) and the level of generalization of the model (e.g. altitude cannot be compared between population in cold environments in the Alps and the Arctic; Pellissier et al., 2013a). These surrogates can be ranked along a scale of "proximality," depending on how close they are to having a causal effect which explains the distribution of the species studied (Austin, 2002, 2007). The different variables can also be classified, depending on their effect or use by the target species, as limiting factors (causing linear or step responses), regulators (modulating the organism physiology, gradual response) or resources (consumed by the organism) (Austin, 2002; Huston, 2002; Austin, 2007; and see Section 4.1).

3.4 The Biotic Environment: Species Interactions, Community Assembly, and Realized Niches

The biotic environment covers all possible forms of interactions among species, either within the same or between different trophic levels. It can also further constrain a species' distribution in space, in time, or along environmental gradients, in addition to the abiotic constraints. For the purpose of simplification, in these examples we often start by considering competition because historically it has been the first biotic factor discussed by the proponents of the environmental niche concept (e.g. Hutchinson). However, interactions between trophic levels (e.g. predator–prey, plant–pollinator, or plant–herbivore) or functional groups (e.g. host–parasite, symbiosis) can be as important in many instances (e.g. Broitman et al., 2009; Pellissier et al., 2013d; Lira-Noriega and Peterson, 2014). Take the fundamental niche of species as described in Section 2.3. It considers that a species can occupy all locations where the abiotic, environmental conditions allow the species to maintain viable populations. However, we also saw earlier on that species do not usually have uniform responses along environmental gradients, and thus the fitness of their individuals, and potentially their competitive potential, also varies along these gradients. One direct implication of this pattern is that when two species co–occur at a same location and compete for the same resource ultimately the less fit species will be excluded, as predicted by the "competitive exclusion principle" (Salisbury, 1929; Gause, 1936). There is extensive evidence of this general principle (e.g. Ellenberg, 1953; Brown, 1971; Grace and Wetzel, 1981; see Lomolino et al., 2010). One piece of evidence, to be further investigated, may be the fact that a number of high-elevation alpine plants can be grown at low-elevation botanical gardens (i.e. in warmer parts of the temperature gradient) in the absence of competition. However, they are never observed at these low elevations in natural systems, probably because they are systematically outcompeted by higher stature plants optimized for low elevations (e.g. Vetaas, 2002). Of course, this example remains speculative for mountain plants, which require some of their environmental conditions to be physically engineered (e.g. watering by gardeners) in order to survive at low elevations, as these may be the actual factors defining their lower limit. The signatures of such competitive exclusion was generally considered to be only observable at local scales (e.g. Pearson and Dawson, 2003), where individuals compete, but evidence has recently shown that the signatures of competitive exclusion can also be detected on larger scales (Gotelli et al., 2010).

When a species is systematically excluded from parts of an environmental gradient (as is putatively the case for alpine plants), the response along these gradients revealed from field observations only gives a partial view of the full physiological (i.e. fundamental) response. The species response from field observations along gradients thus depends on the particular biotic configuration in the area studied. This type of field observation-based response along a single gradient has been called the "realized response" or "ecological response" (Austin et al., 1990; Austin and Gaywood, 1994), in contrast to the "physiological response" or "fundamental response" described above. The differences between these two types of response have also been demonstrated experimentally, as for instance by Ellenberg (1953) with monocultures and mixtures of plants along a gradient of increasing depth of water table. Depending on which part of a fundamental response is excluded by competition, different shapes are obtained for the realized response. Walter (1960; in Ellenberg, 1988), Austin (2002) or Hector et al. (2012) illustrate different cases, such as exclusion of the optimum leading to a bimodal realized response, exclusion of a margin leading to a truncation of the realized response and displacement of the optimum, or exclusion at both margins leading to a similar optimum, but a narrower realized response (Figure 3.4).

Interestingly, it was observed very early on that the fundamental niche of species is more constrained by competition in the milder parts

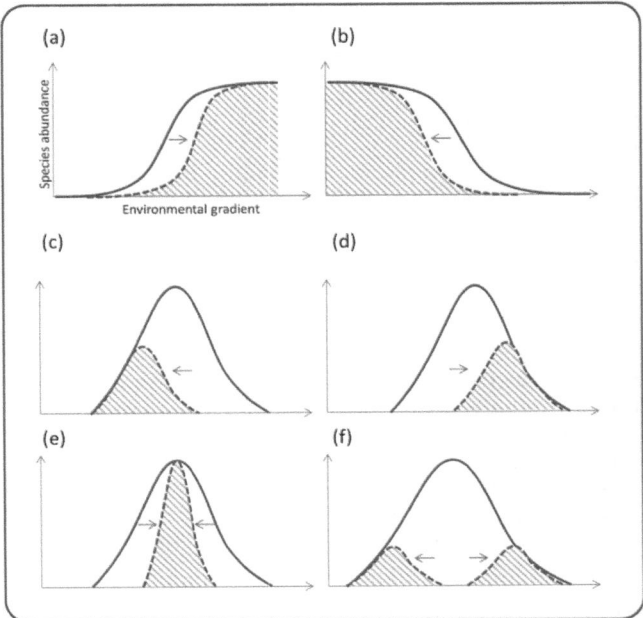

Figure 3.4 The main situation of competitive exclusion constraining the fundamental response curve of a hypothetical species along a hypothetical environmental gradient into the realized response curve observed in natural systems (i.e. niche truncation). These modifications include linear response displacement (a, b), optimum displacement (c, d), narrowing of the niche (e), and optimum removal (resulting in a bimodal response) (f). Modified with permission from Austin and Smith (1989) and Austin (1990).

of environmental gradients that require less physiological adaptation and thus provide better growth conditions (e.g. at the warm end of a niche) than at the gradient edges with higher physical stress (Austin and Gaywood 1994). This resulted in the proposal of a general rule stating that the ranges of species along environmental gradients tend to be limited by physiological tolerance toward the physiologically more stressful edge, and by competitive interactions (i.e. exclusion) toward the physiologically less constraining and more productive parts of environmental gradients (Connell, 1961; Austin et al., 1990; Brown et al., 1996), a hypothesis recently named the "asymmetric abiotic stress limitation" (AASL) hypothesis (see Normand et al., 2009; Meier et al., 2011). This has, for instance, been shown experimentally for cattails along a water depth gradient (Emery et al., 2001) and for chipmunks along an elevation gradient (Heller and Gates, 1971); see Smith and Smith (2015).

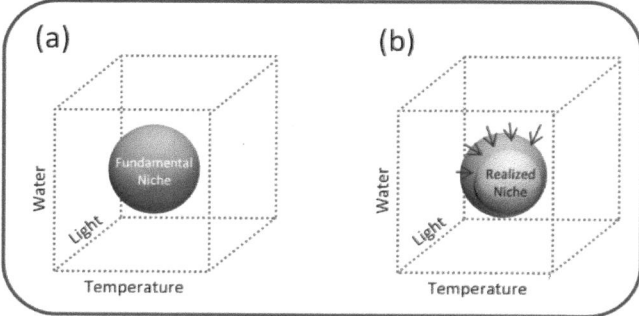

Figure 3.5 Illustration of a three-dimensional environmental niche for a hypothetical species. (a) Fundamental niche representing the envelope of environmental conditions (envelop) within which a species can maintain a viable population in the absence of competitive interactions. (b) Realized niche, representing a subset of the fundamental niche constrained by negative interactions (e.g. competition) with one or several other species. With more than three dimensions, the niche can no longer be represented graphically and is called a hypervolume.

The distinction between the two types of niche responses along single gradients leads to the same distinction at the level of the whole environmental niche, which was termed the "realized niche" by Hutchinson (1957) or the "ecological potential" by Ellenberg (1953), describing a subset of the fundamental niche (or physiological potential) constrained by competition with one or several other species (Figure 3.5).

We have now seen that biotic interactions constrain the fundamental niche thus forming the realized niche. Numerous different biotic interactions can affect the predictability of a species at a site from environmental predictors only (Araújo and Guisan, 2006; Sutherst et al., 2007; Elith and Leathwick, 2009; Kissling et al., 2012; Wisz et al., 2013). These biotic interactions may either be negative, by excluding a species from sites that are a priori environmentally suitable (i.e. within its fundamental niche; see above; e.g. competition) or facilitate a species at sites that appear environmentally unsuitable based on measured average site conditions (i.e. depending on the scale of measurement; Pellissier et al. (2010)).

Examples of positive interactions (i.e. facilitation; Boucher et al., 1982; Callaway, 1995; Stachowicz, 2001; Bruno et al., 2003) include commensalism, mutualism (e.g. non-symbiotic, Pellissier et al., 2012a; or symbiotic, Pellissier et al., 2013c), biotic engineering (i.e. a species improving the micro-habitat conditions for another), for instance forest understory species that cannot grow in plain light benefiting from the shade from the surrounding tree canopy (Linder et al., 2012; Nieto-Lugilde et al., 2015), some plant species benefiting from fungi (i.e. mycorrhizae) or bacteria (i.e. nodules) that capture atmospheric phosphorus or nitrogen in nutrient-depleted soils (e.g. Defossez et al., 2011; Pellissier et al., 2013c), or parasites requiring the presence of their host (e.g. Olwoch et al., 2003). Examples of negative interactions are more common and include competitive exclusion (e.g. Leathwick and Austin, 2001; Hakkarainen et al., 2004; Meier et al., 2010; Pellissier et al., 2010; Meier et al., 2011; Meier et al., 2012), predation (incl. herbivory; e.g. Pellissier et al., 2012b), or parasitism when the host is sufficiently affected by its parasite to be removed from entire parts of its range (e.g. Olwoch et al., 2003). Many of the previous examples of interactions often involved pairs of species (or functional groups of species), but interactions also naturally take place within more complex biotic frameworks such as food webs and large interaction networks (Schoener, 1989; Polis et al., 1997; Bascompte et al., 2003; Sargent and Ackerly, 2008; Ings et al., 2009). From a geographic perspective, the interactions derived from these interactive systems can be used as predictors of individual species distributions (Gravel et al., 2011; Pellissier et al., 2013d; Albouy et al., 2014). The way biotic interactions influence the presence (positive interactions), absence (negative interactions) or abundance (both) of a given species therefore also influences the final composition of a community (Weiher and Keddy, 2001; Lortie et al., 2004; Ricklefs, 2008; Guisan and Rahbek, 2011; Kissling et al., 2012; Wisz et al., 2013), as will be discussed later (Section 4.3). Quantifying these interactions into assembly rules is therefore required in order to understand how communities assemble, and — if the quantified rules allow — to ultimately predict assemblages (Weiher and Keddy, 2001; Fraser and Keddy, 2005; Ricklefs, 2008; Guisan and Rahbek, 2011; Hortal et al., 2012).

3.5 Further Discussion of the Realized Environmental Niche and Other Related Niche Concepts

We have mainly relied so far on Grinnell's and Hutchinson's concept of the environmental niche. However, several other concepts exist. Leibold (1995), in a review, and later on Chase and Leibold (2003), present two opposing views of the niche: (i) determined by the environmental *requirements* of species, i.e. consistent with Hutchinson's definition of the environmental niche; or (ii) determined by the *functional role* a species plays within a food chain and the *impact* it has on its environment (mainly the resources it consumes), i.e. consistent with Elton's (1927) definition of the niche. Hence, the former is embedded within an autecological and physiological view of the niche (*environmental niche*; see e.g. Austin, 1992), whereas the latter opts for the trophic levels and food web theory perspective (*trophic niche* as termed by Elton; see Austin et al., 1990; Silvertown, 2004). The two concepts are both relevant in nature, yet are often considered independently in the contemporary literature.

The habitat models literature addressed in this book, mostly focuses on the environmental niche concept. However, part of the trophic niche is nevertheless implicitly included in the restriction of the fundamental environmental niche into the realized niche and recent developments have linked food webs (and thus trophic relations) with habitat distribution modeling, demonstrating that the two concepts are intertwined (Pellissier et al., 2013d; Albouy et al., 2014). After the early work by Grinell, Hutchinson and Elton, further contributions to the niche concept were made by MacArthur (1968, 1972) and colleagues, resulting in numerous refinements, complications, and clarifications, published in the interim (e.g. Colwell and Futuyma, 1971; Vandermeer, 1972; Whittaker et al., 1973; Austin et al., 1990; Leibold, 1995; Pulliam, 2000; Chase and Leibold, 2003; Kearney and Porter, 2004; Silvertown, 2004; Wiens and Graham, 2005; Araújo and Guisan, 2006; Kearney, 2006; Leibold and McPeek, 2006; Soberón, 2007; Pearman et al., 2008a).

One proposed refinement – still not generally accepted – was to differentiate between α and β niches (and associated α and β traits) according to the scales at which measurements are made and at which niche occupancy operates. According to this distinction, the α-niche would represent the realized niche of a species, which would correspond to the niche built from environmental measurements made at the local community scale at which interactions among species occur (Pickett and Bazzaz, 1978; Silvertown, 2004). It would thus represent field observations and abiotic physico-chemical measurements at the level of conditions occupied by individuals of a species with respect to other individuals of the coexisting species in a community, assuming that differences between species measured at this scale should facilitate the species' coexistence (Silvertown, 2004; Silvertown et al., 2006). In contrast, the β-niche would correspond to the niche built from measurements at the level of mean habitat conditions within a community (e.g. mean annual temperature of a vegetation plot), and was initially defined as measures making it possible to differentiate environmental conditions between communities along gradients (Silvertown et al., 2006).

Although clearly of interest and worth further developments, these two niche measurement levels have been so far developed by only a few authors, and therefore should be treated with caution before applying them too widely. When considering these concepts in an HSM context, most analyses and models built by relating species occurrences to attributes of environmental variables therefore represent calibrations of the β-niche, while the α-niche can only be approached through local studies with exhaustive sampling of whole communities within plots and measurements of the physical environment at a sub-pixel or micro-scale (M. D'Amen and A. Guisan, unpublished). It can therefore be assumed that we will mostly address issues related to the β-niche in this book, but we will refer to the "niche" throughout for the purposes of simplification. With the latest advances in very high-resolution data coupled with intense field sampling (e.g. 1 or 2 m; Lassueur et al., 2006; Pradervand et al., 2014), however, we may get closer to the α-niche in the future (D'Amen and Guisan, unpublished).

Another recently-defined concept concerns the "potential niche". This represents the part of the fundamental niche that is actually available to species as constrained by the realized environment (Jackson and Overpeck, 2000; Ackerly, 2003). Not all the possible combinations of a set of environmental variables that make up the fundamental niche actual exist in a study area (e.g. Austin et al., 1990), or possibly even on Earth (see also Section 6.3 for some examples). In this context, another useful term is that of the "realized environment," which describes the combination of all the environmental conditions that do exist in the study area (see Guisan et al., 2014). The potential niche thus slightly differs from Hutchinson's niche in that it is constrained by what is actually available to the species in the area studied. Thus, in practice, when fitting the realized niche from field observations, one implicitly takes into account the constraints of the potential niche, provided that the sampling is representative of the whole study area. The realized niche can thus be given a more complete working definition as the fundamental niche constrained by biotic interactions, dispersal, but also the realized (available) environment. This is the definition we will use throughout the book when referring to the realized niche. It should be noted that we find the term "potential niche" rather misleading in the context of HSMs, since we refer to the model-based, projected distribution of a species as the "potential distribution", thus accepting that some parts of this mapped distribution are actually not colonized by the species due to dispersal constraints, random effects, local peculiarities not captured in the predictors, etc. However, the potential distribution of a species refers to the realized niche, not to the fundamental or potential niche.

Finally, we should remember at this stage that the original definition of the niche implies a fitness component, i.e. within the niche the species should be able to maintain positive population growth rate. In practice, observations in the field may include sink populations under conditions outside of the fundamental niche of a species (i.e. unsuitable habitats) but which are maintained through high propagule pressure (i.e. dispersal) from nearby source populations under conditions inside the fundamental niche (Pulliam, 2000). A direct corollary of this is that quantifying the realized niche from empirical field observations requires conducting a thorough sampling of the local populations and measurements need to include fitness parameters so that inappropriate sink populations can be identified and excluded from the dataset before quantifying the niche (Pellissier et al., 2013b). If this is not possible (e.g. if only occurrences or presence–absence data are available), then only an approximation of the realized niche is obtained, which may suffice to produce appropriate predictions in many cases (e.g. raw predictions of climate change effects) but not in others (e.g. conservation actions for sensitive species). In these cases, habitat suitability maps derived from presence–absence data may not provide information on population demography but rather on the species' frequency in those habitats (Thuiller et al., 2014b). Most of the predictions of species distribution found in the literature are indeed based on such approximations of the realized niche. Accidentally including a few sink populations will not drastically change the estimation of the realized niche if a fairly high number of samples from fit populations have been collected.

To sum up, the realized niche fitted in HSMs can be viewed as the intersection of conditions suitable to the species in terms of: (i) the abiotic conditions within the available environment (i.e. the potential niche), (ii) the biotic conditions, and (iii) the accessible conditions (i.e. within the colonizable or accessible range; see Section 4.2) (Pulliam, 2000; Soberón, 2007; Soberón and Nakamura, 2009). Furthermore, HSMs often fit this realized niche by relating presence-only or presence–absence observations to average environmental conditions within the modeled cells, and thus tend to ignore population fitness and to fit the niche at the β-level. Future HSM research should tend to use population fitness data more systematically, and therefore sampling such data in the field should be encouraged.

4 · From Niche to Distribution: Basic Modeling Principles and Applications

In this chapter, we describe and illustrate the principles for modeling the distribution of suitable habitats for a given species. However, as discussed in Chapter 1, habitats for other biological entities can also be modeled using the same approach, these may include intraspecific levels (e.g. subspecies, haplotypes), supra-specific levels (e.g. functional groups, communities, ecosystem types), or features that are transversal across species (e.g. species traits, genes or alleles, etc.).

Here, we address how to fit the niche of the modeled entity from field observations (4.1), how the fitted species' niche can then be projected into geographical space to predict species distributions (4.2), how individual single species predictions can be assembled into community-level predictions (e.g. species richness; 4.3), and finally the possible applications of these models and their predictions in ecology, biogeography and evolution (4.4).

4.1 From Geographical Distribution to Niche Quantification

We saw in Chapter 3 that each site on Earth is characterized by a set of environmental conditions, which define a specific habitat inhabited, or not, by a community of species (Kearney, 2006). Habitats are sometime characterized *per se*, as distinguishable units (e.g. discrete vegetation units), but most often they are simply described based on a set of variables that express various characteristics of the environment, such as climate, land use, and soil for plants, or climate, food resources, and habitat structure for animals, at one or a series of locations. Quantifying a species' realized environmental niche (see 3.3 and 3.4), makes it possible to assess the habitat suitability of any geographic unit for which those environmental niche characteristics are known, i.e.

an environmental variable with a spatially explicit coverage, as obtained through and managed in geographic information systems (GIS) and remote-sensing (RS) technologies (see Part II). Once spatially explicit layers (GIS maps) are available for relevant environmental variables defining a species' niche, it is possible to predict the potential distribution of the species' suitable habitats within the study area. This is the basic principle of habitat suitability modeling (or niche-based species distribution modeling if one can expect the whole niche envelope to be captured within the study area). HSMs thus predict a snapshot of the realized environmental niche for a species in geographic space for a given period of time (i.e. defined by the sampling dates and the environmental variables used).

How is the species' realized niche quantified? One simple approach is to take all known localities of the species and for each of them extract the attributes of the environmental variables that define the species' niche (Figure 4.1a). The next step is to relate the species observations to the environmental data in order to obtain a model of the species–environment relationship. When variation in abundances and/or known absences of the species at localities are also available, they can greatly improve the discrimination between suitable and unsuitable habitats, and therefore improve the quantification of the niche (Brotons et al., 2004; Howard et al., 2014), but such model improvement may not always be observed (Pearce and Ferrier, 2001). Such species' occurrence, presence–absence or abundance data are usually related to the corresponding environmental data using simple rules (e.g. expert or logical) or more advanced statistical techniques (Figure 4.1b; see Section 4.4 and Part III). However, many observations of species distributions used to calibrate HSMs and predict species distributions are simple occurrences, i.e. observations of species presence with no information of absence, often originating from heterogeneous databases of natural history collections (Graham et al., 2004a; see Chapter 7). In the best case scenario, these might be presence–absence data from an atlas or designed scientific surveys, or even abundance data or population fitness measurements, but these scenarios are surprisingly infrequent in the HSM literature. Observations are thus most commonly based on simple observations, and are rarely associated with field measurements of abundance or population fitness, and thus do not provide any measure of these. Without population fitness measurements, such data can potentially reflect some of the suboptimal sink situations described in Chapter 3, which might in turn affect the habitat suitability and niche estimations (Pulliam, 2000).

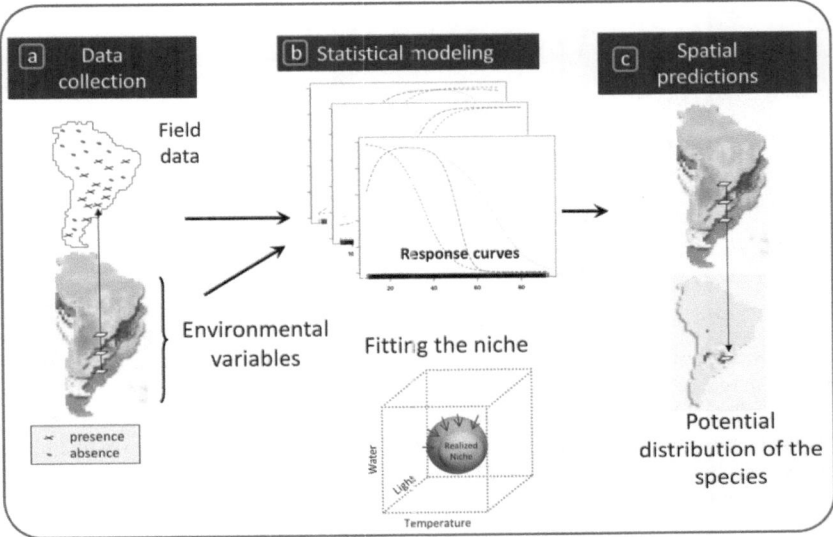

Figure 4.1 Principle of habitat suitability modeling, illustrated on a study area representing South America. (a) Field observations are collected and geo-referenced and the attributes of a set of environmental maps covering the area are extracted for each of them (see Part II). (b) Species observations are related to the environmental values using statistical approaches (Part III) to fit species response curves to each environmental predictor, in order to quantify the envelope of suitable habitat conditions, which represents the realized environmental niche of the species or of part of it if the study area does not encompass all the environmental conditions suitable for the species (see Chapter 3; the arrows suggest competitive exclusion of part of the fundamental niche from other species). (c) The fitted habitat suitability model is then used to combine the initial environmental maps and come up with a spatial projection of the model in geographic space (Part V), corresponding to the species' potential distribution. (*A black and white version of this figure will appear in some formats. For the color version, please refer to the plate section.*)

Whether absences are used or not is an important question because, as we will see later (Part III), it can condition which modeling technique is used to fit and predict the niche. However, any useful information is important when modeling species distribution. So, if absence data, for example, are available and reliable then they should be used. The modeling techniques used to fit the niche may differ (Part III). The niche may be simply captured using a series of Boolean rules (e.g. min and max along gradients, see Figure 4.2a), but in most cases it is based on statistical functions (Figure 4.2b; see also Part III). In general, the functions describing the species distribution along each environmental variable

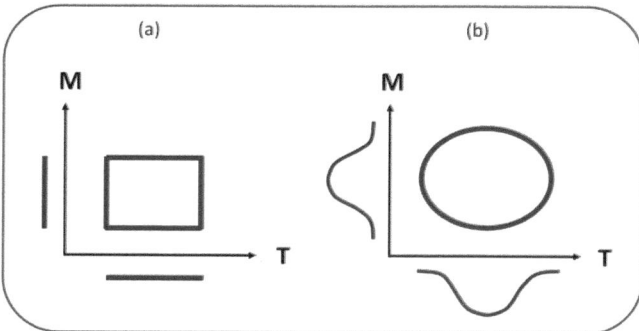

Figure 4.2 Illustration of how modeled species' responses to environmental predictors (e.g. temperature T and moisture M) in the form of simple ranges of tolerance lead to "box-like" (e.g. BIOCLIM; see Part III) representations of the niche (a), whereas smoothed unimodal response curves lead to more realistic ellipse-like niche representations (b; here using a minimum value of the combined responses to define the envelope, e.g. at predicted probability = 0.1; see Part III).

in these models are called "species response curves" (Austin et al., 1994). These response curves therefore either represent simple rectilinear "box-like" envelopes resulting in simple binary (inside/outside the niche) predictions (Figure 4.2a), or more realistic representations of the niche (e.g. unimodal) based on more gradual (and often more complex) responses in environmental space (Figure 4.2b), resulting in continuous index of habitat suitability or probability of species occurrence (usually from low to high, e.g. [0–1] or [0–100]; see Part III and Merow et al. (2014) for discussion of simple versus complex response curves and associated models).

A crucial step in habitat suitability modeling is the acquisition of spatially explicit environmental variables (i.e. maps) at the right resolution, which are sufficiently accurate to be used to determine a species niche as close to its ecophysiological needs as possible. As previously seen (3.3), environmental variables (or predictors; see Part II) can exert direct or indirect effects on species, which can be expressed as a gradient ranging from proximal to distal predictors (Austin, 2002; Huston, 2002; Austin, 2007; Mod et al., 2016). These are ideally chosen to reflect the three main types of influences on the species (Guisan and Zimmermann, 2000; Guisan and Thuiller, 2005): (i) regulators (or limiting factors), defined as factors controlling a species' metabolism (e.g. low temperatures); (ii) disturbances, defined as all types of perturbations affecting environmental systems (natural or human-induced); and (iii) resources, defined as all compounds that can be consumed by organisms (e.g. radiations for plants, water, prey for animals, nutrients for plants). All these variables are considered to be proximal predictors and are the preferential option when building HSMs. However, there are many other variables available that exert an indirect, rather than a direct, effect on species distribution. One extensively discussed example of an indirect (or distal) predictor is elevation (see Körner, 2007), which indirectly affects species distributions through its impact on numerous co-varying variables, such as solar radiation, mean annual temperature, precipitation, or topography (e.g. frequency of steep slopes). Unfortunately, the correlation with these more proximal variables is not constant across the globe, and therefore elevation (or other indirect variables) can only produce meaningful predictions when applied to relatively small regions where the conditions are unchanging (no climate or land-use change, for example).

In practice, the environmental variables available as spatial layers or maps usually vary in their degree of proximality (Austin, 2002, 2007; Mod et al., 2016). Furthermore, it can be difficult to evaluate how proximal a variable really is, because its effects are species specific (e.g. not all species are limited by low temperatures) and physiological causality can only be assessed experimentally (available only for a limited number of species). Therefore, habitat suitability modeling largely depends on prior ecological and ecophysiological knowledge, be it experimental or theoretical, about factors that physiologically and ecologically determine species distributions (e.g. Woodward, 1987; Jones, 1992 for plants). The more detailed our knowledge of the physiological drivers of species distributions, the more precisely we can identify what types of environmental maps are necessary, which will ultimately make our predictive ability more accurate.

As the most proximal variables (e.g. soil physical and chemical properties for plants) are often not available, at least in a spatially explicit way, the most common solution is to use topographic descriptors (least proximal) or environmental variables that represent averages or sums of daily or monthly values (e.g. annual mean air temperature, growing degree days; see Chapter 5) that are biologically meaningful and thus expected to provide a sufficiently good approximation of the true proximal variables (Mod et al., 2016). The degree of proximality of the predictor variables used to quantify the niche thus also conditions the degree of accuracy that can be expected when using the resulting model to make predictions (Chapter 5). However, this does also depend on the scale of the study (Austin and Van Niel, 2011).

Therefore, because of its correlative nature and due to the lack of proximality of some predictors, a statistical approach to niche quantification and habitat suitability modeling can only support previously evidenced biogeographical theories and hypotheses, or suggest new ones. It does not constitute definitive proof of a theory in its own right. However, it does provide a simple and efficient way of making biogeographic predictions (see below).

4.2 From the Quantified Niche to Spatial Predictions

After fitting the realized environmental niche (Figure 4.1b), or part of it (i.e. any envelope of suitable habitats) in a model, the next question we need to ask is: how can the quantified niche be projected into geographic space to obtain the potential distribution of a species (Figure 4.1c)? In order to answer this question, we need to transfer the statistical model, which contains a quantification of a species' response in environmental space, back to the geographic space in which the species was originally observed. To do this, one can apply the statistical functions, fitted in environmental space, to any position in geographic space using the environmental conditions at these locations from the corresponding maps. This makes it possible to express the suitability of the habitat for the species modeled in a spatially explicit way across the whole study area (Figure 4.1c). For spatial predictions, we therefore need to assign a habitat suitability value to every geographic cell in the area by applying the models to the spatial layers representing the niche variables used in the model. Since all the approaches to quantifying a species' niche result in these types of envelopes or more refined niche quantifications in environmental space, they all can yield spatial predictions. However, the exact way of doing it can differ enormously and depends on the modeling technique used (Parts III and V).

We will see later, in Part V, that many statistical models include a function for predicting a suitability value at any new location using the environmental variables available for this location. The most straightforward way of predicting the species' habitat suitability is to use this function to make predictions for all cells of the study area. To do this the function needs to be able to use a pile (or stack) of environmental maps corresponding to the variables used as predictors in the HSM (Figure 4.1c). This will yield a final habitat suitability value for each cell. If the environmental values for a set of species presence–absence locations are first entered into a statistical model as the input and then directly predicted back to a stack of environmental layers, without visualization in environmental space, then we can consider this to be a blind approach. A better way of understanding how a species' niche response has been fitted in a model is thus to first check the prediction process in environmental as opposed to geographic space alone. This can be done by projecting the habitat suitability into the environmental space in which the niche was originally fitted.

So far, we have discussed cases where models have been used to make predictions in the same area and under the same environmental conditions as those used to fit the models. However, there are many instances where predictions need to be made for separate areas or for different time periods, potentially with different environmental conditions, thus "projecting" the model and niche in space and time (Pearson and Dawson, 2003; Heikkinen et al., 2006; Pearman et al., 2008a; Thuiller et al., 2008; Elith et al., 2010; Bellard et al., 2012). Projecting a model to a different area can, for instance, be useful when assessing the invasive potential of an exotic species on a new continent, with the model being fitted in the native range of species and then projected worldwide or to specific areas (Peterson, 2003; Thuiller et al., 2005b; Broennimann et al., 2007; Fitzpatrick et al., 2007; Petitpierre et al., 2012; Guisan et al., 2014). Being able to project a model to another time period can, for instance, be useful when assessing the potential impact of future climate and land-use changes on species (Iverson and Prasad, 1998; Peterson et al., 2002a; Meyneeke, 2004; Simmons et al., 2004; Thomas et al., 2004; Thuiller et al., 2005a; Araújo et al., 2006; Engler et al., 2011b) and on biodiversity (Buisson et al., 2008; Thuiller et al., 2011; Pio et al., 2014; Thuiller et al., 2014a; Thuiller et al., 2014c). We can also assess our capacity to predict the future by testing how successfully we can project models back in time or from the past to the present (Araújo et al., 2005a; Pearman et al., 2008b; Maiorano et al., 2013), or couple HSM projections to the past using phylogeographic or phylogenetic studies (e.g. Hugall et al., 2002; Carnaval et al., 2009; Schorr et al., 2013).

How these models can be used to derive predictions under scenarios of climate and/or land-use change will be explained and treated in greater depth in Part V. As we will see, it is actually done in the same way as for predicting a model under current environmental conditions in the same area, but the model is applied to a distinct set of input environmental maps (see Part V), either for a distinct study area (for instance, the invaded range for an exotic species) or for a distinct time period according to certain environmental change scenarios (e.g. in land use or climate).

4.3 From Individual Species Predictions to Communities

We have introduced how spatial predictions can be made for individual species, under current or future conditions, or in one or several study areas. However, in many instances, several species are modeled simultaneously and the next step is to infer community or diversity patterns from the individual species models. Here, we discuss how this can be done.

Numerous studies to date have looked at changes in species composition (i.e. species turnover) in space or time. For example, measurements of species turnover are often generated in studies assessing future threats from climate change (e.g. Thuiller et al., 2005a; Engler et al., 2011b). These metrics are comparable to those used to measure changes in community composition (i.e. similarity coefficients; e.g. Soerenson) or richness (beta-diversity) between survey units (see Legendre et al., 2005). Such metrics usually compare the number and identity of shared species to the total pool of species between the two communities or regions being contrasted. It is thus a measure of species difference. Caution should therefore be taken and explanations given when applying such measurements to surveys with partial taxonomic coverage, as is the case when using incomplete atlas data (e.g. *Atlas Florae Europaeae*, Jalas and Suominen, 1972–1996; Thuiller et al., 2005a).

In cases where species are exhaustively surveyed for one taxonomic group within spatial units small enough to still correspond to a meaningful community (e.g. small pixels, field quadrats; their upper meaningful size will depend on the group studied), modeling a large set of species opens up perspective for community analyses based on assembling single species predictions (Ferrier and Guisan, 2006). By cumulating single species predictions (i.e. stacked species distribution models, S-SDMs;

Dubuis et al., 2011), one can attempt to reconstruct patterns of species richness and community properties. This is an emerging field (Ferrier and Guisan, 2006; Guisan and Rahbek, 2011; Kissling et al., 2012; Wisz et al., 2013; D'Amen et al., 2015b) and there are therefore still relatively few examples available (e.g. Guisan and Theurillat, 2000; Iverson and Prasad, 2001; Peppler-Lisbach and Schroder, 2004; Parviainen et al., 2009; Thuiller et al., 2011; Civantos et al., 2012; Mateo et al., 2012; Buisson et al., 2013; Pottier et al., 2013; Thuiller et al., 2014a; Thuiller et al., 2014c; D'Amen et al., 2015a).

The important question regarding community assembly is: to what extent can single species models integrate biotic interactions, site history and human disturbances that condition species assemblages, or how many additional constraints need to be included in a formal assembly process, for instance to limit the number of species possibly co-occurring in a given modeled unit from the potential pool of species predicted to occur there by the HSMs (Guisan and Rahbek, 2011; D'Amen et al., 2015a). This "spatially explicit species assemblage modeling" (SESAM; Guisan and Rahbek, 2011; Pottier et al., 2013; D'Amen et al., 2015a) approach provides a way of testing theories and hypotheses about how communities assemble. It opens up additional perspectives for HSM applications to address basic and applied questions in community and ecosystems ecology.

4.4 Main Fields of Application

In recent years, predictive modeling of species distribution based on estimating habitat suitability (HSMs) has become an increasingly important tool for addressing a range of issues in ecology, biogeography, evolutionary biology, conservation biology, and climate change impact research. HSMs can be used to answer both fundamental and applied questions, such as identifying the drivers of species distributions, testing biogeographic hypotheses, assessing niche conservatism, anticipating biological invasions, or assessing global change impacts on species distribution and diversity. They are also increasingly applied to entities transversal to species, such as functional traits, infraspecific taxa, or genes, and they are now coupled with other tools and data from other disciplines, in order to answer questions in phylogeography, phylogenetics, population genetics, population dynamics, and other fields.

Table 4.1 *Some possible uses and applications of HSMs.*

Type of use	Examples of studies
1. Quantifying the environmental niche of species and its changes in time and space	Austin et al. (1990); Westman (1991); Vetaas (2002); Wharton and Kriticos (2004); Luoto et al, (2006); Broennimann et al. (2012); Guisan et al. (2014)
2. Disentangling the environmental drivers determining species ranges	Duckworth et al. (2000); Anderson et al. (2002); Leathwick (2002); Normand et al. (2009); Boulangeat et al. (2012a); Gallien et al. (2015)
3. Relating HSMs to species characteristics or population demography	Huntley et al. (2004); Thuiller et al. (2010); Kharouba et al. (2013); Thuiller et al. (2014b)
4. Testing evolutionary hypotheses in biogeography	Leathwick (1998); Anderson et al. (2002); Wiens and Graham (2005); Moloney and Jeltsch (2008); Evans et al. (2009); Boucher et al. (2012); Broennimann et al. (2014c)
5. Assessing species invasion and proliferation	Beerling et al. (1995); Dirnbock et al. (2003); Peterson (2003); Thuiller et al. (2005b); Peterson et al. (2008b); DeVaney et al. (2009); Gallagher et al. (2010); Gallien et al. (2012); Petitpierre et al. (2012); Broennimann et al. (2014a)
6. Assessing the impact of climate, land use and other environmental changes on species distributions	Beaumont and Hughes (2002); Peterson et al. (2002a); Meyneeke (2004); Simmons et al. (2004); Thomas et al. (2004); Thuiller et al. (2005a); Araújo et al. (2006); Broennimann et al. (2006); Pio et al. (2014); Thuiller et al. (2014a); Thuiller et al. (2014c)
7. Suggesting unsurveyed sites with a high potential of occurrence for rare or new species	Raxworthy et al. (2003); Engler et al. (2004); Edwards et al. (2005); Guisan et al. (2007b)
8. Supporting appropriate management plans for species recovery and mapping suitable sites for species restoration	Pearce and Lindenmayer (1998); Côté and Reynolds (2002); Regan et al. (2008)
9. Supporting conservation planning and reserve selection	Ferrier et al. (2002); Araújo et al. (2004); Kremen et al. (2008); Alagador et al. (2011); Meller et al. (2014)

Table 4.1 (*cont.*)

Type of use	Examples of studies
10. Modeling community and ecosystem properties from individual species predictions	Guisan and Theurillat (2000); Ferrier et al. (2002); Olden (2003); Peppler-Lisbach and Schroder (2004); Algar et al. (2009); Guisan and Rahbek (2011); Pottier et al. (2013)
11. Detecting and anticipating disease spread and outbreaks	Peterson et al. (2002b); Peterson (2003, 2006); Estrada-Peña and Venzal (2007); Meentemeyer et al. (2008); Williams et al. (2008); de Oliveira et al. (2013); Yañez-Arenas et al. (2014); Zhu and Peterson (2014)
12. Incorporating habitat suitability into landscape (meta-)population dynamic assessments	Ferrier et al. (2002); Larson et al. (2004); Binzenhofer et al. (2005); Anderson et al. (2009); Brook et al. (2009); Dullinger et al. (2012); Naujokaitis-Lewis et al. (2013); Boulangeat et al. (2014)

As the number of papers on, and topics addressed using, HSMs has increased exponentially in recent decades (Guisan et al., 2013), it would be impossible to list them all. Instead, we have provided some examples in Table 4.1 and refer readers to review papers such as (Guisan and Zimmermann, 2000; Guisan and Thuiller, 2005; Thuiller et al., 2008; Elith and Leathwick, 2009; Zimmermann et al., 2010; Guisan et al., 2013; Thuiller et al., 2013) or complementary books (Scott et al., 2002; Franklin, 2010a; Peterson et al., 2011).

5 · *Assumptions Behind Habitat Suitability Models*

In this chapter, we present and discuss the main assumptions behind HSMs of species distribution under current conditions and within the same study area. By representing simplified models of the real world, HSMs are based on strong assumptions that are implicitly considered and should be reviewed before the models and their predictions can be used to answer basic and applied questions. We successively address the theoretical (Section 5.1) and methodological (Section 5.2) assumptions that are implicit when applying these models. Meeting all these assumptions is rarely feasible and in the two next sections we will address the possible implications for the models and their use of failure.

This chapter does not tackle the additional assumptions required when projecting HSMs in space and time, such as whether the ecological niches fitted by these models are fully captured and assumed to be stable in time and space. These additional assumptions behind HSM projections are addressed in Part V (Chapter 17), together with other related aspects.

5.1 Theoretical Assumptions

The use of HSMs implies a number of important theoretical assumptions (Guisan and Thuiller, 2005; Araújo and Guisan, 2006; Elith and Leathwick, 2009; Franklin, 2010a; Peterson et al., 2011). The three most important assumptions for applications of models in the present time are: (i) the species–environment relation needs to be considered to be at equilibrium (or pseudo-equilibrium); (ii) all important environmental predictors required to capture the desired niche of the modeled species are assumed to be available at the resolution relevant for the organism being modeled, (iii) species observations (simple occurrences, frequencies, abundance, etc.) need to be suited to the later use of the model to answer the initial aims of the study. We present and discuss these in the next paragraphs.

(i) **Species–environment equilibrium assumption:** Species data are usually sampled over a limited period of time and therefore only reflect a snapshot view of the species–environment relationship. A practical working postulate is to assume that the modeled species is in pseudo-equilibrium with its environment (Franklin, 1995; Guisan and Theurillat, 2000). This is one of the most important assumptions implicit in this type of models (see Guisan and Zimmermann, 2000; Guisan and Thuiller, 2005; Elith and Leathwick, 2009; Franklin, 2010a; Araújo and Peterson, 2012). Under this assumption, the models capture the realized environmental niche of species and then project it elsewhere or into a different time period. It is, therefore, expected that the species–environment relationship will not change in space or time. In other words, the species is expected to have colonized most of its suitable habitats in the studied area. However, there are obvious circumstances in which this assumption does not hold, for instance during biological invasions (see Thuiller et al., 2005b; Broennimann et al., 2007; Petitpierre et al., 2012; Guisan et al., 2014) or when species are still recolonizing a territory after major environmental changes (Svenning and Skov, 2004; Normand et al., 2011). Many invasive species are not in equilibrium with their environment in the invaded range, and should thus preferably be modeled using data from their native range (Peterson, 2003; but see Robertson et al., 2004; Gallien et al., 2010) or from both the native and the invaded ranges (Broennimann and Guisan, 2008; Beaumont et al., 2009; Gallien et al., 2010; Gallien et al., 2012). In the case of post-glacial recolonization, Svenning and Skov (2004) measured limited range filling (RF) – calculated as the realized/potential range size ratio – for many European tree species (RF < 50% for 36/55 species), their results suggested that many of these species still appear to be strongly controlled by dispersal constraints since post-glacial expansion, and thus might not be in full equilibrium with their environment throughout their whole range (see also Normand et al., 2011). Hence, using models that fit the observations too closely might lead to underestimating the true potential range of the species. Interestingly, although equilibrium is a required assumption for projecting a species in space or time, surprisingly few critical considerations have been raised in the recent literature about how close a given species really is to being in equilibrium with climate, and how long it would take to reach a new equilibrium, e.g. after environmental change (but see Araújo and Pearson, 2005). Nevertheless, limited RF does not imply that a species' niche cannot be captured from its current distribution. For instance, if all the possible environmental combinations that make the niche of a species are represented by species ocurrences, then the niche can be fitted successfully (Guisan et al., 2012; Guisan et al., 2014). Hence, limited RF does not necessarily hamper proper modeling of the species' environmental niche and this can be assessed using different methods, for example by comparing the realized environment where the species occurs at the time of the study (for example, using the extent of the range colonized so far) with the realized environment in the whole area of interest (e.g. Europe). However, limited RF may in some cases also result in limited niche filling (Guisan et al., 2012; Guisan et al., 2014), meaning that the realized niche of the species, as illustrated by the current state of colonization at a given time, is smaller than the one that can be expected once the whole territory is (re)colonized. Only fitting part of the realized niche in the area will then logically result in underestimated range sizes. Other factors can also affect the extent to which the realized niche can be quantified, such as positive or negative biotic interactions affecting a target species (Araújo and Guisan, 2006), which questions the stability of the modeled realized niche in space or time (Wiens and Graham, 2005; Pearman et al., 2008a; Guisan et al., 2014). Niche stability in space and time is another important assumption associated with HSMs, and is discussed in Chapter 17 (in Part V, model projections).

(ii) **Availability of all important predictors for the niche being captured:** An absence of important predictors when modeling species leaves us with unexplained variance (Austin and Van Niel, 2011; Mod et al., 2016). This may also bias the quantification of the climate niche (Bertrand et al., 2012) and/or translate into added spatial autocorrelation in the model residuals if a missing predictor is itself spatially autocorrelated (Diniz-Filho et al., 2003; Guisan and Thuiller, 2005; Dormann et al., 2007). Therefore, discussions of model predictions and their use to test theories and hypotheses should always clearly refer to the model being used and which predictors it includes. Important predictors that are unavailable should be identified prior to model fitting and implications anticipated to ensure successful predictions and avoid drawing spurious conclusions. Using a partial set of predictors might be acceptable if it is clearly stated that the study intends to consider only a subset of the environmental niche, for instance the climatic niche (e.g. Petitpierre et al., 2012).

(iii) **Appropriateness of species observations:** Whether species observations are appropriate to fit a model can only be determined if we know what the model will ultimately be used for. For instance, identifying potential locations of population persistence requires estimations of population fitness at observation sites (Keith et al., 2008; Heinrichs et al., 2010; Fordham et al., 2012). However, most species observation data do not account for population fitness (e.g. if simple presence–absence is available). If this is the case, one cannot exclude sink populations (i.e. outside the species' fundamental niche, where mortality is greater than fecundity and populations cannot maintain viable populations without constant immigration; see Pulliam, 2000). If the models are calibrated from data that include sink populations, this may seriously mislead some further applications, for instance if the predictions are used to guide conservation decision-making (Guisan et al., 2013). This issue has implications regarding the applicability of the models. Ideally, habitat suitability should be based on measurements of population fitness at each geographic location (Guisan and Thuiller, 2005), but this would prevent the use of most of the data available in natural history collections (Graham et al., 2004a), as these usually do not contain such information. However, depending on the type of organisms, sink populations may be difficult to detect when modeling over large areas or at coarse resolution (e.g. 10 km resolution). Nevertheless, this remains a potentially important issue, especially when species are modeled at fine spatial resolution.

5.2 Methodological Assumptions

The use of HSMs also implies a number of important methodological assumptions, often less explicitly stated (but see Guisan and Zimmermann, 2000), which complement the theoretical assumptions discussed in Section 5.1. These are: (iv) the statistical modeling methods need to be appropriate for the data being modeled, (v) predictors need to be measured without error, (vi) species data need to be unbiased, (vii) the species observations used to fit the models need to be independent. These assumptions are discussed in the rest of this section.

(iv) **Appropriateness of the statistical methods:** Different types of response variables require different types of statistical models (Guisan and Zimmermann, 2000). For instance, semi-quantitative data require very specific modeling techniques, such as ordinal regressions (Guisan and Harrell, 2000). Quantitative data are easier to model and there are numerous techniques available for doing so (see Part III) but, different types of quantitative responses will still require different types of statistical models. Counts of species or of individuals usually require specifying Poisson, negative binomial, or other probability distribution functions (PDF) for discrete positive values, a requirement that can be met using several modeling techniques, such as generalized regressions (e.g. Vincent and Haworth, 1983) or boosted regression trees (e.g. Thuiller et al., 2006; see Part III). Binary response variables, such as species' presence–absence data, require binomial PDF and logistic transformation (e.g. the link function in GLMs and generalized additive models (GAM); see Part III) and represent by far the data for which the largest range of modeling techniques is available in ready-to-use packages. Techniques for binary response variables are also the main ones presented in this book. Once a statistical method has been chosen, it is assumed to be the right one for the data in hand. Failure to identify the correct method can lead to errors and uncertainty in the predictions (e.g. Guisan et al., 2002).

(v) **Predictors measured without error:** This issue is rarely assessed in studies of SDMs, although errors are an inherent factor in each GIS predictor layer (McInerny and Purves, 2011; Marion et al., 2012). Although it is essentially impossible to guarantee zero errors in mapped environmental predictors, estimates of spatial distribution of errors could be associated with each layer and used to calculate spatial uncertainty in the model predictions (Barry and Elith, 2006; Van Niel and Austin, 2007). There is still a need for a proper method for combining errors from the different environmental variables in the model. This might be achieved using Bayesian approaches that can account for prior knowledge and uncertainty in environmental layers (McInerny and Purves, 2011; Keil et al., 2014).

(vi) **Unbiased species data:** HSMs attempt to quantify the environmental niche of species through models, and therefore the data used to do so need to include all possible environments that represent suitable habitats for the species modeled, at least within its colonizable range (Soberón, 2007). Therefore, the species data need to be unbiased. Bias is likely to arise when the chosen sampling design lacks a random component or when the data are gathered without employing a designed sampling strategy (i.e. subjective sampling) (Graham et al., 2004a; Albert et al., 2010). The latter typically results in data being clustered in more accessible areas, for instance along communication axes (Kadmon et al., 2004), or in some habitats being preferentially sampled or outside others being left out based on prior knowledge or the observer's judgment (Edwards et al., 2006). Any bias can potentially lead to partial niche quantification and thus to models that fail to identify all suitable habitats of a species, with obvious consequences for the spatial predictions (Thuiller et al., 2004b; Albert et al., 2010). Several studies have tested or accounted for the effect of bias in species data in HSMs (Stockwell and Peterson, 2002a; Delisle et al., 2003; Kadmon et al., 2004; Johnson and Gillingham, 2008; Loiselle et al., 2008; Phillips et al., 2009; Lahoz-Monfort et al., 2014).

(vii) **Independence of species observations:** When species observations are not independent (e.g. if spatially autocorrelated), the effective number of degrees of freedom used in many statistics associated with the models, e.g. for model selection, no longer corresponds to the apparent number of observations (Legendre, 1993; Legendre et al., 2002; Crase et al., 2014). Therefore, some statistics may be inaccurate or even wrong (Crase et al., 2014). There is no simple way of checking whether observations are independent, especially in biological systems where species interact with each other and disperse into neighboring sites. It is possible to assess spatial autocorrelation in the model residuals, but this only informs on spatial patterns, not on the processes behind these patterns. This means no one can know for certain if the data are dependent due to biological processes or if the observed patterns simply result from hidden spatially clustered important environmental variables that affect species distributions (Guisan and Thuiller, 2005; Crase et al., 2012). If dependence is suspected, for example when spatial autocorrelation is detected, how can it then be corrected in the model? In situations where spatial autocorrelation cannot be removed by subsampling the data to ensure a minimum distance between observations (i.e. the simplest solution, but infeasible if sample size is small), several other solutions have been proposed for HSMs, including spatial eigenvector mapping, autoregressive models, and generalized estimating equations (Dormann et al., 2007; Crase et al., 2012). If spatial autocorrelation is not corrected, this might cause additional problems if the HSMs are projected into the future. If spatial autocorrelation is not accounted for, projections under future climates result in higher range shifts compared to models that explicitly account for spatial autocorrelation, although under current conditions both approaches generate similar range patterns (Crase et al., 2014). Although this issue deserves more attention, we will not develop this aspect any further as it goes beyond the scope of this book. We refer readers to Dormann et al. (2007), and to the other references previously mentioned, for details on their implementation.

PART II · Data Acquisition, Sampling Design, and Spatial Scales

This part gives a broad overview of general data preparation and preliminary analysis steps. Chapter 6 covers data acquisition from existing sources for species and environmental data; it introduces spatial analyses in R that usually are carried out in GIS or RS environments. It discusses issues of pre-selecting variables for model building explains how to analyse and avoid correlation structures among variables, and discusses statistical accuracy vs. ecological explanation in predictor variables. While all this traditionally has been managed within a GIS, this part demonstrates the required steps can be done in R by providing a number of examples. This part introduces the main databases of environmental predictors and explains how digital elevation models (DEM) and RS can be used to derive ecologically more meaningful predictors. Issues related to species data are dealt with in Chapter 7. It explains how to prepare one's own sampling, discusses issues of sample size, prevalence, and spatial autocorrelation. It specifically introduces algorithms to generate designed sampling using regular or random design elements; and discusses and compares presence–absence vs. presence-only sampling. Chapter 8 addresses issues of ecological scale, namely resolution and extent aspects in the spatial, temporal, and thematic realm. This chapter is partly theoretical, highlighting the effects of scale and extent on habitat suitability modeling, but it also includes practical solutions for scaling data to appropriate common resolution and extent.

6 · *Environmental Predictors: Issues of Processing and Selection*

Predictive habitat suitability modeling usually requires extensive access to resources from a GIS, and one of the difficulties in producing effective modeling lies in the need to combine statistical and GIS modeling. This difficulty has already been addressed, mostly in reference to the limited availability of statistical functions in GIS environments (Guisan and Zimmermann, 2000). While for large spatial data structures and extensive databases this limitation still exists to some extent, there have been major improvements over the last decade, primarily in terms of the functionality of standalone packages for habitat suitability modeling (e.g. Maxent, OpenModeller, GRASP) and the ever-developing statistical environment R. Many R packages can now deal with different types of spatial data, and although most GIS functionalities have been included in several packages, they often use different formats to store spatial objects. However, the field of spatial data handling and analysis in R is rapidly evolving, and this will make GIS access obsolete for many analytical paths. Here, we introduce some of these new functionalities primarily using the `raster` package. Although under continuous development, it offers many of the well-known raster functions usually used in ArcGIS or other GIS software, and it is therefore very well suited to numerous aspects of exploratory analyses and spatial modeling in R. Other packages used here are `maptools`, `sp`, `rasterVis`, and `rgdal` from CRAN.

6.1 Existing Environmental Databases

There are a large number of available datasets ranging from the global[1] to the local scale, and we will not attempt to provide an exhaustive list here.

---

[1] http://worldgrids.org/doku.php?id=source_data and http://freegisdata.rtwilson.com/index.html

Most readers will have their own databases anyway. However, there are a number of datasets that are very commonly used, so we feel it is helpful to list some of these data sources, as they are likely to be useful when used in combination with one's own data. These have been organized according to the nature of the data. Please note that these links change very frequently, and might already be out of date when this book goes to print.

6.1.1 Digital Elevation Data

Digital elevation data (DEM), a three-dimensional representation of a terrain's surface, is very useful for deriving altitude, slope, and aspect, for example. There are two datasets used extensively at global scale. The first is the relatively dated GLOBE[2] dataset (Hastings et al., 1999), also known as GTOPO30, which is available in geographic projection at a resolution of 30 arc seconds (~1 km resolution at the equator). This dataset provides a full global coverage, but is not very precise. Nonetheless, it is perfectly sufficient for most global modeling approaches. A similar dataset including bathymetry data is available as ETOPO1[3] (Amante and Eakins, 2009) at a resolution of 1 arc minute (roughly 2 km at the equator) in geographic projection.

For finer scale analyses, most researcher use their own data, yet the globally available 90m (30m for the US) dataset SRTM[4] is a helpful data source for many, especially since it offers seamless availability for all areas between 60°N and 60°S. This dataset is also available in geographic projection as 1° (lat/lon) tiles. It has been updated and improved compared to version four, mostly with regards to gap filling.

Another interesting and comparably recent source is the ASTER-DEM, available from the Jet Propulsion Lab.[5] It was derived from multiangular ASTER space-borne imaging. It is available seamlessly for areas between 83°N and 83°S, and thus overcomes the data gap for northern latitude in the SRTM dataset. The spatial resolution is 30 m, and the data is made available in 1° tiles globally. Although the data is available at even finer resolution than the SRTM, it is not necessarily more precise. So we advise checking the accuracy carefully when it is applied to small spatial extents, and to resample the grids at a coarser resolution, if necessary.

---

[2] www.ngdc.noaa.gov/mgg/topo/gltiles.html
[3] www.ngdc.noaa.gov/mgg/global/global.html
[4] http://srtm.csi.cgiar.org/
[5] http://asterweb.jpl.nasa.gov/gdem.asp

There are many other national datasets, and most of these have restricted access or require registration. Consistent DEM datasets are very important in order to generate predictors for scaling, e.g. scaling coarser resolution climate data to finer resolutions. Most DEM data are available in ESRI grid format. Other useful formats are GeoTIFF or NetCDF. Global data are usually available in the WGS84 geographic coordinate reference system, while regional applications often have the spatial data transformed to projections that correct for angles or area. The latter is important for many ecological applications.

6.1.2 Climate Data

There are several large-scale datasets available, and depending on the nature of the analysis, researchers can choose the one dataset that is appropriate for their analysis. Worldclim[6] is probably the most widely used global climate dataset for ecological analyses (Hijmans et al., 2005). It is available in geographic projection at a spatial resolution of 30 arc seconds (~1 km), but also at coarser resolutions (2.5, 5, and 10 arc minutes). In order to project the climate extrapolation spatially, the SRTM dataset (see above) was used and resampled to the resolution and spatial registration of the GTOPO30 DEM where available, while the latter was used elsewhere. Worldclim maps are based on a large number of climate stations using long-term (1950–2000 in general, but 1961–1990 for most stations) monthly mean climate information for precipitation (47,554 stations), maximum (24,542 stations), and minimum (14,930 stations) temperature. It is generally available as version 1.4. Just recently, version 2.0 was released for beta testing (beta release 1, June 2016). Due to the globally uneven distribution of climate stations, the mapping uncertainty varies substantially in space. The climate mapping method is based on the ANUSPLIN package, which uses thin-plate smoothing splines (Hutchinson, 1995). In addition to basic climate parameters such as monthly mean, minimum, and maximum temperature and precipitation, this dataset provides a set of 19 so-called bioclimatic variables (bioclim), which are supposed to be more biological relevant than the original monthly climate layers, from which the bioclim variables are derived. The datasets are primarily made available for the current climate. However, the website also offers datasets for historical data (three time slices for the last interglacial, last glacial maximum, and mid-Holocene), as well as for projected future climates for large numbers of global circulation models (GCM) and scenarios

[6] http://worldclim.org/

originating from CMIP5 (Coupled Model Intercomparison Project 5; Meehl et al., 2009) and the Fifth Assessment Report (IPCC, 2013) of the Intergovernmental Panel on Climate Change (IPCC). Based on the Worldclim baseline (current) climate, the CGIAR (Consultative Group on International Agricultural Research) website operated by the Climate Change, Agriculture and Food Security research program (CCAFS) provides a large array of additional projected future climate datasets for 10-year monthly means.[7] These are available at different resolutions and are based on various downscaling techniques. They currently represent the largest available archive of downscaled climate data for use in ecology, evolution and environmental sciences.

For North America, there are three widely used datasets at similar spatial resolution. The first is the PRISM dataset.[8] This mapping approach is semi-statistical in that certain theoretically derived maximum lapse rates are not exceeded when calculating regressions from station values (Daly et al., 1994). Climate data is available for individual months and years at a 4 km spatial resolution from 1895–2010, while 1981–2010 climatological normals are available at an 800 m and a 4 km spatial resolution. PRISM also provides daily data (at 4 km resolution) since 1981. The DAYMET dataset[9] is based on daily climate interpolations (Thornton et al., 1997; Thornton and Running, 1999) originally available for the 48 conterminous United States and maps daily climate surfaces for minimum and maximum temperature, precipitation, global incoming shortwave radiation, and vapor pressure deficit for the period 1980–1997. The new version v3 now covers all of North America, Puerto Rico, and Hawaii and is available at a 1 km spatial resolution in the NetCDF format and covers the period of 1980–2015 (Thornton et al., 2016). Maps are also made available for monthly and yearly data. Point locations can be sampled on the DAYMET website for any sequence of days. The DAYMET method is based on a distance-weighted regression, where weights follow a Gaussian shape filter, which is adjusted on a yearly basis according to the availability of stations and the spatial density of these stations is also considered. Another set of climate layers is available from the Moscow Forestry Sciences Lab.[10] It is available at 30 arc second resolution and is based on the ANUSPLIN method (as Worldclim), but more emphasis was placed

---

[7] www.ccafs-climate.org/
[8] www.prism.oregonstate.edu/
[9] https://daymet.ornl.gov/
[10] http://forest.moscowfsl.wsu.edu/climate/

on regional climate station density. Products are available for the western United States, for Mexico, and for all of North America, including Mexico, the United States (including Alaska), and Canada. As in Worldclim, a set of 15 derived variables has been computed that is supposed to have more biological relevance. In addition to current climate layers, data are provided for 2–3 SRES scenarios from three GCMs for three time slices of an average of 10 years each representing simulations for the 4th IPCC Assessment Report (IPCC, 2007). Now, the site also offers 2–3 RCP scenarios for 5 GCMs (again three time slices of 10 years each) representing simulations for the 5th IPCC Assessment Report (IPCC, 2013). In addition, an ensemble of 17 GCMs for 3 RCPs is provided in the same temporal resolution as the datasets mentioned above.

While all the products described above are available at approximately 1 km spatial resolution, there are many more available at much coarser resolutions. At the 10' to 0.5° resolution, the most widely used dataset is the one produced by the climate research unit (CRU) of the University of East Anglia.[11] These datasets have coarser spatial resolutions, but provide time-series for both historical (1900–2000) and projected future (2001–2100) climates, where the latter are available for a range of IPCC scenarios and GCMs. The biggest source of projected future climate data is available from the CMIP data portal,[12] which currently hosts the most recent global GCM simulations (CMIP5) that were used for the 5th IPCC Assessment Report (IPCC, 2013). Based on these GCMs, a new effort to physically downscale these GCMs using regional climate models (RCMs) was initiated, known as CORDEX (Coordinated Regional Climate Downscaling Experiment), which offers global coverage for climate data at a 0.44 ° (50 km) spatial resolution.[13]

Such data are frequently used in combination with higher resolution climate data such as Worldclim in order to provide downscaled projections of future climate at a regional scale (e.g. Engler et al., 2009; Randin et al., 2009). While the CRU website provides a consistent set of data (variables, IPCC scenarios, GCMs, spatial resolution), the different versions of global GCM runs in the CMIP database are less consistent with regards to extent, spatial resolution, or temporal coverage. The CORDEX database is more consistent, as all projections are generated at the same spatial resolution. These datasets are usually stored in the

---

[11] www.cru.uea.ac.uk/cru/data/hrg/
[12] http://cmip-pcmdi.llnl.gov/index.html
[13] www.cordex.org/

NetCDF format, which is optimized for storing the complex data structures produced by GCMs or RCMs.

6.1.3 Land Cover/Land-Use Data

There are a large number of global land and vegetation cover datasets. While the different datasets were developed independently from each other and for a range of different purposes, more modern products are often provided with several classification schemes, and therefore include more or less classes for the same basic product. Some of these products are regularly updated, while others represent the land cover classification for input data from a certain time period. In this case, this data can only be used meaningfully if explicit reference is made to the time period in question.

Land-use is often reclassified from land cover datasets, and there are fewer products available. However, both types of datasets are important and useful for a range of large to regional scale applications. Both land-cover and land-use change are primary threats to regional biodiversity (Sala et al., 2000; Thuiller, 2007), and it is vital that information on these patterns and processes is included when assessing and modeling species patterns in space and time (Thuiller et al., 2014a). We will discuss herein a number of datasets of interest at regional or global scale, but this list is by no means exhaustive.

One of the early datasets was the International Geosphere Biosphere Program (IGBP) classification of land cover containing 17 classes (Belward et al., 1999). It represents a classification of data collected on a daily basis between April 1992 and March 1993 by the Advance Very High Resolution Radiometer (AVHRR) scanner, a satellite operated by the National Oceanic and Space Administration (NOAA). The first version of this dataset was released in 1997 as version 1.2, and an updated classification of the same original data was later made available as version 2.0.[14] This basic 17-item legend is now continued with a product from the MODIS sensor on board the TERRA satellite, available yearly since 2000 at 500 m and 0.05° spatial resolution.[15] A similar land cover product was generated using AVHRR data and a classification scheme of 12 classes globally (Hansen et al., 2000),[16] available at 1 and 8 km, as well as at 1° spatial resolution. MODIS data can be obtained from an array of

---

[14] http://landcover.usgs.gov/globallandcover.php
[15] https://lpdaac.usgs.gov/dataset_discovery/modis/modis_products_table
[16] www.landcover.org/data/landcover/

different gateways,[17] and the so-called collections (processing schemes) have to be selected very carefully, since these differ slightly with respect to spatial and temporal resolution, as well as classification method (Friedl et al., 2010).

The European Space Agency (ESA) also provides a global land cover product. This Climate Change Initiative (CCI) land cover product (also known as GlobCover) was generated from time-series of ENVISAT MERIS images. The dataset covers the majority of the globe (75°N to 56°S, excluding Antarctica) at 300 m spatial resolution (Bartholome and Belward, 2005) and is now available for three time slices representing *c.* 2000, 2005, and 2010.[18] The classification follows the 22-class levels of the Land Cover Classification System (LCCS).

Other land cover products are based on much higher resolution sensor data such as Landsat. One example is the National Land Cover Database (NLCD) (Homer et al., 2012), a Landsat-based, 30 m resolution land cover database for the USA. NLCD provides land surface information such as thematic class, percent impervious surface, or percent tree canopy cover. NLCD data are available for several time steps (1992, 2001, 2006, and 2011) and are free for download from the MRLC website.[19] Also the Gap Analysis Program (GAP) methodology (Scott et al., 1993) has produced a large series of state-wise or regional land cover products (e.g. Lowry et al., 2007), which are very detailed and often used for ecological applications, especially in biodiversity assessment and animal conservation planning (Scott et al., 2001; Pearlstine et al., 2002). The criticism that conservation decisions for animals should not rely on land cover data alone has been raised since error rates are considered too high (Schlossberg and King, 2009). However, in many instances, they are likely to provide important information on local-scale filtering between climatically suitable areas, because of their structural and contextual relevance. Another product often used in Europe is the Coordinated Information on the European Environment (CORINE) land cover (also known as CLC).[20] This combined land cover/land-use dataset is built on a hierarchical legend, which distinguishes >50 classes at finest scales. It comes in two spatial resolutions, 100 m and 250 m. Three datasets exist, one for 1990, one for 2000, and one for 2006, the latter

---

[17] https://lpdaac.usgs.gov/data_access
[18] http://due.esrin.esa.int/page_globcover.php
[19] www.mrlc.gov/
[20] www.eea.europa.eu/data-and-maps/

covering more countries of the European Union than the first. There is also a "change product" available, where land cover/use change has been explicitly mapped.

A slightly different group of interesting land cover datasets is available from what are termed land cover continuous fields (also called fractional cover or sub-pixel classification; see also the NLCD dataset above). These maps usually originate from coarse resolution satellite images (MODIS, AVHRR, etc.) where associating a single discrete land cover class to a pixel is difficult due to the small-scale structure of land cover in many parts of the world (DeFries et al., 1997, 1999; Hansen et al., 2002; Schwarz and Zimmermann, 2005). On the other hand, such coarser resolution satellite data are usually available at a much higher temporal resolution, which enhances the capacity for classification due to the multi-temporal nature of the data. In order to maximize the usefulness of this spatially coarser resolution data, the pixels are not assigned to a discrete class, but instead the fraction of each class per pixel is mapped (Ju et al., 2003). Strictly speaking, in spectral terms, such cover fractions do not map the fraction of a land cover class, but the fraction of the elements that make this land cover class: e.g. trees, grass, impervious (Hansen et al., 2002; Schwarz and Zimmermann, 2005). Such global data can be downloaded from the MODIS data website (see link above). Another source of information for global continuous field datasets can be obtained from the global land cover facility (GLCF),[21] both for the MODIS-based as well as AVHRR-based products. These datasets are usually available at 250 m (MODIS MOD44B product) and 1 km (AVHRR) spatial resolution. A downscaled version using Landsat images is available as "Landsat Tree Cover" from the same website. It is available at 30 m spatial resolution globally, and specifically improves estimates in agricultural areas (Sexton et al., 2013). However, such datasets can easily be generated for one's own purposes anywhere on the globe with high accuracy (Schwarz and Zimmermann, 2005), and it can also be easily applied directly to other sensors, such as Landsat (Mathys et al., 2009).

Often though, we are interested in land-use rather than in land cover. This is more difficult to map, since it involves interpreting human use of what can objectively be seen from the above, e.g. mapped grassland might be a meadow or a pasture. It might be hard to distinguish between the two, since the difference is the use and not the cover. Furthermore, grasslands can originate from agricultural use, and would naturally revert

[21] www.landcover.org/data/

to forest by means of succession if this use is stopped, while other grasslands are permanent because the conditions are not suitable for forests. However, when projecting any model into the future, we should be aware that climate is not changing independently of changes in human land-use. In fact, human actions result in a modification of the radiative forcing, some of which is direct e.g. burning fossil fuels, but another important factor is land-use change. Thus, human activities including land-use change are the cause of global climate change, and therefore future projections of species or biodiversity may suggest that we include the effects of associated land-use change. Some aspects of land-use are included to a greater or lesser extent in most of the land cover products described above. At a global scale, consistent land-use products are most often available at very coarse spatial resolutions, which are generally too coarse to be used effectively in HSMs. We will therefore not discuss these products here. There are regional or local products available, which are suitable for use in HSMs.

6.1.4 Borders, Political Units, and Other Vector Data

There are several data sources available from which users can access and download free spatial vector datasets. One source is the Global Administrative Areas database (GADM)[22] that contains the spatial data of the world's administrative areas, such as countries, and lower level subdivisions, such as provinces. The data are available in several formats such as shapefile, ESRI geodatabase, RData, and Google Earth kmz. Data can be downloaded either by country or for the whole world.

Another source of vector data is Natural Earth,[23] which also includes countries, disputed areas, first-order admin (e.g. departments, states). Data on populated places, urban polygons, parks and protected areas, and water boundaries are also available with different levels of detail.

A source for European-oriented (but also global) datasets is the ESPON database.[24] It provides access to regional, local, urban, neighborhood, and historical data, but also to indicators and tools that can be used for European territorial development and policy formulation at different geographical levels. The data included in the ESPON database is mainly from European institutions such as EUROSTAT and EEA and is aimed at a wide range of users (researchers, policy makers, stakeholders).

---

[22] http://gadm.org/
[23] www.naturalearthdata.com/
[24] http://database.espon.eu/db2/

Historical data can be obtained from the CShapes database.[25] This database contains worldwide historical maps of country boundaries and capitals in the post–World War II period together with dates of when these changes occurred. The CShapes database is available as a shapefile format or as a package for R (CShapes) and can be used for a number of spatial analyses.

Finally, a very useful database for marine information is Marine Regions.[26] It is an integration of the VLIMAR Gazetteer and the VLIZ Maritime Boundaries Geodatabase and includes marine boundaries, fishing zones, ecological classifications of marine ecosystems and names such as seas, sandbanks, seamounts, and ridges. The geographic cover of the database is global.

6.2 Performing Simple GIS Analyses in R

In this chapter, spatial data visualization, processing, and analysis are introduced using functions and packages available in R. Each reader may have their own favorites and can combine these with the functions and commands presented here. We will primarily use regional to global scale datasets that are freely available on the internet. This chapter primarily deals with raster data, which are often used in spatial habitat suitability modeling. The reader learns how to adjust resolution and extent of raster layers, how to re-project and recalculate them, or how to stack and overlay them with point observation data (field samples). Other sections introduce how to generate contours from raster layers or how to use lines or polygons in combination with raster data. In a final section, the generation of a consistent data structure for habitat suitability modeling is introduced.

6.2.1 Introduction

We will make use of the GIS layers available in ESRIs grid and vector formats as well as GeoTIFF, which are the most widely distributed formats for interacting statistics with GIS. In order to prepare for further statistical analyses, simple spatial operations and analyses are introduced, such as building spatial datasets, importing grids, vectors and points, developing new and partly DEM-derived raster layers, intersecting points with grids, or building simple SDMs and predicting the

---

[25] http://nils.weidmann.ws/projects/cshapes

[26] www.marineregions.org/about.php

modeled habitat suitability and associated spatial uncertainties for the whole spatial study area.

In order to carry out these analyses in R, several spatial R packages need to be installed and loaded. Since some of these packages are still under development, it is important to use the right version. The required packages are: `maptools`, `sp`, `rgdal`, and `raster`. Installing the packages is generally comparatively straightforward. Once these packages are installed, they can be loaded as follows:

```
> library(sp)
> library(rgdal)
> library(raster)
> library(dismo)
> library(maptools)
```

Furthermore, and for all future analyses, it is important to make sure that R is directed to the correct working directory:

```
> setwd("PATH/data/")
```

All data loaded in subsequent examples are indicated in the R commands relative to this path.

In addition to the preparation in R, all the required spatial data have to be prepared and are stored, preferably in one major folder. Preparations ideally include converting all raster and vector layers to a common projection and map extent/resolution (for grids). As raster layers can cause memory limitation problems, they should be read as integer grids where possible. Preparation thus may also include a conversion to integers. Floating grids are often best multiplied by a factor of 10 or 100 prior to being converted to integers in order to keep thematic resolution. Most GIS and RS data used in subsequent analyses are available and loaded directly from online sources, while some are available from this book's website.[27] Please download and unpack the data from the book's website locally on your computer. The data are generally stored in subfolders according to their nature and thematic similarity.

6.2.2 Loading the Data and Initial Exploration

Loading data into R using the raster package commands can be done very easily with the `raster()` command. Load those grids you want to use later in your exercises. Here we provide the commands for four

[27] www.unil.ch/hsdm

layers, namely: temperature isothermality (bio3), temperature annual range (bio7), mean temperature of coldest quarter (bio11) and annual precipitation (bio12). The data is loaded from the basic subfolder termed "raster" within the set path (see above).

```
> bio3    <-raster("raster/bioclim/current/grd/bio3")
> bio7    <-raster("raster/bioclim/current/grd/bio7")
> bio11   <-raster("raster/bioclim/current/grd/bio11")
> bio12   <-raster("raster/bioclim/current/grd/bio12")
```

These grids are now loaded into R and we can access them directly. In order to evaluate the imported datasets, we can use GIS-type commands that give us this information, similar to the "describe" command in ArcInfo:

```
> bbox(bio7); ncol(bio7); nrow(bio7) ; res(bio7)
          min       max
s1  -180.00000  180.00000
s2   -57.49999   83.50001
  [1] 240
  [1] 94
  [1] 1.5 1.5
```

Wordclim data have been loaded at 1.5° lat/lon resolution, which roughly corresponds to a 167 km spatial resolution at the equator.

We then load the GTOPO30 global DEM. As seen in Section 6.1.1, this dataset is available at 30 arc seconds, translating into roughly 1.85 km at the equator. We also assign a projection to this raster, without much explanation here as this topic is treated later (Section 6.2.8):

```
> elev <- raster("raster/topo/GTOPO30.tif")
> projection(elev) <- "+proj=longlat +datum=WGS84 +ellps=WGS84
    +towgs84=0,0,0"
```

If we now compare the elev and the bioclim grids (e.g. bio7), it becomes obvious that they do not have the same spatial extent, pixel resolution or number of rows and columns:

```
> bbox(elev); ncol(elev); nrow(elev); res(elev)
    min       max
s1 -180  180.00002
s2  -60   90.00001
  [1] 43200
  [1] 18000
  [1] 0.008333334 0.008333334
```

Specifically, we see that elev has many more rows and columns than bio7, and also the lower-left coordinate, the extent, and the pixel size differ. They have neither the same resolution nor extent.

6.2.3 Resampling, Spatial Alignment, and Indices

In order to generate a clean data structure for modeling and combining all grids into one stack of grid layers for our study area (this facilitates subsequent analyses) we first resample all grids with deviating resolution and extent (here `elev`) with reference to a master grid (here `bio7`). In this way, we resample the `elev` grid from 30 arc second resolution to a 1.5 ° resolution, and align the new `elev` grid to the lower-left corner of the `bio7` grid, which serves as a master for this operation. Two methods are available within the `resample()` command to adjust the resolution, namely nearest neighbor (`method="ngb"`) and bilinear (`method="bilinear"`). The first is fast and picks the value at the central location of the new grid from the old grid with no interpolation. This method is usually recommended for categorical data. The bilinear method is much slower, as it interpolates from all neighboring cells to calculate the new cell values. It is usually preferred for continuous data. For the purposes of simplicity, here this resampling is carried out using the nearest neighbor method (`method="ngb"`).

```
> elev1<-resample(elev,bio7,method="ngb")
```

In order to accurately upscale continuous grids to much larger pixel sizes it is advisable to use the `aggregate()` function, which extends the size of the pixel by a multiple of the original pixel size. This method is much faster than bilinear resampling, although it still interpolates among the aggregated pixels, e.g. by using the mean function (`fun=mean`) as argument. This is done by first aggregating to a resolution close to 1.5°, if this value cannot be obtained by a direct multiple of the original cell, and then interpolating to the exact resolution and extent using the `resample()` command.

Now all raster layers are converted to the same spatial structure and can be visualized. For the plotting, the `terrain.colors()` option represents the default color theme in the raster package. The other three standard themes, `topo.colors()`, `rainbow()` and `heat.colors()`, can also be used.

```
> plot(bio12,main="Bio.12")
> par(mfrow=c(3,1))
> plot(elev1,col=rev(topo.colors(50)),main="Elevation")
> plot(bio3,col=heat.colors(100),main="Bio.3")
> plot(bio11,col=rainbow(100),main="Bio.11")
> par(mfrow=c(1,1))
```

This creates three maps with conserved x- and y-axis ratios. In a regular plot, such axis ratio conservation can be achieved using the option

(a)

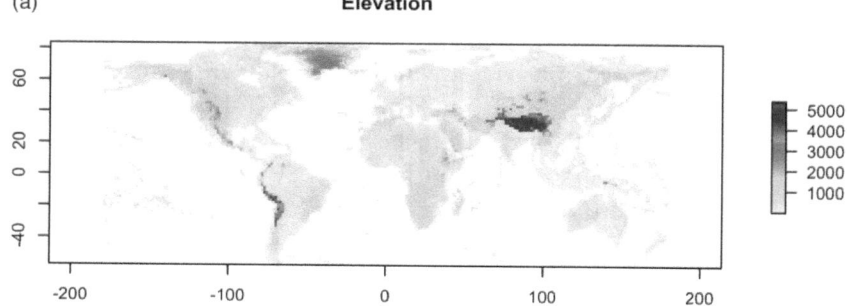

(b)

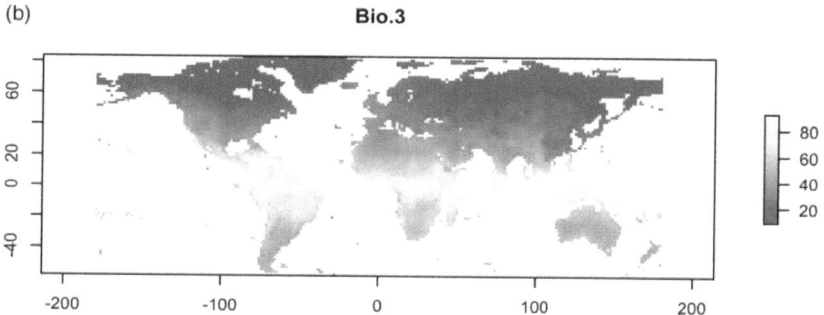

(c)

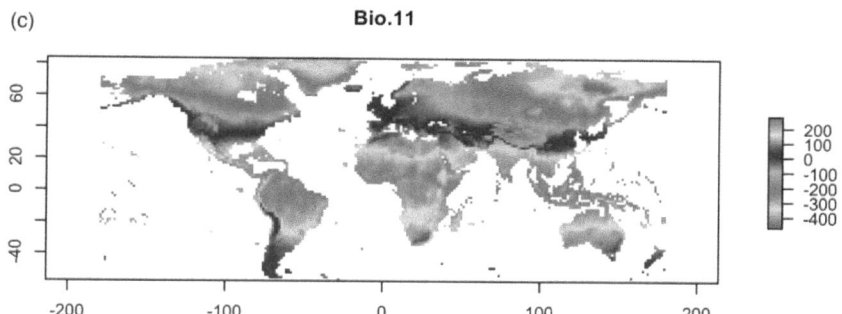

Figure 6.1 Illustration of three maps using the raster package: (a) elev1 = altitude, (b) bio3 = isothermality (×100), and (c) bio11 = mean temperature of the coldest quarter (×10). (*A black and white version of this figure will appear in some formats. For the color version, please refer to the plate section.*)

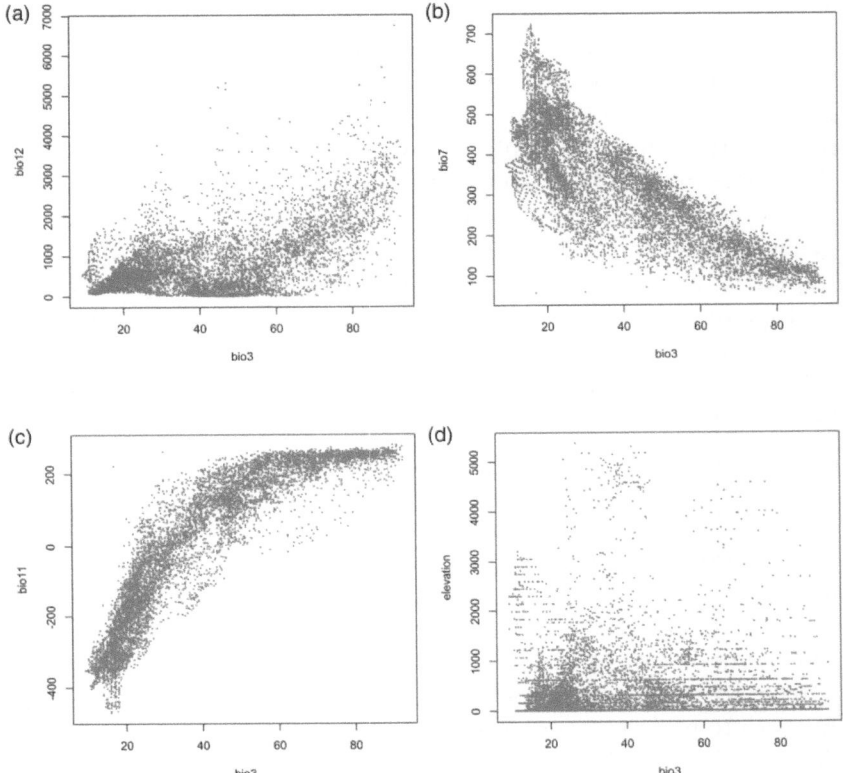

Figure 6.2 Illustration of the pairwise correlation structure of three bioclim and one elevation grid.

(asp=1), but in raster this is not necessary. In the last sequence of commands we have used three different color schemes in one plot window by using the par() command (Figure 6.1), which is a very flexible instrument to define plot parameters.

The raster package provides an easy way of observing the correlation between rasters. The same plot() command in raster can be used to visualize the shape of correlations among raster layers (Figure 6.2):

```
> par(mfrow=c(2,2))
> plot(bio3,bio12,xlab="bio3",ylab="bio12",col="gray55")
> plot(bio3,bio7,xlab="bio3",ylab="bio7",col="gray55")
> plot(bio3,bio11,xlab="bio3",ylab="bio11",col="gray55")
> plot(bio3,elev1,xlab="bio3",ylab="elevation",col="gray55")
> par(mfrow=c(1,1))
```

6.2.4 Working with Contours and Lines

Other types of spatial data can also be loaded in R for habitat suitability modeling, namely shapefiles that represent points, lines and polygons. Several packages are useful for reading and writing shapefiles, but some of them require detailed knowledge to do so. We prefer to use the simple shapefile() command from the raster package, which can be used to read points, lines and polygons. Often, however, we generate lines or points in R, and then we may want to store these (especially lines and polygons) as shapefiles. Here, we use the rgdal package to do so.

First, we return to the elev1 raster object, which we have created. The contour() function can then be used to represent selected elevation (Figure 6.3).

```
> plot(elev1)
> contour(elev, nlevels=7, levels=c(0, 500, 1000, 1500, 2000,
3000, 4000, 5000), add=T, labels="", lwd=.2)
```

Of course, the contour function can also be used to generate probability isolines from SDM output rasters, or other isolines from any numerical or continuous raster surface. It is thus a very practical command for adding lines to a raster to better visualize the results.

In order to visualize these contours in more detail, we select South America to plot these contours. We do so by cropping the global elevation model to a smaller extent and then plotting the same contours to the graph as done above, but for the elev_sa elevation raster (Figure 6.4).

```
> elev_sa<-crop(elev, extent(-85,-30,-60,15))
> elev_na<-crop(elev, extent(-188,-50,15,90))
> plot(elev_sa, main="Elevation Contours South America")
> contour(elev_sa, nlevels=7,
levels=c(0, 500, 1000, 1500, 2000, 3000, 4000, 5000),
add=T, labels="", lwd=0.2)
```

Next, we need to store the global contour line as a spatial object, and finally write it into a line shapefile. To do this, we use the rasterTo-Contour() command in the raster package.

```
> iso<-rasterToContour(elev, nlevels=7, levels=c(0, 500, 1000,
1500, 2000, 3000, 4000, 5000))
> writeOGR(iso, dsn="vector/globe", layer="isolines",
"ESRI Shapefile", check_exists=T, overwrite_layer=T)
```

Essentially, the command for creating contour lines is very similar to drawing them on an existing plot. We then store the lines as shapefiles using the writeOGR() command from the rgdal package. We now

Elevation Contours

Figure 6.3 Elevation contours displayed over digital elevation model. (*A black and white version of this figure will appear in some formats. For the color version, please refer to the plate section.*)

South American Elevation Contours

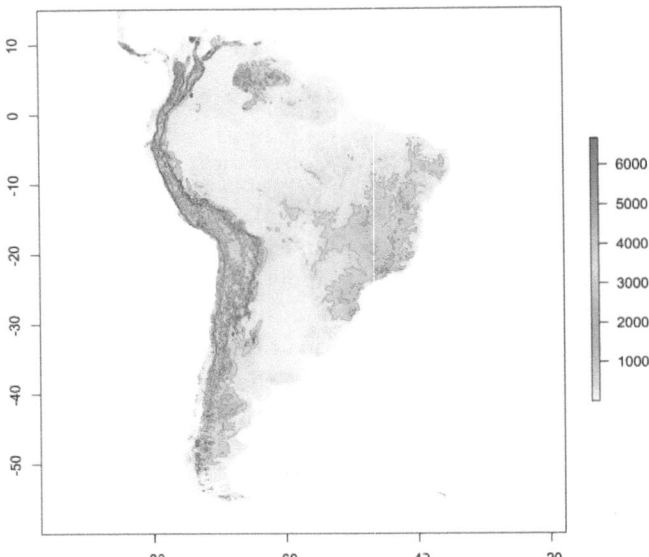

Figure 6.4 Contours displayed over a digital elevation model of South America. (*A black and white version of this figure will appear in some formats. For the color version, please refer to the plate section.*)

can read this (or other) shapefiles using the simple shapefile() command from the raster package.

```
> iso1<- shapefile("vector/globe/isolines")
> plot(elev_na,col=terrain.colors(40),main="Elevation Contours")
> lines(iso1, lwd=0.2)
```

6.2.5 Raster Analyses of Type "Global"

Here, we calculate simple raster operations. These are called "global" operations, because each cell is treated exactly the same way, meaning that the operation is applied equally to all cells. In the first example, we assume that temperature on the one hand is related to elevation, latitude and longitude. If we conduct this type of analysis using ordinary least-square regressions, we might obtain the following output:

```
> summary(reg.lm)

Call: lm(formula = bio11 ~ elev + abs(lat) + abs(lon) + I(lon^2)

Coefficients:
              Estimate    Std. Error  t value  Pr(>|t|)
(Intercept)   4.138e+02   2.146e+00   192.82   <2e-16 ***
alt          -3.624e-02   8.603e-04   -42.13   <2e-16 ***
abs(lat)     -8.216e+00   2.885e-02  -284.82   <2e-16 ***
abs(lon)     -1.794e+00   5.389e-02   -33.29   <2e-16 ***
I(lon^2)      7.122e-03   3.206e-04    22.22   <2e-16 ***
---
Signif. codes:  0 '***' 0.001 '**' 0.01 '*' 0.05 '.' 0.1 ' ' 1

Residual standard error: 58.56 on 8290 degrees of freedom
Multiple R-squared:  0.9191,  Adjusted R-squared:  0.9191
F-statistic: 2.356e+04 on 4 and 8290 DF,  p-value: < 2.2e-16
```

We see that we can express the mean temperature of the coldest quarter (bio11) from a sample of pixels in the climate map as a function of elevation, latitude, and longitude. We now apply this function to each cell of the elevation grid and then plot the resulting grid (Figure 6.5). To implement this calculation, we first need to load the two missing raster files (lat, lon) from TIFF files and then calculate the difference between the observed and the modeled mean temperature of the coldest quarter:

```
> lat<-raster("raster/other/latitude.tif")
> lon<-raster("raster/other/longitude.tif")
> tcold<-4.138e+02 + (-3.624e-02 * elev1) + (-8.216e+00 *
abs(lat)) + (-1.794e+00 * abs(lon)) + (7.122e-03 * lon^2)
> diff_obs_model_temp <-bio11 - tcold
> par(mfrow=c(2,1))
> plot(tcold, col=rev(rainbow(100))[20:100],main="Modelled mean
temperature of the coldest quarter")
> contour(elev, nlevels=7,
levels=c(0, 500, 1000, 1500, 2000, 3000, 4000, 5000), add=T,
labels="", lwd=.3)
```

```
> plot(diff_obs_model_temp, col=rev(rainbow(100))[20:100],main=
"Difference between modelled and observed temperatures")
> contour(elev, nlevels=7, levels=c(0, 500, 1000, 1500, 2000,
3000, 4000, 5000), add=T, labels="", lwd=.3)
> par(mfrow=c(1,1))
```

(a)   **Modeled mean temperature of the coldest quarter**

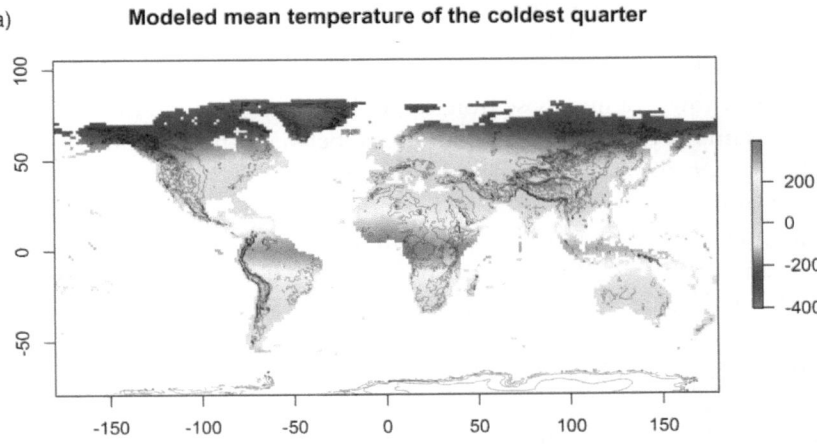

(b)

Difference modeled from observed mean temperature of the coldest quarter

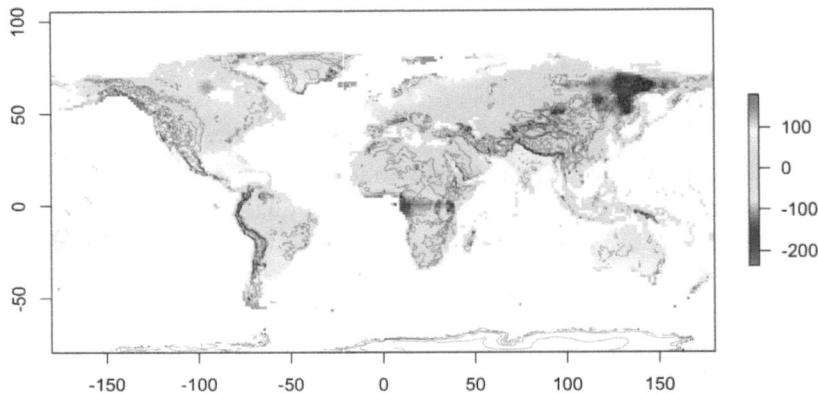

Figure 6.5 Elevation contours displayed over (a) modeled mean temperature of the coldest quarter and (b) difference between modeled and observed temperatures. (*A black and white version of this figure will appear in some formats. For the color version, please refer to the plate section.*)

In the same way, we can perform any mathematical operation between rasters or between rasters and scalars, so that each cell is processed identically over the whole raster extent.

6.2.6 Raster Analyses of Type "Focal" and Terrain Analyses

The next typical raster operation is the "focal" type. We talk about focal analyses when applying neighborhood-based analyses. A moving window is usually run across the whole grid, a function is applied to the cells within the window, and the result of this function is written in the center cell of the window. The window is then moved from one cell to the next row/column-wise from upper left to lower right. In the following example, we calculate the difference between the elevation grid and the average elevation in a moving window of 15 × 15 cells. The calculation of the average elevation in a window (i.e. the focal analysis) means to apply a smoother to the elevation, and we compare this smoothed elevation to the original elevation. Therefore, if the center cell now has a "higher" value than the surrounding 15 × 15 cells, then this cell must be a positioned on a ridge or peak. Likewise, if the center cell has a considerably lower elevation than the surrounding cells, then it must be positioned in a toe slope or gully. If the difference is negligible, then the center cell of the moving window must be either a flat plain or an even slope. Such analyses can be used to classify the terrain. Here, we carry out this analysis on a *c.* 10 km DEM of South America. In order to do so, we first crop a *c.* 1 km DEM called `elev` (see above) to the extent of South America and then we aggregate this temporary grid by an aggregation factor of 10.

```
> tmp<-crop(elev, extent(-85,-30,-60,15))
> elev_sa10<-aggregate(tmp, fact=10, fun=mean, expand=TRUE,
na.rm=TRUE)
```

Next, we need to define the focal window including the weights. In this analysis, we give equal weight to all cells in a quadratic 15 × 15 cell window. The window weights matrix is created as follows so that all weights have a value of 1:

```
> w <- matrix(rep(1,225), nr=15, nc=15)
```

Finally, we calculate the focal operation over the 10 km DEM using the `w` window weights and we subtract this focal analysis result from the original elevation grid. We call the resulting raster `TopEx`, representing topographic exposure with negative values representing sinks and valleys and positive values representing ridges and peaks.

Topographic Exposure over South America

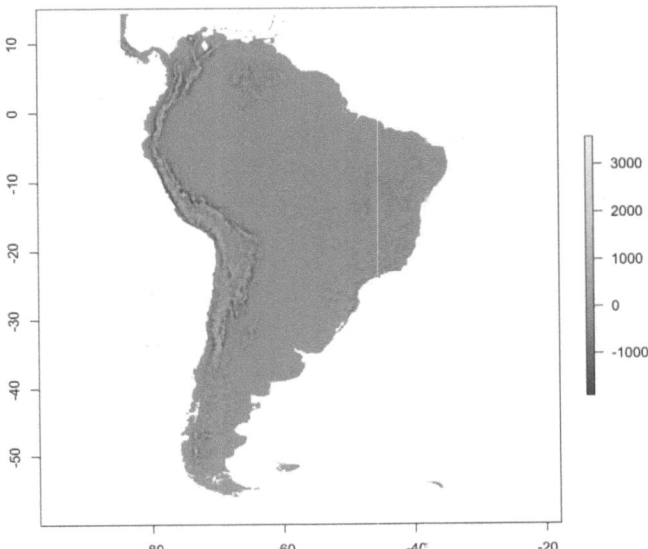

Figure 6.6 Topographic exposure over South America calculated from a focal analysis.

```
> TopEx <- elev_sa10 - focal(elev_sa10, w=w, fun=mean, na.rm=T)
```

To plot this raster, we first define a color scale that is optimal for terrain data, be it solar radiation or topographic exposure data. Then we apply this color scheme to the plot so that positive values are shown in light shades and negative values in dark shades (Figure 6.6).

```
> topography.c <-colorRampPalette(c("dodgerblue4",
"lemonchiffon", "firebrick3"))
> plot(TopEx,col=topography.c(100),
main="Topographic Exposure over South America")
> contour(elev_sa10, nlevels=7,
levels=c(0, 500, 1000, 1500, 2000, 3000, 4000, 5000),
add=T, labels="", lwd=.1)
```

The default option in the focal analysis requires that values be only calculated for those windows that have all 15 × 15 (= 225) cells available with numerical information. Since that would remove some of the marginal area along the coast, we set na.rm=T. This setting removes unavailable or missing values (NAs) from the focal computations and allows calculating an output for windows with fewer than 225 cells with numerical values in our case.

Terrain analyses constitute a special kind of focal analysis. This type of analysis can be used to calculate slope, aspect, or to shade a DEM. Here, we prepare to shade the original *c.* 1 km DEM called `elev` using the `hillShade()` command.

```
> slope <- terrain(elev, opt='slope')
> aspect <- terrain(elev, opt='aspect')
```

These two maps are now used to shade the terrain model `elev`, from which slope and aspect are derived. Here, we set the altitude angle to 30° and the azimuth angle to 315°, in order to illuminate the terrain model.

```
> hillshade <- hillShade(slope, aspect, 30, 315)
```

The hill-shaded raster is stored here as a TIFF file for later use, it can then be read from here again in a new R session.

```
> writeRaster(hillshade, "raster/topo/hillshade.tif",
overwrite=T)
> hillsh<-raster("raster/topo/hillshade.tif")
```

Such maps can now be used as background images for illustrating thematic layers. We illustrate this with the example of North America, first by cropping the global hillshade image to the North America extent, and then by overlaying the elevation map (`elev`) in a semi-transparent manner. In order to do this, we redefine the plot extent as above for North America. The argument `alpha=0.5` adds semi-transparency, making the superimposed layer 50% transparent. In this way, the underlying hillshade is partly visible.

```
> plot_extent<-extent(-124,-66,24,50)
> hillsh_na<-crop(hillshade, extent(-188,-50,15,90))
> plot(hillsh_na, col=grey(0:100/100), legend=FALSE, axes=F,
ext=plot_extent)
> plot(elev_na,col=terrain.colors(100),alpha=.5,add=T,
ext=plot_extent)
```

In the next plotting example, we create our own color palette, mimicking the elevation color ramp of ArcGIS. We first generate a function for this color palette, and then we generate 100 color values.

```
> dem.c<-colorRampPalette(c("aquamarine", "lightgoldenrodyellow",
"lightgoldenrod", "yellow", "burlywood", "burlywood4", "plum",
"seashell"))
> cols<-dem.c(100)
```

Since we cannot assign an `alpha` value in this function, we simply paste the transparency values for 60% transparency to the BinHex color code. For the alpha values 0.0, 0.3, 0.4, 0.5, 0.6, and 1.0, we add the values "00", "4D", "66", "80", "99", "FF", respectively (with alpha = 0.0 being

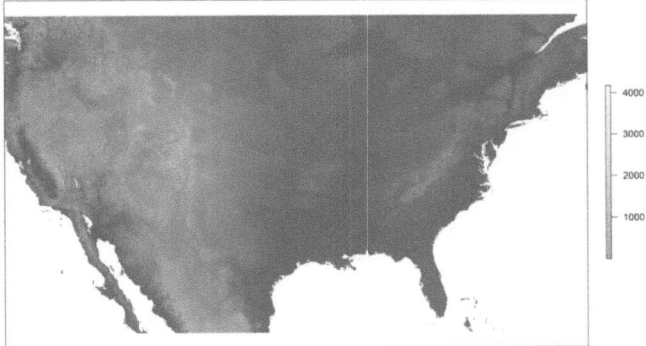

Figure 6.7 Elevation map draped semi-transparently over hill-shaded DEM.

fully transparent and alpha = 1.0 being intransparent). Finally, we plot the map with this color ramp as described above and an alpha value of 0.4 (Figure 6.7). We plot this map again at a smaller extent than the hillsh_na raster, using the same extent as above.

```
> cols<-paste(cols,"66",sep="",collate="")
> plot(hillsh_na, col=grey(0:255/255), legend=FALSE, axes=F,
    ext=plot_extent)
> plot(elev_na, col=cols,add=T, ext=plot_extent)
```

There are numerous other possible analyses. Only a few examples have been provided here. Read for example the manuals for the raster, maptools, rgdal, and sp packages in order to find out more. Again, for large datasets, the GIS functionality may be somewhat slow. However, with medium to small datasets, R offers an extremely flexible and powerful way of combining statistical modeling with GIS functions.

6.2.7 Stacking Grids to a Grid Stack

We often prepare environmental data in a raster format (grid) and sample these grids with our observational data points (*x*/*y*-coordinates) in order to build an environmental database for our field data. We can do this very easily in R by first stacking all the grids together (stack() command from raster package) and then sampling the whole stack with one command (extract() from raster package):

```
> world.stk <- stack(elev1,bio3,bio7,bio11,bio12)
```

We can easily summarize the information in a grid stack by typing the summary command: summary(world.stk)

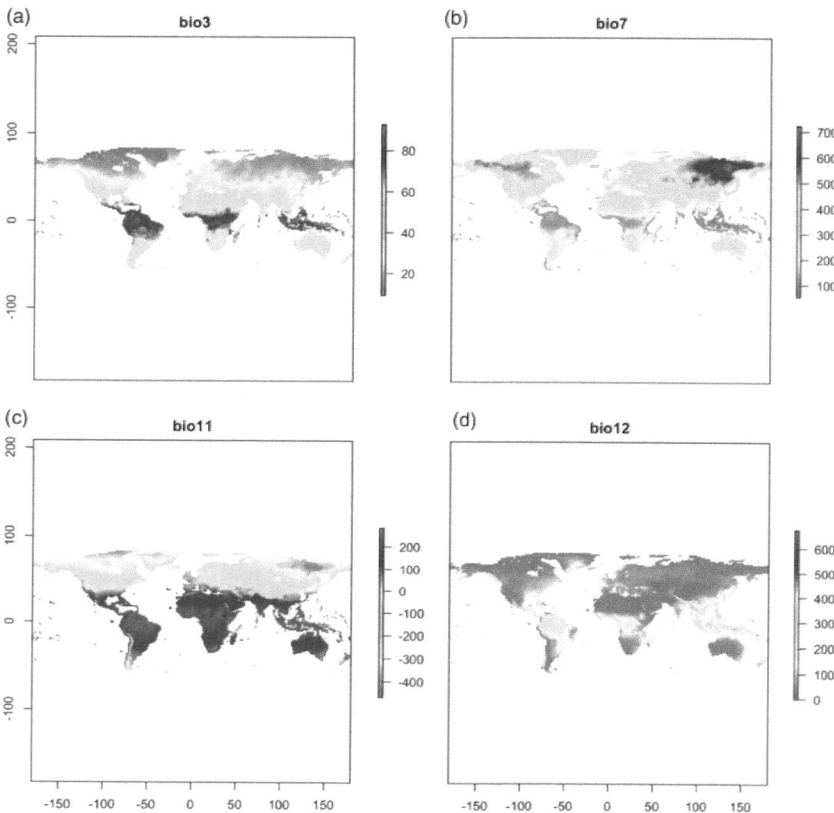

Figure 6.8 Visualization of all elements in a stacked grid. (*A black and white version of this figure will appear in some formats. For the color version, please refer to the plate section.*)

Single grids from a grid stack can be selected using the $ selection (e.g. `world.stk$bio7`). In addition, we can obtain a very rapid overview of a stack by mapping its content (Figure 6.8). This is only recommended if the stack contains a small number of grids. Here, we plot the stack for layers 2 to 5, thus not plotting the first layer, i.e. elevation:

```
> plot(world.stk[[2:5]], col=rainbow(100,start=.0,end=.8))
```

Working with grid stacks has several advantages. It enables sampling all elements of a stack from a point file using a single command (see below), or one can process all stacked grids equally with one single command. One example of this latter option is changing the cell size (spatial resolution) of all grids in our stack. Assuming we want to convert our 1.5° stack of rasters (our `world.stk`) to a lower resolution (3 × 3 degree), we need to aggregate the rasters by a factor of 2 (storing the mean of four cells into larger 3° cells).

```
> world.stk3 <- aggregate(world.stk,fact=2, method="mean")
```

When plotting this stack, we can immediately see that the cells are larger than in the original stack. This represents an upscaling procedure (with respect to grain size).

6.2.8 Re-Projecting Grids

Rasters (or vector type files) assembled from various sources might not always be found in the same projection. Overlaying these environmental files with a set of sample points for modeling purposes or creating a stack of raster files requires all the files to be part of the same projection. We have seen previously how to assign a projection. Here we work through an example of the steps that need to be taken when reading a raster from a different projection.

Here, we read long-term (30-year normal) annual climate data for precipitation and temperature from the PRISM project (see Section 6.2.1).

```
> prec_yearly_usa <- raster("raster/prism/prec_30yr_normal_
annual.asc")
> tave_yearly_usa <- raster("raster/prism/tave_30yr_normal_
annual.asc")
```

If we check extent and projection, then we realize: first, that the extent has coordinates similar to the lon/lat WGS84 projection and second, that the projection is already set, but that the datum is different:

```
> extent(prec_yearly_usa)
> projection(prec_yearly_usa)
 [1] "+proj=longlat +datum=NAD83 +no_defs +ellps=GRS80
+towgs84=0,0,0"
```

The datum and the ellipsoid of the projection differ from the ones used when reading the Worldclim data (bio3, etc.). In order to superimpose these maps correctly with the points, vectors and rasters that use the WGS84 datum and ellipsoid, we need to re-project the PRISM rasters to the same WGS84 datum and ellipsoid. It is also interesting to see that the raster already has the projection information, despite being read from an ascii file. This information is attached, because in the same folder of PRISM climate files, there are also associated *.prj files available with the same name as the *.asc ascii files. The raster environment recognizes these files, and reads the projection information from them. Removing these *.prj files from the directory results in reading un-projected raster files.

In the following example, we re-project the two rasters to a lon/lat WGS84 projection, and in doing so, we adjust the resolution to 0.025°. We could also keep the original resolution of 0.00833333°, yet this would take considerably longer.

```
> prec_yearly_usa_wgs <- projectRaster(prec_yearly_usa,
res=0.025, crs="+proj=longlat +datum=WGS84 +no_defs +ellps=WGS84
+towgs84=0,0,0", method="bilinear")
> tave_yearly_usa_wgs <- projectRaster(tave_yearly_usa,
res=0.025, crs="+proj=longlat +datum=WGS84 +no_defs +ellps=WGS84
+towgs84=0,0,0", method="bilinear")
```

The new output raster now has a 0.025° spatial raster resolution. This translates to *c.* 2.5 km in a metric projection. The original 0.00833333° resolution would project to *c.* 800m cell size if projected to a metric projection such as Albers equal area (aea) or Lambert azimuthal equal area (lazea). It is important to note that the lon/lat geographic coordinate system does not conserve area or angles, and therefore cannot be used to perform any area- or distance-based calculations. For such analyses, all layers should be projected to equal area-based projections, for example. An overview of PROJ4 projections that represents the basis for the projection definitions in R are available online.[28]

6.2.9 *Importing and Overlaying Species Data*
One frequently used GIS analysis is to import and overlay field sampled species distribution data with environmental predictor layers to later model their habitat suitability (Part III onward). Here, we import a set of distribution data for *Pinus edulis* L., downloaded from GBIF (see Section 6.1.1) in July 2014.

```
> pinus_edulis<-read.table("tabular/species/pinus_edulis.txt",
h=TRUE,sep=",")
```

Such a file is read as a data frame object, which includes two columns that represent coordinates (lat and lon). However, the file is not of class spatial. We can check this with the class() command.

```
> class(pinus_edulis)
```

Next, we define the object to be of class spatial, and therefore we need to assign which columns represent the x- and the y-coordinates.

```
> coordinates(pinus_edulis) <- c("lon", "lat")
```

---

[28] http://geotiff.maptools.org/proj_list/

When comparing the `pinus_edulis` object with the original version, there are no longer any coordinates visible in this object. These are stored separately and can be retrieved using the following command:

```
> coordinates(pinus_edulis)
```

When checking again using the `class()` command, the object is now of class `spatial`. However, we also see that no coordinate system has yet been defined for this spatial object. So we then need to define it:

```
> projection(pinus_edulis) <- "+proj=longlat +datum=WGS84 +no_
defs +ellps=WGS84 +towgs84=0,0,0"
```

This finalizes the transformation of a simple data frame object into a spatial object. Such objects require both the columns that represent the coordinates, and a projection to be defined. Numerous datasets are available in the simple Lon/Lat geographic projection, which is a simple coordinate system. More advanced projections are needed to ensure that either the angles or areas are correctly represented.

As an alternative to reading the *Pinus edulis* data from a file, we can derive these distribution points directly from online sources such as the Global Biodiversity Information Facility (GBIF). This can be done from within R using the `dismo` package.

```
> library(dismo)
> pinus_edulis <- gbif('pinus', 'edulis', download=T,
geo=T, sp=T, removeZeros=T)
```

Using the option (`sp=T`), we ensure that the generated file is converted directly into an object of class `spatial`, and not only into a tabular data frame object. The new spatial object has just one column (next to the invisible coordinates), namely the download date. We just give it this name:

```
> names(pinus_edulis)[1]<-"dwnld.date"
```

If we want to download more variables, we need to set respective arguments. We can now save this spatial data object in a text file for later use. We do this by converting it to a comma-separated CSV file.

```
> write.table(data.frame(pinus_edulis@coords, pinus_edulis@data),
"tabular/species/p_edulis.txt", sep="," ,row.names=FALSE)
```

An almost identical method would be to use the `write.csv()` command. This directly generates comma-separated files, but requires more care when reading the data back if using commands other than `read.csv()`, such as the `read.table()` command, for example. Both methods are very similar, but attention must be paid to the way row names are written and imported.

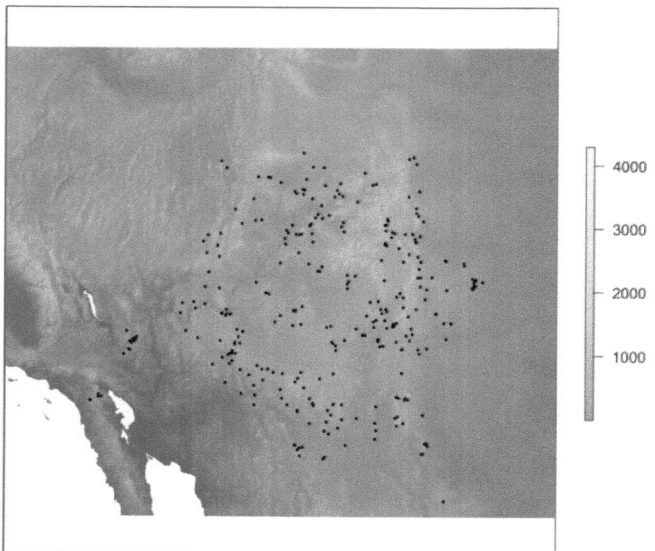

Figure 6.9 Downloaded distribution points of *Pinus edulis* from GBIF.

```
> write.csv(data.frame(pinus_edulis@coords, pinus_edulis@data),
"tabular/species/p_edulis.csv")
```

Next, we want to extract climate data from the `world.stk` object used above (Section 6.2.7) so that we overlay several stacked, bioclimatic raster objects with one single command using the `bilinear` interpolation method for a set of points.

```
> pts.clim<-extract(world.stk, pinus_edulis, method="bilinear")
> pin_edu.clim<-data.frame(cbind(coordinates(pinus_edulis),
pts.clim, pinus_edulis@data))
```

The `extract()` command generates a data frame object (`pts.clim`) containing climate data, which is merged with the coordinates (`coordinates(pinus_edulis)`) and the data (`pinus_edulis@data`) from the `pinus_edulis` spatial object. The resulting data frame is now no longer a spatial object, but we can easily generate such a structure again:

```
> coordinates(pin_edu.clim)<-c("lon","lat")
```

Here, we now plot this file over North America in order to visualize the presence of *Pinus edulis* as downloaded from GBIF using a predefined extent that is smaller than the hillshade created above (Figure 6.9, note that this figure is printed in gray).

```
> map.ext<-extent(-120,-100,30,44)
> plot(hillsh_na, col=grey(0:100/100), legend=FALSE, axes=F,
ext=map.ext)
> plot(elev_na,col=cols,add=T,ext=map.ext)
> plot(pinus_edulis, pch=16, cex=.5, add=T)
```

6.2.10 Generating a Uniform Spatial Data Structure for Modeling and Analysis

We demonstrate how a simple data structure is prepared for species distribution modeling and how we can project a simple model using the GIS functionality in R. The whole of Part III will be devoted to deepening our understanding of many different statistical modeling approaches, while Part V is meant to introduce the projections in space and time in more detail. Here, we generate a first simple data structure describing the distribution of the red fox (*Vulpes vulpes*), one of the species used later in the examples in Part III. A data structure is imported that contains a training or calibration (cal.txt) and a validation or evaluation (eva.txt) dataset, both created by regularly sampling the global range of the species. We then overlay these datasets with four bioclim variables from the world.stk raster stack, and fit a very simple GLM, see Chapter 10). We then simply plot this model for the purposes of illustration, but we do not test it against the evaluation dataset (which will be treated in Part IV), and we do not further explore the spatial (or temporal) projections (which will be covered in Part V).

In our example, we use *Vulpes vulpes* and four bioclimatic predictors to build a simple GLM-based distribution model. We process the model from a simple data frame and chose not to transform this data frame into an object of type spatial. We also perform a stepwise variable reduction and then use the function ecospat.adj.D2.glm() from the ecospat library to calculate the adjusted D^2 values (adj.D2) (see Guisan and Zimmermann, 2000 for details).

```
> pts.cal<-read.table("tabular/species/cal.txt")
> pts.eva<-read.table("tabular/species/eva.txt")
> plot(pts.cal[which(pts.cal$VulpesVulpes==0),1:2],pch=15,cex=.3,
col="grey50",xlab="Longitude",ylab="Latitude")
> points(pts.eva[which(pts.eva$VulpesVulpes==0),1:2],pch=15,
cex=.3, col="grey85")
> points(pts.cal[which(pts.cal$VulpesVulpes==1),1:2],pch=16,
cex=.4, col="firebrick3")
> points(pts.eva[which(pts.eva$VulpesVulpes==1),1:2],pch=16,
cex=.4, col="seagreen3")
```

In this way, the calibration and the evaluation points are plotted in maroon and sea green, respectively, for presence points (Figure 6.10). The absence points for the two datasets are presented in darker and lighter shades of gray.

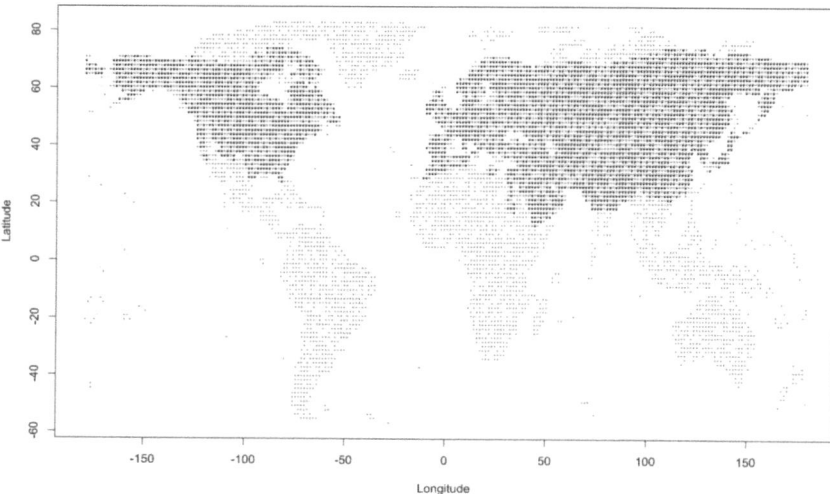

Figure 6.10 Distribution of presence and absence points sampled from a global range map of *Vulpes vulpes*. Presence calibration (red) and evaluation (green) points reflect the species' range, while the absence points of the calibration (darker gray) and the evaluation (lighter gray) datasets reflect areas of absence for the species. (*A black and white version of this figure will appear in some formats. For the color version, please refer to the plate section.*)

We then overlay the calibration and evaluation points with the world.stk object using the extract() function from the raster package. Here, we concatenate two commands. First, we extract the four grid layers from the world.stk raster stack using the extract() command (layers 2 to 5). This simply generates a matrix of bioclim variables for each of the coordinates from the pts.cal object, the coordinates for the extraction are found in the first two columns. Second, we combine column eight from the original calibration and evaluation data frames with the extracted bioclim variables, so that these are joined at the end. This eighth column contains the distribution data for *V. vulpes*. By combining the overlaid matrices with the original data frames, we generate two new data frames:

```
> pts.cal.ovl<-cbind(pts.cal[,8],extract(world.stk[[2:5]],
pts.cal[,1:2]))
> pts.eva.ovl<-cbind(pts.eva[,8],extract(world.stk[[2:5]],
pts.eva[,1:2]))
```

Now, we first need to remove the NA values using the na.omit() command and at the same time convert the object to a data frame. Second, we need to assign the species (*V. vulpes*) a name in the first column. The first omits points that were sampled from the coarse global distribution map, but were actually located in the sea according to the bioclimatic variables. It is also easier to operate with data frames when modeling species distributions. We therefore apply the data.frame() command, which converts the numerical matrix into a data frame.

```
> pts.cal.ovl<-data.frame(na.omit(pts.cal.ovl))
> pts.eva.ovl<-data.frame(na.omit(pts.eva.ovl))
> names(pts.cal.ovl)[1]<-"Vulpes.vulpes"
> names(pts.eva.ovl)[1]<-"Vulpes.vulpes"
```

We have now completed our data preparation and can fit a simple GLM object.

A simple GLM model is fitted to illustrate GIS capability in R. Here, we fit a four-parameter model (bio3, bio7, bio11, and bio12) with both linear and quadratic terms, and we perform simple stepwise bi-directional parameter selection in order to optimize this model.

```
> vulpes.full <- glm(Vulpes.vulpes~bio3+I(bio3^2)+bio7+I(bio7^2)+
bio11+ I(bio11^2)+bio12+I(bio12^2), family="binomial",
data=pts.cal.ovl)
> vulpes.step <- step(vulpes.full, direction="both", trace=F)
```

Next, we load the ecospat library to be able to use the ecospat. adj.D2.glm() function, which calculates the adjusted D^2 calibration strength of a GLM according to Weisberg's (1980) formula (see Guisan and Zimmermann, 2000).

```
> library(ecospat)
> ecospat.adj.D2.glm(vulpes.full)
 [1] 0.6141926
> ecospat.adj.D2.glm(vulpes.step)
 [1] 0.6142384
```

We find that both models give roughly the same adjusted D^2 value, with a slightly higher value for the stepwise-optimized model. The (unadjusted) D^2 would be higher for the full model, and slightly lower for the stepwise-optimized model. However, the adjusted D^2 considers the number of parameters and the number of observations used, and thus penalizes the stepwise-optimized model, which has the linear term of the bio7 variable removed, less:

```
> summary(vulpes.step)
Call:
glm(formula = Vulpes.vulpes ~ bio3 + I(bio3^2) + I(bio7^2) +
    bio11 + I(bio11^2) + bio12 + I(bio12^2), family = "binomial",
    data = pts.cal.ovl)
```

```
Deviance Residuals:
    Min      1Q   Median      3Q      Max
-2.9355  -0.2531   0.0000  0.3655   3.8624


Coefficients:
              Estimate Std. Error  z value Pr(>|z|)
 (Intercept) -7.596e+00  7.330e-01 -10.362  < 2e-16 ***
bio3          3.711e-01  4.003e-02   9.269  < 2e-16 ***
I(bio3^2)    -6.517e-03  5.446e-04 -11.967  < 2e-16 ***
I(bio7^2)     3.084e-05  1.233e-06  25.009  < 2e-16 ***
bio11         5.277e-03  7.390e-04   7.141 9.25e-13 ***
I(bio11^2)   -2.346e-05  2.582e-06  -9.084  < 2e-16 ***
bio12         2.694e-03  2.776e-04   9.702  < 2e-16 ***
I(bio12^2)   -8.662e-07  1.228e-07  -7.052 1.76e-12 ***
---
Signif. codes:  0 '***' 0.001 '**' 0.01 '*' 0.05 '.' 0.1 ' ' 1

(Dispersion parameter for binomial family taken to be 1)

    Null deviance: 8178.2  on 5899  degrees of freedom
Residual deviance: 3151.1  on 5892  degrees of freedom
AIC: 3167.1

Number of Fisher Scoring iterations: 8
```

Projecting this model to space is now a straightforward process using the GIS functionality in R. We simply predict to the `world.stk` grid stack, which we have used to overlay the calibration points over the climate (predictor) layers. We then generate a map (Figure 6.11) of the projected global distribution of *V. vulpes*, and we add the calibration points that we have used to fit the GLM to this model projection. We take these points from the `pts.cal` object, which has *x*- and *y*- coordinates.

```
> vulpes.map<-predict(world.stk,vulpes.step, type="response")
> plot(vulpes.map, col=rev(heat.colors(10)),
main="Predicted distribution: Vulpes vulpes")
> points(pts.cal[which(pts.cal$VulpesVulpes==1),1:2], pch=15,
cex=.25)
```

We will not further test this model here, as the whole of Part IV is devoted to model evaluation. Many more GIS functions are available in R, which can be used for habitat distribution modeling of species. The basic steps introduced so far should now offer a good basis from which readers can further explore this functionality.

6.3 RS-Based Predictors

In this section, we introduce how to download, load, and visualize RS data and how to carry out simple GIS-like analyses using functions and

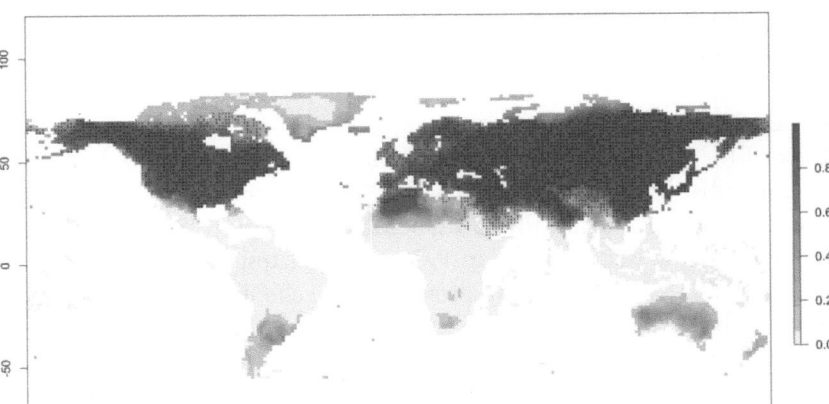

Figure 6.11 Map of the projected global distribution of *Vulpes vulpes* contrasted with the observed distribution points used to calibrate the simple four-parameter GLM model. (*A black and white version of this figure will appear in some formats. For the color version, please refer to the plate section.*)

the available packages in R. Each reader may have his/her own preferred RS datasets, which they can combine with other geospatial data. We will only superficially address the issue of RS data here. Most importantly, this section does not aim to introduce the basics of RS data processing, such as georegistration, relative or absolute atmospheric correction, or cloud masking, for which we refer readers to specialized books (e.g. Normand et al., 2013). Instead, it constitutes a simple introduction to loading and re-calculating remotely sensed data in order to combine these with other GIS layers for powerful statistical analyses (Carlson et al., 2014; Pottier et al., 2014). For this purpose, we use a smaller-scale dataset, which improves the handling and visualization of RS products.

6.3.1 Introduction
RS has a lot to offer biogeographers and macroecologists (Kerr and Ostrovsky, 2003). The careful processing of RS data requires specialist knowledge, and this is not the subject here. However, RS data is increasingly available in pre-processed formats, and can be used like any other form of GIS data, if prepared carefully. One of the biggest advantages of RS data is that it informs us objectively, usually with full coverage of a larger study area, about the state of the Earth's surface at a specific point in time. There are many different systems such as passive optical or active

LiDAR and radar RS. Here, we will primarily work with passive optical sensors. Once the dataset has been selected, we have to retrieve the information that best explains the patterns we are interested in.

6.3.2 Data Sources

One of the most widely used data sources for RS imagery is Landsat.[29] The Landsat Program is a series of Earth-observing satellites (EOS) managed by NASA and the US Geological Survey (USGS).[30] The first Landsat satellite was launched in 1972 and the latest satellite, Landsat 8 in February 2013. It continues to provide an excellent RS dataset of almost 40 years by now.

The Landsat 8 satellite collects data for the entire Earth every 16 days with an 8-day difference from Landsat 7. The data collected are available for download free of charge within 24 hours of retrieval.[31]

Landsat data is available as a 4–11 band product, covering the globe with approximately 180 × 180 km single images (called scenes) every 16 days. Most bands have a spatial pixel resolution of roughly 30 × 30 m, depending on latitude. The whole Landsat archive is now available for free, and can be accessed over the internet. A useful archive is the one hosted by the USGS.[32].

```
LandSat8 OLI bands:

    * Band 1        0.43-0.45       Coastal aerosol
    * Band 2        0.45-0.51       Blue
    * Band 3        0.53-0.59       Green
    * Band 4        0.64-0.67       Red
    * Band 5        0.85-0.88       Near IR
    * Band 6        1.57-1.65       SWIR 1
    * Band 7        2.11-2.29       SWIR 2
    * Band 8        0.50-0.68       Panchromatic
    * Band 9        1.36-1.38       Cirrus
    * Band 10       0.52-0.90       Thermal Infrared (TIRS) 1
    * Band 11       0.52-0.90       Thermal Infrared (TIRS) 2

LandSat7 ETM+ bands:

    * Band 1        0.45-0.52       Blue
    * Band 2        0.53-0.61       Green
    * Band 3        0.63-0.69       Red
    * Band 4        0.75-0.90       Near IR
```

[29] http://landsat.usgs.gov/
[30] http://landsat.gsfc.nasa.gov/
[31] http://landsat.usgs.gov/landsat8.php
[32] http://landsat.usgs.gov/band_designations_landsat_satellites.php

```
* Band 5        1.55-1.75       Mid-IR
* Band 6        10.4-12.5       Thermal IR
* Band 7        2.09-2.35       Short Wave IR
* Band 8        0.52-0.90       Panchromatic
```

The moderate resolution spectroradiometer (MODIS) provides another frequently used dataset that is also available free of charge. It scans the whole Earth every 1–2 days (depending on latitude) and two satellites – TERRA and AQUA – host the same MODIS sensor. This means a morning and an afternoon scene can be obtained for every day. The sensor has a much larger swath (width of the scanned path) than Landsat, scans a total of 36 different bands and has a pixel resolution of 1 km for 29 bands, except those that mimic Landsat bands, which are available at 500 m resolution (except for the RED and NIR bands that are available at 250 m). One of the big advantages of MODIS is that the data are processed automatically and made available "ready to use," along with a suite of derived variables.

The MERIS sensor is a programmable, medium-spectral resolution, imaging spectrometer on board the ESA's environmental research satellite ENVISAT. MERIS has 15 spectral bands that have a programmable width and location in the visible and near-infrared spectral range (390 nm to 1040 nm). MERIS is a pushbroom instrument that has a 68.5° field of view around the nadir, with a swath width of 1150 km. This allows the instrument to collect data for the entire planet every 3 days in equatorial regions while polar regions are visited more frequently due to the convergence of orbits. The spatial resolution of the data collected by MERIS is 300 m at the nadir (full resolution product) while most common products are generated at 1200 m resolution (reduced resolution product) by aggregating the data to 1200 m.

One of the widest ranging data warehouses is NASA's Reverb, which provides an interface for discovering, accessing, and using EOS data.[33] It allows users to order data from many different sensors and satellite missions. All data holdings stored in the so-called Land Processes Distributed Active Archive Centers (LP DAAC) of the USGS are available here. Data are classified by target application or field, such as atmosphere, cryosphere, land, ocean, and solar. Reverb is probably the best place to order MODIS data, whereas data from the Landsat sensor is best ordered from the Landsat Mission USGS website.[34]

---

[33] http://reverb.echo.nasa.gov/reverb/
[34] http://landsat.usgs.gov/Landsat_Search_and_Download.php

Furthermore, several RS datasets can be downloaded from the ESA's portal,[35] primarily acquired from European satellites like Envisat.

6.3.3 Importing, Resampling, and Grid Stacking

The applicable commands are similar to those used for loading and analysis of GIS data. Here, we assume that all data is available in georegistered and atmospherically corrected TIFF format. We work with bands 1–5 and 7, and leave out the thermal infrared band 6 and the panchromatic band 8, as both are available at different spatial resolutions. We load an image located over north-eastern Switzerland (Path = 194/Row = 027) and we first load a shapefile encompassing the canton of Zurich, in order to crop the image to the shape, so as to reduce the size of the loaded data.

```
> Cantons <-readShapePoly("vector/swiss/Swiss_Cantons.shp")
> Zurich<-Cantons[Cantons$NAME=="ZUERICH",]
```

Next, we load all six Landsat bands.

```
> band1_blue<-raster("raster/landsat/L7_194027_2001_08_24_B10.
TIF")
> band2_green<-raster("raster/landsat/L7_194027_2001_08_24_B20.
TIF")
> band3_red<-raster("raster/landsat/L7_194027_2001_08_24_B30.
TIF")
> band4_nir<-raster("raster/landsat/L7_194027_2001_08_24_B40.
TIF")
> band5_swir1<-raster("raster/landsat/L7_194027_2001_08_24_B50.
TIF")
> band7_swir2<-raster("raster/landsat/L72194027_2001_08_24_B70.
TIF")
```

The next step is to crop the six Landsat bands to the canton of Zurich.

```
> band1_blue_crop<-crop(band1_blue,extent(Zurich))
> band2_green_crop<-crop(band2_green,extent(Zurich))
> band3_red_crop<-crop(band3_red,extent(Zurich))
> band4_nir_crop<-crop(band4_nir,extent(Zurich))
> band5_swir1_crop<-crop(band5_swir1,extent(Zurich))
> band7_swir2_crop<-crop(band7_swir2,extent(Zurich))
```

We then stack all six images into one compound multilayer brick.

```
> L7_010824<-brick(band1_blue_crop, band2_green_crop,
band3_red_crop,band4_nir_crop,
band5_swir1_crop,band7_swir2_crop)
```

Of course, there is a more economical way of executing the same three commands!

---

[35] https://earth.esa.int/web/guest/data-access

(a) (b) (c)

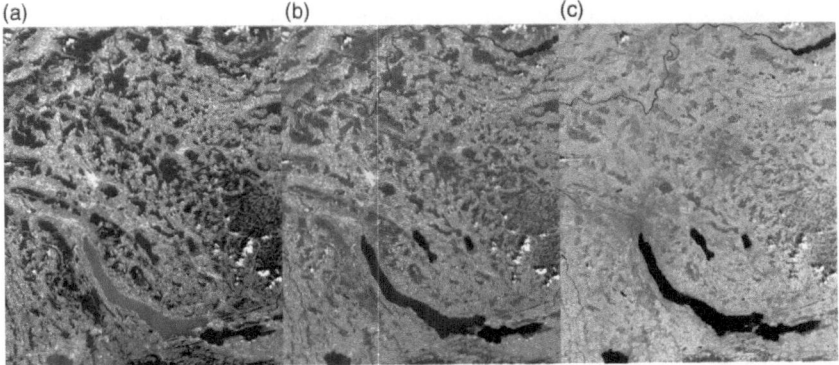

Figure 6.12 Illustration of three-color composites from Landsat bands of the area of Zurich in Switzerland. (*A black and white version of this figure will appear in some formats. For the color version, please refer to the plate section.*)

```
> tmp<-stack(band1_blue,band2_green,
band3_red,band4_nir,band5_swir1, band7_swir2)
> L7_010824<-crop(tmp, extent(Zurich))
```

Finally, we assign layer names and check the naming.

```
> names(L7_010824)<-c("band1_blue","band2_green","band3_red",
"band4_nir","band5_swir1","band7_swir2")
> names(L7_010824)
 [1] "band1_blue" "band2_green" "band3_red" "band4_nir" "band5_
swir1" "band7_swir2"
```

An RGB color composited image from Landsat is usually plotted with bands 3 (red), 2 (green), and 1 (blue), while a false color composite is usually generated from bands 4, 3, and 2. In some cases, bands 7, 5, and 4 (note that band 7 is the sixth band in our stack or brick) are plotted to visualize Landsat scenes. See Figure 6.12 for three examples of representing color composites:

```
> par(mfrow=c(1,3))
> plotRGB(L7_010824,3,2,1,stretch="lin")
> plotRGB(L7_010824,4,3,2,stretch="lin")
> plotRGB(L7_010824,6,4,3,stretch="lin")
> par(mfrow=c(1,1))
```

Currently, there are R packages available that can be used to perform simple to more sophisticated RS analyses. The landsat and RStoolbox packages have a number of functions for pre-processing and analyzing Landsat, MODIS, ASTER and other remote-sensing data, such as generating most commonly used vegetation indices. Furthermore, other packages such as MODISTools and MODIS can be used to download, import

Table 6.1 *Summary of some of the important indices that are often used in RS-based applications.*

Vegetation index	Equation	Reference
Structural indices		
Normalized difference vegetation index (NDVI)	$NDVI = (R_{NIR} - R_{red}) / (R_{NIR} + R_{red})$	Rouse et al. (1974)
Simple ratio index (SR)	$SR = R_{NIR}/R_{red}$	Rouse et al. (1974)
Soil adjusted vegetation index (SAVI)	$SAVI = (1+L)(R_{800} - R_{670})/(R_{800} + R_{670} + L)$ $(L = 0.5)$	Huete (1988)
Normalized difference water index (NDWI)	$NDWI = (R_{860} - R_{1240})/(R_{860} + R_{1240})$	Gao (1996)
Tasseled cap transformation		Kauth and Thomas (1976)

and process MODIS data directly within R. Finally, other packages like `hyperSpec` or `hsdar`, currently in the early stages of development, can be used for processing hyperspectral RS data.

6.3.4 Data Processing for Ecological Analyses

We will now briefly introduce the recalculation of simple indices, which are often used in RS applications (e.g. Pottier et al., 2014). Once the images are processed, these recalculations correspond to simple GIS-type analyses. We will briefly present the following four indices, namely: the normalized difference vegetation index (NDVI), the simple ratio (SR), the soil adjusted vegetation index (SAVI), and the normalized difference water index (NDWI). Each of these indices enhances certain features, which are not clearly visible in the available bands (see Table 6.1). Therefore, the calculation of such indices is also termed "signal enhancing," since it makes certain signals hidden in the band data accessible to analysts. In addition, we apply the tasseled cap transformation that is designed to analyse and map vegetation and urban development changes detected by various satellite sensor systems. This transformation can be used to monitor crops, analyse and map vegetation for forestry, carbon sequestering and more. However, in order to apply the tasseled cap transformation we need to convert the digital number (DN) values of the downloaded Landsat scene to at-sensor reflectance. We can do this using the `radiocorr()` function in the `landsat` package in R.

```
> NDVI <- (L7_010824$band4_nir -L7_010824$band3_red)/(L7_
010824$band4_nir + L7_010824$band3_red)
> NDWI <- (L7_010824$band4_nir -L7_010824$band5_swir1)/(L7_
010824$band4_nir + L7_010824$band5_swir1)
> SR <- L7_010824$band4_nir / L7_010824$band3_red
```

Calculating SAVI and the Tasseled Cap requires At-Sensor or Surface reflectance. In this example, we will calculate At-Sensor reflectance using the landsat package in R.

The landsat package does not currently accept raster format objects, so we have to convert the data to SpatialGridDataFrame.

```
> L7_010824_sp <- as(L7_010824, "SpatialGridDataFrame")
> L7_010824_sp1 <- L7_010824_sp[1]
> L7_010824_sp2 <- L7_010824_sp[2]
> L7_010824_sp3 <- L7_010824_sp[3]
> L7_010824_sp4 <- L7_010824_sp[4]
> L7_010824_sp5 <- L7_010824_sp[5]
> L7_010824_sp7 <- L7_010824_sp[6]
```

Now we can calculate At-Sensor reflectance with the radiocorr() function from the landsat package using the apparentreflectance method. Several parameters are required for the function to run properly, including "gain" and "offset" which are needed to convert the DN to radiance. Most of this information can be found in the metadata provided with the Landsat data or can easily be found using the help associated with the function or in the literature.

```
> library(landsat)
> L7_010824_refl_sp1 <- radiocorr(L7_010824_sp1, Grescale=0.76282,
Brescale=-1.52, sunelev= 48.29, edist=ESdist("2011-08-24"),
Esun=1957, method="apparentreflectance")
> L7_010824_refl_sp2 <- radiocorr(L7_010824_sp2, Grescale=1.44251,
Brescale=-2.84, sunelev= 48.29, edist=ESdist("2011-08-24"),
Esun=1826, method="apparentreflectance")
> L7_010824_refl_sp3 <- radiocorr(L7_010824_sp3, Grescale=1.03988,
Brescale=-1.17, sunelev= 48.29, edist=ESdist("2011-08-24"),
Esun=1554, method="apparentreflectance")
> L7_010824_refl_sp4 <- radiocorr(L7_010824_sp4, Grescale=0.87258,
Brescale=-1.51, sunelev= 48.29, edist=ESdist("2011-08-24"),
Esun=1036, method="apparentreflectance")
> L7_010824_refl_sp5 <- radiocorr(L7_010824_sp5, Grescale=0.11988,
Brescale=-0.37, sunelev= 48.29, edist=ESdist("2011-08-24"),
Esun=215, method="apparentreflectance")
> L7_010824_refl_sp7 <- radiocorr(L7_010824_sp7, Grescale=0.06529,
Brescale=-0.15, sunelev= 48.29, edist=ESdist("2011-08-24"),
Esun=80.67, method="apparentreflectance")
```

Now that we have reflectance values (ranging from 0 to 1) we can calculate indices such as SAVI, TSAVI or tasseled cap. Tasseled cap is calculated using the `tasscap()` function from the `landsat` package. Here we choose the option `sat=7` since the satellite data we are using are from the Landsat ETM (Landsat 7) sensor. Finally, we can extract the three newly generated rasters named "brightness," "greenness," and "wetness," which represent band recombinations, and provide ecologically more meaningful information than the raw bands.

```
> L7_010824_tc <- tasscap("L7_010824_refl_sp",  sat = 7)
> L7_010824_Brightness <- raster(L7_010824_tc[[1]])
> L7_010824_Greenness <- raster(L7_010824_tc[[2]])
> L7_010824_Wetness   <- raster(L7_010824_tc[[3]])
```

In order to estimate SAVI we need to define the parameter "L" that varies according to the amount of green vegetation or green vegetation cover: in regions with dense vegetation, $L=0$ and in areas with no green vegetation, $L=1$. Generally, $L=0.5$ works well in most situations and is the default value used.

```
> L <- 0.5
> SAVI <- ((raster(L7_010824_refl_sp4) -
raster(L7_010824_refl_sp3))/(raster(L7_010824_refl_sp4) +
raster(L7_010824_refl_sp3) + L) )* (1+L)
```

Finally, we plot the results of our simple analyses using the 250 m resolution hillshade available for the study area (Figures 6.13 and 6.14). For the second set of graphs, we specifically design a new color palette `ygb.c`, using the `colorRampPalette()` command, which allows us to assign colors to a palette. We then assign the number of color shades to be generated and the command interpolates between the assigned core colors.

```
> hill_250m_utm <- raster("raster/topo/hill_250m_utm.tif")
> par(mfcol=c(2,2))
> plot(hill_250m_utm,col=grey(0:100/100), legend=FALSE, axes=F,
ext=extent(SAVI), main ="NDVI")
> plot(NDVI,col=rev(terrain.colors(20,alpha=0.6)),add=T)

> plot(hill_250m_utm,col=grey(0:100/100), legend=FALSE, axes=F,
ext=extent(SAVI), main ="SR")
> plot(SR,col=rev(terrain.colors(20,alpha=0.6)),add=T)

> plot(hill_250m_utm,col=grey(0:100/100), legend=FALSE, axes=F,
ext=extent(SAVI), main ="NDWI")
> plot(NDWI,col=rev(topo.colors(20,alpha=0.6)),add=T)
```

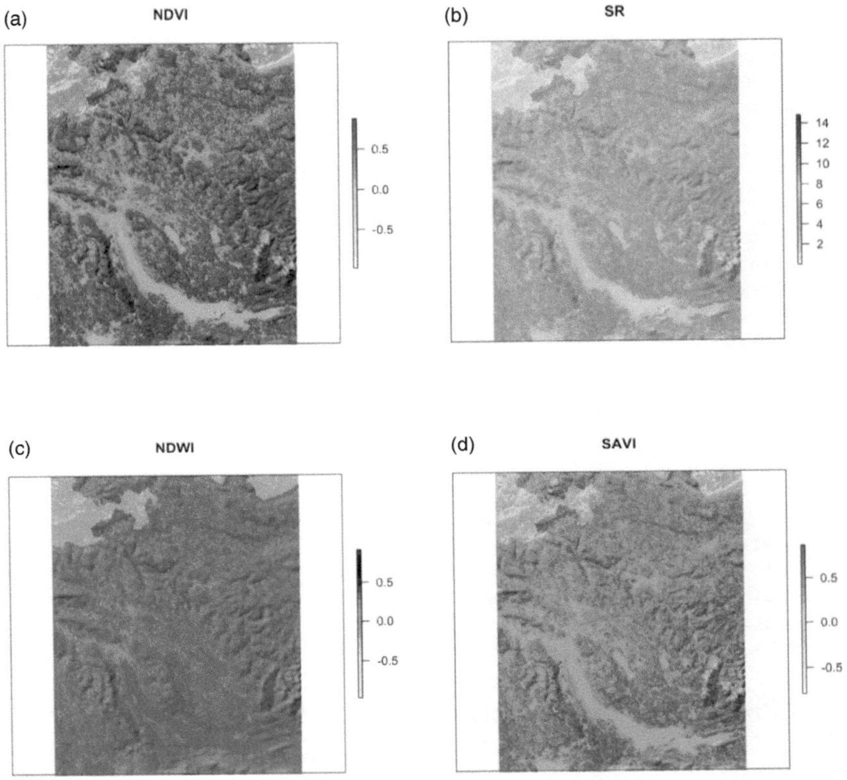

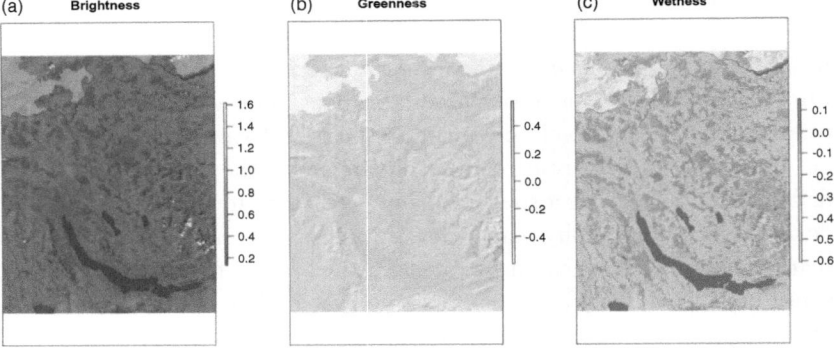

Figure 6.13 Visualization of four band–derived indices: (a) NDVI, (b) SR, (c) NDWI, and (d) SAVI. (*A black and white version of this figure will appear in some formats. For the color version, please refer to the plate section.*)

Figure 6.14 Visualization of the three tasseled cap indices: (a) brightness, (b) greenness, and (c) wetness. (*A black and white version of this figure will appear in some formats. For the color version, please refer to the plate section.*)

```
> plot(hill_250m_utm,col=grey(0:100/100), legend=FALSE, axes=F,
ext=extent(SAVI), main ="SAVI")
> plot(SAVI,col=rev(terrain.colors(20,alpha=0.6)),add=T)
> par(mfcol=c(1,1))

> ygb.c <-colorRampPalette(c("yellow","#7FFF7F","forestgreen",
"deepskyblue4","#00007F"))

> par(mfcol=c(1,3))
> plot(hill_250m_utm,col=grey(0:100/100), legend=FALSE, axes=F,
ext=extent(SAVI), main ="Brightness")
> plot(L7_010824_Brightness,col=rev(paste(ygb.c(20),"B3",
sep="")),add=T)

> plot(hill_250m_utm,col=grey(0:100/100), legend=FALSE, axes=F,
ext=extent(SAVI), main ="Greenness")
> plot(L7_010824_Greenness,
col=rev(terrain.colors(20,alpha=0.6)),add=T)

> plot(hill_250m_utm,col=grey(0:100/100), legend=FALSE, axes=F,
ext=extent(SAVI), main ="Wetness")
> plot(L7_010824_Wetness,
col=rev(topo.colors(20,alpha=0.6)),add=T)
> par(mfcol=c(1,1))
```

6.4 Properties and Selection of Variables

In this section, we discuss issues relating to the choice of variables and what this means for modeling species' habitat suitability. Since the predictive variables for spatial habitat suitability modeling need to be made available as maps in the form of raster or vector layers (or data frames), it is important to first resolve a number of issues when selecting these variables for modeling.

6.4.1 Accuracy vs. Mechanistic Explanation of Predictors

On the one hand, we need our predictor maps to be accurate, and on the other we want them to closely relate to the true drivers of species ranges, i.e. there is a mechanistic relationship between the predictor and the distribution of our target species. Many predictors are either modeled or derived from topography maps (elevation, see Section 6.2.3). This means that deriving more mechanism-oriented variables (e.g. relative humidity, vapor pressure deficit, frequency of frost days, or plant available soil moisture) requires advanced spatial modeling that naturally includes error propagation (Barry and Elith, 2006). Such maps become more "relevant" and more "mechanistically meaningful" for species modeling, while accumulating a higher rate of propagated error compared to using simple topography maps such as elevation, slope, or aspect. However, using simple topographic variables only produces accurate models if we project our maps to the same, comparably small study, area used to calibrate the models, and if we assume no change in environmental conditions.

One example which clearly illustrates this is the question whether one should use elevation or temperature to model species' ranges. Most species show distinct elevation patterns, despite the fact that elevation (units of meters or feet) has no direct, physiological, or otherwise mechanistic effect on species' ranges (Körner, 2007). However, changes in physiologically and ecologically important variables strongly correlate with changes in elevation, and these variables do have a more direct effect on species and can therefore be more meaningfully used for range delimitation. One clear example is temperature, but other variables are also strongly, and more or less linearly (positively or negatively), correlated with elevation, including precipitation, global radiation, wind speed, and potential evapotranspiration. Other variables often show a hump-shaped relationship with elevation, such as cloudiness. All these variables can affect species' ranges in one way or another, and more or less directly. So using annual mean temperature usually has almost the same predictive power as elevation, since the two are very strongly correlated, but using temperature makes a model more applicable to other areas or time periods. Yet temperature is derived from elevation (and other spatial covariates such as longitude or latitude, distance to lakes, etc.) often using geostatistical methods (Daly et al., 1994; Hutchinson, 1995; Thornton et al., 1997), and therefore includes more propagated error compared to elevation or other simple derivatives. New approaches to mapping temperature from field measurements, topography, and imagery use predictive models instead of geostatistical interpolations (Pradervand et al., 2014).

If the study area becomes (too) large, then indirect or distal variables such as elevation or aspect (see Section 4.1 or Guisan and Zimmermann, 2000) become ineffective as predictors, since they have no mechanistic effect on species distribution, but are only locally correlated with such variables. At such larger scales, more direct variables become more effective predictors (Guisan and Hofer, 2003), despite the higher level of uncertainty associated with them. This is due to their higher relevance and the fact that they are better correlated with the mechanisms (e.g. climatic control) that drive the species range patterns than indirect variables (such as elevation). For instance, the climatic treeline elevation changes very little within a small region of say 20 × 20 km, if there are no strong climate gradients in this study area. Using elevation as a predictor thus might even better explain the treeline position than temperature. Across the globe, treeline elevation varies between 0 and 4800 m a.s.l., while the summer mean temperature only varies within the very small range of error in soil temperature of 0.8°C for the globe (6.7±0.8°C; see Körner and Paulsen, 2004). Therefore, soil temperature is a good predictor of treeline position.

This shows that, as far as possible, we need to avoid using indirect variables, unless our study area is small, and our goal is simply to generate highly accurate predictions of current patterns under current environmental conditions. If we used indirect predictors, we would not be able to interpret what drives the spatial distribution of our species. In turn, if we are interested in interpreting our habitat suitability model with regards to the likely drivers of spatial patterns, then we should select more direct and resource (also termed "proximal") variables that have a known mechanistic or direct impact on processes that shape ranges (Austin, 2002, 2007). Such variables are also more likely to remain important under changing environmental conditions and/or when projecting models to new areas (see Part V).

Such considerations are not only relevant for climate variables, but hold for any type of predictor variable, including remotely sensed products or variables related to soils, habitats, or geology. When using RS imagery, one can simply use raw individual bands, irrespective of whether they represent information with close links to mechanistic processes or not (i.e. "distal" variables). RS imagery can also be used to retrieve vegetation structure, or water or nutrient status of the vegetation surface (more "proximal" variables).

6.4.2 Correlation, Collinearity, and Variance Inflation

Most statistical techniques will struggle to successfully fit a stable model if the predictor variables are highly cross-correlated, as this results in multicollinearity issues (see Part III). When GIS was first developed as a tool for spatial modeling, researchers simply obtained and used whichever map was available for modeling. Nowadays, we have a vast array of available maps, be climatic, topographic, or remotely sensed. Since many maps are derived from, or modeled with, a DEM, many of the derived maps reveal strong correlation with elevation, but also correlate with each other. It is therefore important to assess pairwise correlation between variables prior to use in any model fitting. Here, we illustrate how correlations among variables can be visualized. We use a predefined ascii file of

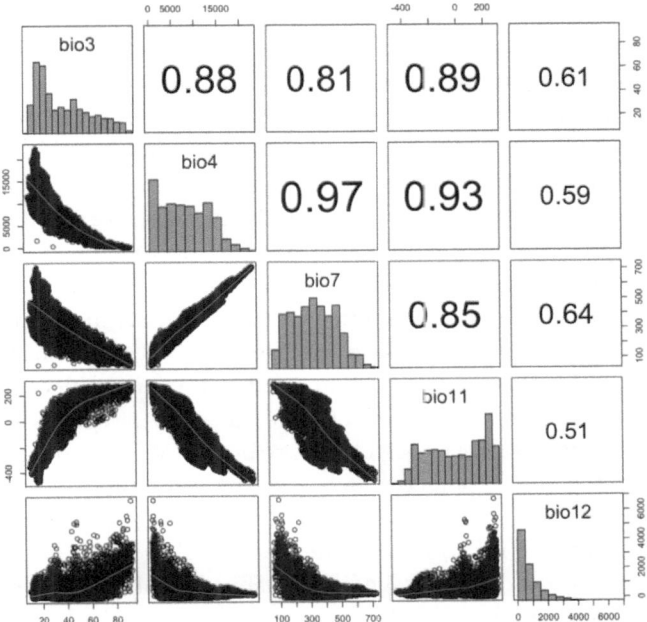

Figure 6.15 Correlation and distribution plot of five bioclim predictor variables. Positive and negative correlations are illustrated in black and red (shown as dark gray), respectively. Weaker correlations are displayed in smaller fonts. The distribution of each variable is illustrated on the diagonal.

bioclim variables (bioclim_table.csv) and the function `ecospat.cor.plot()` from the `ecospat` library, which is based on the `pairs()` command in R. This perfectly illustrates the correlation structures and variable distributions (Figure 6.15, note that it is printed in gray).

```
> data<-read.csv("tabular/bioclin/current/bioclim_table.csv",
header = TRUE, sep = ",")
> ecospat.cor.plot(data[,4:8])
```

This overview obtained using the `ecospat.cor.plot()` function provides us with some interesting information, regarding the five bioclimatic Worldclim[36] variables used. First, bio4 (temperature seasonality) and bio7 (annual temperature range) are strongly correlated ($r = 0.97$) at global scale. Bio4 is also strongly correlated with bio11 (mean temperature of coldest quarter). Furthermore, we see that bio12 in particular shows considerable skew, while bio7 seems more normally distributed.

---

[36] www.worldclim.org

One way of reducing multicollinearity problems is to remove variables that are too strongly correlated. There is no consensus as to what is an acceptable threshold from a statistical point of view. Numerous papers using distribution models refer to a threshold of $r = 0.8$ as recommended by Menard (2002), while others refer to a threshold of 0.7 as recommended by Green (1979). However, there is sometimes a hidden correlation structure that is not clearly visible in the pairwise correlation analysis. It can only be detected with a variance inflation factor (VIF) analysis (Harrell, 2001; Guisan et al., 2002, 2006b; Hair et al., 2006). VIF estimates the severity of the effect of multicollinearity, by measuring the extent to which variance in a regression increases due to collinearity compared to when uncorrelated variables are used. VIF tests are recommended, especially when numerous variables are added to a regression, as they detect the variables' linear correlation structure. We can do so, for example, using the `data` object which we have just imported in combination with the `vif()` command from the `usdm` package. Several packages contain a variance inflation test. The `usdm` package has several advantages for our purposes, as it is designed for habitat suitability model testing. It can be used to test for both data frames and raster stack objects, for example. It also offers options for testing which variable combination remains below a given correlation threshold.

```
> library(usdm)
> vif(data[,4:8])
  Variables       VIF
1      bio3  6.813542
2      bio4 63.384660
3      bio7 32.810217
4     bio11 11.786373
5     bio12  2.168148
```

We can see that almost all the variables are above a value of 10.0. Usually values from 5 to 10 are considered as critical for multi-variable correlation. Some authors suggest that VIF values of up to 20 can be accepted, but we do not recommend going above 10. Specifically, we see that bio4 has a very high VIF value. We next test what happens if we remove bio4 from our analyses, the same variable that showed extremely high correlation with several other variables in Figure 6.15.

```
> vif(data[,c(4,6:8)])
  Variables      VIF
1      bio3 5.866959
2      bio7 4.678556
3     bio11 6.881024
4     bio12 1.933955
```

We can now see that all values are near or below a VIF value of 6, which we can consider as acceptable for further analyses. We would obtain a similar result if we let the usdm package do a stepwise elimination of highly inflating variables using the vifstep() command, which applies a default threshold of 10.

```
> vifstep(world.stk[[2:5]])
```

We actually get a very similar result when using the world.stk object (only layers 2–5, excluding elevation), which has the same variables stacked as a raster stack. The values are not identical, because the stack and the data frame do not have exactly the same dimension or resolution.

```
> vif(world.stk[[2:5]])
   Variables      VIF
1      bio3 5.815840
2      bio7 4.576193
3     bio11 6.935333
4     bio12 1.931942
```

We can now check which variables remain when only a certain level of correlation is accepted, say r = 0.7.

```
> vifcor(data[,4:8], th=.7)
3 variables from the 5 input variables have collinearity
problem:

bio4 bio11 bio7

After excluding the collinear variables, the linear correlation
coefficients range between:
min correlation ( bio12 ~ bio3 ):  0.6021647
max correlation ( bio12 ~ bio3 ):  0.6021647

---------- VIFs of the remained variables --------
   Variables      VIF
1      bio3 1.568879
2     bio12 1.568879
```

In our case, we can see that only bio3 and bio12 remain at a correlation threshold of 0.7 (which we can also see from Figure 6.15). However, vif() does not calculate in the same way as bivariate correlations. VIF is based on the square of the multiple correlation coefficients resulting from regressing a predictor variable against all other predictor variables. It therefore detects multicollinearities that cannot always easily be detected with a simple pairs() scatterplot correlation.

More generally speaking, we might ask why we should be concerned about correlations. On the one hand, some statistical methods will fail to correctly fit the influence of the selected variable and thus return biased coefficients. This might strongly influence the importance of variables in the models. It can also be problematic when a model is applied to a changing climate, if the temperature variable has a biased regression coefficient, for example (Dormann et al., 2013). Another, and perhaps even bigger, problem is the fact that the correlation structure among climate (or other) variables as a whole might change significantly in the future. Having too many highly correlated variables included in a regression model (Part III) will tightly constrain the simulated distribution of a species to this correlation structure, and possibly result in false projected extinction rates, should this variable correlation structure change in the future. It is therefore safer and more prudent to avoid selecting highly correlated variables for habitat suitability modeling.

6.4.3 Variable Pre-Selection

Given the wealth of spatial data available today, selecting the right variables for a model is somewhat arbitrary. We have dozens if not hundreds of candidate variables available, and therefore need to decide which ones should be included in a model. This is known as the pre-selection of variables, as most statistical methods will then either down-weight the variables that are unimportant for a certain species (e.g. shrinkage rules), or use statistical variable selection procedures to remove these unimportant variables (e.g. stepwise selection). In order to come up with a pre-selection of variables, we can either use statistical or conceptual reasoning. Statistical reasoning would involve evaluating whether there are any variables that contribute so little to explaining the spatial distribution of a target species that they are insignificant or not much different from a random variable. Common practice to produce a statistically based pre-selection is to select the most important variables, and from these produce a selection that is not too strongly correlated and which passes a VIF test.

Alternatively, one could start with conceptual reasoning and ecological theory (Austin, 2002, 2007). We might, for example, ask: "*what is primarily constraining the niche of a species, and thus its spatial distribution?*" We can use results from experiments, or more conceptual considerations (see Part I). It still makes sense here to check the correlation structure because when designing a conceptually sound variable pre-selection, we may still pre-select some highly correlated variables (e.g. summer mean temperature and minimum temperature of the coldest month), because they both seem equally important for many species.

If we include both (which we do not recommend in most cases), we need to accept the risk of associated multicollinearity problems with this variable choice. Overall, the pre-selection of variables, and the consideration of correlation or accuracy vs. proximal influence is an open, and still poorly explored field. However, from the content discussed above, it seems clear that a priori using a larger set of predictors (such as all 19 bioclimatic variables of Worldclim) without further considering correlation or their relationship to known drivers of the spatial distribution of species is not an appropriate strategy for the purposes of habitat suitability modeling.

One understudied and underrepresented aspect is the transformation of variables. Some methods primarily expect the response (dependent) variable to follow certain distributions, and several model families have been developed to cope with non-normal distributions. However, it is often beneficial for the analysis to transform predictor. The correlation between variables bio7 and bio12 is e.g. quite a bit smaller if the left-skew is treated with log-transformation.

```
> cor(data$bio12,data$bio7)
 [1] -0.6439925

> cor(log(data$bio12+.001),data$bio7)
 [1] -0.4410228
```

We see quite a large change in correlation (r = -0.44 with log-transformation, r = -0.64 without transformation for bio12), yet we do not see a significant increase in predictive power in all cases where we transformed predictor variables. In essence, transformation of variables has to be tested individually. Note that a small value (0.001) had to be added to the bio12 variable in order to avoid errors from log(0) computations.

7 · Species Data: Issues of Acquisition and Design

When preparing a HSM, we first need to assemble an appropriate species presence–absence dataset. This is done either by collecting our own data in the field using an appropriate sampling design, or extracting our own database from existing large databases, atlases or museum data. In this chapter, we provide an overview of some large databases useful for habitat suitability modeling and how such data is imported and visualized in R. When using existing databases, we have no control over the design. It is therefore important in this case to check the distributions of spatial structures in the biological data and in the residuals from the modeling exercises. It is also possible to generate our own sampling design using a suite of rules and design elements, in order to appropriately sample space and environment with regard to the specific questions being asked. This chapter deals with these issues and provides some of the tools and functions used to this end. Finally, we discuss the benefits and risks of using presence–absence vs. presence-only data, and we give examples of how to generate pseudo-absence datasets. This last section also includes a discussion about how RSF relate to habitat suitability modeling.

7.1 Existing Data and Databases

Nowadays, there are numerous datasets and databases that can be used for habitat suitability modeling. Some of them are open access, while others have restricted access only. One of the most widely used databases for larger scale to global distributions of species is the GBIF.[1] It is immediately obvious when querying and downloading from such databases, that ongoing improvements to web access and web databases mean data are available more rapidly and more easily. Yet it is also clear, that there are

---

[1] www.gbif.org

Table 7.1 *Examples of databases that store species distribution data.*

General	GBIF	www.gbif.org
General	Map of Life	https://mol.org/
General	LifeMapper	lifemapper.org/
General	IUCN Red List	www.iucnredlist.org/
Herps	HerpNET	herpnet.org/
Mammals	MaNIS	vertnet.org/
Marine species	OBIS	www.iobis.org/
Amphibians	AmphibiaWeb	http://amphibiaweb.org/
Birds	ORNIS	http://ornisnet.org
Birds	Bird Life	www.birdlife.org/
Plants	Atlas Flora Europaea	www.luomus.fi/en/ database-atlas-florae-europaeae/
Plants	BIEN	http://bien.nceas.ucsb.edu/bien/
Central America	REMIB	www.conabio.gob.mx/remib_ingles/ doctos/remibnodosdb.html?
Brazil	SpeciesLink	http://splink.cria.org.br/

gaps in the collections and these gaps might cause difficulties later on when analysing the data.

The rapid growth of web databases such as GBIF has made access to data much simpler (e.g. Table 7.1) Yet this does not necessarily mean that the data can be used without restriction. Extractions from large, community databases such as GBIF need to be treated with caution for the following reasons: (i) uncertainty in species identification, (ii) low or unknown accuracy of sample location, (iii) lack of design, (iv) incomplete or uneven spatial coverage of the true distribution of a species, or (v) spatial autocorrelation in sample locations.

The issues related to species identification cannot be easily resolved, and are not addressed in this book. The second issue related to uncertainty in sampling location is covered in Chapter 8. The lack of design is a third, serious issue for all analyses that attempt to derive a probabilistic habitat suitability estimate from large datasets. If lack of design is an issue, one might consider resampling existing databases in order to increase the level of design (Broennimann and Guisan, 2008; Veloz, 2009; Anderson and Raza, 2010; Hijmans, 2012; Syfert et al., 2013; Mateo et al., 2015). Such resampling can only improve, but not fully remove the design bias inherent to such large datasets (see Section 7.4 for some suggestions). The fourth issue relating to the lack of coverage cannot be easily overcome, and its effects are treated in Section 8.2. The fifth issue is spatial autocorrelation, which is an inherent property of spatially structured, ecological data. This is partly dealt with in the following section. Finally, the sixth issue is sample size. When data is derived from large databases, one only has limited control over the sample size of a species. The available data can be considered as "presence-only" or the samples of presences of species other than the target species can be considered as "pseudo-absences" and thus allow the production of presence and absence data. However, it is likely that these derived datasets contain considerable problems in terms of prevalence (Phillips et al., 2009) and design, and may require further subsampling or design efforts with regards to "pseudo-absence" selection (see Barbet-Massin et al., 2012; Mateo et al., 2015).

7.2 Spatial Autocorrelation and Pseudo-Replicates

Spatial autocorrelation expresses the amount of spatial structure and pattern in geographically sampled data. The environmental or biological variables measured in space are not usually distributed randomly, and therefore inherently contain spatial structure. Indeed, both the response variable (e.g. presence–absence) and the predictors might contain some sort of spatial structure. If we have – and we almost always have – spatial autocorrelation in our data then the statistical analyses may require careful treatment. Spatial autocorrelation points to ecological or environmental processes that are influenced by space such as metapopulation processes. On the other hand, it may pose problems for the statistical analyses, as most standard statistical methods expect the individual observations in a dataset to be fully independent from each other. If, however, we do find spatial autocorrelation, then these observations may not be independent from each other, but represent observations that are influenced in the same way or direction by a specific process. It therefore represents a form of pseudo-replication. This problem is particularly severe in small datasets, or if the non-dependence reduces the effective number of observations (and resulting number of degrees of freedom) to critical value (see Thibaud et al., 2014).

Pseudo-replication arises when we consider the same observation twice as two separate samples. For example, when we sample tree species along environmental gradients, and we sample the same individual tree twice (i.e. as two ramets of the same site), then the two samples are not independent from each other. With regards to sampling presences along spatial and environmental gradients, we may also face the problem of pseudo-replication if we measure species presences from samples taken in overly close proximity. This means that we measure the presence or absence of the species under the same set of historical and demographic conditions more frequently than in other regions or locations. Such processes may typically cause spatial autocorrelation, and avoiding it is a design problem, while solving it is an analytical problem.

When we fit a habitat suitability model for a species based on observed presences, we aim to explain the spatial distribution pattern by means of environmental predictors. This means that if the set of predictors properly explains the spatial distribution, we should no longer have spatial autocorrelation in the residuals of our model. If this is the case, then we know that our set of predictors was suitable for explaining the spatial structure of our response variable. If, however, the model residuals are (still) spatially structured, this reveals that our model was unable to explain the spatial patterns of our response variable from environmental predictors alone, which means that there are hidden intrinsic or extrinsic (additional) factors that are responsible for this remaining spatial pattern. In a conventional regression analysis with spatially autocorrelated data, the estimates of the regression parameters for the used predictor variables may be affected, since the model tries to explain these patterns using the available predictors without accounting for the hidden factors. If additional predictors affecting the unexplained structure in the residuals are used in the regression, then the fitted parameters of the other predictors change. In summary, it means that we might over- or under-estimate the importance of the predictors in a regression if we do not correct for spatial autocorrelation found in residuals. The view presented here indicates that although informative per se, checking for spatial autocorrelation in the response variable (presence patterns of species) is not a priority. Rather, spatial autocorrelation should be checked in the residuals after fitting a statistical model (see Legendre, 1993).

Here, we present a simple method for detecting spatial autocorrelation in model residuals. Furthermore, we will discuss different views on whether effects of spatial autocorrelation should or should not be corrected (and when), and refer to other published work on correcting for spatial autocorrelation in regression models. Note that post-correction for spatial autocorrelation still remains an unresolved issue, as it remains unclear: (i) how to correct the number of degrees of freedom and (ii) how to best correct model parameters (e.g. regression coefficients) from indices of spatial autocorrelation.

When fitting a habitat suitability model for a given species, we can test for spatial autocorrelation in different ways. First, we can evaluate globally whether there is any spatial autocorrelation in the residuals. We can do this using a Moran's I test. This involves deriving a distance matrix from all observations, and then testing the distance effect against the residuals. Here, we use the `ape` library and the stepwise-optimized GLM model of *V. vulpes* from Section 6.2.10. As coordinates, we take the x- and y-coordinates from the `pts.cal` object, which we used to derive the calibration dataset. The order and dimension of the two objects are still the same. We generate an inverse-distance matrix with the diagonals set to 0 for the Moran's I test on the GLM model residuals (i.e. `vulpes.step$residuals`).

```
> library(ape)

> xy <- pts.cal[,1:2]
> dists <- as.matrix(dist(xy))
> dists.inv <- 1/dists
> diag(dists.inv) <- 0
> Moran.I(vulpes.step$residuals, dists.inv)

$observed
 [1] 0.01506913

$expected
 [1] -0.0001695203

$sd
 [1] 0.0003468404

$p.value
 [1] 0
```

We learn from this example that the p-value for testing for spatial auto-correlation is highly significant ($p < 0.05$). We therefore find spatial autocorrelation in the residuals. Next, we will want to plot the spatial correlation structure against distances between our observations. Samples separated by a short distance should have greater similarity (and thus correlation) than samples separated by a larger distance. We evaluate this distance dependence using a Mantel correlogram in the `ncf` package. This package makes it easy to plot a spatial (Mantel) correlogram. This is done by first extracting the residuals from the GLM object, and then randomly selecting 500 points from the residuals and from the x- and y-coordinates (`xy` object from the previous example). This information is needed in the `correlog()` command in the `ncf` package. We store the result of this command in the `spat.cor` object. We can either plot this object directly by typing "`plot(spat.cor)`", or we can extract the necessary information, and make a neater plot (see Figure 7.1)

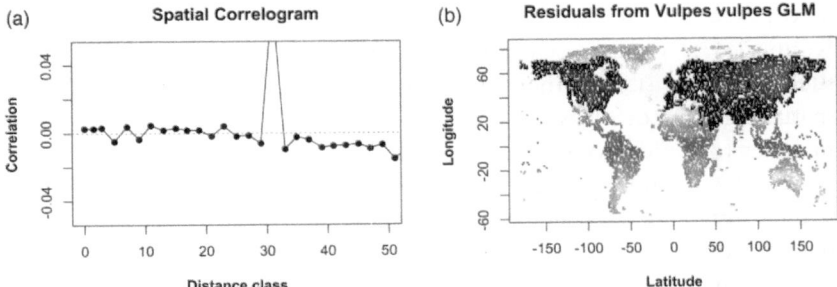

Figure 7.1 Spatial correlation (a) and spatial patterns (b) of model residuals for the *Vulpes vulpes* stepwise-optimized GLM model. The correlogram reveals a low correlation at a short distance. (*A black and white version of this figure will appear in some formats. For the color version, please refer to the plate section.*)

```
> library(ncf)
> rsd<-vulpes.step$residuals
> rnd<-sample(1:length(rsd),500,replace=T)
> spat.cor<-correlog(xy[rnd,1],xy[rnd,2],rsd[rnd],increment=2,
resamp=10)
```

We then plot the correlogram and in addition the spatial residuals in a map for illustration.

```
> par(mfcol=c(1,2))
> plot(spat.cor$mean,spat.cor$ccrr,ylim=c(-.005,.005),
xlim=c(0,50), pch=16,col="firebrick3", ylab="Correlation",
xlab="Distance class", main="Spatial Correlogram", font.lab=2)
lines(spat.cor$mean,spat.cor$corr,
col="firebrick3") abline(0,0,col="grey50",lty=3)
> plot(xy[order(rsd),], pch=15, col=rev(heat.colors(5900)),
cex=.3, main="Residuals from Vulpes vulpes GLM ", xlab="Latitude",
ylab="Longitude", font.lab=2)
> par(mfcol=c(1,1))
```

Residuals from binomial GLMs are naturally asymmetric, with larger values where presence is simulated and lower values where absence is simulated. The highest residuals usually occur at the range edge of species, as is clear in Figure 7.1b.

The debate around spatial autocorrelation is complex (e.g. Hawkins, 2012a, b; Kühn and Dormann, 2012). On the one hand, it is clear that autocorrelation arises when important drivers that shape the geographic patterns of species are not included. However, including them can have a detrimental effect on the assessment of variable importance or even the slope of the calibrated effect of a variable in a model (Kühn, 2007). There are numerous publications available which explain how to remove or incorporate the effects of spatial autocorrelation in models (e.g. Carl and Kühn, 2007; Dormann et al., 2007; Kühn et al., 2009). Broad spatial trends such as those visible in the residuals in Figure 7.1 can be removed using spatial eigenvector maps (SEVM) or similar techniques built on residuals, in order to avoid strong interactions with the other predictors (Maggini et al., 2006; Kühn et al., 2009). This is not currently standard practice in habitat suitability modeling, but we agree that neglecting such effects can have an adverse impact on the modeling and the conclusions drawn from these analyses. Regardless of whether the spatial trends are removed or not, we will always be faced with a lack of certainty regarding the importance and trends in analysed variables, due to the correlative nature of regression-type analyses. Careful selection of a smaller number of important variables helps to avoid these problems. Running sensitivity analyses to estimate the relative effect size of spatial autocorrelation in models, compared to other factors (see subsequent sections) is also helpful. In some cases spatial autocorrelation has a negligible effect on HSM predictions (Thibaud et al., 2014), but see also Guillera-Arroita et al. (2014) for some issues related to the model setup in the aforementioned publication.

7.3 Sample Size, Prevalence, and Sample Accuracy

Numerous factors can affect the predictive power of HSMs. First of all, to fit a sound model of habitat suitability using several environmental predictors requires sufficient species observation data (i.e. sample size). Due to their correlative nature, HSMs require sufficient information on both the presence and the absence data to fit reliable curves. Many publications have reviewed the effect of sample size on the quality of HSMs (e.g. Stockwell and Peterson, 2002b; Kadmon et al., 2004; Hernandez et al., 2006; Guisan et al., 2007b; Wisz et al., 2008; Jimenez-Valverde et al., 2009; Thibaud et al., 2014). Most studies agree that model accuracy metrics decline severely if fewer than 30 presence observations are available, while sample size effects usually become less critical above 50 presences. There are differences among statistical methods in terms of how sensitive they are to small sample size, and some publications claim that methods such as Maxent, or other complex methods, can cope with small sample sizes (e.g. Pearson et al., 2007; Thibaud et al., 2014). However, if this is done with too few observations, multiple and independent predictor variables cannot be fitted in a probabilistic way, and such over-fitted models can cause severe errors when projected to new areas or to changed environmental conditions, as such models heavily constrain the fitted surfaces to current correlation structures of the predictor variables in the study area (Randin et al., 2006). The rule of thumb is to have at least ten presence observations available per environmental predictor used (Harrell et al., 1996; Guisan and Zimmermann, 2000). We also advise defining the minimum number of presence points per species at between 20 and 50 observations (depending on the complexity of the models to be built; Merow et al., 2014).

Below, we illustrate the effect of sample size and prevalence on the quality of habitat models using the global *V. vulpes* dataset and three predictor variables from Worldclim. We first split the dataset into sets of presences and absences, and define resampling schemes within each set separately. In the "sample size" test, we randomly resample equal numbers of points from both presence and absence points to increasing total numbers, while in the "prevalence" test, we take all absences, and add randomly resampled and increasing numbers of presences to the absence dataset. In this example, we only use GLM as a statistical method. In the example code presented below, we only illustrate the prevalence test. The sample size test can be easily reconstructed from the example of the prevalence test. The full (eval.f) and the stepwise-optimized model (eval.s), both predicted against the complete "Sp.Env" dataset (not the reduced one), are tested by resubstitution and the dataset that has been extracted by resampling is also tested with cross-validation. For each dataset containing different characteristics regarding sample size and prevalence three evaluation metrics (Kappa, AUC, and TSS; see Chapter 15, Part IV) are calculated and the final results plotted on a summary graph (Figure 7.1). First, we need to load two libraries, ecospat and PresenceAbsence. The data preparation is starting from the pts.cal.ovl object used in Chapter 6:

```
> library(ecospat)
> library(PresenceAbsence)
> vulvul.pa<-cbind(pts.cal.ovl,runif(dim(pts.cal.ovl)[1],1,100))
> names(vulvul.pa)[6]<-"srt"
> vulvul.pa<-vulvul.pa[order(vulvul.pa$srt),]
> vulvul.p<-cbind(vulvul.pa[which(vulvul.pa[,1]==1),])
> vulvul.a<-cbind(vulvul.pa[which(vulvul.pa[,1]==0),])
```

Next, the data for resampling and quality testing is prepared.

```
> ybl<-c(10,20,40,60,80,100,125,150,200,250,300,400,600,800,1000
,1500, length(vulvul.p$srt))
> pr.qual<-data.frame(matrix(data=NA,nrow=length(ybl),ncol=12))
> names(pr.qual)=c("adjD2.f","adjD2.s","AUC.f","AUC.s","AUC.x",
"Kappa.f", "Kappa.s","Kappa.x","TSS.f","TSS.s","TSS.x","Prev")
> pr.qual[12]<-ybl/(ybl+length(vulvul.a$srt))
```

Once everything is ready, we run the loop and generate the prevalence test.

```
>for (i in 1:length(yb1)){
 paok<-rbind(vulvul.p[1:yb1[i],],vulvul.a)
 rownames(paok)<-1:dim(paok)[1]
 # Full Model
 paok1f<-glm(Vulpes.vulpes~bio3+I(bio3^2)+bio11+I(bio11^2)+bio12+
 I(bio12^2),family="binomial",data=paok)
 paok0<-predict(paok1f,paok,type="response")
 pr.qual[i,1]<-ecospat.adj.D2.glm(paok1f)
 tmp1 <- data.frame(1:length(paok0),paok[,1],paok0)
 names(tmp1) <- c("ID","Observed","Predicted")
 pr.qual[i,3]<-auc(tmp1)$AUC
 pr.qual[i,6]<-ecospat.max.kappa(paok0,paok[,1])[[2]][1,2]
 pr.qual[i,9]<-ecospat.max.tss(paok0,paok[,1])[[2]][1,2]

 # Stepwise optimized model
 paok1s<-step(paok1f,direction="both",trace=F)
 paok0<-predict(paok1s,paok,type="response")
 pr.qual[i,2]<-ecospat.adj.D2.glm(paok1s)
 tmp1 <- data.frame(1:length(paok0),paok[,1],paok0)
 names(tmp1) <- c("ID","Observed","Predicted")
 pr.qual[i,4]<-auc(tmp1)$AUC
 pr.qual[i,7]<-ecospat.max.kappa(paok0,paok[,1])[[2]][1,2]
 pr.qual[i,10]<-ecospat.max.tss(paok0,paok[,1])[[2]][1,2]

 # Xval procedure
 paok1x<-ecospat.cv.glm(paok1s)
 tmp1 <- data.frame(1:length(paok0),paok[,1],paok1x$predictions)
 names(tmp1) <- c("ID","Observed","Predicted")
 pr.qual[i,5]<-auc(tmp1)$AUC
 pr.qual[i,8]<-ecospat.max.kappa(paok1x$predictions,paok[,1])
[[2]][1,2]
 pr.qual[i,11]<-ecospat.max.tss(paok1x$predictions,paok[,1])[[2]]
[1,2]
}
```

Finally, we plot all model quality results from the prevalence test (Figure 7.2). Note that the sample size plot is also presented in the same figure, but not given as code example here.

```
> plot(pr.qual$Prev,pr.qual$Kappa.f,ty="l",lwd=5,col="#00FF00B4",
ylim=c(0,1.0),xlim=c(0,.5),xlab="Sample size",
ylab="Model  Quality",main="Prevalence Effects")
> points(pr.qual$Prev,pr.qual$Kappa.s,ty="l",lwd=5,col="#00CD00B4",
lty=3)
> points(pr.qual$Prev,pr.qual$Kappa.x,ty="l",lwd=5,col="#008B00B4",
lty=2)
> points(pr.qual$Prev,pr.qual$TSS.f,ty="l",lwd=5,col="#ADD8E6B4")
> points(pr.qual$Prev,pr.qual$TSS.s,ty="l",lwd=5,col="#9FB6CDB4",
lty=3)
```

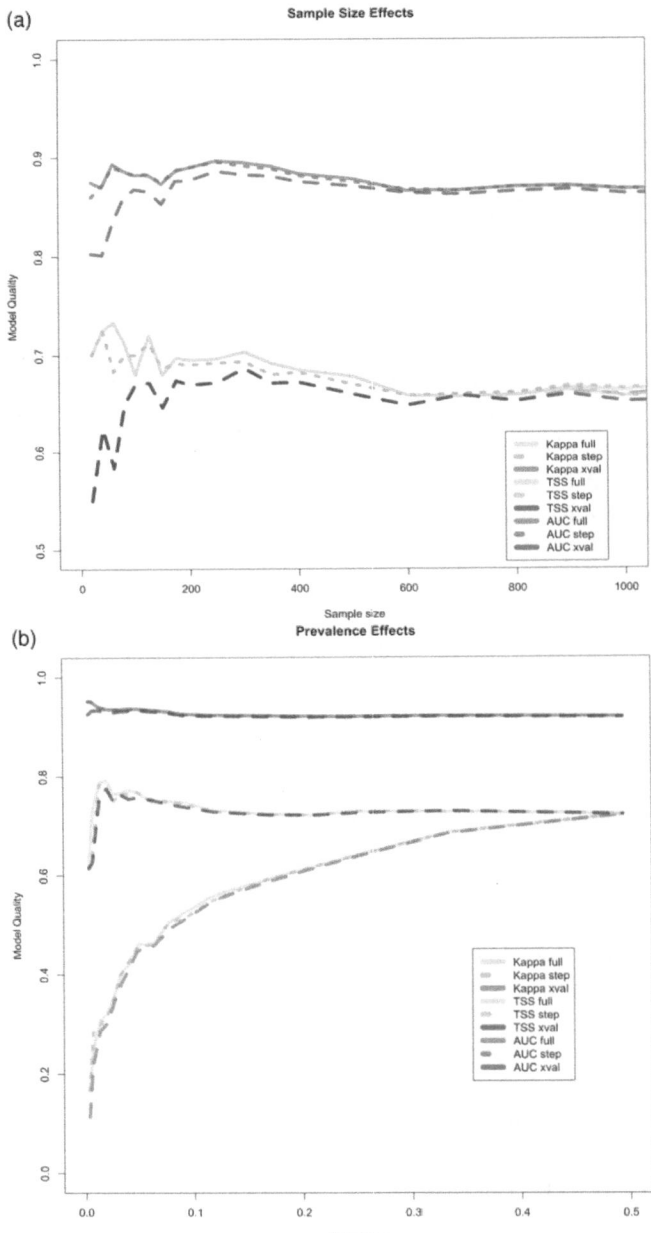

Figure 7.2 (a) Sample size and (b) prevalence effects on model accuracy in the global *Vulpes vulpes* dataset. Low prevalence has a strong effect on cross-validated Kappa and TSS accuracies. (*A black and white version of this figure will appear in some formats. For the color version, please refer to the plate section.*)

```
> points(pr.qual$Prev,pr.qual$TSS.x,ty="l",lwd=5,col="#0000FFB4",
lty=2)
> points(pr.qual$Prev,pr.qual$AUC.f,ty="l",lwd=5,col="#EE2C2CB4")
> points(pr.qual$Prev,pr.qual$AUC.s,ty="l",lwd=5,col="#CD2626B4",
lty=3)
> points(pr.qual$Prev,pr.qual$AUC.x,ty="l",lwd=5,col="#8B1A1AB4",
lty=2)
> legend(.355,0.45,c("Kappa full","Kappa step","Kappa xval",
"TSS full", "TSS step", "TSS xval", "AUC full", "AUC step",
"AUC xval"), lty=c(1,3,2,1,3,2,1,3,2),lwd=c(8),col=c("#00FF00B4",
"#00CD00B4", "#008B00B4","#ADD8E6B4","#9FB6CDB4","#0000FFB4",
"#EE2C2CB4", "#CD2626B4","#8B1A1AB4"))
```

It becomes clear, that unbalanced prevalence and low sample size (below 50 sample points) in particular, reduce the accuracy of the models. Also, both TSS and Kappa clearly respond to reduced prevalence and sample size in the cross-validation exercise. This is the only semi-independent test in this example. AUC is said to be largely unaffected by sample size or prevalence, which has previously been considered as an advantage (McPherson et al., 2004). However, our example shows that it is actually a disadvantage when the metric is insensitive to low sample size or prevalence. It tends to provide overly optimistic model quality estimates in these cases (Lobo et al., 2008), since comparably low sample sizes do not seem to significantly affect model quality. However, we should be aware that in this example, the sample size was always balanced (equally low number of presence and absence points). Most datasets with a low sample size often also have very low prevalence, due to a much larger number of known (or assumed pseudo-) absences compared to known presences. In order to avoid negative prevalence effects, it is recommended that presence and absence points are weighted inversely proportional to the relative fraction on the total sample size. An even more rigorous test would be to evaluate the model against the evaluation dataset (pts.eva.ovl). Chapter 15 addresses model evaluation in more detail.

7.4 Sampling Design and Data Collection

Sampling design for collecting field data has long been a crucial research and teaching topic. Several demonstration studies have highlighted the importance of design in HSMs (Austin, 1987; Hirzel and Guisan, 2002; Kadmon et al., 2004; Austin et al., 2006; Edwards et al., 2006; Phillips et al., 2009). However, often we lack the tools required to combine different sampling approaches into one design. Here, we present several methods that can be implemented in R for a range of purposes.

There is no universal sampling design that fits all needs. We may therefore want to explore different strategies and their implications for drawing inferences from these (Albert et al., 2010). Most designs have either a random or a regular design component (or a combination of the two). A design can be set up to sample the geographical space (x-/y-coordinates), the topographical space (z, aspect, slope), or the environmental space (temperature, moisture, radiation, etc.). The goal of a study should guide us in defining the space we are to sample. Here we explore different design approaches, and we demonstrate how these can be implemented in R.

7.4.1 Preparing Stratifications for Spatial Sampling Design

Let's assume we start a new sampling campaign in North America and we want to design a suite of sampling strategies that explore the region for a range of purposes. Before we start, we want to generate an environmental stratification (environmental space) since we will use it later on. We do this for North America by combining temperature isothermality (bio3) and yearly precipitation sum (bio12). We build classes of similar bio3 and bio12 by reclassifying each of these two variables into nine evenly spaced classes, then we recombine the two into a single map. Each map code then represents similar bio3 and bio12 conditions. In order to reclassify our raster layers, we use a function from the ecospat library, called `ecospat.rcls.grd()`. In addition, we load the library `classInt`.

```
> library(ecospat)
> library(classInt)
```

Before classifying the bio3 and bio12 rasters, we crop the global bioclim rasters from Section 6.1.3 to the extent of the conterminous lower 48 states of the United States. For this we load a shapefile of the states of the United States and extract the lower 48 states:

```
> usa <- shapefile("vector/usa/USA_states.shp")
> usa_contin <- usa[usa$STATE_NAME != "Alaska" & usa$STATE_NAME != "Hawaii", ]
```

With the second command, we select all states that are neither Alaska nor Hawaii, thus representing the conterminous United States. Next we need to generate an empty raster, to which we then rasterize the state polygons using the DRAWSEQ field:

```
> empty_raster <- raster(bio3)
> usa_raster <- rasterize(usa_contin,empty_raster,field="DRAWSEQ")
```

Next, we mask and crop the global bio3 and bio12 rasters to the United States extent:

```
> bio3.us <- crop(mask(bio3, usa_raster), extent(usa_contin))
> bio12.us <- crop(mask(bio12, usa_raster), extent(usa_contin))
```

Now, we are ready to apply the reclassification procedure. We do so, by generating nine classes, using the ecospat.rcls.grd() function applied to bio3 by assigning nine classes evenly spaced across the value range of bio3. We then repeat the same process for bio12 and recombine the two classified grids by adding them up after having multiplied bio12 by a factor of ten. This is a trick for generating a single raster of unique values originating from several reclassified raster layers consisting of values that always add up to unique numbers. We use decimal classifications for this purpose:

```
> B3.rcl<- ecospat.rcls.grd (bio3.us,9)
> B12.rcl<- ecospat.rcls.grd (bio12.us,9)
> B3B12.comb <- B3.rcl+B12.rcl*10
```

We now have generated two reclassified raster layers, one with class numbers running from 1 to 9 and the other with numbers of 10, 20, … 90. Then we summed the two layers so the new values are now unique, meaning that we can trace back their origin. For example, a number of 56 means that it originates from class 5 (code 50) of bio12 and class 6 of bio3. No other combination of classes would generate this code.

Next we plot the histogram of class frequencies, and we plot the stratification map with the rainbow color scheme (Figure 7.3). This includes checking the range of possible values for plotting the histogram (cspan). In order to better visualize the classes, we also allocate random colors to the map, and we click on five locations with the mouse to read values from the map (shown after you click five times at any location on the map). The randomly allocated colors are stored in a variable called yb, a name that has no ecological or R-specific meaning:

```
> cspan<-maxValue(B3B12.comb)-minValue(B3B12.comb)
> yb<-rainbow(100)[round(runif(cspan,.5,100.5))]
> par(mfcol=c(1,2))
> hist(B3B12.comb,breaks=100,col=heat.colors(cspan),
main="Histogram values")
> plot(B3B12.comb,col=yb,main="Stratified map", asp=1)
> click(B3B12.comb,n=5,type="p",xy=T)
> par(mfcol=c(1,1))
```

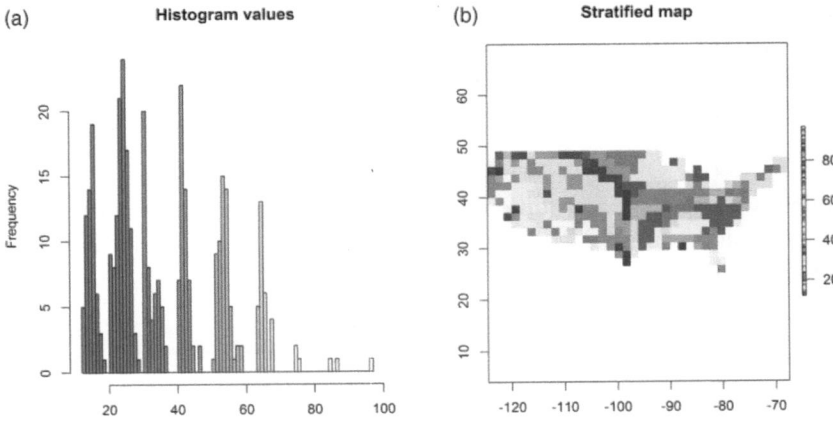

Figure 7.3 (a) Histogram of the frequencies of the classes, and (b) stratification map with the rainbow color scheme used to identify the strata from the environmental stratification of the study region. (*A black and white version of this figure will appear in some formats. For the color version, please refer to the plate section.*)

7.4.2 Spatial Sampling Design Using Built-In Functions in "rgdal"

We now develop our first set of spatial sampling designs using the built-in functions in the rgdal package. We can choose between four different designs, and here we will explore all of these. There are two spatial designs available, a "random" and a "regular" type. Both sample the x/y geographic space either randomly or regularly. The regular design identifies the optimal spacing of regular point distances over a grid, given the total number of sample points defined by the user. In addition, there are two variations to these basic spatial sampling designs, named "stratified" and "non-aligned". The stratified spatial sampling represents a spatial stratification. It allocates a user-defined number of spatial "blocks" by subdividing the x- and y-axes into regular strata, then allocates the user-defined number of samples evenly across the strata and samples randomly within cells. The "non-aligned" sampling adds some random noise to the regular sampling so that the sample points are no longer fully aligned compared to the regular sampling.

Before we start to apply these designs, we need to convert the raster layer used here into the required format for the rgdal package. From now on, we will use the reclassified bio3/bio12 grid as the reference map. This represents our study area, but here is only used to generate the extent of the study area, while we will use the reclassified values later on. It needs to be converted to the SpatialPixelsDataFrame format for the rgdal package. In order to distinguish it from our original raster format, we will call this paok from now on (an artificial name with no ecological meaning). This is different from the raster format in the raster package:

```
> paok <- as(B3B12.comb, "SpatialPixelsDataFrame")
```

Next, we apply the four different sampling designs previously described. Each design consists of 100 sample points that are allocated according to different design strategies:

```
> s.rand<- spsample(paok,n=100,type="random")
> s.strt<- spsample(paok,n=100,type="stratified",cells=3)
> s.regl<- spsample(paok,n=100,type="regular")
> s.nona<-spsample(paok,n=100,type="nonaligned")
```

It is clear to see that these designs are arranged in a sequence from fully random to fully regular. The four graphs are plotted as follows (Figure 7.4):

```
> par(mfcol=c(2,2))
> plot(B3B12.comb,main="random",col=rev(terrain.colors(25)))
> points(s.rand, pch=3, cex=.5)
> plot(B3B12.comb,main="stratified",col=rev(terrain.colors(25)))
> points(s.strt, pch=3, cex=.5)
> plot(B3B12.comb,main="nonaligned",col=rev(terrain.
colors(25)))
> points(s.nona, pch=3, cex=.5)
> plot(B3B12.comb,main="regular",col=rev(terrain.colors(25)))
> points(s.regl, pch=3, cex=.5)
> par(mfcol=c(1,1))
```

We have now developed four different designs, and each of these datasets can be used to sample the environmental layer stacks as done previously. Once this is done, we can check how the different designs affect the retrieved distribution layers sampled with the respective design. We can compare this distribution with the known truth (when plotting the distribution of the raster layers).

7.4.3 Random, Environmentally Stratified Sampling Design
We now want to generate a simple environmentally stratified design. This is a method implemented in the ecospat library. We will use the stratification raster called "B3B12.comb" we developed earlier.

The basic method for environmentally stratified sampling is to use a classified stratum map to drive the sampling. We want to sample randomly in each stratum, but we also want to sample each stratum. This design represents a combination of regular (each stratum with regular sampling along environmental gradients) and random (spatially random within

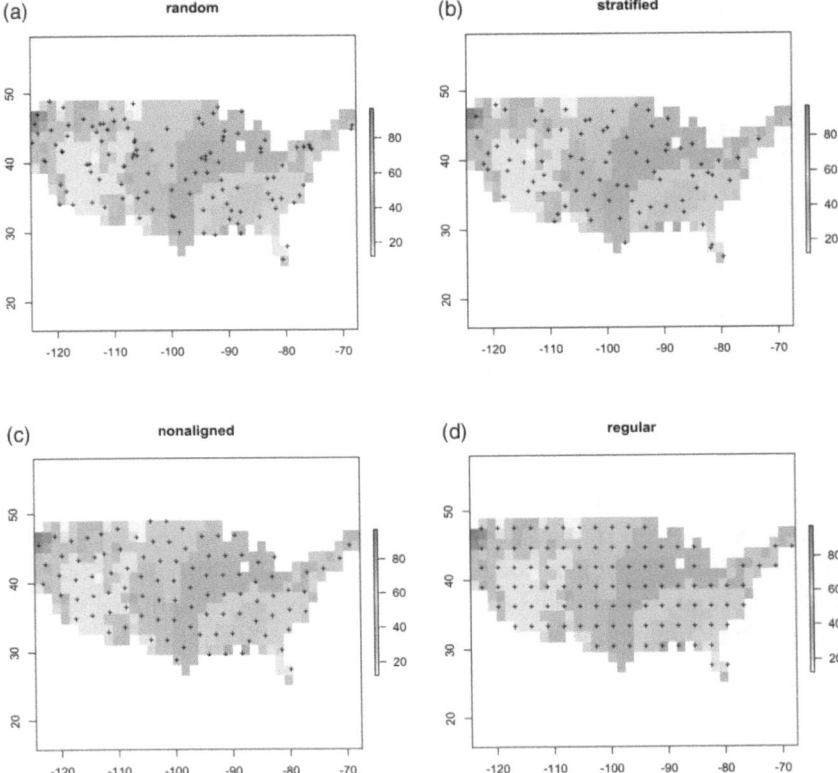

Figure 7.4 Spatial sampling designs: (a) random, (b) stratified, (c) non-aligned, and (d) regular. The color shades of the background indicate the identifier of the strata from the environmental stratification of the study region. (*A black and white version of this figure will appear in some formats. For the color version, please refer to the plate section.*)

polygons of each stratum) sampling elements, and is therefore a comparably complex design. However, it is well suited to sampling the gradients that we consider to be relevant for our study (Albert et al. 2010).

There are two basic approaches to random stratified sampling (Figure 7.5). Most commonly, we sample an equal number of points per stratum. In this way, we give equal weights to each stratum irrespective of its spatial extent (equal number variant). The second approach would be to sample each stratum with points selected proportional in number to the total area of each stratum (proportional variant). With the proportional approach, we accept that the strata have different spatial dimensions and we give respective weights to their frequency in the landscape. Compared to a fully spatially random sampling, the proportional

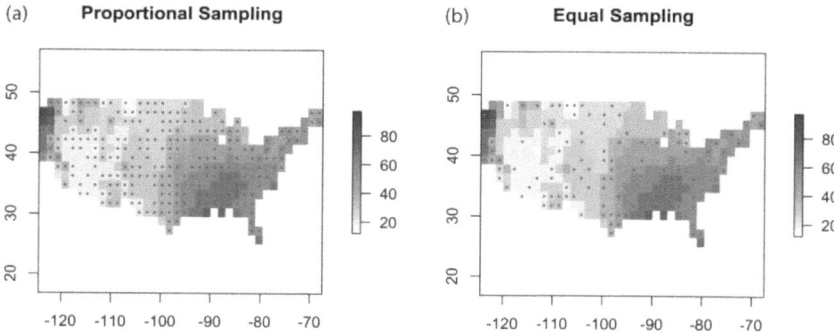

Figure 7.5 Sample points according to our random, environmentally stratified design. The color shades of the background indicate the identifiers of the strata from the environmental stratification of the study region. (*A black and white version of this figure will appear in some formats. For the color version, please refer to the plate section.*)

approach still has a lower risk of failing to detect rare strata. With both strategies, even the rarest strata will still be chosen with a few points, (proportional) or with an equal number of points as other strata (equal), unless they are too small (see Figure 7.6). The equal number variant assigns the same number of points, which – depending on the number of available classes – can result in slightly different numbers than those originally given. The proportional variant assigns the numbers according to the number of pixels available in the stratification grid. If the proportion of a stratum results in less than one sample being allocated to this class, then this class is not sampled.

We apply the equal, and then the proportional, allocation of sample points according to our random (environmentally) stratified design using the `ecospat.recstrat_regl()` and the `ecospat.recstrat_prop()` functions. These functions take the stratification grid and the total number of points to allocate per strata as arguments.

Finally, we plot 150 points from the two designs over the whole study area (Figure 7.5). We can see that the proportional design reveals a very similar distribution as is available in the study area, while the equal design (even numbers per stratum) generates a more uniform distribution of strata.

```
> envstrat_equ<- ecospat.recstrat_regl(B3B12.comb,150)
> envstrat_prp<- ecospat.recstrat_prop(B3B12.comb, 150)

> par(mfcol=c(1,2))
> plot(B3B12.comb,main="Proportional Sampling", col=rev(terrain.
colors(25)))
```

```
> points(envstrat_prp$x,envstrat_prp$y,pch=16,cex=.4,col=2)
> plot(B3B12.comb,main="Equal Sampling",
col=rev(terrain.colors(25)))
> points(envstrat_equ$x,envstrat_equ$y,pch=16,cex=.4,col=2)
> par(mfcol=c(1,1))
```

When evaluating the class distribution of the two sampling approaches, we see that the two designs sample the classes very differently. While the equal number sampling generates a very even distribution (with few deviations from the equal number because these are extremely rare classes which have insufficient pixels to be sampled), the proportional sampling closely mimics the overall proportion of pixels per class as previously plotted (Figure 7.6, note that this figure is printed in gray).

```
> par(mfrow=c(2,1))
> barplot(table(envstrat_prp$class),col="firebrick",
main="Proportional point allocation")
> barplot(table(envstrat_equ$class),col="slategray4", main="Equal
point allocation")
> par(mfrow=c(1,1))
```

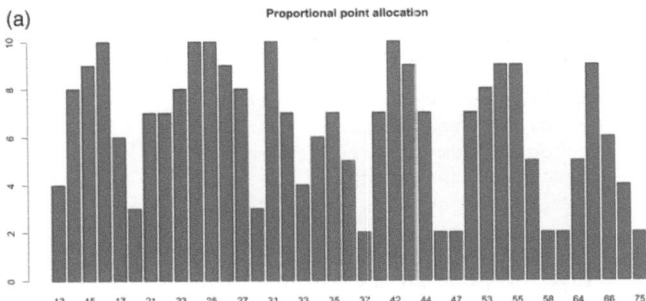

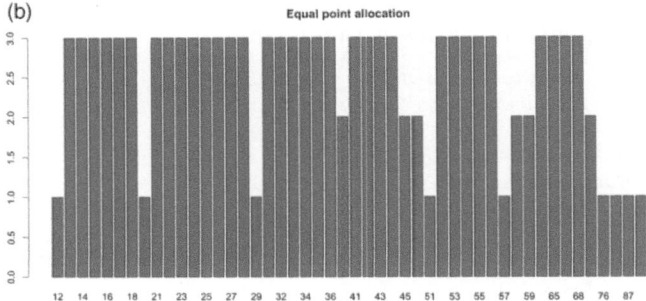

Figure 7.6 Illustration of the class distribution between the two sampling designs. The x-axis labels indicate the identifier of the strata from the environmental stratification of the study region.

7.4.4 Sampling Designs Along Linear Features

We have so far developed six different spatially and environmentally oriented sampling designs, which all cover the whole study area. Some ecological questions will require sampling along linear features, and we will therefore briefly mention this approach as well. For example, you may need to sample along rivers or transects from the sea to inland, or you might ask how a species has invaded along roads. All cases require specifically sampling along linear elements, not across large landscapes as a whole.

The three designs we demonstrate here are available in the `rgdal` package, and resemble the first four spatial designs discussed. The first is a "random" design, where points are allocated randomly at distance D from the origin of the linear path. The "regular" design dissects the whole length of the linear element into even distance classes, and allocates points accordingly. Finally, the "stratified" sampling is again a spatially stratified sampling approach, which dissects the line into blocks of equal distance, and samples randomly within these. The "non-aligned" sampling does not make any sense here, and is therefore not available.

As an example, we load the DEM of the globe, upscale the spatial resolution to 10 km, and crop it to the extent and the mask of the conterminous lower 48 states used in the examples in Section 7.4.1.

```
> dem_globe <- raster("raster/topo/GTOPO30.tif")
> dem_usa <- crop(dem_globe, usa_contin)
> dem_usa_10km <- aggregate(dem_usa, 10, fun=mean)
> empty_raster <- raster(dem_usa_10km)
> usa_raster <- rasterize(usa_contin, empty_raster,
field="DRAWSEQ")
```

We then mask the DEM and generate a contour of the 1000 m altitude band throughout the United States, as illustrated in Figure 7.7. This contour line is now ready to use to design different line sampling strategies by allocating sample points along this elevation contour.

```
> dem_usa_masked <- mask(dem_usa_10km, usa_raster)
> iso_1000m<-rasterToContour(dem_usa_masked, nlevels=1,
levels=c(1000))
> plot(dem_usa_masked,col=dem.c(100),
main="Elevation Contours at 1000m")
> lines(iso_1000m, lwd=1.3, col=2)
```

Now, we generate three point samples of size 150 following a linear design according to three different methods:

```
> l.rand<-spsample(iso_1000m,n=150,type="random")
> l.strt<-spsample(iso_1000m,n=150,type="stratified", cells=50)
> l.regl<-spsample(iso_1000m,n=150,type="regular")
```

Elevation Contours at 1000m

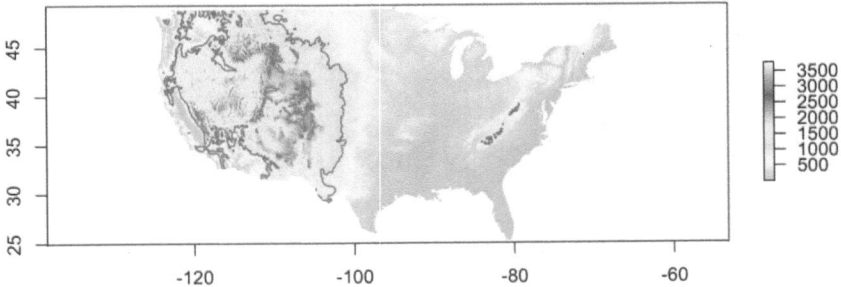

Figure 7.7 Elevation contour lines (1000 m) plotted in the study area. (*A black and white version of this figure will appear in some formats. For the color version, please refer to the plate section.*)

We have now generated our linear sampling along the elevation contours. The three sampling designs are plotted as follows, and we can see the different designs reflected in the points we have sampled (Figure 7.8).

```
> par(mfcol=c(3,1))
> plot(dem_usa_masked,col= gray.colors(100),
main="Random Sampling")
> points(l.rand, pch=3, cex=.5)
> plot(dem_usa_masked,col=gray.colors(100),
main="Stratified Sampling")
> points(l.strt, pch=3, cex=.5)
> plot(dem_usa_masked,col=gray.colors(100),
main="Regular Sampling")
> points(l.regl, pch=3, cex=.5)
> par(mfcol=c(1,1))
```

When discussing linear sampling as a method, we need to bear in mind that this approach can be very powerful if applied carefully to a well-targeted study. However, linear sampling also arises when sampling along roads is done for simple convenience (for a review see Albert et al. 2010). The three plots above neatly illustrate, how much of the "environment" or "geographical space" is left unsampled. Linear designs are not appropriate to sample larger geographical areas.

7.5 Presence–Absence vs. Presence-Only Data

Depending on the datasets used, many users or scientists make a clear distinction between "presence–absence" vs. "presence-only" datasets.

(a)

Random Sampling

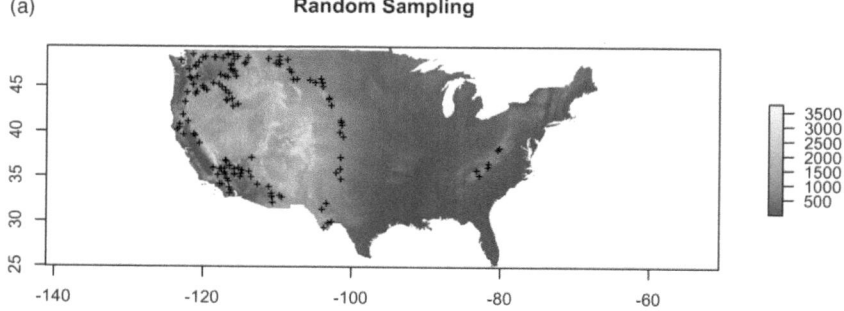

(b)

Stratified Sampling

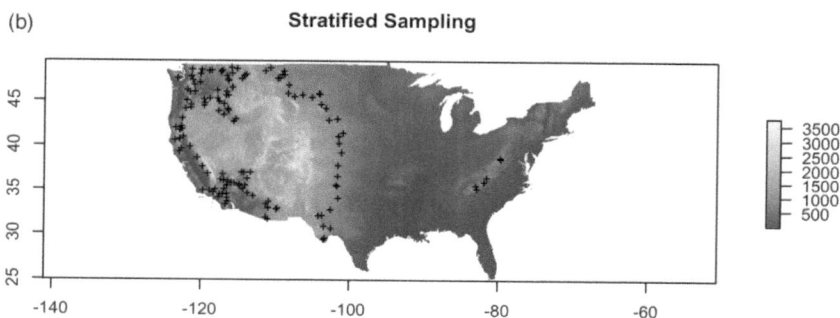

(c)

Regular Sampling

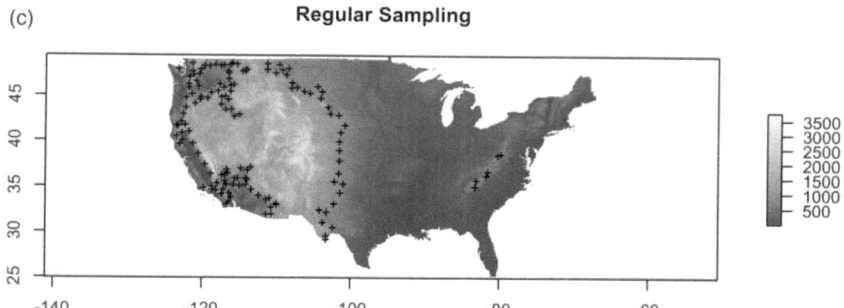

Figure 7.8 Three linear sampling designs applied along elevation contours in the study region: (a) random, (b) stratified, and (c) regular.

While a designed sampling campaign makes it possible to collect a complete set of presence and absence points that fully represent the probabilistic distribution of a species along sampled gradients, this cannot easily be obtained from museum-type collections of species presences. The frequently used GBIF database does not contain "observed absences," but only "observed presences." Building probabilistic HSMs from such data is therefore complicated (Graham et al., 2004a). This is mainly because there is no design available. Instead, the data for a given species may originate from a large collection of partly designed, usually biased observations of presence, and overall, the data have very heterogeneous sampling intensities across the globe (Figure 7.9; Meyer et al., 2015).

Over time, GBIF and other similar databases have amassed a huge number of records. However, the sampling bias has not decreased, but rather increased in severity, since the same sampling gaps still exist (e.g. Northern Africa, Russian Federation, Eastern Mongolia, or tropical regions; see Figure 7.9), while the already well-sampled regions have compiled even larger numbers of observations. This causes severe problems when inferring the habitat suitability of widely distributed species, because nothing is known about their distribution in under-sampled areas. Inferring habitat suitability is thus slightly more straightforward for species that are primarily distributed within the well-sampled regions. However, we still face regional differences in sampling intensity (Meyer et al., 2015), which influence any statistical models derived from the data (Graham et al., 2004a; Phillips et al., 2009).

Different statistical approaches can be used to predict species distributions from presence-only data (see Part III). Among the possible methods, some use information from the entire study area, either called pseudo-absences or background data (e.g. regression techniques) and combine those data with the existing presence data, while others do not require such additional data (i.e. presence-only technique). We refer here to Part III for explanations of these techniques. In principle, the terms background data and pseudo-absences are equivalent, but there are many different ways of generating them. The most common way is to randomly sample a large set of localities in the study area (i.e. random pseudo-absences or background). However, in some cases the sites corresponding to presences can be removed from the random set of pseudo-absences or background, or some sampling bias can be corrected (Phillips et al., 2009; Barbet-Massin et al., 2012). The geographic

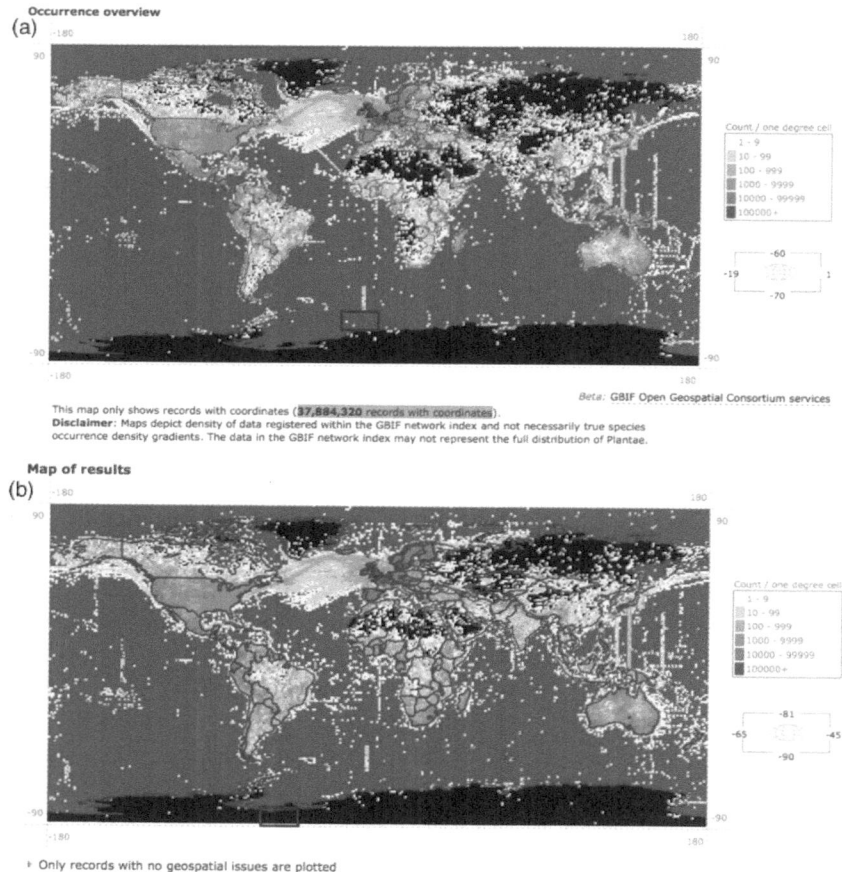

Figure 7.9 Sampling intensity of plant samples in GBIF: (a) 37.8 Mio records as of October 26, 2009; (b) 77.8 Mio records as of February 12, 2013. It becomes evident that the collection intensity varies considerably in space as it is measured by density per 1 × 1 degree cells, which will cause bias in the predictions when the data is used without resampling. Over the period 2009 to 2013 the spatial sampling bias has not decreased visibly. (*A black and white version of this figure will appear in some formats. For the color version, please refer to the plate section.*)

or environmental extent (see also Chapter 8) over which these pseudo-absences are sampled has also been shown to potentially affect model performance (VanDerWal et al., 2009; Barbet-Massin et al., 2012). Approaches using predictions from a preliminary model to stratify the selection of pseudo-absences (background) have also been proposed (e.g. Engler et al., 2004). However, this approach has the disadvantage (compared to simple random selection) of amplifying the bias if present in the initial species data (Wisz and Guisan, 2009; Barbet-Massin et al., 2012).

We suggest that both terms (pseudo-absence or background) can be used interchangeably when general strategies not involving the species' ecology are used (e.g. simple random or random stratified, with or without bias correction). We suggest preferentially using pseudo-absences when target group (i.e. using presences of related taxa in a same dataset as pseudo-absences; Phillips et al. 2009) or model-based (see above) approaches are used. Pseudo-absence is a more general term, and will be used as the term of choice hereafter.

There are numerous methods for selecting pseudo-absences in a given study area or for a given purpose (e.g. Engler et al., 2004; Chefaoui and Lobo, 2008; Phillips et al., 2009; Lobo and Tognelli, 2011; Barbet-Massin et al., 2012; Hanberry et al., 2012). We do not aim to provide a full overview here, which is best found in the relevant literature (see above). For each study or purpose, scope and data required to meeting the stated aim need to be defined. For example, in their study assessing the invasion potential and the extent to which invasive alien species are still spreading or have reached a climatic equilibrium compared to their native range, Gallien et al. (2012) applied a weighting to pseudo-absences generated from a random sampling process. Furthermore, Phillips and co-workers (2009) argue that it is advantageous to constrain the sampling of pseudo-absences based on ecological reasoning, or in a way that accounts for sampling bias, rather than simply sampling the whole background environment, possibly over an unrealistically large geographic area. Both cases illustrate that adding prior ecological or statistical considerations when sampling pseudo-absences can be important, and may have a significant effect on the models derived from the presence observations.

Here, we stress two key factors for reducing subsequent problems with a statistical method that uses presences in combination with pseudo-absences. First, it is important to constrain the study area to a realistic realm. Using a large but unrealistic extent to sample the pseudo-absences has been shown to have a potentially adverse effect on the model and predictions (VanDerWal et al., 2009; Elith et al., 2010). Second, sampling pseudo-absences randomly is the strategy with the least assumptions and should be used by default if there are no strong arguments in favor of a different, more taxon-specific approach (Barbet-Massin et al., 2012). Constraining the sampling to areas where the species has not been observed to prevent conflicting presence and absence observations is likely to increase initial bias (Wisz and Guisan, 2009) and may result in over-predictions of the species ranges (Hanberry et al., 2012).

8 · *Ecological Scales: Issues of Resolution and Extent*

Scale is an important issue in ecological applications; many patterns emerge only at a specific scale, and are perceived as noise or constants, or even go undetected at other scales (Chave, 2013). When preparing the data, one common problem is that the datasets we intend to use do not always match in terms of scale or resolution. We therefore need to scale spatial data to a common spatial extent and resolution in order to obtain a consistent modeling environment (see e.g. Section 6.2 for examples). This is definitely the easiest approach, but there are also other approaches, when a hierarchy of scales is included and implemented in a habitat suitability modeling exercise (see e.g. Lomba et al., 2010; Gallien et al., 2012; Fernandes et al., 2014).

Scaling is an important aspect of modeling, and it is therefore important to recall its conceptual foundation. Scaling is basically the process of translating information from one scale to another (King, 1991; Wu, 1999, 2004). A prerequisite for scaling models is usually to first adopt a concept of how the processes and components involved interrelate and how they can be ordered. Hierarchy theory provides a conceptual framework for this ordering (O'Neill et al., 1986; Allen and Hoekstra, 1992; Schneider, 2001). Depending on the nature of the processes or patterns, scaling can be a complex undertaking (Lischke et al., 2007). When carrying out habitat suitability and species distribution modeling the scaling is comparably simple. There are basically two important aspects to take into consideration: the effect of resolution and the effect of extent.

When considering resolution in a spatial context, we immediately think of "spatial resolution." For a grid, this naturally refers to its cell size. However, for the same grid it can also mean to what extent the thematic information (the map content) is resolved (e.g. how many and to what extent habitat units are distinguished or what units for temperature are used), or to what extent the map resolves temporal information. For our purposes, we will therefore distinguish three major levels of resolution: spatial, temporal, and thematic. Scaling often involves either the choice of a specific resolution, or to scale the data to appropriate or common resolution, by up- or downscaling. While some of these latter processes can be quite tricky, we only use simple examples that can be easily implemented in a GIS/statistical framework (see Chapter 6).

The extent scale can also be analysed on the three basic levels: spatial, temporal, and thematic. Extent effects are more challenging to deal with when modeling the suitability of habitats. What is the appropriate extent for solving a problem? Answering this question is by no means straightforward. On the one hand, we can easily say that once we work in a study area, then the spatial extent is defined. In many instances, however, the questions we ask cannot be answered from within the extent of the study area alone. If our study area is comparably small and encompasses relatively little topographic variation, then habitat suitability modeling under strong climate change scenarios by 2100 will drive most of the species' potential habitats out of the study area (and we can be fairly certain about this aspect), but we lack any certainty about what other species may actually find suitable habitats in this same study area by 2100. This is simply because these species cannot be observed within the current extent. Not even for those existing within the study area, we can be sure to fit all their niche limits appropriately, as they likely extend far beyond the study area. In such cases, HSMs do not reflect the full environmental constraint that limits their distribution.

In the following section, we will discuss how far questions of resolution and extent are relevant in the preparation, analysis and modeling of spatial ecological data.

8.1 Issues of Resolution

8.1.1 Spatial Resolution

Basically, all studies modeling habitat suitability and species distributions, encounter the problem of spatial resolution, both at the level of the species distribution data (dependent variable when modeling) and at the level of spatial predictors. We will briefly evaluate the techniques and principles relevant to the spatial resolution of our data.

Species data usually come either as point measurements or as raster cells from atlas data. The data may be classified as presence–absence or come in any abundance, frequency, proportion or importance scale. When available as raster information – the *Atlas Florae Europaeae* (Jalas

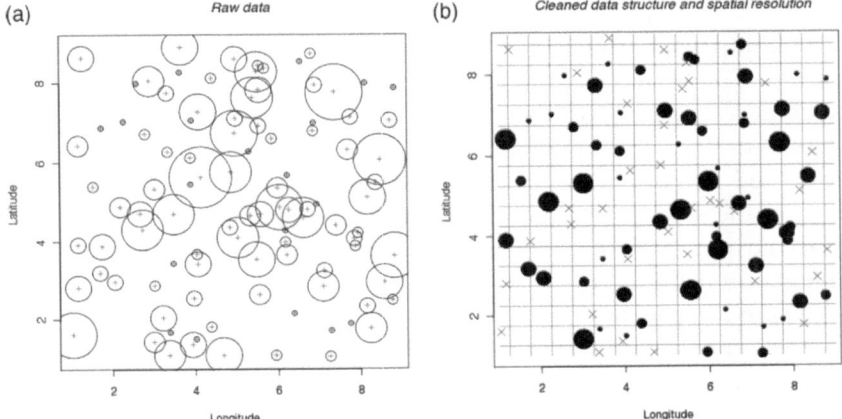

Figure 8.1 (a) Species occurrence data with positional uncertainty illustrated by circles around the noted location (+) mapped along artificial "latitude" and "longitude" coordinates. (b) Remaining points with accepted positional uncertainty and the associated spatial resolution of the raster (gray lines) used for calibration and prediction. The excluded species occurrences are shown with a red cross.

and Suominen, 1972–1996) is one example – then the spatial resolution is defined by the nature of the data (e.g. 50 km). When data are available as point (or small plot) level information (e.g. GBIF data or sampling plots), then the required spatial resolution is not immediately clear. A single point observation is conceptually infinitely small or as small as the size of the plant or animal of concern. However, we cannot assume that the point coordinates are accurate with no error. How large is this error? Some coordinates may be as imprecise as ±1 km or even only be available at the resolution of 0.1°. It makes a huge difference for our subsequent analyses if the coordinates are highly precise (errors < 5 m) or highly imprecise (errors > 10 km). In the latter case, it is difficult to relate our species information to topographic or environmental site variations, since the low precision means the information from a high-resolution surface of environmental predictors, such as climate or topography, cannot be retrieved accurately from the point coordinates. We therefore need to ascertain that the spatial resolution used at the predictor variable level fits with the uncertainty of the position coordinates for the point samples. A virtual example is given in Figure 8.1. Assume varying positional accuracy for the coordinates at each point (illustrated as circles around the position center in panel a). Since we do not want to work with overly high levels of uncertainty, we may want to remove a set of points where the positional uncertainty exceeds a given threshold. In our artificial example, we set the raster resolution to a value of 0.25 (indicated by gray lines in panel b) and remove all the points with greater spatial uncertainty (indicated by the red crosses in panel b). Alternatively, a coarser resolution for elevation could be used with an increased number of points. However, this may not increase the accuracy of the species–environment relationship, but rather introduce a bias, since coarser resolutions blur the environmental variability contained in a higher resolution grid.

There are no objective rules for making decisions on the exclusion or inclusion of points with regards to their positional accuracy. It seems important to reflect on the aim of the analysis before taking such decisions. How much topographic variation and smaller-scale ecological gradient is required for our analysis? Our decision needs to consider these questions in combination with the consequences on sample size when excluding points (see Section 7.3). The result is essentially a trade-off between keeping as many points as possible to increase the sample size, and removing as many points as possible to increase the spatial accuracy, the spatial resolution, and with this the accuracy of the species–environment relationship in the habitat suitability model.

While such decisions are relatively easy to make when the positional accuracy is known, it is more difficult to select a subset of useful data when no or only limited information on positional accuracy is available. If we have additional information at hand for given points, such as elevation, slope, or aspect, then this would be sufficient to at least use digital topography to test which points actually fit the field observed topography. In this case, we ignore the question of whether the field derived positional accuracy or the positional and topographic accuracies of the DEM we use actually cause more serious problems. Our primary aim is to ascertain that the field and digital topography match. We can test for this by overlaying each point in a GIS with DEM-derived elevation, slope and aspect. Most points will show slightly different values for these three variables. Again, we need to define what errors we will accept in our modeling. Imagine that we have a point where the field protocol says it is a steep south-facing slope, and the GIS database says it is a moderately steep north-facing slope. Whatever the reason for this mismatch, we would obviously attach all the environmental variables relating to north-facing slopes to the point in our database, while in fact we know that the point, and in particular the growth conditions for the species we have observed at this location, are significantly different. By deciding how

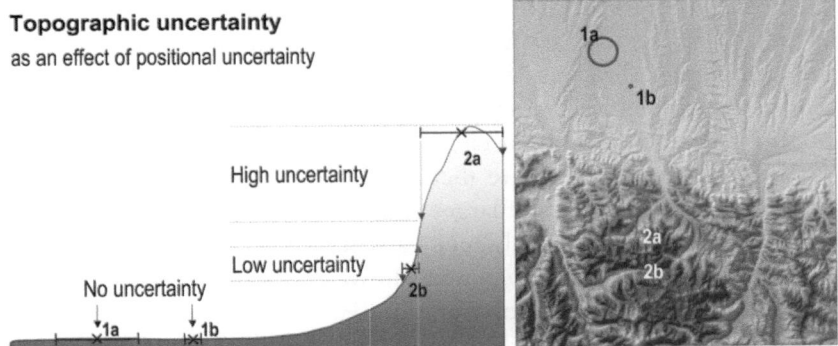

Figure 8.2 Relationship between positional and topographic uncertainty. Positional uncertainty matters especially where there are strong topographic and/or environmental gradients (2a/2b), and do not significantly influence our modeling where there are no such steep gradients visible (1a/1b). (*A black and white version of this figure will appear in some formats. For the color version, please refer to the plate section.*)

much topographic error is acceptable, we decide how well we are able to differentiate between different topographic situations. In essence, this relates to the thematic resolution (Section 8.1.3), as it determines how well we can actually differentiate along topographic axes. Another positive effect of this method for optimizing spatial scale effects is that we can accept larger positional errors in topographically homogenous terrain, since even comparably large positional uncertainties will not affect the linking of GIS-related "ecology." Figure 8.2 illustrates the link between positional and topographic uncertainty.

Most predictor layers are now available as raster maps. One major inconsistency that needs to be addressed at the beginning of a modeling study is that all raster maps have the same spatial resolution, the same lower-left coordinate (alignment), and the same spatial extent. Section 6.2.3 provides example code for resampling and aligning raster layers so that they perfectly match with a master grid that serves as a reference for all mapping. Working with differing resolutions may cause uncontrolled interpolation effects, and these are not easy to deal with if not treated explicitly (McPherson et al., 2006).

Spatial resolution, both at the level of the dependent and independent variable, has an effect on model fit and accuracy. In fact, it impacts more than just accuracy. First, coarser resolution data do not necessarily result in poorer model fit and accuracy (Mitchell et al., 2001; Guisan et al., 2007a; Luoto et al., 2007). Some studies have shown an improvement in predictive accuracy when higher resolution variables are used (Lassueur et al., 2006), yet this is primarily the effect of substituting climate for topography variables, not a direct effect of the resolution per se. Another study demonstrated that when scale differences between local point observations and used raster cells of predictors are too large, the models become less accurate (Seo et al., 2009). Specifically, predictor variables with grain sizes of larger than 10 km fitted to point observations of species presence–absence caused model accuracy to decrease significantly in this study. In general, however, models that use dependent and independent coarse resolution data show very high model fit, with AUC values often above 0.9. However, and this is ecologically problematic, they do not meaningfully fit climate–species relationships, only relationships on very broadly averaged information. Most species are not uniformly distributed within large cells, and most large cells are not flat. This means that the true climate–species relationship is often obscured in coarse scale models. Despite coarse-grained data generally fitting HSMs successfully and producing very good test statistics, it remains rather uncertain which habitat drivers really are responsible for determining the distribution of species. In fact, it has been shown that coarse resolution HSMs tend to predict high extinction rates under projected future climates that do not match with projections when finer grained data are used (Randin et al., 2009). Such finer grained information obviously identifies suitable habitats that are smoothed away at the coarse grain.

8.1.2 Temporal Resolution

Temporal resolution is rarely properly considered in the context of modeling habitat suitability and species distributions. We usually collect a set of distribution data, and take "current" climate and environmental predictor variables to relate the distribution to these predictors. However, there might be a strong mismatch between the observed distribution and the drivers (predictors) responsible for shaping this distribution. We distinguish between three different aspects, which we discuss here briefly. The three aspects all relate to the problem of capturing the niche of a species when overlaying observed presence and absence points with climate and other environmental predictors and then fitting statistical models: (i) Does "current" climate refer to the period responsible for the observed presence of a species? (ii) Is there a time lag in a species' response to changes in environmental conditions or does the species respond instantaneously to these changes? And (iii) is it possible to observe all elements of a species' niche under current climate conditions or can different (non-analog) environments in the past also contributed to capturing a species' niche?

The first issue relates to the question of whether the time of observation fits the time of the "current" climate and environmental data as drivers of this distribution. If we have, say, observations from the period of 1970–1980, then basic climate normals data for the period of 1961–1990 might be perfectly fine as these cover the basic sampling period. However, this is not necessarily a sound assessment of all species and all circumstances. A species might be very long lived, as is the case for trees, and the climate which explains their distribution is likely to date from much earlier. This is especially true if we assume that the regeneration niche determines spatial distribution much more than the adult niche. We then face the problem that trees of between 80 and 150 years of age at the time of the aforementioned period would have been influenced by climate between 1820 and 1900. We can usually still fit the models successfully when using more recent climates. However, we might get a biased estimate of their climate niche if we only consider current climate to explain their distribution. Presence may rather be an expression of persistence than of a viable population niche for some tree species in some locations (Bell et al., 2014). This is not always easy to detect and relates back to the definition of the niche (Part I). While slow migration in response to changing environments may result in incomplete RF (see the discussion of the second issue), no migration or extremely slow migration may result in a current range pattern that cannot be explained well by current, but rather (and better) by past, climate and environmental conditions. Support for this view has, for example, been demonstrated for reptiles and amphibians in Europe, where the late glacial 0°C isotherm is a better predictor of narrow-range species than the current 0°C isotherm, whereas the latter better explains wide-range species (Araújo et al., 2008). In addition, Quaternary climate stability was found to better explain current species richness in reptiles and amphibians than current climate stability in the same study. Other species are very short lived, and respond very directly to fluctuating climate characteristics that are not correctly expressed with 30-year climate normals. Annual plants, marine phyto- or zooplankton all respond on a much faster timescale (resolution). While annuals primarily respond to spring and summer climates, and thus show high levels of annual fluctuation, marine plankton has an even shorter response time, basically from a few weeks to a few months at best. Here, each observation has to be related to the environmental (marine) conditions for that precise moment (week, month). Otherwise, the temporal resolution does not match. The best way of coping with this problem is to use primarily ecological and biological reasoning when selecting climate and environmental data for model-fitting purposes. This means that we need to identify the proper temporal resolution (what level of temporal detail) and extent (how far back in time, what time-span) of our predictors when planning a study. In fact, testing several alternative datasets may help to explore the best solution for a given dataset. Nowadays, there are large numbers of different climate datasets available spanning across larger time periods.

The second issue relates to the question of whether species actually track changing climate and environmental conditions successfully and steadily across the whole range. While fast dispersing species may follow this (pseudo-) equilibrium assumption (see Part I), many other species do not. There is ample evidence in the literature that there have been time lags in the re-adjustment to Quaternary climate change, for example. This evidence comes from the paleo-literature (Davis, 1989), as well as from species modeling studies (Svenning and Skov, 2004, 2007) and from recent observations (Gehrig-Fasel et al., 2007; Delzon et al., 2013; Lenoir and Svenning, 2015). Current climate change seems to shift with such fast velocities that many species are struggling to re-adjust (Foden et al., 2007). However, this may not yet have resulted in significant discrepancies between potential and actual ranges.

Another issue is the fact that many species do not fill their ranges everywhere they would find suitable habitats. This is primarily thought to originate from historical effects such as incomplete migration since the last Ice Age (Svenning and Skov, 2007) or regional changes in competitive interactions after environmental changes due to different migration and adaptation patterns in different species (Lenoir et al., 2010). For example, some studies have shown that the same trees that show limited RF actually seem to fill their niche dimension even across large, disjunct ranges (Randin et al., 2013). Although, filling is stronger at the cold (physiologically constraining) than at the warm limit, where the latter is thought to be more driven by biotic interactions than by physiological constraints (Meier et al., 2010, 2011). All this points to a general issue that may cause significant bias to our sampling and data, even if we perfectly design an environmentally sound sampling scheme. It may (and does) turn out that even if we properly sample gradients following a sound design, we might sample areas that have unfilled ranges for a given species. We then tend to infer these areas as a proof of unsuitable habitat for this species, when in fact it is not. Or we simply find significant noise in the calibration of the species–environment relationship, due to the fact that the species was not always there where the habitat was suitable. There is no best practice for coping with this problem. One solution is to test to what degree differently complex models do predict the observed range of a species, and whether over- or under-prediction rather only concern areas in the core or also at the edge of the range (Merow et al., 2014). Over-prediction in areas that represent the core of the ecological niche is likely to represent limited RF issues. Such an over-predicting, usually simple, model might then give relatively poor model accuracies in cross-validation or independent tests, but might actually be better than statistically evaluated, due to bias in the test data.

The third question has only recently been addressed. Many species, irrespective of whether they exhibit small or large ranges, do not find all of the environment they could tolerate under contemporary environmental conditions. This specifically concerns the interaction of the multiple environmental variables that, together, make up the species' niche. Former times may have exposed the species to different combinations of climates that have no modern analogs. This means that the contemporary range of a species may not give the full picture of its realized niche of a species. Maiorano et al. (2013) demonstrated that using climate and distribution data from the past 13 000 years revealed better niche model fits than data from a single 1000-year time slice. This suggests that the niche in this example was better captured with a multi-temporal analysis of the species' distribution. Such a model is likely to be much better suited to projecting the likely future habitat suitability of the same species under climate change than if it was fitted only from one time period. This is because the species will most certainly also face non-analog climates in the future, as was reported for the past (Williams and Jackson, 2007; Williams et al., 2007). As previously discussed, this third issue is a particular problem for projecting species through time, be it into the past or into the future. There is only limited experience to date on how to best cope with this problem. One way is to use either ensemble modeling techniques or to use models that are not overly complex. Ensemble modeling integrates statistical models of different complexities and different statistical properties (e.g. how much interaction among variables is considered or how a species' flexible responses along environmental gradients are fitted) when projecting a species through time (see Section 17.3). This ensures that several possible projections are considered. Conceptually, more complex models ensure more accurate fitting of a set of training data, but usually at the cost of overfitting and constraining the response too tightly to current correlations among predictor variables. Less complex models or calibrations tend to less accurately fit the models to observations, but tend to have improved transferability. This has been shown for space (Randin et al., 2006), but is also likely to hold for transferability through time (but see Araújo et al., 2005a).

8.1.3 Thematic Resolution

Thematic resolution is another scale issue that is rarely considered or discussed in habitat suitability and species distribution modeling studies. Thematic resolution concerns the extent to which a predictor variable resolves the thematic content it intends to represent. We can represent precipitation in units of millimeters, centimeters, decimeters, or meters, for example. While we may not be able to recognize the difference between millimeters and centimeters, this is clear for decimeters or meters when representing maps in these classified units. Representing precipitation as classified integer unit in meters would be considered a very coarse thematic resolution. Of course, we can always represent any climate variable as real rather than as integer numbers. This cannot, however, be done easily with all predictors, notably categorical variables. So we need to distinguish two issues that we will discuss in more detail below: (1) which thematic resolution is meaningful from an ecological viewpoint; and (2) which thematic resolution is meaningful from a statistical viewpoint?

The first question is not always easy to answer. Of course, we want to have high thematic resolution for all the predictors used. Regarding habitat units, geology, or soil information, we are often left with comparably coarse or ecologically uninformative classifications. A time-classified geologic stratification does not translate directly into ecologically meaningful classes. We therefore often have numerous classes that have more disadvantages than advantages for habitat suitability modeling, because we may not have sufficient observations of presence and absence for each of these strata. Here, we suggest reclassifying all strata that exhibit similar ecological properties, irrespective of age of the tectonic stratum. In this way, we generate a coarser temporal, but more a meaningful, thematic stratification. We can for example collapse sedimentary rock based on their clay and calcium content, which have direct impact on soil development, soil pH, and nutrient availability. In the end, we have a much smaller number of classes (thus lower thematic resolution), but these are more useful for modeling habitat suitability, and this also directly translates into easier sampling protocols and statistical modeling procedures (see the second point below). Generally speaking, we should always aim for a thematic resolution that is meaningful for the purposes of habitat suitability modeling. For example, representing precipitation in 0.1 mm may not be a good choice, since plants or animals do not directly respond to differences at this scale. Storing the maps in integer units of millimeters or centimeters will help in two ways: first, they are units that are more directly meaningful to organisms, and, second, it reduces the size of the stored map data. All categorical variables need to be reconsidered in the same way before starting a modeling exercise. If we have a habitat map available with, say, 50 classes then we need to ask ourselves: are all 50 classes important for explaining the spatial distribution of our target species, or can we safely lump some classes together (especially those that have no or a similar meaning for a target species). Simplifying our classification, not only has statistical benefits (see below), but we also generate a more parsimonious model, a model that is simpler, less complex, and easier to explain. Often, however, we face the opposite problem. We have only very coarse classes available, and we know from ecological reasoning that we should use a finer thematic resolution for our modeling exercise. In such cases, we need to decide if we make the effort to post-process the coarse categories and refine them from other sources, or if we keep them as they are, and then discuss the results in the light of this known limitation. The latter is often done where no (or insufficiently resolved) soil or land cover information is available. Refining classes requires considerable work. We can for example, refine an existing habitat classification (say five classes: forest, grassland, wetlands, unvegetated, urban) by means of ecological post-processing using either RS, or field data, or by simply classifying these classes into altitudinal belts or based on other prior knowledge. Reclassifying from RS data can be very tedious, while simple reclassifications from altitude or other forms of prior knowledge come with the risk of generating circular reasoning. Regarding numerical variables, we can consider the transformation of variables as an issue related to thematic resolution. For example, if we log-transform a precipitation gradient, then we add finer resolution to low values and coarser resolution to high values of precipitation. Such a transformation can be ecologically meaningful, since low precipitation values tend to have higher species turnover compared to high precipitation values.

The second question aims to answer how much thematic resolution is actually statistically meaningful. If we have answered the first question

Random stratified design

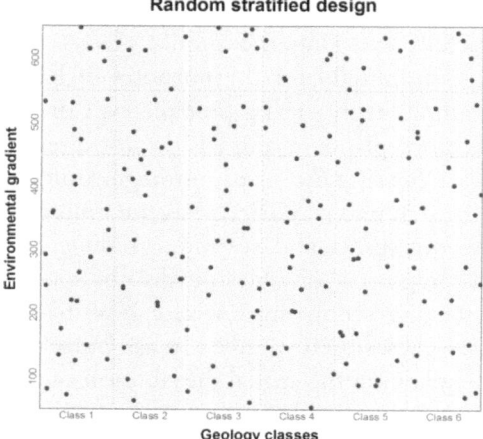

Figure 8.3 Illustration of a random stratified design along two variables, one being a nominal class (here geology) and one being a numerical environmental variable (here classified into classes of 100 units). The nominal variable has six classes, and the numerical gradient has also been classified into six classes. Within each class combination, a randomly allocated set of five observations has been chosen.

(on ecological reasoning), and if our sampling is perfect (sampling all thematic layers perfectly), then there are no real statistical limitations. However, we are usually confronted with thematic resolution problems after the fieldwork has been completed. From a statistical viewpoint, we need to have sufficient observations along each gradient (of numerical predictors) or within each thematic class (of nominal predictors). These are a prerequisite for building models that represent sound probabilistic responses of our target species to these variables. In fact, sufficient observations are needed for each possible combination of variables and categorical classes. Very often, we realize that we have no or only insufficient observations for some rare nominal classes (e.g. geological, soil, habitat units). In such cases, we cannot make meaningful predictions for these rare nominal classes. Similarly, when planning a stratified random sampling design we might ask ourselves if we will really be able to sample each gradient combination for each available nominal class variable (see Figure 8.3)? If our answer is no, then we need to recombine such rare classes with more frequent ones, preferably with classes that have similar ecological meaning for a species' distribution.

If, on the other hand, we have large numbers of classes available for a nominal predictor, then we first need to reconsider, which classes can actually be used in a statistically meaningful way. Similarly, as previously discussed, we may want to collapse several nominal classes that are too rare or that have very similar ecological meaning to other classes. This is not only important to reduce complexity for the sake of parsimony, but also has an impact from a statistical viewpoint. Each nominal class will use one degree of freedom in a regression-based model, so numerous nominal classes result in a large loss of degrees of freedom. We can only reduce this effect by collapsing nominal variables to fewer classes.

8.2 Issues of Extent

Extent is an important issue in habitat suitability and species distribution modeling. It primarily concerns the question: do my observation data represent the full or only a part of the full geographic range of my target species? Depending on the goal of the study, it can be very important to capture the full range of a species in order ascertain (at least under current climate conditions, see Section 8.1.2 for more discussion of this issue) if the full niche of a species is captured. Fitting models only using observation data from a subset of the whole range means it is not possible to fit the full niche, and it is likely that some important components of the niche will be overlooked. If the aim of the study is to locally predict habitat suitability under current environmental conditions (e.g. for conservation planning or species management purposes), then using only part of the range from local sampling at high spatial resolution is acceptable (Lyet et al., 2013). If, however, the goal is to project species habitat suitability to future climates, to test ecological hypotheses, or to project species habitat suitability to other, more distant, regions then, as far as possible, we need to capture the full niche of our target species (see Part I for discussion of this issue).

The effect of limited niche coverage on HSMs has been tested on numerous occasions, mostly in the context of climate change projections (Thuiller et al., 2004b; Barbet-Massin et al., 2010). These studies have shown that data from a subset of the current range do not represent all the conditions the species finds suitable throughout its whole range. Fitting a model to these incomplete datasets might cause high levels of bias when the models are projected through time or space (Thuiller et al., 2004b; Barbet-Massin et al., 2010). This limitation heavily constrains what the models fitted from such data can be used for. In one study that focused on evaluating the predictive performance of HSMs at the range margins of three widespread African vertebrates, the authors demonstrated that models fitted against fine (1 km) or coarse (10 km) resolution regional predictors performed better at the range margins than models fitted across a whole species range from coarse resolution (10 km) predictors (Vale et al., 2014). In another study, however, models fitted against coarse resolution (10 km) predictors capturing the whole Iberian range of amphibians predicted much better at regional scale than regionally fitted models, if the coarse resolution models were applied to high resolution (1 km) regional predictors (Suarez-Seoane et al., 2014). We have already discussed (Section 8.1.1) the fact that scaling from fine to coarse spatial resolution can obscure the real climate–species relationship (and thus the niche quantification). However, the example of Suarez-Seoane et al. (2014) suggests that applying a habitat suitability model fitted from coarse resolution predictors from the whole species range (thus representing a blurred niche shape) results in good performance at regional scales, due to the fact that it captures the whole niche and not only parts of it. This was, however, only true if this simple form of downscaling was applied. We argue that this is a valid way of scaling models fitted from the whole range to smaller regions. It is often applied even across continents, e.g. when fitting models at a coarser scale (e.g. 30' lat/lon), and then predicting at a finer scale (e.g. 10' lat/lon), done e.g. by Thuiller et al. (2005a). Indeed, it has been shown that this type of scaling is appropriate, as long as it does not exceed a ratio of 1:10 in scaling (Bombi and D'Amen, 2012). This approach is therefore a useful way of scaling models from the whole range of a species that can only be fitted from coarse resolution data to the finer resolution useful at regional scales.

If the focus is on regional predictions, then we often observe the phenomenon described by Vale et al. (2014, see above): range margins are partially poorly predicted from models fitted from the whole range, and regionally fitted models often perform better at these marginal locations. This may be for several reasons. On the one hand, the niche shape might be quite complex, and the training data might not always be sufficient to capture regional peculiarities in the niche that do not have sufficient probabilistic support from the whole range. On the other hand, such regional peculiarities can (and probably often do) represent adaptations to special ecological conditions at this range margin location or they might reflect different effects of dispersal and biotic interactions. Boulangeat et al. (2012a) developed a hierarchical habitat suitability model that could be used to test the differential effects of limited dispersal and biotic interactions on species distribution. This model allows for testing, to what degree the factors environment, dispersal or biotic interactions contribute to constraining the range margins of species. Of course, such margins can also represent locations where the range is insufficiently filled, which can also cause deviations from the margin conditions observed elsewhere. However, the local adaptation of populations across the full extent of a species' range is interesting both from a habitat suitability perspective and from range extent perspective, and it is difficult to comprehensively deal with. Several authors have started to study this effect, and they found quite surprising discrepancies between models fitted for infraspecific taxa (e.g. subspecies, genetic clades) that are then aggregated to the "whole species" and models fitted to the species data as a whole (Estrada-Peña and Thuiller, 2008; Pearman et al., 2010; D'Amen et al., 2013; Serra-Varela et al., 2015). From these studies, it becomes evident, that some infraspecific taxa or genetic clades: (i) show quite significant differences in niche structure and habitat preference, (ii) do not necessarily fill all the environmental space of the niche fitted from all intraspecific taxa jointly (species-level model), (iii) have only partially significant overlap in suitable habitats, and (iv) reveal obvious adaptations to more marginal environments than the species-level model. These results indicate that there may also be valid ecological reasons for fitting models from parts of the species' range only. If such a region is selected from genetic clustering, then fitting the models to observations from this region will capture the specific ecological niche of this genetic population, which may differ from that of populations from other parts of the niche. We expect to see more research in this area, bridging the gap between phylogeography and niche modeling in the coming years. Such studies will make an important contribution to better understanding the effects of extent and local adaptation.

Issue of temporal and thematic extents have partly been addressed already in relation to the treatment of temporal and thematic resolutions, and we therefore do not expand on these issues further.

PART III · Modeling Approaches and Model Calibration

This part covers the different statistical modeling approaches that can be used to predict habitat suitability for species or other biological entities. It does not aim to be exhaustive as this would require a book in itself. Rather, it aims to present the modeling techniques that (i) are most commonly used and (ii) are implemented in R or can be easily called from R. Numerous alternative or complementary approaches can be found in Guisan and Zimmermann (2000), Elith et al. (2006), Franklin (2010a), and Maher et al. (2014), for instance. As we have already seen in Part I, selecting the appropriate modeling approaches is ultimately based upon the ecological questions the researcher would like to address, and the availability and accuracy of data to fit the models.

With the development of new powerful statistical techniques, the use of HSMs in ecology has increased rapidly (see Part I, Box 2.1). These models are static and probabilistic in nature, since they statistically relate the distributions of populations, species, communities or biodiversity to their contemporary environment. A wide array of models has been developed to cover research areas as diverse as evolutionary biology, macroecology, biogeography, functional ecology, conservation biology, global change biology, and habitat or species management (see Guisan and Zimmermann, 2000; Guisan and Thuiller, 2005; Thuiller et al., 2008; Elith and Leathwick, 2009; Franklin, 2010b; Peterson et al., 2011; Guisan et al., 2013; see Section 4.4 and Table 4.1).

In practice, ecological models can be separated into three main types: descriptive, explanatory and predictive. This terminology is also sometime used to distinguish between different biogeographic approaches (e.g. Blondel and Aronson, 1995). In the modeling literature, most discussions compare the respective strengths and drawbacks of predictive versus explanatory models (e.g. Mac Nally, 2000; Austin, 2002; Guisan et al., 2002), with little attention paid to descriptive models.

A descriptive approach aims to explore the links between the response variable (e.g. species occurrence) and other potentially explanatory variables, of which there may be many, in order to select, for instance, just one significant subset. Multivariate statistical analyses are commonly used for these types of explorations (e.g. ter Braak, 1986; Dray et al., 2003).

The explanatory (or hypothetico-deductive) approach uses prior knowledge of the system to derive a set of testable hypotheses that can be confirmed or invalidated by estimating model parameters. In other words, in ecology this means analyzing the effects of one or more environmental variables on a response variable and assessing their explanatory power (i.e. in terms of deviance explained). Here, using the appropriate statistical model for a given type of data is of prime importance as inference testing is explicitly used to discuss the significance and importance of environmental descriptors (Leathwick and Austin, 2001).

Finally, when using a predictive approach (see Peters, 1991; Côté and Reynolds, 2002), describing or explaining the data might be of less importance, the aim of the modeling being spatial prediction, not interpretation. Basically, it tries to optimize the quality of estimators and predictors in order to, for instance, minimize the sum of errors or maximize a spatial statistic of accuracy (Johnson and Omland, 2004). This is usually obtained from parsimonious models that include a limited number of explanatory variables that are not strongly correlated (see Chapter 4), but if only the predictive ability matters here, a model might be satisfactory even if it cannot be properly interpreted (e.g. black-box algorithms). This is because a model may be right but the theory might not yet be available to interpret it (e.g. lack of knowledge about the species' ecology). The "best" model – i.e. the one obtaining the best fit – can lead to biased parameters in favor of a smaller variance. In this sense, the right model is not the one that best explains the ecological data from a deviance reduction or R^2 point of view measured on the training data, because this type of model is likely to have numerous explanatory variables and associated collinearity problems. The right model is the one which produces the most accurate predictions on independent or cross-validated data (Merow et al., 2014).

Nowadays, with the increase in computer power, numerical environmental and biological data (see Part II) and tremendous progress in statistics, the difference between descriptive, explanatory and predictive models lies mainly in the level of confidence associated with the tools used, and the way in which model characteristics and specific information about species ecology is retrieved. A useful model (i.e. on which explains a good proportion of variance or deviance) is (i) statistically appropriate for the data being modeled (i.e. no violation of statistical assumptions) and (ii) easy to interpret and tested against independent data so that it can be used to make predictions.

As a result, conflicting issues can arise in the field of biogeography when both explanatory and predictive power of a model are required (Mac Nally, 2000). Although we should ultimately aim for explanatory models that simultaneously offer good predictive performance, in practice, there is a trade-off between explanatory and predictive power (Guisan and Zimmermann, 2000). Indeed, although this trade-off seems implicit in biogeography, not all statistical methods are capable of dealing with it effectively (Merow et al., 2014).

In this part of the book, instead of following a data type structure (e.g. presence, presence–absence, abundance, percentage cover), we decided to follow a model-technique structure. Statistical models can be roughly discriminated between using basic principles (regression vs. tree-based approaches, parametric vs. non-parametric approaches), which then require specific data structure or link functions. The examples presented focus on presence-only or presence–absence data, but many of the models (except those presented in Sections 9.1 and 9.2) can accommodate abundance or proportional data with very little modification to the model options (e.g. family, link).

When modeling species distributions, the user is usually confronted with the difficult choice of which statistical algorithm to select. We will see later on that most of the best-known algorithms have pros and cons and there is no ultimate algorithm to answer every possible question in ecological modeling. Rather, each algorithm has its own strengths and weaknesses. Since the emergence of R (R Core Team 2014), most of the algorithms available for analyzing and predicting species distributions can be run jointly and comparatively with the same data and on the same platform (see Moisen and Frescino, 2002; Thuiller, 2003; Thuiller et al., 2003a; Brotons et al., 2004; Segurado and Araújo, 2004; Elith et al., 2006; Meynard and Quinn, 2007; Tsoar et al., 2007; Maher et al., 2014 for examples of such comparative analyses). In addition, several packages have been developed to make the best of the different algorithms implemented in R (e.g. packages biomod2 (Thuiller et al., 2009) or dismo (Hijmans et al., 2013). The recent literature recommends using combinations of data, algorithms, models and predictions, taking advantage of this possibility to use different algorithms on the same platform (Thuiller, 2004; Araújo and New, 2007; Marmion et al., 2009; Meller et al., 2014). Combining models is by no means a new idea. Model selection and multi-model inference have long been discussed in ecology (Burnham and Anderson, 2002; Johnson and Omland, 2004) and it has been proposed that predictions from the same algorithm (e.g. GLM) should be averaged and run over some competitor models or over all subsets of the available variables. The advantage of multi-model inference is that the inference can be made from more than one single "best" model, by extending the concept of likelihood of the parameters given the model and data, to a concept of the likelihood of the model given the data, the latter being the core of Bayesian statistics. This idea has also percolated into the species distribution modeling world. Modelers have started to average outputs from different algorithms to get the best out of them and analyse the uncertainty around the mean. There are various ways of creating such ensemble models, which are discussed in Chapter 14 (see also Marmion et al., 2009).

Finally, Part III does not cover aspects of spatial autocorrelation in the data and collinearity in the environmental variables, which have already been partly addressed in Part II (see also Legendre, 1993; Dormann, 2007a; Dormann, 2007b; Dormann et al., 2013).

9 · Envelopes and Distance-Based Approaches

9.1 Concepts

Presence-only approaches are the simplest and oldest methods available, usually based on very simple rules and assumptions (see Box 2.1). They are particular in that they deal with presence-only data with no need to create any background or pseudo-absence data. They can roughly be separated into two categories – envelopes (e.g. BIOCLIM, HABITAT) and distance-based approaches (e.g. ENFA, DOMAIN, Mahalanobis distance) – which will be developed in the next two sections. In Chapter 20, we will briefly introduce point-process models that have been recently been introduced into the field of species distribution modeling and address most of the criticisms of traditional presence-only approaches (see below). Since point-process models are still quite new in the field and do not specifically model species presence, but rather species density (i.e. species records per area), we decided to not detail them in this edition. However, Maxent is a specific case of point-process models which is fully developed in Chapter 13.

9.2 Envelope Approaches

There are two types of envelope approaches – geographic and environmental. Geographic envelopes are models that focus on the geographic distribution of a species or population. They usually define the "extent of occurrence" of a species as the area contained within the shortest continuous geographic boundary (e.g. a convex hull) and are typically the approach used by the IUCN for monitoring changes in species ranges and deriving threat status (IUCN, 2001). Different refinements have been made by manipulating the hull to remove potential outlier populations and to provide more conservative estimates of species' ranges (Burgman and Fox, 2003). This approach, however, does not require any independent variables supposed to constrain and explain the range of species since it relies solely on geography. Environmental envelopes, on the contrary, are rather more elaborate as they are based on the potential environmental drivers of species distributions (see Part I). The pioneering approach, BIOCLIM, defines the ecological niche of a species as the n-dimensional bounding box (i.e. minimal rectilinear envelope) that encloses all the records of the species in the environmental space defined by n pre-selected variables (Busby, 1991). In a way, this is similar to Hutchinson's view of the realized niche except that it only considers presence data and does not provide an estimate of habitat suitability. The BIOCLIM-type approach has the advantage of having been the first model to predict the geographic distribution of a species more than 25 years ago, when the use of computer technology in ecology was still in its infancy (revisited in Booth et al., 2014). Busby (1991) developed the BIOCLIM model to map plant species distributions in Australia. Holdridge (1967) had already applied the same approach to ecosystems, and Box (1981) to plant functional types. The rectilinear envelope is defined in the environmental space by means of the most extreme (minimum and maximum) records of the species along each selected environmental variable. In order to reduce the sensitivity of model predictions to outliers, species records can be sorted along each variable and only the records that lie within a certain percentile range of these environmental gradients (e.g. 5–95%) can be used for model construction. In this way, the model is less sensitive to outliers (i.e. sink populations) or to the detrimental effects of thematic uncertainty (see Chapter 6). The `biomod2` package proposes a flexible function – species range envelope (SRE) –, which essentially reproduces the original BIOCLIM with the possibility of applying different percentiles (Figure 9.1). The `dismo` package also provides more refinement to produce continuous probability maps.

```
> library(biomod2)
> ## Load the species and environmental datasets
> mammals_data <- read.csv("tabular/species/mammals_and_bioclim_
table.csv", row.names = 1)
> head(mammals_data)
   X_WGS84  Y_WGS84 ConnochaetesGnou GuloGulo PantheraOnca
1   -86.25 82.75001                0        0            0
2   -84.75 82.75001                0        1            0
3   -83.25 82.75001                0        1            0
4   -81.75 82.75001                0        1            0
5   -80.25 82.75001                0        1            0
6   -78.75 82.75001                0        1            0
```

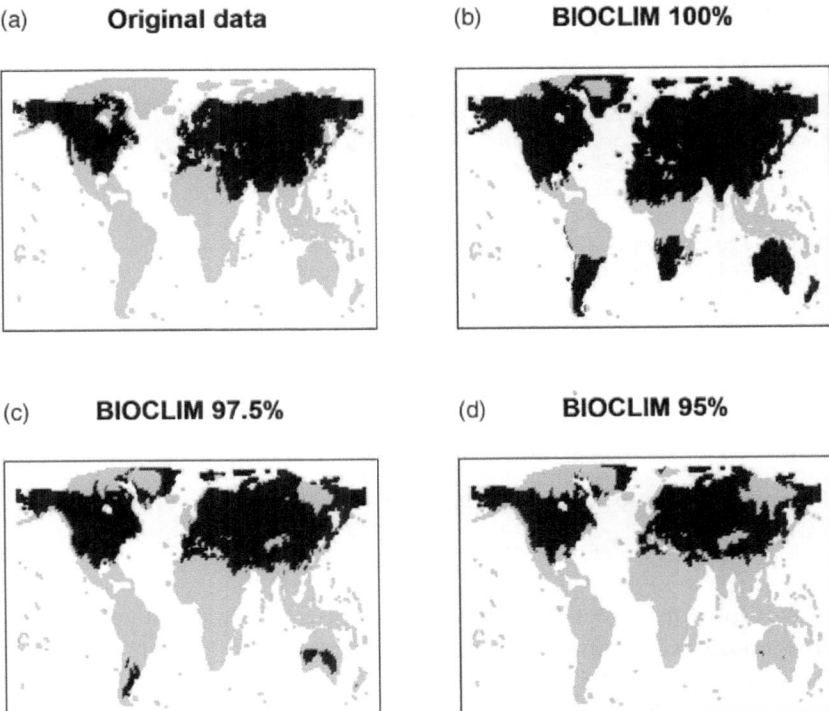

Figure 9.1 Observed and potential distribution of the red fox using a rectilinear envelope model (sre function in the biomod2 package). The potential distributions differ by the use of different percentiles to delineate the envelope. In all maps, black = presence, light gray = absence.

```
  PteropusGiganteus TenrecEcaudatus VulpesVulpes      bio3      bio4
1                 0               0            0  0 0 11.66666 16213.33
2                 0               0            0  0 0 11.33333 16200.00
3                 0               0            0  0 0 11.33332 15981.33
4                 0               0            0  0 0 0 11.50000 15726.50
5                 0               0            0  0 0 0 11.00001 15559.50
6                 0               0            0  0 0 0 11.16667 15305.00
      bio7      bio11      bio12
1 491.6667 -389.6667   75.00004
2 489.3333 -399.9999  101.99969
3 483.3334 -396.0001  108.00011
4 478.6667 -388.6667  103.66661
5 472.0000 -396.8333  138.49980
6 464.5000 -395.1666  151.16643
```

```
> pred_BIOCLIM <- sre(Response = mammals_data$VulpesVulpes,
Explanatory = mammals_data[, c("bio3", "bio7", "bio11", "bio12")],
NewData = mammals_data[, c("bio3", "bio7", "bio11", "bio12")],
Quant = 0)
> pred_BIOCLIM_025 <- sre(Response = mammals_data$VulpesVulpes,
Explanatory = mammals_data[, c("bio3", "bio7", "bio11",
"bio12")], NewData = mammals_data[, c("bio3", "bio7", "bio11",
"bio12")], Quant = 0.025)
> pred_BIOCLIM_05 <- sre(Response = mammals_data$VulpesVulpes,
Explanatory = mammals_data[, c("bio3", "bio7", "bio11",
"bio12")], NewData = mammals_data[, c("bio3", "bio7", "bio11",
"bio12")], Quant = 0.05)
> par(mfrow = c(2, 2))
> level.plot(mammals_data$VulpesVulpes, XY = mammals_data[, c("X_
WGS84", "Y_WGS84")], color.gradient = "grey", cex = 0.3,
show.scale = F, title = "Original data")
> level.plot(pred_BIOCLIM, XY = mammals_data[, c("X_WGS84", "Y_
WGS84")], color.gradient = "grey", cex = 0.3,
show.scale = F, title = "BIOCLIM 100%")
> level.plot(pred_BIOCLIM_025, XY = mammals_data[, c("X_WGS84",
"Y_WGS84")], color.gradient = "grey", cex = 0.3,
show.scale = F, title = "BIOCLIM 97.5%")
```

We note that predictions from SRE using 100 percent of the data erroneously predict the southern hemisphere as being suitable for the red fox. Using the core 95 percent quantile allows for more accurate prediction of the southern hemisphere, but at the cost of underestimating the distribution in Russia. Generally speaking, such over- and under-predictions highlight the relatively low predictive accuracy of SRE (Elith et al., 2006). Indeed, it assumes independent rectilinear bounds and that all variables are known, and it will cause over-prediction when not enough variables are included and under-prediction with too many (or even spurious) variables (Barry and Elith, 2006). This approach, although quite simple, should thus be used with parsimony and care. However, it does give a quick rough estimate of the habitat suitability of a given species without much effort. It does not expect the predictor variables to be uncorrelated, and it can map the distribution using many different variables at the same time.

Several refinements of environmental envelopes were later developed, including DOMAIN (Carpenter et al., 1993), but we will not discuss them here (and note that some of them are partly distance approaches; see Section 9.3). These approaches are no longer routinely applied and the few comparative analyses that have tested their predictive accuracy have revealed only moderate to weak performance (Elith et al., 2006;

Tsoar et al., 2007) compared to more modern techniques (Elith et al., 2006), BIOCLIM has the advantage of being the simplest and can thus be used as a baseline prediction.

The first major shortcoming of rectilinear envelopes is that they assume the relationship between the presence of a given species and any given variable is binary. In other words, a single presence record under an extreme climatic condition at the edge of a species' range, for example, has the same weight as thousands of presences recorded in the core of the range. As previously mentioned, this can be dealt with by adjusting the percentile of the data included. Nonetheless, as we have seen in our example, strongly reducing the percentile can also lead to the exclusion of relevant range information. The second major problem is that every explanatory variable modeled is apportioned the same weight when constructing the complete species model, and that explanatory variables are treated as independent (Barry and Elith, 2006). This highlights the importance of carefully selecting the variables (Austin et al., 1990). In BIOCLIM, even if 100 variables were selected, they would all be used with the same weight and thus all contribute equally to defining the multidimensional envelope for the given species. Such a highly constrained model might prove highly accurate in defining the current extent of a given species, but it would expectedly perform relatively poorly when used to project the distribution of the species in space and time. This can be tested by playing with different sets of variables, running one's own variable selections within a cross-validation scheme (see Part IV) and then evaluating the predictive power of the different (over-fitted or not) models.

9.3 Distance-Based Methods

Distance-based approaches are refined alternatives to simple envelope approaches. Instead of building on rectilinear discrimination, they are usually built on the distance between the environmental centroid of the study area and that defined for the species (Guisan and Zimmermann, 2000; Elith and Leathwick, 2009). This approach is meant to overcome some of the limitations previously discussed such as variable selection and variable importance, which can be used to calculate the axes of the environmental space. Various approaches have been proposed, such as those based on principal component analysis (PCA-sp; Robertson et al., 2001), Mahalanobis distance (MD; Farber and Kadmon, 2003) or the ecological niche factor analysis

(ENFA; Hirzel et al., 2002a) approaches. ENFA is discussed in more detail below.

ENFA, initially proposed to map the distribution of mammals in Switzerland (Hausser, 1995), calculates a measure of habitat suitability based on the analysis of marginality (to what extent a species' mean of the environmental space differs from the global environmental mean across the whole study area, known as background in ENFA) and environmental tolerance (to what extent a species' variance in environmental space differs from the global environmental variance). A threshold of suitability value can then be applied to determine the boundaries of the ecological niche (Hirzel et al., 2002a). In this way, ENFA measures the ecological niche that is actually occupied by a given species by comparing its distribution in the ecological space (i.e. a species' distribution) with the distribution of the environment across the whole study area (i.e. the global distribution) (Hirzel et al., 2002a). As ENFA takes into account background it is not a presence-only method in the strict sense of the word, but rather a presence-background data approach. We prefer to put ENFA in this category for the sake of simplicity and as a natural extension of the other distance-based approaches.

With respect to the definition of Hutchinson's niche, a species' marginality indicates the species' niche position (i.e. niche optimum), while the environmental breadth negatively correlates with a species' specialization. A generalist species, a species that tolerates a large range of environmental conditions, will have a large estimated niche breadth, and vice versa. Environmental niche breadth usually strongly correlates to other ecological niche dimensions such as functional traits (Thuiller et al., 2004c) or sensitivity to environmental changes (Broennimann et al., 2006).

More specifically, ENFA performs a factor analysis with orthogonal rotations to: (i) transform the predictor variables into a set of uncorrelated factors (as in principal component analysis), and (ii) construct the axes in a way that accounts for all the marginality of the species on the first axis, and then minimizes species' ecological tolerances along all following axes (see Hausser, 1995). ENFA has been fully implemented in a standalone package called BIOMAPPER[1] (Hirzel et al., 2002b) but

---

[1] www.unil.ch/biomapper/

can also be found in the adehabitatHS package in R (Calenge and Basille, 2008).

```
> library(adehabitatHS)
> library(pROC)
```

Using the same data as for BIOCLIM.

ENFA starts by performing a PCA on the environmental variables to find the major axis of variation

```
> pc <- dudi.pca(mammals_data[, c("bio3", "bio7", "bio11",
"bio12")], scannf = FALSE, nf = 2)
```

The data are then ready to be used in ENFA to define "ecologically more meaningful" axes:

```
> en <- enfa(pc, mammals_data$VulpesVulpes, scan = FALSE)
```

adehabitatHS has interesting features which can be used to display the major axes of variation and the description of the niche (Figure 9.2).

```
> par(mfrow = c(2, 2))
> barplot(en$s)   # the specialization diagram
> scatterniche(en$li, mammals_data$VulpesVulpes, pts = T)
# plot the niche
> s.arrow(cor(pc$tab, en$li))   # meaning of the axes
```

The scatterniche plot (Figure 9.2b) represents the environment used by the species of interest (in gray) against the global environment, here represented by the environmental conditions for the whole world. The major axis of variation is mainly determined by bio3, bio7, and bio11, whereas the second is mostly influenced by bio12.

ENFA generates the environmental suitability of a species by using the MD between any presence (representing a site) and the centroid of the environmental niche of the species. ENFA thus produces habitat suitability values as distances, and not as values between 0 and 1. In order to compare ENFA results to those from BIOCLIM, for instance, we need to transform the ENFA values into binary presence–absence information (Figure 9.3). Here we used a function from the pRoc package called roc(), which balances the percentage of presence and background data (here assuming they represent non-suitable areas) (see Part IV, Chapter 13 for more details on the different ways of transforming habitat suitability maps into presence–absence data).

(a)

(b)

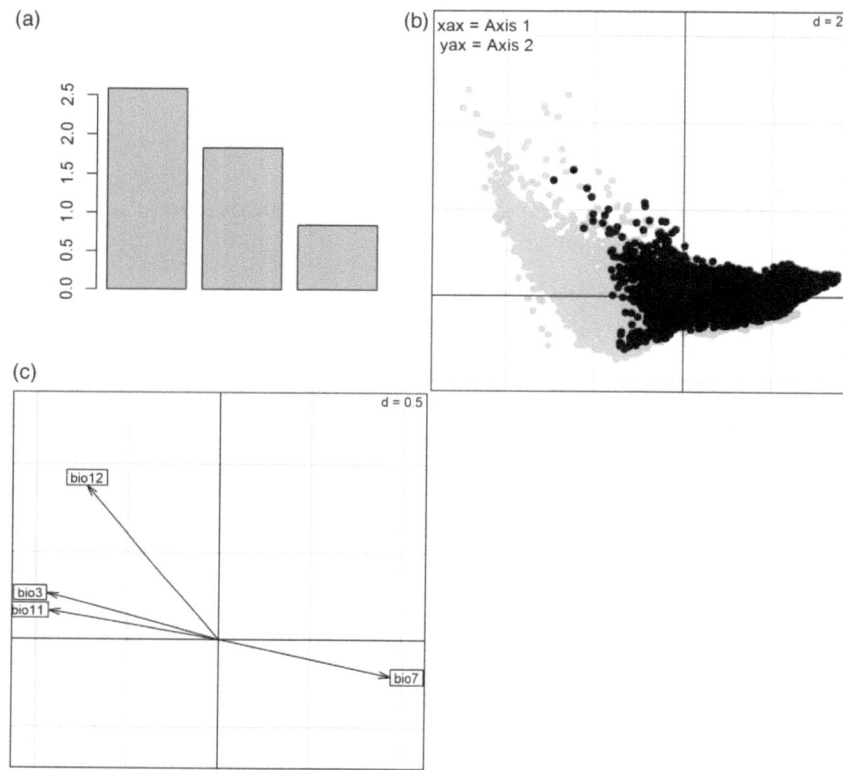

(c)

Figure 9.2 Ecological niche description of the red fox (function enfa() in the package adehabitatHS)

```
> par(mfrow = c(2, 2))
> level.plot(mammals_data$VulpesVulpes, XY = mammals_data[,
c("X_WGS84", "Y_WGS84")], color.gradient = "grey",
cex = 0.3, show.scale = F, title = "Original data")
> level.plot(en$li[, 1], XY = mammals_data[, c("X_WGS84",
"Y_WGS84")], color.gradient = "grey", cex = 0.3,
show.scale = F, title = "ENFA")
> roc_enfa <- roc(mammals_data$VulpesVulpes, en$li[, 1])
> threshold_enfa <- coords(roc_enfa, "best",
ret = c("threshold"))
> Pred01 <- as.numeric(en$li[, 1] > threshold_enfa)
> level.plot(Pred01, XY = mammals_data[, c("X_WGS84",
"Y_WGS84")], color.gradient = "grey", cex = 0.3,
show.scale = F,
title = "ENFA binary")
```

(a) **Original data**

(b) **ENFA**

(c) **ENFA binary**

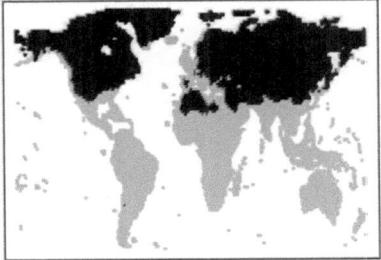

Figure 9.3 Observed and potential distribution of the red fox modeled using ENFA. The potential distribution is either expressed along a scale of habitat suitability values (light= low suitability to dark = high suitability), or in a binary form picturing presence–absence (black = presence, light gray = absence).

As shown in the two previous sections, these two approaches – BIOCLIM and ENFA – enable us to make predictions of potentially suitable habitats based on relatively limited assumptions and using fairly simple algorithms. So far, we have not provided measures of the predictive accuracy of the different models (but see Part IV). However, these methods have been extensively compared in isolation (Hirzel et al., 2006; Pearce and Boyce, 2006; Tsoar et al., 2007; Ward et al., 2009) or against methods using either presence and absence or pseudo–absence data (Brotons et al., 2004; Hirzel et al., 2006). ENFA and BIOCLIM generally have lower predictive accuracy than standard methods using presence and absence data, which is logical given that they use less information about the distribution of the data points. In the examples presented so far, ENFA generally performs best. It is also an interesting approach for representing the estimated niche in the ecological space (assuming the PCA axes are meaningful) and to derive simple niche characteristics such as niche position and niche breadth (see the outlying mean index approach for a natural extension to multi-species niche characterization; Dolédec et al., 2000; Thuiller et al., 2004c).

10 · *Regression-Based Approaches*

10.1 Concepts

Regression-based approaches are by far the most commonly used in ecology and other disciplines, and particularly in habitat suitability modeling (Guisan et al., 2002). They usually rely on robust statistical theories (e.g. sum of squares, maximum likelihood) and are treated in detail in textbooks.

Regression relates a response variable (e.g. presence–absence, abundance, biomass) to a set of pre-selected environmental predictors (e.g. climate, land use, resource). The predictors can be used as untransformed environmental variables or, in order to prevent multicollinearity in the data, as orthogonal components derived from the environmental variables through multivariate analyses. As seen in section 6.4.2, one diagnostic to test for multicollinearity is the VIF (Montgomery and Peck, 1982; see Part II) and its derivation to test for various combinations of variables. The classical ordinary least-square (OLS) linear regression approach (often simply called linear model, LM) is theoretically valid only when the response variable is normally distributed (i.e. Gaussian) and the variance does not change as a function of the mean (homoscedasticity). In other words, homoscedasticity relates to the specific case in which the error term (i.e. the random effect in the relationship between the predictors and the response variable) is constant across all values of the predictor variables. GLMs constitute a more flexible family of regression models, which allow the response variable to follow other distributions and non-constant variance functions to be modeled. In GLMs, the combination of predictors (the linear predictors) is related to the mean of the response variable through a link function. Using such link functions makes it possible to both transform the response to linearity and maintain the predicted values within the original range of values allowed for the response variable. By doing so, the GLMs can handle Gaussian (e.g. biomass), Poisson (species abundance, species richness), binomial (e.g. presence–absence), or gamma distributions – with link functions set to identity, logarithm, logit, and inverse respectively, for example.

If the response shape is not a linear function of predictors, a transformed (higher-order polynomial) term of the latter can be included in the model (Hastie et al., 2009). This type of regression is called a polynomial regression. Second order polynomial regressions simulate unimodal symmetric responses (e.g. a hypothetical bell-shaped relationship between species abundance and a given environmental variable; Austin, 1985), whereas third-order or higher terms make it possible to simulate skewed and bimodal responses, or even a combination of both. Fitting complex curves should, however, be done carefully, since there is then a high risk of obtaining undesired shapes for the resulting response curve outside of the calibration range of the model (e.g. rising again exponentially outside the possible range of the species, which will make problem when projecting the model on these values; Thuiller et al., 2004b). See Merow et al. (2014) for a discussion of fitting simple versus complex response shapes.

Alternative regression techniques for relating the distribution of biological entities to environmental gradients are based on non-parametric smoothing functions of predictors. GAMs are commonly used to implement non-parametric smoothers in regression models (Wood, 2006; Wood et al., 2015), making them semi-parametric approaches. This technique applies smoothers independently to each predictor and additively calculates the component response.

The major difference between GLMs and their extensions (e.g. GAM, MARS, BRUTO) thus lies in the choice of model-driven versus data-driven response shapes. Indeed, to properly use a GLM, one should have some expectation regarding the shape of the response variable along the predictors. When a highly limiting factor is expected, a linear relationship could be sufficient, whereas when a unimodal response along a wide continuous gradient is expected, a bell-shaped (quadratic) curve is required (see Part I). If there is no expectation regarding shape, then various shapes would need to be tested, which could become tedious when several predictors are used together. Data-driven approaches, such as GAM, are slightly more flexible in this regard, but other choices have to be decided up front (e.g. type of smoother, degrees of freedom). We will see later what impact the decision to select one approach over another can have on the predictions.

Table 10.1 *Examples of commonly used distributions, associated families and links for GLM. A classical ecological example is also given.*

Distribution	Family or/and usefulness	Link	Example
Normal	Ordinary linear model	Identity	Biomass (usually log-transformed)
Poisson	Log-linear model	Log or square root	Species richness
Binomial	Logistic regression, probit	Logit or probit	Presence–absence
Gamma	Alternative to lognormal model	Log or inverse link	Species abundance distribution
Negative binomial	Account for overdispersion	log	Frequency count data

10.2 Generalized Linear Models

As touched on earlier, GLMs generalize OLS regression by allowing the linear model to be related to the response variable via a link function, and by allowing the magnitude of the variance for each measurement to act as a function of its predicted value.

There are a number of distributions in addition to the normal distribution that leads to a GLM (Table 10.1).

The linear predictor determines the mean of the response (McCullagh and Nelder, 1989a). It is unbounded, but the mean of some of these distributions (e.g. binomial) is restricted. The mean is supposed to be a (monotone) function of the linear predictor and the inverse of this function is called the inverse-link function. As stated above, this function ensures that the reversely transformed predictions remain within the original scale of the response variable. Users need to define the link before running any models (see Table 10.1).

If we go back to our red fox species modeled in Chapter 9, a simple GLM with a number of predictor variables can be easily implemented:

```
> glm1 <- glm(VulpesVulpes ~ 1 + bio3 + bio7 + bio11 + bio12,
data = mammals_data, family = "binomial")
> glm2 <- glm(VulpesVulpes ~ 1 + poly(bio3, 2) + poly(bio7,
2) + poly(bio11, 2) + poly(bio12, 2), data = mammals_data,
family = "binomial")

> library(biomod2)
> par(mfrow = c(2, 2))
```

```
> level.plot(mammals_data$VulpesVulpes, XY = mammals_data[,
c("X_WGS84", "Y_WGS84")], color.gradient = "grey",
cex = 0.3, show.scale = F, title = "Original data")
> level.plot(fitted(glm1), XY = mammals_data[, c("X_WGS84",
"Y_WGS84")], color.gradient = "grey", cex = 0.3,
show.scale = F, title = "GLM with linear terms")
> level.plot(fitted(glm2), XY = mammals_data[, c("X_WGS84",
"Y_WGS84")], color.gradient = "grey", cex = 0.3,
show.scale = F, title = "GLM with quadratic terms")
```

The two models *glm1* and *glm2* mostly differ in terms of the hypotheses used regarding the shape of the relationship between all variables and the presence of the species. In *glm1*, one assumes that linear predictors are sufficient, in *glm2* one expects quadratic relationships, (i.e. non-symmetric, unimodal or sigmoidal relationships). The poly function in glm2 is an effective way of dealing with correlation between x and x^2 and provide a more flexible response (i.e. non-symmetric unimodal) that simply uses $x + I(X)^2$ in the formula.

In this particular example, the spatial distributions of the probability of occurrence from the two different models appear rather similar at first glance (Figure 10.1). However, let's examine how the modeled responses differ in environmental space by analysing the response curves of the species along the environmental gradients fitted in the models (Figure 10.2).

There are several ways of visualizing the response curves of a species for the different models. One possibility is to use a function in the biomod2 package, which implements the evaluation strip method proposed by Elith et al. (2005). This method has the advantage of being independent of the algorithm used. For building the predicted response curves, n-1 variables are set as constants to a fixed value (mean, median, min or max, i.e. fixed.var.metric argument) and only the remaining one (remaining two for three-dimensional response plots) varies across its whole range (given by Data). The variations observed and the curve thus obtained shows the sensibility of the model to that specific variable (Figure 10.2.).

```
> library(ggplot2)
> ## create the response plot
> rp <- response.plot2(models = c("glm1", "glm2"),
Data = mammals_data[,
c("bio3", "bio7", "bio11", "bio12")],
show.variables = c("bio3", "bio7", "bio11", "bio12"), fixed.var.
metric = "mean",
plot = FALSE, use.formal.names = TRUE)
> ## define a custom ggplot2 theme
```

(a) **Original data**

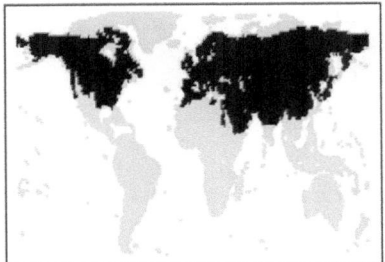

(b) **GLM with linear terms**

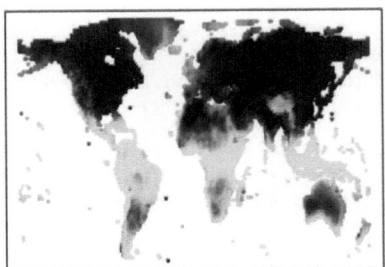

(c) **GLM with quadratic terms**

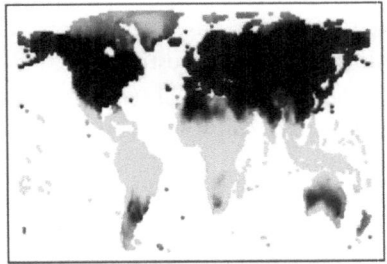

Figure 10.1 Observed (black = presence, light gray = absence) and potential distribution of species Sp290 modeled by different GLM differing by the complexity of the parameters (linear, quadratic, and second-order polynomials). The gray scale of predictions (b, c) shows habitat suitability values between 0 (light, unsuitable) and 1 (dark, highly suitable).

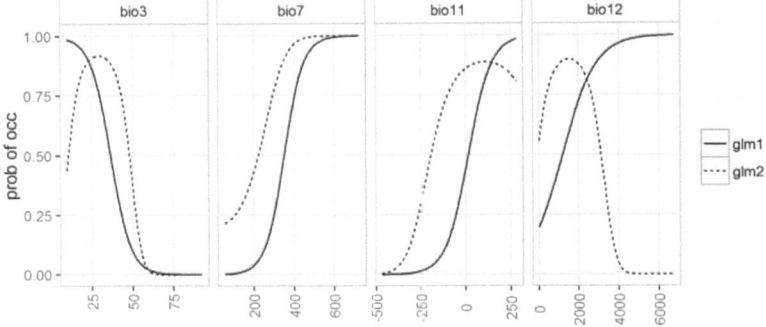

Figure 10.2 Response curves of model glm1 (linear terms) and glm2 (quadratic terms). Plotted are the probabilities of occurrence in function of the bioclimatic variables.

```
> rp.gg.theme <- theme(legend.title = element_blank(),
axis.text.x = element_text(angle = 90, vjust = 0.5),
panel.background = element_rect(fill = NA,
colour = "gray70"), strip.background = element_rect(fill = NA,
colour = "gray70"), panel.grid.major = element_line(colour =,
"grey90"), legend.key = element_rect(fill = NA,
colour = "gray70"))

> ## display the response plot
> gg.rp <- ggplot(rp, aes(x = expl.val, y = pred.val,
lty = pred.name)) + geom_line() + ylab("prob of occ") + xlab("")
+ rp.gg.theme + facet_grid(~expl.name, scales = "free_x")
> print(gg.rp)
```

It is interesting to note that although the response curves differ between the models, the spatial predictions remain relatively similar. This reminds us that slightly different models can yield very similar predictions. This shows the importance of producing these plots in order to analyse the model and decide whether the estimated relationships meet expectations. In this respect, Merow et al. (2014) analysed the pros and cons of simple versus complex response shapes when calibrating SDMs.

Obviously, when one has no idea what the a priori importance of each variable might be and which should be included in the model, there is a need for some sort of variable selection. Stepwise regression – backward, forward, or both – is a traditional method for examining the relative importance of each derived variable to explain presence–absence or abundance of species. Usually, stepwise regressions are based on the Akaike information criteria (AIC; Akaike, 1974) or its Bayesian derivation (BIC, see Chapter 12), but other information criteria also exist (Johnson and Omland, 2004). In both backward and forward stepwise regressions, variables are tested sequentially, and the one producing the lowest AIC or BIC is retained. The method then assesses the contribution of the other variables after accounting for the variable selected. This approach is appealing as it classifies the variables, ranks them based on their contribution to reducing the total AIC, and retains the most parsimonious combination of variables. The backward and forward strategies differ with regards to the starting model. In the latter case, model selection will start from the intercept (null) model including no variable, while, for the former, it will start from the saturated (full) model, including all the initial variables.

Although stepwise regression is certainly appealing and used to be one of the most commonly used means of reducing complexity in regression-like methods, it is often deemed to be a high-variance exercise since the slightest disturbance in the response data can sometimes lead to vastly different subsets of the variables (Johnson and Omland, 2004; Whittingham et al., 2006). This is especially the case when the number of predictor variables is large (over 10) and the variables correlated with each other. We highly recommend, at least, reducing the number of variables first with PCA, VIF analyses, or simple pairwise correlation tests, to ultimately select a series of non-correlated, ecologically relevant variables (see Part II and Dormann et al., 2013).

The last few years have also seen the development of penalized regression and shrinkage rules as alternatives to stepwise regression. Penalizing algorithms such as "lasso" or "ridge" have gained momentum in the statistical literature, but also in the habitat suitability modeling literature (Hastie et al., 2009; Renner and Warton, 2013; and see Chapter 11). Lasso (Tibshirani, 1996, 1997) and ridge (Hoerl and Kennard, 1970; Le Cessie and van Houwelingen, 1992) provide alternative algorithms that shrink the estimates of the regression coefficients toward zero relative to the maximum likelihood estimates. The overarching goal of the penalty (or shrinkage) is to accurately estimate the parameters while avoiding overfitting either due to multicollinearity of the predictors or overly high dimensionality (i.e. too many predictors). The ridge penalty generally leads to many small but non-zero regression coefficients, while the lasso penalty results in few regression coefficients with little shrinkage and the remaining ones shrunk to zero. However, as in any optimization process, one has to decide a priori what criteria should be used to optimally shrink the parameters. This is determined by tuning a shrinkage parameter (usually called λ) that takes values between zero (i.e. no shrinkage, maximum likelihood estimation) and infinity (i.e. infinite shrinkage, all regression coefficients set to zero). The penalized package offers interesting tools to perform lasso and ridge regressions and select the optimal λ by means of cross-validation.

Here, we provide an example of stepwise selection using the `ste-pAIC()` function (in the `MASS` package). Let's start by running an intercept model that will serve as the starting model. Then, the `stepAIC()` function will sequentially add and remove the different variables. There are three important parameters in that function: `scope`, `direction` and `k`. Scope can be used to specify the form of the different variables to be tested (e.g. linear, quadratic, interactions). Direction can be used to specify the direction of the variable selection. Starting with the intercept model means only the forward direction can be used, but one can also use the more advanced both option, combining forward and backward. If glmStart is the saturated model, then backward or both would be the two choices proposed. k is the multiple of the number of degrees of freedom used for the penalty. If the k parameter is set to 2, then AIC is used for variable selection. If k = log(n), then variable selection is based on BIC.

The scope argument is rather tedious to write, but it can be done using the formula function.

```
> library(MASS)
> glmStart <- glm(VulpesVulpes ~ 1, data = mammals_data,
family = binomial)

> glm.formula <- formula("VulpesVulpes ~ 1 + poly(bio3,2) +
poly(bio7,2) + poly(bio11,2) + poly(bio12,2) + bio3:bio7 +
bio3:bio11 + bio3:bio12 + bio7:bio11 + bio7:bio12 + bio11:bio12")

> glm.formula <- formula("VulpesVulpes ~ 1 + poly(bio3, 2) +
poly(bio7, 2) + poly(bio11, 2) + poly(bio12, 2) + bio3:bio7
+ bio3:bio11 + bio3:bio12 + bio7:bio11 + bio7:bio12 +
bio11:bio12")

> glmModAIC <- stepAIC(glmStart, glm.formula, data = mammals_
data, direction = "both", trace = FALSE, k = 2, control = glm.
control(maxit = 100))

> glmModBIC <- stepAIC(glmStart, glm.formula, direction = "both",
trace = FALSE, k = log(nrow(mammals_data)),
control = glm.control(maxit = 100))

> rp <- response.plot2(models = c("glm1", "glm2", "glmModAIC",
"glmModBIC"), Data = mammals_data[, c("bio3", "bio7", "bio11",
"bio12")], show.variables = c("bio3", "bio7", "bio11", "bio12"),
fixed.var.metric = "mean", plot = FALSE, use.formal.
names = TRUE)
> gg.rp <- ggplot(rp, aes(x = expl.val, y = pred.val, lty = pred.
name)) +
geom_line() + ylab("prob of occ") + xlab("") + rp.gg.theme +
facet_grid(~expl.name, scales = "free_x")
> print(gg.rp)
```

We can see now the effects of the variable selection on the retained best model (Figure 10.3).

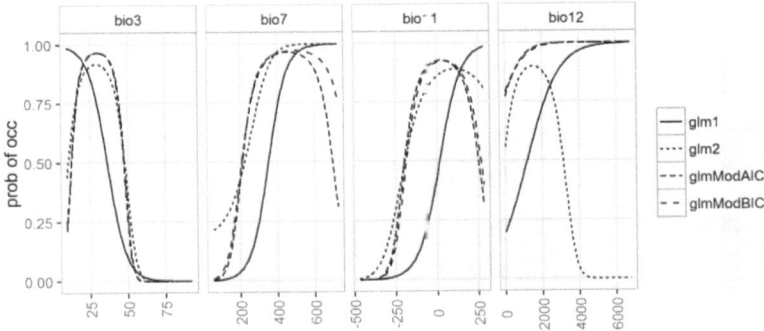

Figure 10.3 Two-dimensional response curves for the different fitted models.

Another way to look at a species' response is to generate and visualize bivariate response curves. Here, we illustrate the example of the best glm selected according to the AIC scores.

```
> rp.2D <- response.plot2(models = c("glmModAIC"),
Data = mammals_data[, c("bio3", "bio7", "bio11", "bio12")],
show.variables = c("bio3", "bio7", "bio11", "bio12"),
fixed.var.metric = "median",
do.bivariate = T, plot = FALSE, use.formal.names = TRUE)
> gg.rp.2D <- ggplot(rp.2D, aes(x = expl1.val, y = expl2.val,
fill = pred.val)) + geom_raster() + rp.gg.theme + ylab("") +
xlab("") + theme(legend.title = element_text()) + scale_fill_
gradient(name = "prob of occ.",
low = "#f0f0f0", high = "#000000") + facet_grid(expl2.name ~
expl1.name, scales = "free")
> print(gg.rp.2D)
```

These bivariate plots allow analysing the joint effects of two variables on the modeled probability of presence (Figure 10.4). For instance, the probability of occurrence is high for high values of bio3 and low values of bio7. When both bio3 and bio7 are both low, the probability of occurrence of the red fox is also low.

The variable rankings can be easily extracted using the anova() function.

```
> anova(glmModAIC)
        Analysis of Deviance Table
        Model: binomial, link: logit
        Response: VulpesVulpes
        Terms added sequentially (first to last)
```

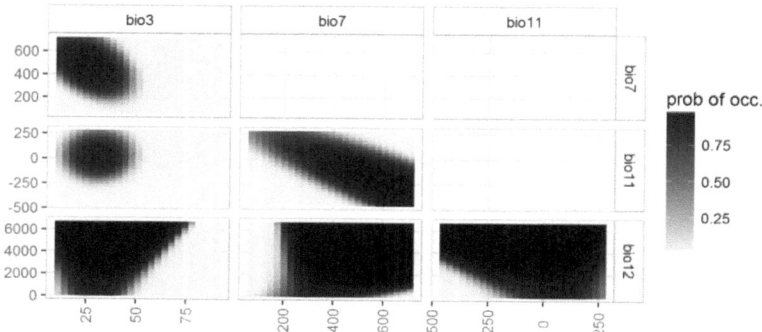

Figure 10.4 Bivariate response curves from the model `glmModAIC` for four predictor variables.

```
Df Deviance Resid. Df Resid. Dev
   NULL                             8541    11839.6
   poly(bio3, 2)    2   5581.8      8539     6257.8
   poly(bio7, 2)    2   1247.0      8537     5010.8
   poly(bio11, 2)   2    299.1      8535     4711.7
   poly(bio12, 2)   2    136.1      8533     4575.7
   bio3:bio7        1    338.3      8532     4237.4
   bio7:bio11       1    129.4      8531     4107.9
   bio11:bio12      1     53.9      8530     4054.0
   bio7:bio12       1     53.0      8529     4001.1
   bio3:bio12       1      3.6      8528     3997.5
```

Variable `bio3` is by far the best explaining variable, followed by `bio7` and `bio11` which strongly influences the distribution of our model species. Although they are less influential than formal variables, adding interaction terms can slightly improve the model's performance.

```
> par(mfrow = c(2, 2))

> level.plot(mammals_data$VulpesVulpes, XY = mammals_data[,
c("X_WGS84", "Y_WGS84")], color.gradient = "grey", cex = 0.3,
level.range = c(0, 1), show.scale = F, title = "Original data")
> level.plot(fitted(glmModAIC), XY = mammals_data[, c("X_WGS84",
"Y_WGS84")], color.gradient = "grey", cex = 0.3, level.
range = c(0, 1), show.scale = F, title = "Stepwise GLM
with AIC")
> level.plot(fitted(glmModBIC), XY = mammals_data[, c("X_WGS84",
"Y_WGS84")], color.gradient = "grey", cex = 0.3, level.
range = c(0, 1), show.scale = F,
title = "Stepwise GLM with BIC")
```

(a) **Original data**   (b) **Stepwise GLM with AIC**

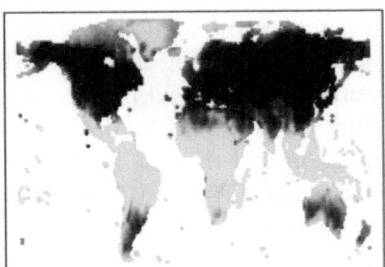

(c) **Stepwise GLM with BIC**

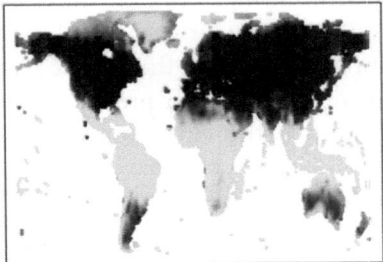

Figure 10.5 (a) Observed (black = presence, light gray = absence) and potential distribution of red fox extracted from (b) glmModAIC and (c) glmModBIC models. The gray scale of predictions shows habitat suitability values between 0 (light, unsuitable) and 1 (dark, highly suitable).

The potential distribution of the red fox does not differ significantly between the two stepwise procedures (Figure 10.5). We would have expected larger differences primarily when using small sample sizes, but not when using big datasets as is the case here.

10.3 Generalized Additive Models

GAMs are techniques designed to capitalize on the strengths of GLMs but which do not require postulating a shape for the response curve from a specific parametric function. GAMs use algorithms called "smoothers" that automatically fit response curves "as closely as possible" to the data given the permitted level of smoothing. GAMs are therefore useful when the relationship between the variables is expected to be of a more complex form, not easily fitted with standard parametric functions of the predictors (e.g. GLM with a linear or quadratic response), or where there is no a priori reason for using a particular shape (Hastie and Tibshirani, 1990). If one wants to remain within a parametric scheme, GAM can also be used in complement to GLM, firstly to explore the general shape of the response function and then to implement it in the best possible way in a GLM (Guisan et al., 2006b). Link and family in GAM are the same as in GLM.

There are now several packages, which can be used to fit GAMs in R (e.g. `gam`, `mgcv`, `gamair`, `GAMBoost`). The `gam` package iteratively fits weighted additive models using backfitting (i.e. iteratively smoothing partial residuals (Hastie et al., 2009). There are different smoothers available, but the most commonly used is the cubic-spline smoother, a collection of polynomials of degree less than or equal to 3, defined on subintervals. A separate polynomial model is fitted in each neighborhood (using a moving window algorithm), thus enabling the fitted curve to connect all the points. Nevertheless, the user has to predetermine the degree of smoothing applied when fitting the curve (or select it through cross-validation). In the SDMs field, researchers have generally used degrees lower than 4, which corresponds roughly to a polynomial of degree 3 (Hastie et al., 2009). Higher degrees will generate more locally complex curves.

The syntax is exactly the same as for a GLM, except that the user needs to specify the smoother (below, a cubic-spline called *s*) and the degree of smoothing (below 2 and 4). Note that the degree of smoothing can change across the variables in a model (i.e. a different smoothing level can be specified for each variable).

```
> if (is.element("package:mgcv", search()))
detach("package:mgcv")  ## make sure the mgcv package is not
loaded to avoid conflicts between packages

> library(gam)
> gam1 <- gam(VulpesVulpes ~ s(bio3, 2) + s(bio7, 2) + s(bio11,
2) + s(bio12, 2), data = mammals_data, family = "binomial")
> gam2 <- gam(VulpesVulpes ~ s(bio3, 4) + s(bio7, 4) + s(bio11,
4) + s(bio12, 4), data = mammals_data, family = "binomial")
```

The `gam` package provides its own function (`plot.gam()`) to extract the response curves, which works in exactly in the same way as the `response.plot2()` function in the `biomod2` package. However, it is important to note that these responses are expressed in the transformed unit (here, the logit scale; Figure 10.6). A very nice feature of `plot.`

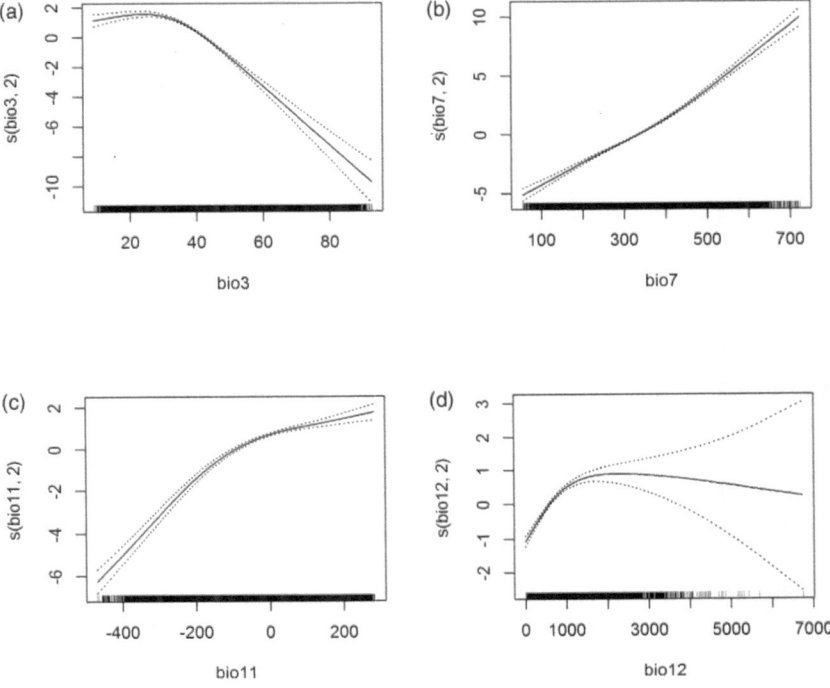

Figure 10.6 Response curves of model gam1 expressed in logit scale (function plot.gam() from the gam package).

gam() is the possibility to include upper and lower point-wise ±2 standard error curves.

```
> par(mfrow = c(2, 2))
> plot(gam1, se = T)
```

We can compare the influence of the degree of smoothing on the response curves expressed in the original unit (between 0 and 1) using the response.plot2() function in the biomod2 package.

```
> rp <- response.plot2(models = c("gam1", "gam2"),
Data = mammals_data[,
c("bio3", "bio7", "bio11", "bio12")],
show.variables = c("bio3", "bio7", "bio11", "bio12"),
fixed.var.metric = "mean", plot = FALSE, use.formal.names = TRUE)
> gg.rp <- ggplot(rp, aes(x = expl.val, y = pred.val, lty = pred.
name)) + geom_line() + ylab("prob of occ") + xlab("") + rp.gg.
theme + facet_grid(~expl.name, scales = "free_x")
> print(gg.rp)
```

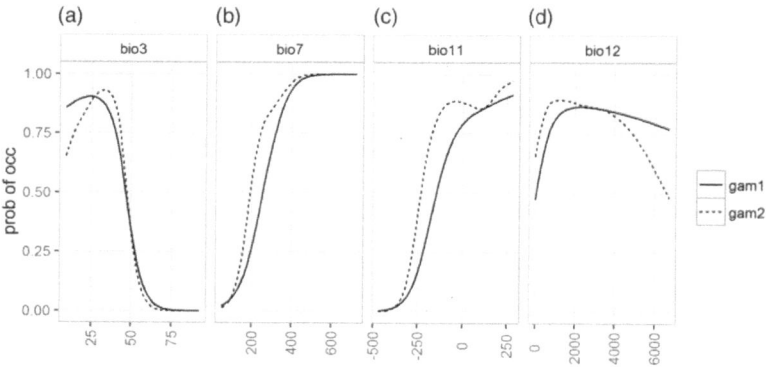

Figure 10.7 Response curves of the gam1 (degree of smoothing = 2) and gam2 (degree of smoothing = 4) models.

Note that the response curves are quite similar to those obtained from the GLMs (Figure 10.7). Therefore, it is clear that the degree of smoothing has a relatively small effect in this example. However, it is important to carefully check the complexity of models. GAMs are data-driven and thus prone to overfitting the data when highly complex smoothers are used. When modeling species distributions for predictive purposes, we do not recommend using degree of smoothing higher than 4 or 5. Users who want to model more complex relationships, e.g. in order to very closely fit and predict the calibration data, may use a higher degree of smoothing, but at the cost of reduced generalization (Merow et al., 2014).

Similarly to a GLM, the gam() function supports various options for variable selection using stepwise procedures or shrinkage rules. These are implemented in the same way as in a GLM. It is also possible to use a custom function for the scope argument from the biomod2 package (function.scope()). Here we will illustrate the use of the stepwise procedure with another function called step.gam() (note however that the stepAIC() function also works for gam() and can be implemented in the same way as previously shown for GLM).

```
> gamStart <- gam(VulpesVulpes ~ 1, data = mammals_data,
family = binomial)
> gamModAIC <- step.gam(gamStart, biomod2::::.scope(mammals_
data[1:3,
c("bio3", "bio7", "bio11", "bio12")], "s", 4), trace = F,
direction = "both")
```

In the step.gam procedure, we will test a single degree of smoothing (here 4), which represents the maximum degree achieved by the model.

(a) **Original data**

(b) **Stepwise GAM with AIC**

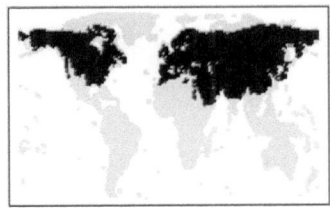

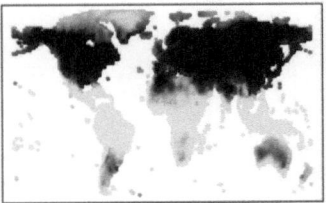

Figure 10.8 (a) Observed (black = presence, light gray = absence) and (b) potential distribution of *Vulpes vulpes* extracted from gamModAIC. The gray scale of prediction shows habitat suitability values between 0 (light, unsuitable) and 1 (dark, highly suitable).

In practice, when the observed relationship is linear, the GAM will also fit a linear relationship even if the degree has been pre-set to 4. An alternative would be to test for different degree of smoothing using c(2,3,4) instead of 4.

The spatial prediction can easily be displayed and compared with the observed distribution (Figure 10.8).

```
> par(mfrow = c(1, 2))

> level.plot(mammals_data$VulpesVulpes, XY = mammals_data[,
c("X_WGS84", "Y_WGS84")], color.gradient = "grey",
cex = 0.3, level.range = c(0, 1), show.scale = F,
title = "Original data")

> level.plot(fitted(gamModAIC), XY = mammals_data[, c("X_WGS84",
"Y_WGS84")], color.gradient = "grey", cex = 0.3, level.
range = c(0, 1), show.scale = F,
title = "Stepwise GAM with AIC")
```

Alternatively, the mgcv package provides a slightly different version of GAM. Smooth terms are implemented through penalized regression splines with smoothing parameters selected through generalized cross-validation or AIC in the mgcv package, or regression splines with fixed degrees of freedom, as in the gam package. The most interesting feature is the possibility to explore interactions between variables through multidimensional smoothers using penalized thin-plate regression splines (isotropic) or tensor product splines (when an isotropic smooth is inappropriate) (Wood, 2006).

The default syntax in mgcv is very similar to the gam package except that the user does not have to specify the degree of smoothing, which is automatically defined by means of internal cross-validation.

```
> if (is.element("package:gam", search())) detach("package:gam")
## make sure the gam package is not loaded to avoid conflicts
> library(mgcv)
> gam_mgcv <- gam(VulpesVulpes ~ s(bio3) + s(bio7) + s(bio11) +
s(bio12), data = mammals_data, family = "binomial")
> ## see a range of summary statistics
> summary(gam_mgcv)

Family: binomial
    Link function: logit

Formula:
    VulpesVulpes ~ s(bio3) + s(bio7) + s(bio11) + s(bio12)

Parametric coefficients:
                Estimate Std. Error z value Pr(>|z|)
    (Intercept)  -1.3421     0.5091  -2.636  0.00839 **
    ---
    Signif. codes:  0 '***' 0.001 '**' 0.01 '*' 0.05 '.'
0.1 ' ' 1

Approximate significance of smooth terms:
              edf Ref.df Chi.sq  p-value
    s(bio3)  5.888  6.589 491.77  < 2e-16 ***
    s(bio7)  6.601  7.358 732.09  < 2e-16 ***
    s(bio11) 8.740  8.968 347.37  < 2e-16 ***
    s(bio12) 7.013  8.022  56.49 2.34e-09 ***
    ---
    Signif. codes:  0 '***' 0.001 '**' 0.01 '*' 0.05 '.' 0.1 ' ' 1

R-sq.(adj) =  0.712   Deviance explained = 65.8%
    UBRE = -0.51977  Scale est. = 1          n = 8542
> gam.check(gam_mgcv)
```

The mgcv package provides a lot of summary statistics that can be very useful when carefully examined (see gam.check()). Additionally, response curves can also be plotted using the internal functions of mgcv (Figure 10.9).

```
> plot(gam_mgcv, pages = 1, seWithMean = TRUE)
```

This makes it possible to compare the response curves from the mgcv implementation of GAM to those from the gam package (Figure 10.10).

```
> rp <- response.plot2(models = c("gam1", "gam2"),
Data = mammals_data[,
c("bio3", "bio7", "bio11", "bio12")],
show.variables = c("bio3", "bio7", "bio11", "bio12"),
fixed.var.metric = "mean", plot = FALSE, use.formal.names = TRUE)
> gg.rp <- ggplot(rp, aes(x = expl.val, y = pred.val,
lty = pred.name)) + geom_line() + ylab("prob of occ") + xlab("")
+ rp.gg.theme + facet_grid(~expl.name, scales = "free_x")
> print(gg.rp)
```

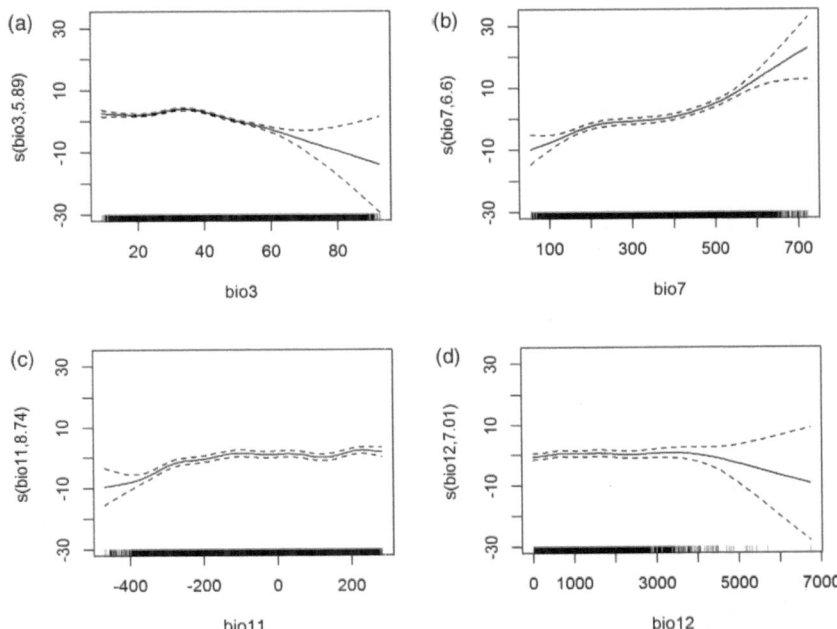

Figure 10.9 Response curves of model gam_mgcv plotted using the internal function of mgcv().

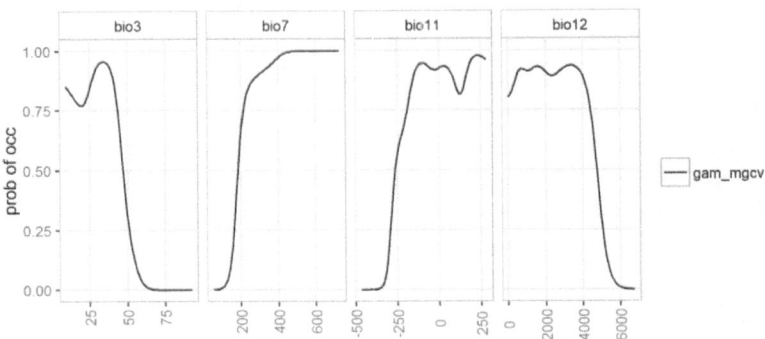

Figure 10.10 The response curves from the model calibrated with the mgcv package (gam_mgcv).

(a) **Original data**   (b) **GAM with mgcv**

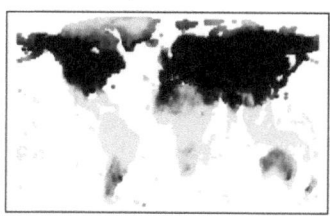

Figure 10.11 (a) Observed (black = presence, light gray = absence) and (b) potential distribution of *Vulpes vulpes* extracted from the gam_mgcv object. The gray scale of predictions illustrates habitat suitability values between 0 (light, unsuitable) and 1 (dark, highly suitable).

Despite these slight differences between models calibrated with the gam algorithm from gam package (gam1, gam2 and gamModAIC) and the model calibrated with the mgcv package (gam_mgcv), the resulting spatial predictions (see Part V) are similar to those obtained from the gam package (Figure 10.11).

We can see that for this particular species (*V. vulpes*), all the models we have seen so far (except SRE and ENFA) have yielded quite similar potential distributions. We will later learn about (Part IV) different ways of testing the predictive accuracy of different models in order to obtain quantitative metrics and compare their predictive power.

10.4 Multivariate Adaptive Regression Splines

Like GAM, multivariate adaptive regression splines (MARS) constitute a more flexible regression technique than GLM, as they also do not require any assumptions to be made about the underlying functional relationship between the species and the environmental variables. Instead of using a predefined shape, such as polynomial functions in GLMs, MARS fits piecewise functions that together can accommodate nonlinear responses. In this sense, it is quite similar to GAM and the smoothed functions. Knots define the breaks between segments and different regression lines with different slopes are thus fitted between each pair of knots, while the full fitted function is constrained to have no breaks or abrupt steps. Generalized cross-validation is used to assess the effect of adding or removing knots. Backward and forward variable selection is also possible, as in GAM and GLM.

MARS is implemented in R in both the mda and earth package. Here, we use the earth package, which provides additional functions that are not available in mda.

Very few parameters are required to fit a MARS model. One important parameter concerns the maximum interaction degree, which determines whether interactions between variables are fitted or not. This is set to one by default, but more complicated response curves are likely to be required in certain instances. In the following examples, we thus use both a degree of 1 (no interactions) and 2 (pairwise interactions).

```
> library(earth)
> Mars_int1 <- earth(VulpesVulpes ~ 1 + bio3 + bio7
+ bio11 + bio12, data = mammals_data, degree = 1,
glm = list(family = binomial))
> Mars_int2 <- earth(VulpesVulpes ~ 1 + bio3 + bio7
+ bio11 + bio12, data = mammals_data, degree = 2,
glm = list(family = binomial))
> ## print the summary of objects
> Mars_int1
      Earth selected 14 of 15 terms, and 4 of 4 predictors
      Termination condition: Reached nk 21
      Importance: bio7, bio11, bio3, bio12
      Number of terms at each degree of interaction: 1 13
(additive model)
      Earth GCV 0.08460021    RSS 718.0938    GRSq 0.6615926
RSq 0.6636498

 GLM null.deviance 11839.56 (8541 dof)    deviance 4267.856 (8528
dof)    iters 11
> Mars_int2
      Earth selected 18 of 21 terms, and 4 of 4 predictors
      Termination condition: Reached nk 21
      Importance: bio7, bio3, bio11, bio12
      Number of terms at each degree of interaction: 1 4 13
      Earth GCV 0.07349056    RSS 621.379    GRSq 0.7060321    RSq
0.7089503

 GLM null.deviance 11839.56 (8541 dof)    deviance 3625.926 (8524
dof)    iters 25 did not converge
```

From the summary statistics in earth(), we can visualize that the r-square of Mars_int2 ($R^2 = 0.71$) is slightly better than the simpler version (Mars_int1, $R^2 = 0.66$).

```
> summary(fitted.values(Mars_int1))
      VulpesVulpes
    Min.   :-0.5111
```

```
1st Qu.: 0.0582
Median : 0.5105
Mean   : 0.4920
3rd Qu.: 0.8625
Max.   : 1.2911
```

There is no in-built `fitted.value` function in MARS. If a user wants to extract the predictions the `predict` function needs to be used and the `type` argument "response" employed to make sure the predictions are converted to the appropriate scale.

```
> pred_Mars_int1 <- predict(Mars_int1, type = "response")
> summary(pred_Mars_int1)
      VulpesVulpes
   Min.   :0.0000001
   1st Qu.:0.0240546
   Median :0.5087504
   Mean   :0.4920393
   3rd Qu.:0.9383953
   Max.   :0.9996739
> pred_Mars_int2 <- predict(Mars_int2, type = "response")
> summary(pred_Mars_int2)
      VulpesVulpes
   Min.   :0.0000
   1st Qu.:0.0227
   Median :0.4513
   Mean   :0.4920
   3rd Qu.:0.9731
   Max.   :1.0000
```

One interesting feature of the `earth` package is that it can be used to plot the distribution of observed presences and absences across classes of predicted values. This allows us to clearly visualize whether species presences are located within high values of predictions, and conversely whether absences are distributed in areas with low prediction values (Figure 10.12).

```
> plotd(Mars_int1, hist = T)
```

Figure 10.12 clearly shows the discriminatory power of MARS for this species, as the probability values are always extremely low when the species is absent and reciprocally high when the species is present. The optimal threshold for transforming the probability values into presence–absence is also easily identifiable in this case (around 0.5–0.6).

We can also see the difference between the two MARS models, which differ in terms of the amount of interaction between variables (one

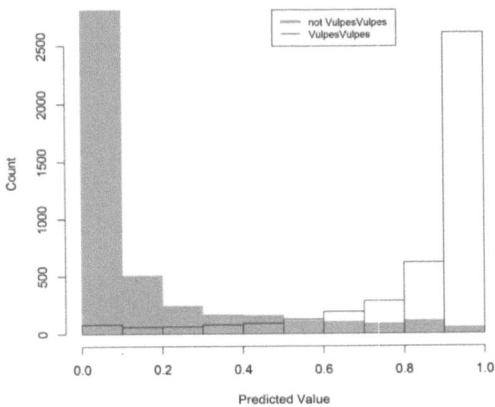

Figure 10.12 The distribution of the predicted values from MARS for both the presence and absence of *Vulpes vulpes*.

model with no interaction, one with pairwise interactions). This can be represented visually by plotting the fitted probabilities against each other (Figure 10.13).

```
> plot (pred_Mars_int1, pred_Mars_int2,
xlab = "MARS with max inter degree 1",
ylab = "MARS with max inter degree 2")
```

Although the two MARS models have similar predictions at and close to 0 and 1, we can see fairly high variability in the predicted probabilities at intermediate values (Figure 10.13). For instance, there are points where the probability of occurrence is close to 1 in MARS with no interactions, whereas it is close to 0 when predicted by MARS with 2 degrees of interactions. This highlights that those kinds of choices are not without consequences, so need to be very carefully evaluated.

The spatial predictions of the two versions of the model look relatively similar overall, except for few regions such like Greenland and some parts of the southern hemisphere (Figure 10.14)

```
> par (mfrow = c(2, 2))
> level.plot (mammals_data$Vulpes_vulpes, XY = mammals_data[,
c("X_WGS84", "Y_WGS84")], color.gradient = "grey", cex = 0.3,
level.range = c(0, 1), show.scale = F,
title = "Original data")
```

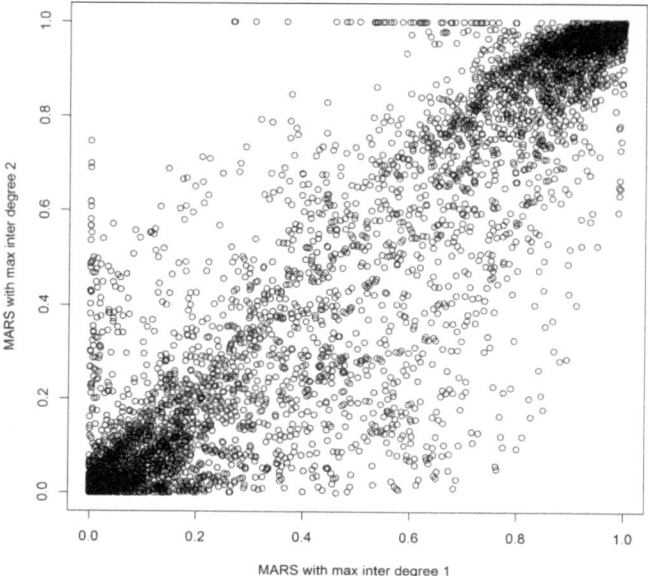

Figure 10.13 Differences between probability of occurrence between a MARS model with a maximum of 1 degree of interaction and a MARS model with a maximum of 2 degrees of interaction.

```
> level.plot(pred_Mars_int1, XY = mammals_data[, c("X_WGS84",
"Y_WGS84")], color.gradient = "grey", cex = 0.3, level.
range = c(0, 1),
show.scale = F, title = "MARS with interaction degree 1")
> level.plot(pred_Mars_int2, XY = mammals_data[, c("X_WGS84",
"Y_WGS84")], color.gradient = "grey", cex = 0.3, level.
range = c(0, 1),
show.scale = F, title = "MARS with interaction degree 2")
```

The response curves for MARS are not shown here, as they can be extracted using the same function as in GLM or GAM, as shown above.

(a)   **Original data**   (b) **MARS with interaction degree 1**

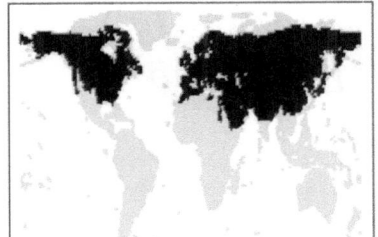   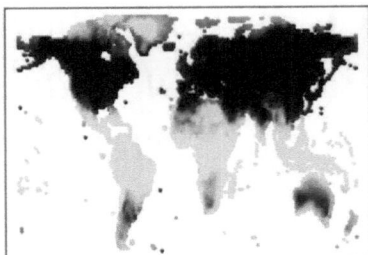

(c) **MARS with interaction degree 2**

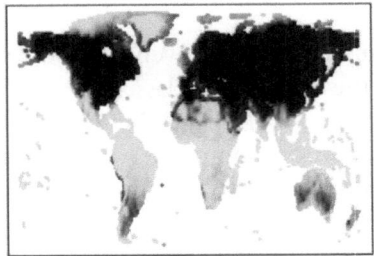

Figure 10.14 (a) Observed (black = presence, light gray = absence) and potential distribution of *Vulpes vulpes* extracted from the (b) MARS 1 and (c) MARS 2 objects. The gray scale of predictions (upper-right and lower-left panels) illustrates habitat suitability values between 0 (light, unsuitable) and 1 (dark, highly suitable).

11 · *Classification Approaches and Machine-Learning Systems*

11.1 Concepts

Classification approaches, recursive partitioning, and even some of the machine-learning approaches rely on the concept of classifying observations into homogenous groups (two or more). It is difficult to trace back to the first application of classification approaches in ecology, as many different implementations were developed to answer different scientific questions. Cluster analysis is the approach most widely used to group observations, based on one or several predictor variables. Clustering is a method of unsupervised learning, and a common technique for statistical data analysis used in many fields, including machine learning, data mining, pattern recognition, image analysis, and bioinformatics. Other examples of methods include supervised approaches, such as discriminant analyses (Hastie et al., 1994), recursive partitioning (Breiman et al., 1984; Quinlan, 1986) neural networks (Ripley, 1996; Franklin, 2010a) or support vector machine (Drake et al., 2006).

These methods have been compared or tested in a number of studies (e.g. Manel et al., 1999a; Loiselle et al., 2003; Thuiller et al., 2003a, 2003b; Lawler et al., 2006; Maher et al., 2014). The main finding is that generally speaking, classification or machine-learning approaches do not provide better results than regression-based approaches, but some of them are easy to understand and allow the models to be represented in a very informative or complementary format (e.g. recursive partitioning), or reveal properties not automatically available from other approaches (e.g. interactions between predictors). We will detail here three different approaches: recursive partitioning, discriminant analysis, and artificial neural networks.

11.2 Recursive Partitioning

Among the different techniques generally categorized as classification approaches, recursive partitioning is one of the most interesting for habitat suitability modeling. First of all, the approach is relatively easy to explain to inexperienced users and the results can be presented in the simple form of a decision tree, where the interactions between variables are visible. Second, recursive partitioning techniques usually form the basis of more complex and powerful techniques such as bagging or boosting (see Chapter 12), although the latter can theoretically be applied to regression approaches as well (James et al., 2013).

Recursive partitioning methods (RP; see Strobl et al., 2009), originally based on automated interaction detection (Morgan and Sonquist, 1963), were first introduced by both Breiman et al. (1984) as "classification and regression trees" (CART) and by Quinlan (1986) as "decision trees." RP approaches are meant to explain the variation for a single response variable (e.g. species presence–absence, biomass, abundance) with one or more explanatory variables. The response variable can be either discrete (classification trees) or continuous (regression trees), whereas the explanatory variables can be of any type, as is the case in most of the approaches so far addressed in this book (e.g. GLM or GAM). Specifying a binary response (e.g. presence–absence) as a factor will lead to a classification tree, whereas leaving it as numeric will lead to a regression tree. In the latter case, however, the final predictions might fall outside of the 0–1 range (unless a bounding function is applied). A decision tree is grown by repeatedly splitting the data, defined at each split (node) by a rule based on a single explanatory variable. At each split the data is partitioned into two mutually exclusive groups. The criteria for segmenting the data are based on either minimizing the classification error rate in the case of a classification tree, or maximizing the inter-class variance in the case of a regression tree. The splitting procedure is then re-applied to each group separately, repeating the same procedure at parallel nodes (i.e. other branches), thus growing the tree iteratively. The key trade-off is to partition the response into homogeneous groups, but also to keep the tree reasonably small in order to avoid overfitting the data through a very complex model. Furthermore, a complete tree will predict each data point perfectly, but will have limited power to predict outside of the training data. So, data-splitting is first performed until an overly large tree is grown (the maximum possible size equals the number of samples or sites). This complex tree is then pruned back to the desired size using specific rules to reduce overfitting. This pruning of the tree is the trickiest part of RP. The goal is to reduce the tree to an optimal size while maintaining enough predictive power to ensure accurate predictions. There are several algorithms for defining rules for pruning. The most common rely on cross-validation, where data-splitting is performed on a subset of data and then the predictive power is evaluated on the remaining data (Breiman et al. 1984).

Each final leaf (or terminal node) corresponds to one or a group of observations, and is typically characterized by either the distribution (discrete response) or mean value (continuous response) of the response variable (e.g. probability of presence). It is predicted by the values of the explanatory variables that define the nodes along the path to the terminal leaf.

Obviously, the way the splits are defined depends on the type of the predictor variables. For continuous variables, a split is defined using values of less than, or greater than, a chosen splitting value. Thus, only the rank order of numeric variables determines a split, and for z unique values there are $z-1$ possible splits. For discrete predictor variables with only two levels (e.g. presence or absence of a substrate type), only one split is possible, with each level defining a group. In the case of multiple levels (e.g. different soil types), any combination of levels can form a split, and for z levels, there are $2^{z-1}-1$ possible splits. Out of all the possible splits of all explanatory variables at a node, the one selected is that which minimizes the classification error rate or maximizes the inter-class variance at this node.

One advantage of RP is that it does not rely on assumptions about the relationship between the explanatory variable and the response variable of interest. Also, it does not expect the dependent variable to follow any specific distribution (as in GLM or GAM models). The approach is thus entirely data-driven.

There are several packages available in R for implementing RP (e.g. `tree`, `rpart`, `party`, `REEMtree`). The examples in this section use the `rpart` package, which offers built-in cross-validation procedures to optimally prune the final tree. The party package offers a nice alternative to `rpart` since it can also deal with conditional inference. In other words, the calculation of variable selection and variable importance is conditioned by the correlation between the variables (Strobl et al., 2008). It therefore accounts for the correlation structure between variables. This feature is very useful when a large set of correlated explanatory variables is used, although this algorithm has proven to be rather slow for large datasets.

There are only few options needed to fit a tree in the `rpart` package. The best way to start a tree model is to use the `rpart.control()` function, which specifies certain important parameters, such as the minimum number of observations that must remain to define a node. The default is 20, which makes sense with an extensive dataset. However, researchers working on restricted datasets with less than 100 points for instance should consider decreasing this parameter to 5 or 10. Another important parameter is the number of cross-validations, which by default is ten. When using large datasets with large numbers of variables, this number can be set slightly higher. In our case, we will adjust the cross-validation to 1000 repetitions and set the minimum number of observations to the default value. One key advantage of `rpart` is the efficiency of its algorithm that makes it possible to run over large data sets and to set a relatively large number of cross-validation runs.

```
> library(rpart)
> RP <- rpart(VulpesVulpes ~ 1 + bio3 + bio7 + bio11 + bio12,
data = mammals_data, control = rpart.control(xval = 10),
method = "class")
```

A print function for `rpart` objects is available (`print(RP)`) which details the different decisions taken during tree construction. However, a graphical representation as shown in Figure 11.1, is more convenient. Observations that satisfy the condition shown for each node go to the left while all others are element of the right branch in each node. The numbers plotted in the leaves (i.e. terminal nodes) are the presence or the absence of the species for the given combination of variables.

```
> plot(RP, uniform = F, margin = 0.1, branch = 0.5,
compress = T)
> text(RP, cex = 0.8)
```

Similarly to other techniques (e.g. GAM, GLM), the spatial prediction of an RP model can be easily obtained using the `predict()` function. Note that `rpart` provides both the presence–absence values and the probabilities of presence. Using the `prob` arguments makes it possible to extract the probabilities of presence (Figure 11.2). The presence–absence data come from a basic transformation of the probabilities of presence using a cutoff at 0.5. This is presented in Figure 11.1. We will see in Part IV that this is not the best binarization approach for all models (i.e. this threshold can be optimized).

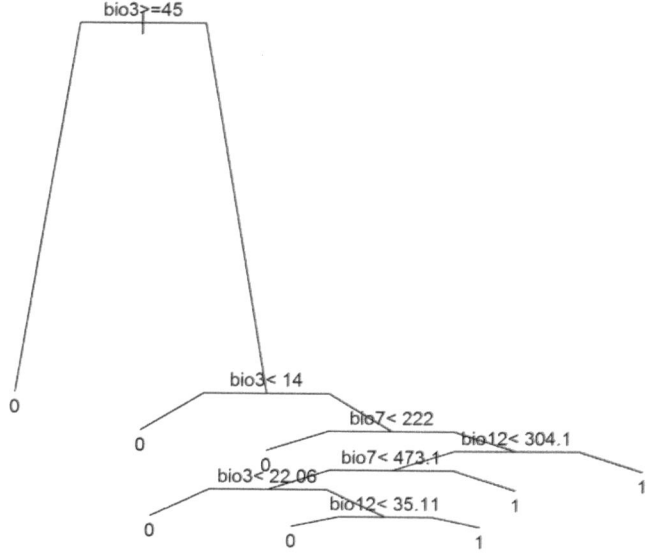

Figure 11.1 Classification tree for *Vulpes vulpes* using the rpart() function.

(a) **Original data**   (b) **Recursive partitioning**

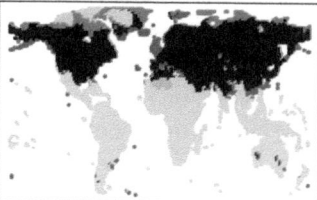

Figure 11.2 (a) Observed (black = presence, light gray = absence) and (b) potential distribution of *Vulpes vulpes* predicted by recursive partitioning. The gray scale of predictions illustrates habitat suitability values between 0 (light, unsuitable) and 1 (dark, highly suitable).

```
> RP.pred <- predict(RP, type = "prob")[, 2]
> par(mfrow = c(1, 2))
> level.plot(mammals_data$VulpesVulpes, XY = mammals_data[,
c("X_WGS84", "Y_WGS84")], color.gradient = "grey", cex = 0.3,
show.scale = F, title = "Original data")
```

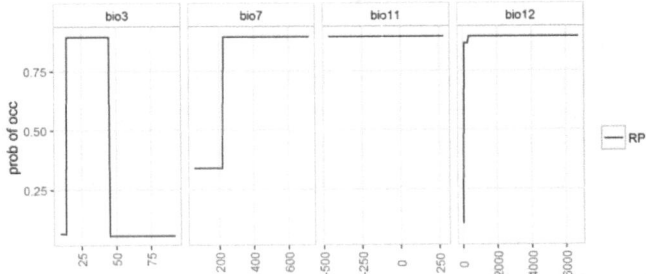

Figure 11.3 Response curves of a recursive partitioning (RP) model.

```
> level.plot(RP.pred, XY = mammals_data[, c("X_WGS84",
"Y_WGS84")], color.gradient = "grey", cex = 0.3,
level.range = c(0, 1),
show.scale = F,
title = "Recursive partitioning")
```

Response curves, based on the evaluation strip method (Elith et al., 2005), can also be extracted for RP in the same way as for GLM or GAM (Thuiller et al., 2009). Partial dependence plots show the response curve of the species along a given predictor variable while accounting for the average effects of all other predictor variables (Figure 11.3). However, such plots do not perfectly represent the effect of each variable in cases where strong interactions between variables have been found or are suspected (Friedman 2001). Since RP methods can easily fit interaction among variables, such plots should thus be interpreted with caution. Alternative plots for such complex methods are discussed in Zurell et al. (2012).

```
> rp <- response.plot2(models = c("RP"), Data = mammals_data[,
c("bio3", "bio7", "bio11", "bio12")],
show.variables = c("bio3", "bio7", "bio11", "bio12"),
fixed.var.metric = "mean", plot = FALSE, use.formal.names = TRUE)
> gg.rp <- ggplot(rp, aes(x = expl.val, y = pred.val,
lty = pred.name)) + geom_line() + ylab("prob of occ") + xlab("")
+ rp.gg.theme + facet_grid(~expl.name, scales = "free_x")
> print(gg.rp)
```

As previously mentioned, RP techniques form the basis of more complex and more powerful approaches, which we will detail in Chapter 12. These alternative approaches are interesting, as it is not always easy to determine the optimal size of a tree. Cross-validation strategies can sometimes produce puzzling findings, supporting different tree sizes that can lead to approximately similar predictive performance and error rates. More generally speaking, we will see later that fitting one tree to the data is a high-variance operation since local optima could lead to non-optimal trees. We will see later that bootstrap aggregations of trees and boosting are interesting alternative responses to the dilemma of selecting an appropriate tree size.

11.3 Linear Discriminant Analysis and Extensions

Discriminant analyses are methods used in statistics and machine learning to classify individuals (e.g. sites, samples, and populations) into groups (e.g. low suitable, moderately suitable, and highly suitable) based on a set of features (e.g. environmental variables), in order to describe them. In comparison to RP that is data-driven, linear discriminant analysis (LDA) assumes that linear combinations of environmental variables can separate these groups. The resulting combination can be used as a linear classifier, or, more commonly, for dimensionality reduction before subsequent classification.

LDA is closely linked to ANOVA and linear regression analyses, which also relate one response variable as a linear combination of other explanatory variables. In the other two methods, however, the response variable could be continuous, while in LDA it is discrete. Logistic (or related link) regressions (Chapter 10, GLM) are closer to LDA, as they also explain a discrete variable. However, these other methods (GLM, GAM, etc.) are preferable for ecological applications, where the explanatory variables are not necessarily assumed to be normally distributed which is a fundamental prerequisite for the LDA method.

Because LDA relies on linear combinations of predictor variables, it is not always relevant for modeling species distributions, for instance. Different extensions have been proposed in the past, notably the flexible discriminant analysis (FDA) proposed by Hastie et al. (1994), that allows the user to replace the linear combination with non-parametric functions such as MARS and BRUTO (see Chapter 10).

FDA is implemented in the mda package and does not require a lot of arguments, except for specifying MARS as the fitting method, for example:

```
> library(mda)
> fda_mod <- fda(VulpesVulpes ~ 1 + bio3 + bio7 + bio11 + bio12,
data = mammals_data, method = mars)
```

A confusion matrix can be extracted with a predefined (arbitrary) threshold of 0.5 to transform probability values into 0 (absence) and 1 (presence).

(a) **Original data**

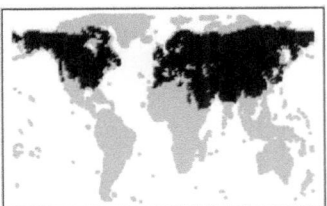

(b) **FDA**

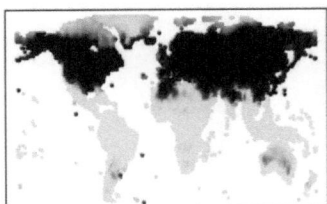

Figure 11.4 (a) Observed and (b) potential distribution of *Vulpes vulpes* predicted by flexible discriminant analysis based on MARS algorithm.

```
> fda_mod$confusion
            true
   predicted    0    1
           0 3866  449
           1  473 3754
   attr(,"error")
   [1] 0.1079373
```

As in RP, however, we will see in Part IV that more objective approaches can be used to define an optimal threshold for binarizing the predictions. There is no built-in function to plot FDA results to date. Because FDA naturally predicts two–class levels (0 or 1), predictions from fda can only be 0 or 1. However, in a similar way to rpart() when using the argument type="response" in the predict() function, using the argument type = "posterior" in FDA will produce a matrix of posterior probabilities that will allow the user to work with the inherent probabilities instead of the classified response (0 or 1) (Figure 11.4). The "posterior" argument returns the probabilities of each sample or site to belong to one of the modeled classes (species' present or absent). In the presence and absence case, we have two classes, which explains why the predict() function returns a matrix with two columns (we have selected the second one in the example below).

```
> FDA.pred <- predict(fda_mod, mammals_data[, c("bio3", "bio7",
"bio11", "bio12")], type = "posterior")[, 2]
> par(mfrow = c(1, 2))
> level.plot(mammals_data$VulpesVulpes, XY = mammals_data[,
c("X_WGS84", "Y_WGS84")], color.gradient = "grey", cex = 0.3,
show.scale = F, title = "Original data")
> level.plot(FDA.pred, XY = mammals_data[, c("X_WGS84",
"Y_WGS84")], color.gradient = "grey", cex = 0.3,
show.scale = F, title = "FDA")
```

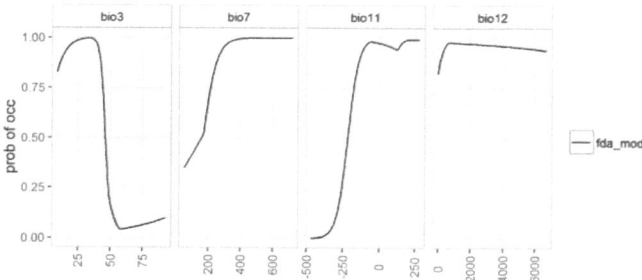

Figure 11.5 Response curve of *Vulpes vulpes* modeled using flexible discriminant analysis.

As with the other algorithms, the evaluation strip method (Elith et al., 2005) makes it possible to extract the response curves from the FDA and visualize the shape of the modeled relationships between the species and its environment (Figure 11.5).

```
> rp <- response.plot2(models = c("fda_mod"),
Data = mammals_data[, c("bio3", "bio7", "bio11", "bio12")],
show.variables = c("bio3", "bio7", "bio11", "bio12"),
fixed.var.metric = "mean",
plot = FALSE, use.formal.names = TRUE)
> gg.rp <- ggplot(rp, aes(x = expl.val, y = pred.val,
lty = pred.name)) + geom_line() + ylab("prob of occ") + xlab("")
+ rp.gg.theme + facet_grid(~expl.name, scales = "free_x")
> print(gg.rp)
```

While interaction is not explicitly modeled in FDA, the bivariate response curves also enable us to visualize how combinations of variables influence the habitat suitability predicted for our target species (Figure 11.6).

```
> rp.2D <- response.plot2(models = c("fda_mod"),
Data = mammals_data[, c("bio3", "bio7", "bio11", "bio12")],
show.variables = c("bio3", "bio7", "bio11", "bio12"),
fixed.var.metric = "median",
do.bivariate = T, plot = FALSE,
use.formal.names = TRUE)
> gg.rp.2D <- ggplot(rp.2D, aes(x = expl1.val, y = expl2.val,
fill = pred.val)) + geom_raster() + rp.gg.theme + ylab("") +
xlab("") + theme(legend.title = element_text()) +
scale_fill_gradient(name = "prob of occ.",
low = "#f0f0f0", high = "#000000") + facet_grid(expl2.name ~
expl1.name, scales = "free")
> print(gg.rp.2D)
```

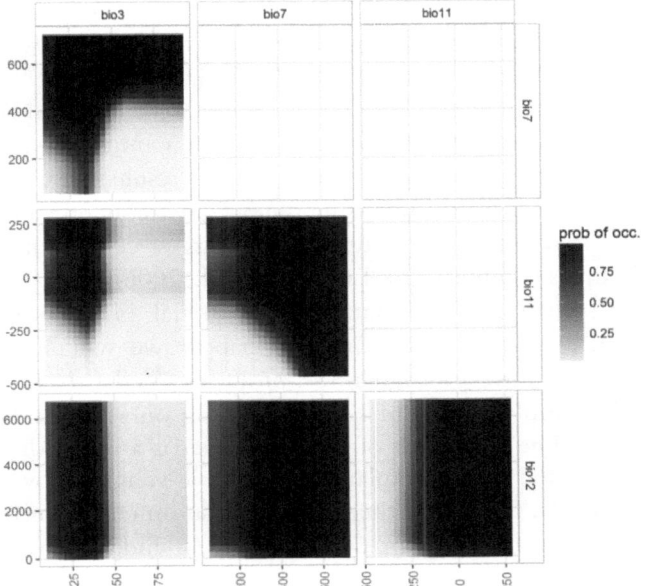

Figure 11.6 Bivariate response curve of *Vulpes vulpes* modeled using flexible discriminant analysis along four predictor variables.

11.4 Artificial Neural Networks

Artificial neural networks (ANNs), sometimes simply called "neural networks" (NNs), summarize a mathematical technique that attempts to simulate the structure and/or functional aspects of biological neural networks in order to classify objects (e.g. sites, samples, populations) and make predictions. ANNs have two basic characteristics, which we explain herein. First, ANNs use sets of adaptive weights, which are numerical parameters fine-tuned with a learning algorithm, in order to link the response to the predictors. These weights basically represent the interconnection between hidden layers, called "neurons." The strengths of these interconnections, initially set to random, are progressively updated throughout the learning process (which is why they are usually called adaptive). Secondly, ANNs accommodate nonlinear relationships between the response variables and the explanatory variable, which make them highly suitable for modeling complex systems. ANNs can handle any type of explanatory variable (e.g. continuous, categorical, Boolean), do not assume normal distribution of the data, and are said to be robust to noise in the data (e.g. false negatives; Dawson et al., 1998). They have not been widely used in ecology although they have generally shown relatively high predictive accuracy (Lek et al., 1996; Manel et al., 1999a; Pearson et al., 2002; Thuiller, 2003). The lack of transparency in ANNs, the difficulty of implementing them correctly, and their inherent stochasticity (different model runs return slightly different results) have no doubt discouraged many scientists from using them more frequently, and partly explains why they were not included in the largest comparative assessment of techniques conducted to date (Elith et al., 2006).

There are now several implementations of artificial NNs in R (e.g. `nnet`, `AMORE`, `neuralnet`). In this chapter, we will illustrate their use, with the `nnet` package. `nnet` proposes "feed-forward" NNs, which are the simplest form of NNs, including only one hidden layer (or neuron). The basic unit of a hidden layer is a block that will sum a set of weighted inputs (explanatory variable weighted by a coefficient, as in a GLM). The block then passes the summed response to a nonlinear function to create an output node (= hidden layer) response. This is done several times in several blocks that give different answers, because the initial weights are random. The ANN algorithm thus "learns" the correct weights by measuring the error between the observed response and the values predicted by the learned model. This error gets passed backward and the feedback algorithm individually increases or decreases those weights proportional to the error at each node. The network then iteratively moves forward, measures the predicted response, then updates and corrects the weights until the errors are minimized by the set criteria.

The important parameters in `nnet` are the number of units in the hidden layer (`size`) and the `weight decay`, which steers the optimization process and avoids overfitting (Venables and Ripley, 2002). As in most machine-learning algorithms, the weight decay "regularizes" (i.e. penalty to complexity) the weights. This is designed to overcome overfitting by keeping the weight estimates small when producing smooth nonlinear functions. This is the same principle as setting the lambda parameter in lasso and ridge regression (see Chapter 10).

The syntax is similar to other regression or classifications functions.

```
> library(nnet)
> set.seed(555)
> nnet.Init <- nnet(mammals_data[, c("bio3", "bio7", "bio11",
"bio12")], mammals_data$VulpesVulpes, size = 2, rang = 0.1,
decay = 5e-04, maxit = 200)
```

Here, we used two hidden layers and a value of 5e-4 for the distance decay in a very arbitrary way, although they do fall within traditionally used ranges (see Venables and Ripley, 2002).

Here, we use a cross-validation procedure to select the optimal size of the hidden layer and weight decay. This cross-validation is implemented in the `biomod2` package, using the `.CV.nnet()` function, as follows:

- Create the ensemble of conditions SIZE × DECAY to be tested.
- Cut the original data into two subparts (training and test) while keeping the species' prevalence constant (to avoid having a test subset with no presence, for instance).
- Train the NNs for each combination of SIZE and DECAY using the training dataset.
- Test the predictive power (here using AUC, see Part IV) using the test dataset.
- Run the procedure several times (here 10×) to account for variability in the subsampling procedure.
- Return the SIZE and DECAY combination, which together gives the best fit (i.e. the highest AUC score).
- Run the final ANN model using the optimal combination.

Interested readers are advised to take a look at the source code for this function by typing:

```
biomod2:::.CV.nnet in R
```

Then we use the "optimal" combination of SIZE and DECAY to develop the final model.

```
> set.seed(555)
> CV_nnet <- biomod2:::.CV.nnet(Input = mammals_data[, c("bio3",
"bio7", "bio11", "bio12")],
Target = mammals_data$VulpesVulpes)
> nnet.Final <- nnet(mammals_data[, c("bio3", "bio7", "bio11",
"bio12")], mammals_data$VulpesVulpes, size = CV_nnet[1, 1],
rang = 0.1, decay = CV_nnet[1, 2], maxit = 200, trace = F)
```

The optimal combination here was:

```
> CV_nnet
       Size Decay
    15    6   0.1
```

Note that due to the inherent stochasticity of NNs, the results may differ slightly each time.

We now visualize the predictions (Figure 11.7):

```
> nnet.Init.pred <- predict(nnet.Init, mammals_data[, c("bio3",
"bio7", "bio11", "bio12")])
> nnet.Final.pred <- predict(nnet.Final, mammals_data[, c("bio3",
"bio7", "bio11", "bio12")])
> par(mfrow = c(2, 2))
> level.plot(mammals_data$VulpesVulpes, XY = mammals_data[,
c("X_WGS84", "Y_WGS84")], color.gradient = "grey",
cex = 0.3, show.scale = F, title = "Original data")
> level.plot(nnet.Init.pred, XY = mammals_data[, c("X_WGS84",
"Y_WGS84")], color.gradient = "grey", cex = 0.3,
show.scale = F, title = "nnet.Init")
> level.plot(nnet.Final.pred, XY = mammals_data[, c("X_WGS84",
"Y_WGS84")], color.gradient = "grey", cex = 0.3,
show.scale = F, title = "nnet.Final")
```

(a)   **Original data**   (b)   **nnet.Init**

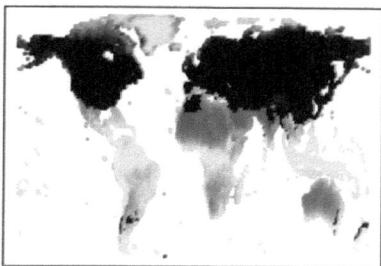

(c)   **nnet.Final**

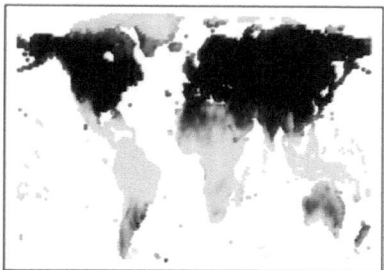

Figure 11.7 (a) Observed (black = presence, light gray = absence) and potential distribution of *Vulpes vulpes* modeled using a neural network algorithm with two different sets of (b) SIZE and (c) DECAY. The gray scale of predictions (upper-right and lower-left panels) shows habitat suitability values between 0 (light, unsuitable) and 1 (dark, highly suitable).

We can see that the SIZE × DECAY optimization has worked relatively well. The ANN model with optimized parameters predicts the true distribution of the species more accurately than the non–optimized one (Figure 11.7).

As for most of the approaches presented in this book, users interested in extracting the response curve for the species can do so using the `response.plot2()` function implemented in `biomod2` (Figure 11.8).

```
> rp <- response.plot2(models = c("nnet.Init", "nnet.Final"),
Data = mammals_data[, c("bio3", "bio7", "bio11", "bio12")],
show.variables = c("bio3","bio7", "bio11", "bio12"),
fixed.var.metric = "mean", plot = FALSE, use.formal.names = TRUE)
> gg.rp <- ggplot(rp, aes(x = expl.val, y = pred.val,
lty = pred.name)) + geom_line() + ylab("prob of occ") + xlab("")
+ rp.gg.theme + facet_grid(~expl.name, scales = "free_x")
> print(gg.rp)
```

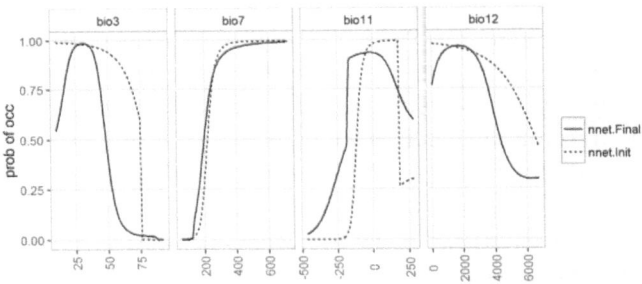

Figure 11.8 Response curve of *Vulpes vulpes* modeled by neural networks. The red lines represent a first model with reasonable but not optimized parameters set for SIZE and DECAY, while the blue line represents the final model with optimized parameters.

12 · Boosting and Bagging Approaches

12.1 Concepts

We have seen that RP methods can be used as alternative approaches to classification (e.g. FDA) and regression techniques (e.g. GLM, GAM) for predicting species distributions. They are not based on assumptions of normality and user-specified model statements as is discriminant analysis (e.g. FDA) and OLS regression. However, as for stepwise regression, the classification into groups can be influenced by local optima or noise in the data. Therefore, there is not one single decision tree that best explain the habitat suitability of a given species, but rather several trees which perform just as accurately when predicting a response. Here we present two different types of technique that have emerged over the last few years and that have been mostly applied to RP, although, in theory, they can be applied to any method. Bagging and boosting are ensemble modeling techniques, for which a classification or regression method is applied to various resampling of the original data set or through a stage-based framework, respectively. The results from each model are then combined (ensembled) using different weighting schemes.

Bagging – a short for bootstrap aggregation – was proposed by Breiman (1996), based on the principle of bootstrapping. In this approach, a large number of bootstrap samples are drawn from the available data (random subsampling with replacement of rows of data), a model (e.g. RP) is applied to each bootstrap sample, and then the results are combined into an ensemble. The final prediction is made either by averaging the outputs of regression tree approaches or by simple voting in the case of classification tree approaches (committee averaging; see Section 17.3.2). This type of procedure has been shown to drastically reduce the associated variance of the prediction (Breiman 2001). This bagging procedure applied to RP together with certain other refinements (see below) has given rise to the well-known random forests algorithm (Breiman 2001). Note that other types of bagged trees methods exist in the machine-learning literature.

Boosting, like bagging, is another ensemble approach developed to improve the predictive performance of models (Ridgeway, 1999; Friedman et al., 2000; Friedman, 2001). However, unlike bagging that uses a simple averaging (in regression trees) or voting (in classification trees) of results to obtain an overall prediction, boosting is a forward stage-wise procedure. In a boosting process, models (e.g. logistic regressions or decision trees) are fitted sequentially to the data. Interestingly, in this approach, model fitting is conducted on the residuals of the previous model(s), at each iteration. This is done repeatedly until a final fit is obtained. There are various ways of conducting this forward procedure and the method can be applied to different model types. Friedman (2001) also proposed the stochastic gradient boosting procedure which improves the quality of the fit and avoids overfitting. Boosted regression trees belong to this category (Elith et al., 2008).

In the next two chapters, we will introduce the use of random forests and boosted regression trees, two special types of bagging and boosting approaches, respectively, that are available in R and have gained momentum in the ecological literature in recent years (Thuiller et al., 2006; Cutler et al., 2007; De'Ath, 2007; Peters et al., 2007; Elith et al., 2008; Leathwick et al., 2008; Pearman et al., 2008b; Thuiller et al., 2009).

12.2 Random Forests

At this point, we assume that the user knows how to implement RP approaches to model species distributions (Chapter 11). One of the trickier aspects of RP is that it is a high-variance process. Indeed, small changes in the chosen variables or small changes in the dataset could lead to very different selected trees. The optimal tree size is also difficult to select. As an example, let's select another species from our dataset. We will use the jaguar (*Panthera onca*) for the example below:

```
> set.seed(555)
> RP.PantheraOnca <- rpart(mammals_data$PantheraOnca ~ bio3 +
bio7 + bio11 + bio12, data = mammals_data,
control = rpart.control(xval = 10), method = "class")
```

For this species, we can see that the fitted tree is slightly more complicated than for *V. vulpes* (Figure 12.1).

```
> plot(RP.PantheraOnca, uniform = F, margin = 0.1, branch = 0.5,
compress = T)
> text(RP.PantheraOnca, cex = 0.8)
```

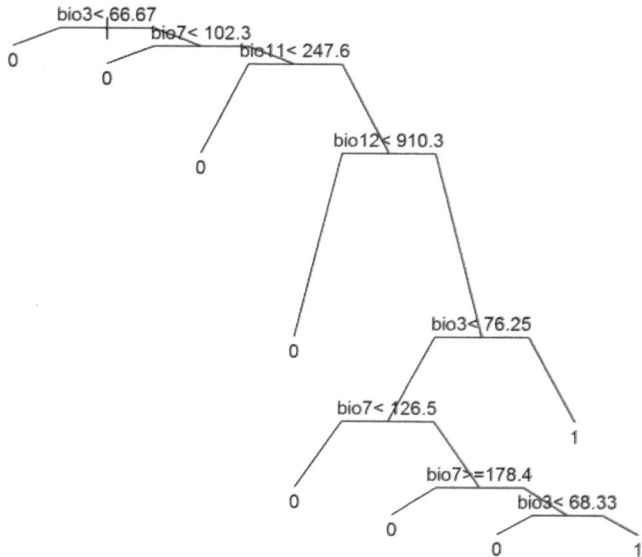

Figure 12.1 Classification tree for *Panthera onca* using the rpart() function.

Ten splits have been selected in the optimized model. How does this value change with different cross-validations runs, for instance? How robust is it to noisy data or small perturbations in the input data? These are fundamental questions one should preferably ask when applying RP approaches, instead of taking the first decision tree as given.

The idea of bagging is to fit several trees to different resampling of the original dataset and then to average the trees from the different subsamples. This is a relatively easy way of generating a naïve bagging approach using a bootstrap procedure. First, the bootstrap samples can be drawn from a multinomial distribution of parameter n (the number of sites or plots) and with the initial probability of drawing a plot from this distribution being equal to $1/n$.

```
> trees <- vector(mode = "list", length = 50)
> n <- nrow(mammals_data)
> boot <- rmultinom(length(trees), n, rep(1, n)/n)
```

We first create a complete tree with no pruning (xval=0) and then use the update function to re-evaluate the initial tree (Full_tree) without altering the weights (i.e. fitting a tree to a bootstrap sample specified by the weights) and store the trees in the list called "trees."

```
> Full_tree <- rpart(mammals_data$PantheraOnca ~ bio3 + bio7 +
bio11 + bio12, data = mammals_data, control = rpart.control(
xval = 0), method = "class")
> for (i in 1:length(trees)) {
trees[[i]] <- update(Full_tree, weights = boot[, i])
}
```

In order to understand the benefit of this approach, it is interesting to look at the structure of the multiple trees. A simple use of the table() function allows us to see which variable has been selected for a set of nodes.

```
> table(sapply(trees, function(x) as.character(x$frame$var[1])))

    bio3
     50
> table(sapply(trees, function(x) as.character(x$frame$
var[3])))

    bio12   bio7
      1     49
> table(sapply(trees, function(x) as.character(x$frame$var[5])))

    bio11 bio12   bio3  bio7
      16    30      3     1
> table(sapply(trees, function(x) as.character(x$frame$
var[10])))

    <leaf>  bio11  bio12   bio3   bio7
       29      3      4      3     11
```

We can see that through the 50 bootstraps, bio3 is always for the first split. When going down the trees, it becomes clear that all the variables could have been selected for a given split. The further we go down the tree, the higher the variability of the selected variables.

The advantage of the bootstrap approach is that one can extract the averaged probability (and the variance) of occurrences across all bootstrap samples.

```
> Pred <- matrix(0, nrow = n, ncol = length(trees))
> for (i in 1:length(trees)) {
# extract the prediction for each of the trees
Pred[, i] <- predict(trees[[i]], newdata = mammals_data[,
c("bio3", "bio7", "bio11", "bio12")], type = "prob")[, 2]
# remove potential predictions with a negative
# weight in the # bootstrap procedure
Pred[boot[, i] < 0, i] <- NA
+ }
> ## calculate the average probability of occurrence (e.g.
> ## habitat suitability)
> Pred.AVG <- rowMeans(Pred, na.rm = TRUE)
```

We can easily compare the performance of our DIY bagging with the initial rpart tree developed with 100 cross-validation runs. Here we use the receiver-operating characteristic (ROC) curve, detailed in Part IV, to compare model performance.

```
> require(pROC, quietly = T)
> roc_AVG <- roc(mammals_data$PantheraOnca, Pred.AVG,
percent = T)
> AUC_AVG <- as.numeric(auc(roc_AVG))
> AUC_AVG
    [1] 90.94637
> roc_RP <- roc(mammals_data$PantheraOnca,
predict(RP.PantheraOnca, type = "prob")[, 2],
percent = T)
> AUC_RP <- as.numeric(auc(roc_RP))
> AUC_RP
    [1] 86.7767
```

The predictive accuracy using the simple bagging approach (AUC=90.9) is much higher than the initial predictive accuracy (AUC=86.8). However, we have not controlled for overfitting here in our simple bagging approach, since we did not prune the initial regression tree at the basis of the bootstrapped model.

Random forests have been developed to check for overfitting by adding some stochasticity to the process of building the trees, but also at each node of each tree (Breiman, 2001). Let's assume that we have N plots or sites and X explanatory variables, each tree is grown based on the follow procedure:

1. Take a bootstrapped sample of N sites at random with replacement. This sample represents the training set for growing the tree.
2. At each node, select x candidate variables randomly out of all X predictors and evaluate the best split based on one of these x variable for the node. The value of x has to be selected beforehand and is kept constant during the forest growing.
3. Each tree is grown to the largest possible extent. There is no pruning.

The number of candidate variables taken randomly at each node is one of the few adjustable parameters to which random forests are somewhat sensitive. However, Breiman (2001) argued that the square root of the number of variables is a good compromise for classification trees and the number of variables divided by three for regression trees (Breiman, 2001). The few tests we have performed, did indeed reveal that x is of limited importance when it more or less follows Breiman's rule of thumb (2001). We will now go back to our initial example (*V. vulpes*) for the purposes of comparison with other algorithms.

```
> library(randomForest)
> RF <- randomForest(x = mammals_data[, c("bio3", "bio7",
"bio11", "bio12")], y = as.factor(mammals_data$VulpesVulpes),
ntree = 1000, importance = TRUE)
```

The importance = TRUE argument makes it possible to estimate the importance of each variable based on a permutation procedure that measures the drop in mean accuracy when the given variable is permuted. Note that in our example we transformed the binary presence–absence into a factor in order to enforce a classification tree. The results would be similar with a numeric response, thus enforcing a regression tree, although some of the final predictions might fall outside the 0–1 range. In that particular case, a logistic transformation as used in GLMs and GAMs, and shown in MARS, solves the problem.

```
> importance(RF)
                0         1 MeanDecreaseAccuracy MeanDecreaseGini
bio3     91.00154  78.45245            137.51393         1823.810
bio7     36.97712 410.76112            164.58416          947.334
bio11    52.24918  61.35427             85.40074         1026.730
bio12   111.97419  90.89983            153.76295          471.509
```

According to the random forests, and when looking at the change in the Gini coefficient (a measure of predictive accuracy; see Part IV), bio3, bio11, and bio7 show up as the most influential variables for the red fox.

The predictions (fitted values if using the training set; see Part IV) are extracted using the predict() function. Because we transformed the 0/1-classified dependent variable into a binary factor, predictions from the random forest return a matrix of class probabilities (probability of 0 and probability of 1). When predicting species distributions, we are interested in the probability of presence (the second column) (Figure 12.2).

```
> RF.pred <- predict(RF, type = "prob")[, 2]
> par(mfrow = c(1, 2))
> level.plot(mammals_data$VulpesVulpes, XY = mammals_data[,
c("X_WGS84", "Y_WGS84")], color.gradient = "grey",
cex = 0.3, show.scale = F,
title = "Original data")
> level.plot(RF.pred, XY = mammals_data[, c("X_WGS84",
"Y_WGS84")], color.gradient = "grey", cex = 0.3,
show.scale = F, title = "RF")
```

(a) **Original data**   (b)   **RF**

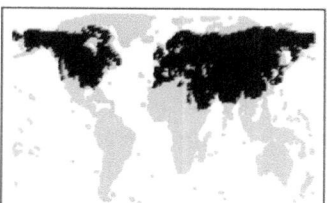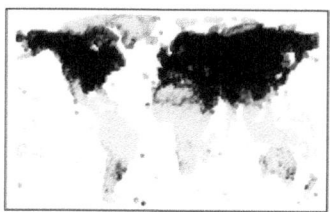

Figure 12.2 (a) Observed (left; black = presence, light gray = absence) and (b) potential distribution (right) of red fox modeled using random forest. The gray scale of predictions illustrates habitat suitability values between 0 (light, unsuitable) and 1 (dark, highly suitable).

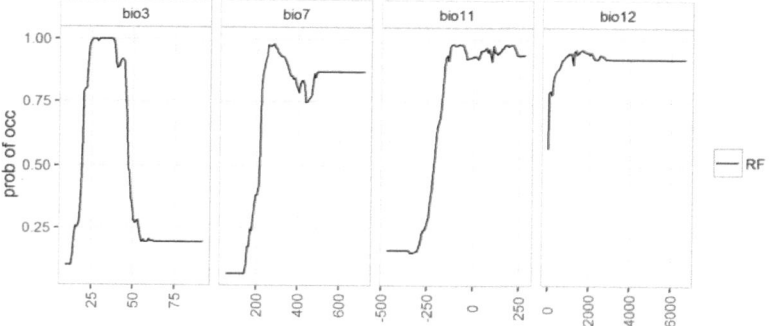

Figure 12.3 Probability of occurrence of red fox modeled by random forest in function of environmental.

It can be informative to look at the modeled response curves to explore the shapes of the species' responses along the predictor variables in more detail (Figure 12.3). We can see from the response curves that random forests are based on RP with sharp steps along the gradients, but that in general the response curves look similar to those extracted from models such as GAM or GLM.

```
> rp <- response.plot2(models = c("RF"), Data = mammals_data[,
c("bio3", "bio7", "bio11", "bio12")],
show.variables = c("bio3",
"bio7", "bio11", "bio12"),
fixed.var.metric = "mean", plot = FALSE, use.formal.names = TRUE)
```

```
> gg.rp <- ggplot(rp, aes(x = expl.val, y = pred.val,
lty = pred.name)) + geom_line() + ylab("prob of occ") + xlab("")
+ rp.gg.theme + facet_grid(~expl.name, scales = "free_x")
> print(gg.rp)
```

12.3 Boosted Regression Trees

As we have seen in the introduction to Chapter 12, when applied to RP (i.e. decision trees), gradient boosting models (also called boosted regression trees, BRT; Elith et al., 2008) provide a very flexible alternative ensemble modeling procedure to bagging. Unlike bagging that averages unpruned trees built on bootstrapped sample data, boosting uses a forward stage-wise procedure that iteratively fits simple trees to the training data, while gradually increasing focus on poorly modeled observations (by fitting residuals to the same predictors again). The general idea is to compute a sequence of very simple trees, where each new tree is fitted to the residuals of the set of trees so far developed. This procedure, also called additive weighted expansions of trees, has been show to improve not only the predictive ability of the model but also the bias and variance of estimates, even when the relationships between the environmental variables and the species are very complex (Friedman et al., 2000; Friedman, 2001; Elith et al., 2006; Elith et al., 2008; Merow et al., 2014). In addition to this process, Friedman (2002) proposed improving the quality of BRT by adding stochasticity. For each consecutive shallow tree, only a random sample of the dataset is used for training and the remaining for testing. This is similar to random forests for this purpose, except that the sampling is without replacement in BRT. Building consecutive trees from a random sub-sample of observations is called stochastic gradient boosting (Friedman, 2002), and has been shown to improve predictive accuracy and obviously to increase the computational speed since the trees are built from smaller fractions than the original datasets (Friedman, 2002). The fraction of data used at each consecutive tree, called the bag fraction, has been suggested to contain 0.5 and 0.75 of the full dataset (Elith et al., 2008). Setting a smaller bag fraction to train the model will result in excessively high prediction variance between runs if the number of trees is low.

·Among the different parameters of importance when fitting BRTs, the number of trees to be fitted, the learning rate, and the interaction depth are all critical. Hereafter, we will define these parameters and discuss their parameterization. The learning rate in BRT regularizes the weight given to the successive trees. Regularization refers to smoothing the model, making it more regular, so as to avoid fitting an overly complex model. It is usually preferable to have a slow learning rate (low weight for each single tree) and a large number of trees, not the other way around. The number of trees and the learning rate are thus closely linked and should be optimally selected using independent data or cross-validation procedures as introduced in Section 12.2. However, it is quite time consuming to perform this n optimization for large datasets. Elith et al. (2008) showed, as part of a case study in New Zealand, that as a rule of thumb a minimum of at least 1000 trees need to be fitted, and found that the appropriate learning rate for fitting at least 1000 trees was a good strategy (learning rate < 0.01). These authors also showed that the amount of trees and the associated learning rate was dependent on the prevalence of the species. Rare species require very slow learning rates compared to more abundant ones. The interaction depth relates to the complexity of the tree (i.e. the number of nodes). There is no single solution for selecting the optimal tree complexity (as for complexity in models in general; Merow et al., 2014). Numbers between 2 and 10 have shown to be a good option (Elith et al. 2008). One has to bear in mind that more complex trees require fewer trees and accommodate faster learning rates, and vice versa. However, ecologists should always strive to fit a large number of trees because it reduces the variation between runs. Additionally, using more than three nodes makes it possible to fit interactions among variables at each stage of the tree construction.

Boosted regression trees are implemented in the gbm package (Ridgeway, 1999). The dismo package proposes some additional features on top of the gbm package, to improve variable selection and offer additional summary statistics (as published in Elith et al., 2008). Elith et al. (2008) proposed a very detailed working guide for boosted regression trees with a case study in which they investigated the combined effects of tree complexity, learning rate, and number of trees on the overall predictive accuracy of BRT models.

For the sake of simplicity, we will stick to the gbm implementation, but we strongly recommend that interested readers take a look at the additional functions in dismo.

A number of important parameters need to be set: n.trees, interaction.depth, shrinkage, bag.fraction and cv.folds.

The parameter n.trees sets the maximum number of trees to be fitted. The different diagnostic tools proposed by generalized boosting model (gbm) will then reduce this number to the "relevant number" of trees. The `interaction.depth` corresponds to the complexity of the fitted trees at each stage (three nodes in our worked example). The `shrinkage` parameter corresponds to the learning rate. The `bag.fraction` corresponds to the random fraction of data used to fit each consecutive tree. Finally, `cv.folds>1` can be used, in addition to the usual fit, to perform a cross-validation and calculate an estimate of generalization error returned in `cv.error`. This is very useful for selecting the appropriate number of trees for predictions.

The function `glm.perf` allows the user to extract the number of relevant trees based on the cross-validation procedure (Ridgeway, 1999). In the example below, we chose a slow learning rate (0.01), a bag fraction of 0.5 and an interaction depth of 3. The optimization will also perform a 10-fold cross-validation to select the appropriate number of trees in light of the tree complexity and learning rate. We set up a large number of initial trees to further check how the improvement in deviance explains change as new trees are added.

```
# Note that this line of code takes quite a bit of time to run.
> GBM.mod <- gbm(VulpesVulpes ~ bio3 + bio7 + bio11 + bio12,
data = mammals_data, distribution = "bernoulli", n.trees = 10000,
interaction.depth = 3, shrinkage = 0.01,
bag.fraction = 0.5, cv.folds = 10)
```

We recommend the user to investigate how the improvement in fit changes as more trees are added (using the gbm.perf function with plot.it=T) (Figure 12.4). This is an important step. The generalized boosting model (GBM) requires a lot of fine-tuning and if the interested analyst wants to further investigate how the learning rate influences the outputs, we recommend conducting tests with different settings, while continuing to work with a large number of trees.

```
> gbm.mod.perf <- gbm.perf(GBM.mod, method = "cv", plot.it = T)
```

From the curve we can see that 1000 trees is not enough to get a reliable and stable model while a model with more than 5000 trees is enough. The user can here consider to either manually select 6000 trees for making predictions or plotting the response curve, or to select the optimal number of trees using the function (gbm.perf) which is here 10 000. For the sake of simplicity, we will here use the output from the gbm.perf function.

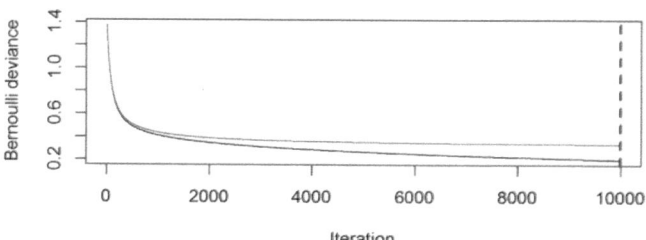

Figure 12.4 Optimal number of iterations (trees) for the GBM object. The *y*-axis represents the error of the model in function of the total number of trees (*x*-axis). The black line represents the error of the calibrated model with all data, while the grey line represents the error from the cross-validation runs.

Like the randomForest package, the gbm package also provides an interesting function, summary.gbm(), which is used to extract the relative importance of each explanatory variable. This function proposes two different ways of estimating variable importance. The first is the relative.influence, which is the default and the same as the one described in Friedman (2001). Interestingly, this approach is very similar to the weight of evidence based on AIC proposed by Burnham and Anderson (2002; see Chapter 12). In BRT, for each consecutive tree, the number of times a variable is selected at each node is weighted by the squared improvement to the model as a result of the split. Therefore, the variables that build more basic splits get higher weights. This procedure is then averaged over all fitted trees, performed for all variables, and scaled in order to obtain the relative influence of all variables summed to 100 (Friedman and Meulman, 2003)

The second choice is the permutation.test.gbm, which is very similar to the one used in random forests since it corresponds to the reduction in predictive performance when the variable of interest is permuted.

```
> summary(GBM.mod, method = relative.influence, plotit = F)
              var    rel.inf
      bio3   bio3 60.864446
      bio7   bio7 19.614715
      bio11 bio11 11.421784
      bio12 bio12  8.099055
> summary(GBM.mod, method = permutation.test.gbm, plotit = F)
           var    rel.inf
      1   bio7 58.403195
      2   bio3 23.549499
      3 bio11 13.012762
      4 bio12  5.034543
```

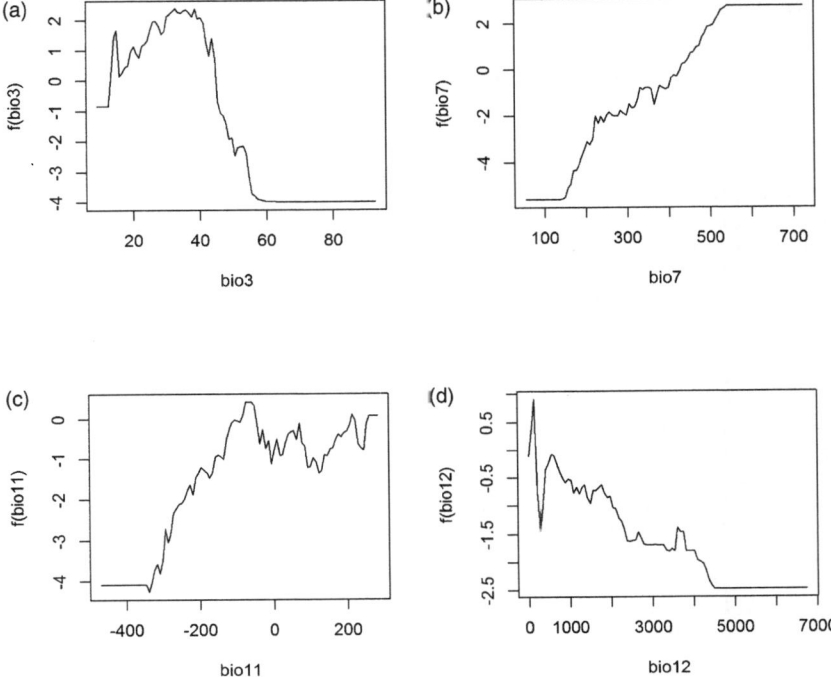

Figure 12.5 Response curves of *Vulpes vulpes* as a function of the explanatory variables built using the plot.gbm function in the GBM package.

We can see for the red fox (*V. vulpes*) example above that the same ranking is obtained using either method, with the species distribution strongly influenced by bio3 and then bio7 and bio11. All the models we have seen so far in Part II provide the same ranking of variable importance for this species.

An additional feature of gbm is the "inner" argument (i.var) used to plot the response curve of species as a function of the environmental variables (Figure 12.5).

```
> par (mfrow = c(2, 2))
> for (i in 1:ncol(mammals_data[, c("bio3", "bio7", "bio11",
"bio12")])) plot(GBM.mod, n.trees = gbm.mod.perf,
i.var = i)
```

However, it should be noted that the scale of the *y*-axis is expressed in the transformed scale (here using presence–absence and using the binomial model family, the scale is logistic).

In case of species' presence–absence data, we can use the `response.plot()` from `biomod2` to plot univariate and bivariate response curves in the probability scale (Figure 12.6).

```
> library("cowplot")
> # Univariate response curves
> rp <- response.plot2(models = c("GBM.mod"),
Data = mammals_data[, c("bio3", "bio7", "bio11", "bio12")],
show.variables = c("bio3", "bio7", "bio11", "bio12"),
fixed.var.metric = "mean",
plot = FALSE, use.formal.names = TRUE)
> gg.rp <- ggplot(rp, aes(x = expl.val, y = pred.val, lty = pred.
name)) + geom_line() + ylab("prob of occ") + xlab("") +
rp.gg.theme + facet_grid(~expl.name, scales = "free_x")
> # Bivariate response curves
> rp.2D <- response.plot2(models = c("GBM.mod"),
Data = mammals_data[, c("bio3", "bio7", "bio11", "bio12")],
show.variables = c("bio3", "bio7", "bio11", "bio12"),
fixed.var.metric = "median", do.bivariate = T,
plot = FALSE, use.formal.names = TRUE)
> gg.rp.2D <- ggplot(rp.2D, aes(x = expl1.val, y = expl2.val,
fill = pred.val)) + geom_raster() + rp.gg.theme + ylab("") +
xlab("") + theme(legend.title = element_text())+
scale_fill_gradient(name = "prob of occ.",
low = "#f0f0f0", high = "#000000") +
facet_grid(expl2.name ~ expl1.name, scales = "free")
> plot_grid(gg.rp, gg.rp.2D, labels = c("(a)", "(b)"), ncol = 1,
nrow = 2, rel_heights = c(1, 2))
```

Predictions from GBM can be obtained with the usual `predict()` function, with a supplementary parameter specifying the number of trees that should be used to make the prediction (Figure 12.7). The best practice is to use the results from the `gbm.perf()` function.

```
> GBM.pred <- predict(GBM.mod, newdata = mammals_data[, c("bio3",
"bio7", "bio11", "bio12")], type = "response",
n.trees = gbm.mod.perf)
> par(mfrow = c(1, 2))
> level.plot(mammals_data$VulpesVulpes, XY = mammals_data[,
c("X_WGS84", "Y_WGS84")], color.gradient = "grey",
cex = 0.3, show.scale = F, title = "Original data")
> level.plot(GBM.pred, XY = mammals_data[, c("X_WGS84",
"Y_WGS84")], color.gradient = "grey", cex = 0.3,
show.scale = F,
title = "GBM")
```

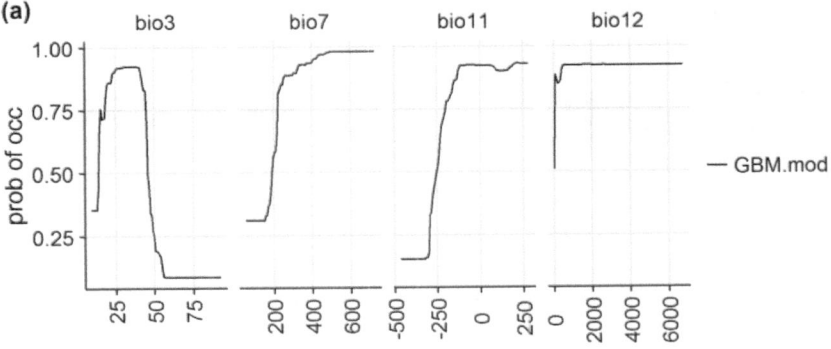

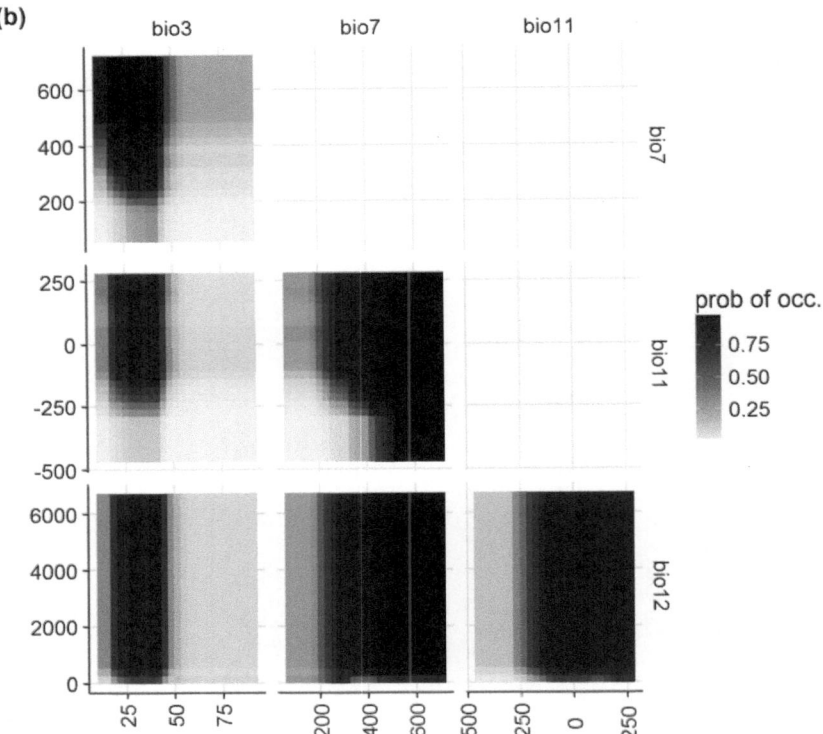

Figure 12.6 Response curves of red fox as a function of one (a) or two (b) explanatory variables at a time.

(a) **Original data**   (b)   **GBM**

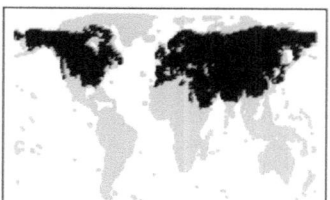 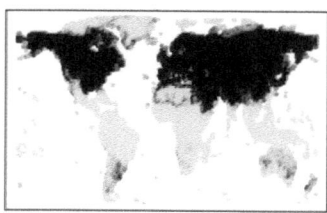

Figure 12.7 (a) Observed (black=presence, light gray= absence) and (b) potential distributions of the red fox modeled using a boosted regression tree approach. The gray scale of predictions shows habitat suitability values between 0 (light, unsuitable) and 1 (dark, highly suitable).

13 · *Maximum Entropy*

13.1 Concepts

In recent years, we have seen a rise in applications using the maximum entropy principle in ecology; for instance, to predict species abundances from functional traits (Shipley et al., 2006, 2011), to predict macroecological patterns (Harte, 2011), or to model species distributions (Phillips et al., 2004, 2006). From a Bayesian perspective, the principle of maximum entropy states that, subject to known constraints, the probability distribution that best represents the data is the one with the greatest entropy, i.e. the one which best reproduces the data. When applying Maxent to presence-only species distribution data, the space within which the Maxent probability distribution is defined encompasses all pixels in the study area (i.e. background information, or quadrature points, see Renner et al. 2015), the pixels representing the distribution of species occurrences constitute the sample points, and their environmental features are the explanatory variables.

The application of the maximum entropy formalism to species distribution modeling was first introduced by Phillips et al. (2004) and is now well-developed in the standalone package Maxent.[1] Although it is not formally implemented in R, we decided to add a short introduction here and present a way of running Maxent from R, so that the Maxent results can be compared with those from other modeling techniques and approaches (see Part IV and Part V). Both `dismo` and `biomod2` can be used to run Maxent in a batch mode. In addition, a maximum entropy R package is currently in development (see Halvorsen et al., 2015). For more information about Maxent, we refer interested readers to Elith et al. (2011), and for its equivalence to GLM and more general discussion of point-pattern process models, Renner et al. (2015) and to Chapter 20 of this book.

---

[1] www.cs.princeton.edu/~schapire/maxent/

Conceptually, Maxent contrasts observed presence data ($y = 1$) to the available environment in a given region (named z, a vector of environmental predictors). Following Elith et al. (2011), we define f(z) as the probability density of predictors across the region and f1(z) as the probability density of covariates across locations within the region where the species occurs. MaxEnt uses the predictors from the occurrence and the background sample to estimate the ratio f1(z)/f(z). The optimization algorithm looks for f1(z) that minimizes the distance from f(z). Indeed f(z) is here seen as a null model for f1(z) since there is no reason to expect the species to prefer any particular environmental conditions in the absence of occurrence data. In the latter case, the best prediction is that the species occupies environmental conditions proportionally to their availability in the region. In MaxEnt, this distance from f(z) is taken to be the relative entropy of f1(z) with respect to f(z).

Since Maxent does not specifically model presence data but rather the density of used environmental conditions, the raw outputs of Maxent are then back-transformed into a logistic outputs to be directly interpreted as a probability of occurrence (for more details, see Elith et al. 2011).

As in regression-based approaches (e.g. GLM), where linear or quadratic terms could be fitted between presence–absence data and the predictors, Maxent also offers the option of using different ways of modeling the relationships between f1(z)/f(z) and the environmental predictors. These are known as features, by default they are all used in the model optimization (unless one or more features are specified), meaning that in most models there will be more features than covariates. There are six feature classes: linear, product, quadratic, hinge, threshold, and categorical (see Elith et al. 2011, Appendix S1). Elith et al. (2011) describe the different features as:

> products of all possible pairwise combinations of covariates, allowing simple interactions to be fitted. Threshold features allow a 'step' in the fitted function; hinge features are similar except they allow a change in gradient of the response. Many threshold or hinge features can be fitted for one covariate, giving a potentially complex function. Hinge features (which are basis functions for piecewise linear splines), if used alone, allow a model rather like a GAM: an additive model, with non-linear fitted functions of varying complexity but without the sudden steps of the threshold features.

In light of these features, Maxent tends to overfit the data if no penalty or regularization is used to down-weight unimportant variables (as in boosted regression trees). Thus, in a similar vein to other penalties for complexity such as Akaike's information criterion (Akaike, 1974), Maxent fits a penalized maximum likelihood model that aims to trade-off model fit and model complexity (Phillips and Dudik, 2008).

13.2 Maxent in R

Maxent takes the sample points or coordinates of observed presences of the species of interest in a comma-separated text file and the environmental variables in grid formats. Here, in our implementation, Maxent directly uses ascii grids to sample the environmental variables for the presence locations of the species, and to define the available space for the Maxent probability distribution. Maxent then creates the background data with a default number of 10 000 randomly selected points across the ascii grids (also called quadrature point, see Renner et al. 2015). If presence–absence data are both available and are reliable, it is generally advisable to use a presence–absence modeling method (as seen in the previous parts of this book), as this makes the models less susceptible to sample selection bias and means they take advantage of all information in the data. In other words, using Maxent with true presence and absence data is not recommended (e.g. Elith et al., 2011; Guillera-Arroita et al., 2014).

First of all, we need to inform Maxent where the species and the grids files are located.

The path to maxent.jar should also be referred.

```
> ## The folders 'book.data' should be in the a directory just a
> ## before your working directory test if the data directory is
> ## well located (i.e. in dirname(getwd()))
> parent.dir <- dirname(getwd())  ## get the name of the
directory where data dir should be
> any(file.exists("data", parent.dir))  ## ok if return TRUE
[1] TRUE
> dir.create("MaxEnt.res")
> MaxEnt.layers.dir <- "../data/bioclim"
> MaxEnt.samples.dir <- "../data/species"
> MaxEnt.out.dir <- "MaxEnt.res"
> MaxEnt.soft.path <- "../data/maxent.jar"  ## the path to
maxent.jar file
> Java.soft.path <- "C:/Program Files (x86)/Java/jre7/bin/java.
exe"  ## the path to java software binaries => to be adapted
according to your computer settings
```

The environmental layers can be listed by:

```
> list.files(MaxEnt.layers.dir, pattern = ".asc", recursive = T)
    [1] "current/ascii/bio11.asc" "current/ascii/bio12.asc"
    [3] "current/ascii/bio3.asc"  "current/ascii/bio7.asc"
    [5] "future/ascii/bio11.asc"  "future/ascii/bio12.asc"
    [7] "future/ascii/bio3.asc"   "future/ascii/bio7.asc"
```

and the species of interest (VulpesVulpes.csv) containing presence-only observations by:

```
> list.files(MaxEnt.samples.dir, pattern = ".csv")
    [1] "mammals_and_bioclim_table.csv" "mammals_table.csv"
    [3] "VulpesVulpes.csv"
```

Then, we call Maxent directly from R in batch mode (see the Maxent manual for further explanations, Elith et al., 2011, and Renner et al., 2015, for additional code):

```
> ## define the shell command we want to execute
> maxent.cmd <- paste0("\"", Java.soft.path, "\" -mx512m -
jar \"", MaxEnt.soft.path, "\" environmentallayers=\"",
file.path(MaxEnt.layers.dir,
"current", "ascii"), "\" samplesfile=\"",
file.path(MaxEnt.samples.dir,
"VulpesVulpes.csv"), "\" projectionlayers=\"",
file.path(MaxEnt.layers.dir,
"current", "bioclim_table.csv"), "\" outputdirectory=\"",
MaxEnt.out.dir, "\"  outputformat=logistic
maximumiterations=500 jackknife visible=FALSE redoifexists
autorun nowarnings notooltips")
> ## run Maxent
> system(command = maxent.cmd)
```

This should normally load Maxent and run it for the species *V. vulpes*.

The command –mx512m gives Maxent 512Mb of RAM. Then, one has to provide Maxent with the location of the environmental layers, the sample file, the output directory, and a few more options. For instance, we ask to obtain the probability of occurrence (transformed from the raw data) instead of the raw data using `outputformat=logistic`. To be able to compare Maxent predictions to those from other models in R, we need to provide Maxent with a projection file in.csv format (bioclim_table.csv). This file contains the coordinates and values of explanatory variables for all grid cells (same data as in the mammal_ data table).

Most of the outputs from Maxent are finally stored in the MaxEnt. out folder:

(a) **Original data**　　　　　　(b) **MAXENT**

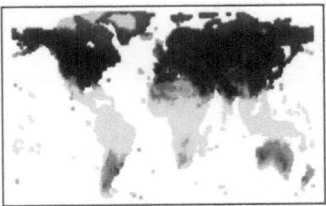

Figure 13.1 (a) Observed (black = presence, light gray = absence) and (b) potential distribution of the red fox modeled using Maxent in batch mode from R. The gray scale of predictions shows habitat suitability values between 0 (unsuitable) and 1 (highly suitable).

```
> list.files(MaxEnt.out.dir)
      [1] "maxent.log"
      [2] "maxentResults.csv"
      [3] "plots"
      [4] "VulpesVulpes.asc"
      [5] "VulpesVulpes.html"
      [6] "VulpesVulpes.lambdas"
      [7] "VulpesVulpes_bioclim_table.csv"
      [8] "VulpesVulpes_bioclim_table_clamping.csv"
      [9] "VulpesVulpes_omission.csv"
     [10] "VulpesVulpes_sampleAverages.csv"
     [11] "VulpesVulpes_samplePredictions.csv"
```

The predictions for the sample points are saved in VulpesVulpes_samplePredictions.csv and predictions for the whole area are stored in VulpesVulpes_bioclim_table.csv.

```
> Maxent.pred_AllFeatures <- read.csv(file.path(MaxEnt.out.dir,
"VulpesVulpes_bioclim_table.csv"))
```

One can then plot the predictions and compare them to the observed data as for the other models (Figure 13.1).

```
> par(mfrow = c(1, 2))
> level.plot(mammals_data$VulpesVulpes, XY = mammals_data[,
c("X_WGS84", "Y_WGS84")], color.gradient = "grey", cex = 0.3,
level.range = c(0, 1), show.scale = F, title = "Original data")
> level.plot(Maxent.pred_AllFeatures[, 3],
XY = Maxent.pred_AllFeatures[, c("X_WGS84", "Y_WGS84")], color.
gradient = "grey",
cex = 0.3, show.scale = F, title = "MAXENT", level.
range = c(0, 1))
```

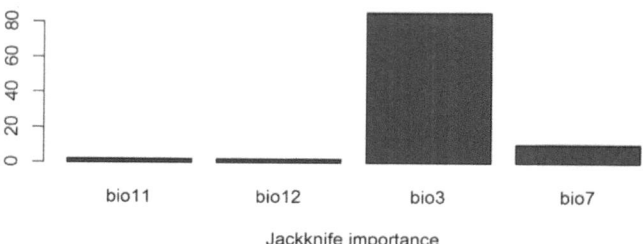

Jackknife importance

Figure 13.2 Heuristic estimate of relative contributions of the four environmental variables to the Maxent model using a jackknife procedure.

All results regarding the predictive accuracy and other results are stored in the maxentResults.csv file.

```
> Maxent.results <- read.csv("MaxEnt.res/maxentResults.csv")
> names(Maxent.results)
```

The importance of each variable obtained using the jackknife procedure can be displayed as a bar plot (Figure 13.2):

```
> par(mfrow = c(1, 1))
> barplot(as.matrix(Maxent.results[1, c(8:11)]), horiz = F,
cex.names = 1, names = sub(".contribution$", "",
names(Maxent.results[1, c(8:11)]))), xlab = "Jackknife
importance")
```

The results obtained from Maxent are relatively similar to those from the other techniques we have looked at so far.

The example run here used the default option which allows all types of feature. This is however worth considering simpler models as we have seen through the entire book (see also Merow et al., 2014).

Here we use hinge features by turning off the other feature types (nonlinear, etc.).

```
> ## define the shell command we want to execute
> maxent.cmd <- paste0("\"", Java.soft.path, "\" -mx512m -
jar \"", MaxEnt.soft.path, "\" environmentallayers=\"",
file.path(MaxEnt.layers.dir, "current", "ascii"), "\"
samplesfile=\"",
file.path(MaxEnt.samples.dir, "VulpesVulpes.csv"), "\"
projectionlayers=\"",
file.path(MaxEnt.layers.dir, "current", "bioclim_table.csv"), "\"
outputdirectory=\"", MaxEnt.out.dir, "\" outputformat=logistic
nowarnings
nolinear noquadratic nothreshold noproduct maximumiterations=500
jackknife visible=FALSE redoifexists autorun nowarnings
notooltips")
> ## run Maxent
> system(command = maxent.cmd)
> Maxent.pred_Hinge <- read.csv("MaxEnt.res/VulpesVulpes_bioclim_
table.csv")
```

(a) **MAXENT - all features**   (b) **MAXENT - hinge feature**

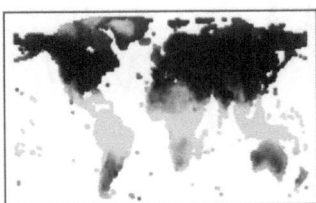

Figure 13.3 Comparison between the potential distribution of the red fox modeled using (a) Maxent with all features (by default) and (b) Maxent with only the hinge feature selected. The gray scale of predictions shows habitat suitability values between 0 (unsuitable) and 1 (highly suitable).

The predictions can then be plotted and compared to the initial model with all features and the one with only the hinge feature (Figure 13.3).

```
> par(mfrow = c(1, 2))
> level.plot(Maxent.pred_AllFeatures[, 3],
XY = Maxent.pred_AllFeatures[, c("X_WGS84", "Y_WGS84")],
color.gradient = "grey",
cex = 0.3, level.range = c(0, 1),
show.scale = F,
title = "MAXENT - all features")
> level.plot(Maxent.pred_Hinge[, 3], XY = Maxent.pred_Hinge[,
c("X_WGS84", "Y_WGS84")], color.gradient = "grey", cex = 0.3,
show.scale = F, title = "MAXENT - hinge feature", level.
range = c(0, 1))
```

As we can see in Figure 13.3, the maps are almost exactly the same. The hinge function that behaves similarly to a GAM is enough to predict the distribution of the red fox. Interested readers could also take a look at the jackknifing results that are also the same.

This result underlines that Maxent, like any other modeling technique, needs tuning to ensure it is correctly parameterized. This could be done in a semi-automatic fashion once the criteria are clear (e.g. the best fit with the simplest model, for instance).

14 · *Ensemble Modeling and Model Averaging*

So far, we have seen that HSMs can be implemented with a large range of statistical tools. This raises the question of which one(s) to use and how? There is no simple answer to this question, but it has fueled more than ten years of comparative analyses comparing, for example, regression-based versus tree-based algorithms (Thuiller et al., 2003a; Meynard and Quinn, 2007) or model-based versus machine-learning based (Manel et al., 1999a; Segurado and Araújo, 2004), parametric versus non-parametric algorithms (Thuiller et al., 2003a; Segurado and Araújo, 2004), and all the other types of model contests. With just a few exceptions (e.g. Maher et al., 2014), the main conclusion has been that presence–absence models usually work better (Brotons et al., 2004), that the most recently proposed approaches to HSM such as boosting or bagging tend to offer higher predictive performance (Elith et al., 2006), but this also usually depends on the context, data bias, and resolution (Elith and Leathwick, 2009), and that better predictive performance at model calibration usually comes at the expense of model transferability to new regions or to new conditions (Randin et al., 2006).

One way of selecting a model from the plethora of existing algorithms is to simply select the best one for the data, based on one or a set of predictive performance metrics (Thuiller, 2003, see Part IV). When modeling a large number of species, one model can be selected per species, resulting in different models selected for different species. The advantage of this solution is that the predictive performance metric selects the best model for the user, but it does make it more difficult to compare models across species. An alternative to the strict selection of one single model is to use an ensemble of models (e.g. fitted with different techniques, or with different sets of predictors) and to derive a general prediction from all (or a part) of them. The rationale behind using and ensembling several models is that two or more models may have very similar predictive performance even when they contain different environmental predictors and/or yield vastly different spatial predictions, making it difficult to know which of the equivalent candidate models to use. Furthermore, the "best" model may not necessarily be the best one for predictions in a different area or under new conditions, or some models it may be more sensitive (than one or the other models) to sampling bias, which might also reduce model transferability (Randin et al., 2006). We therefore usually only know which model performs best given the available dataset, but not how it might perform in a different region (for which there is no data available) or in a different time period (where the environmental conditions and relationships between predictors may change). Ensemble modeling is particularly powerful in these situations, as it maps both the main trend (i.e. mean, median, or some other percentile) and the overall variation (and thus uncertainty) across all models. Many other aspects of models can also be "ensembled," such as the measured importance of variables or the modeled response curves.

The concept of ensemble modeling is relatively recent. It was first introduced through the averaging of several models fitted with a single modeling technique, as popularized by Burnham and Anderson (2002) in the case of regression models such as GLM or GAM in their seminal book *Model Selection and Multi-model Inference: A Practical Information-theoretic Approach*. The main idea behind multi-model inference (MMI) is to avoid selecting the best model and instead relying on multiple candidate models (chosen based on theory or another decision-making process) with no variable selection and then weighting each model based on an information criterion to obtain the weighted sum of parameters and outputs. This approach thereby retains all information (all models, their weights, and their predictions) and draws inference in a very informative, probabilistic way. Such an approach was initially put forward as an alternative to stepwise procedures. Indeed, with an increasing number of predictor variables, the number of candidate models increases exponentially, making the selection of a single best model increasingly difficult, as several combinations of variables (possibly fully distinct) may give similar model fit (e.g. measured with the AIC or BIC criterion) although only one would be selected as the "optimal" solution in a stepwise model selection approach. This is rather problematic if the output of such stepwise selection procedures is then used to infer causal relationships between the biological response and the selected environmental variables or to support environmental management or conservation (species reintroduction, for example) (Guisan et al., 2013). Ensemble modeling avoids selecting one single best model, and thus eliminates (or at least limits) model selection bias but also provides a relative measure of each predictor's importance throughout all candidate models (weight of evidence, Burnham and Anderson, 2002).

Here, we will briefly explain MMI by focusing on one information criteria – the AIC – although alternative criteria could also be used, such as the BIC (Link and Barker, 2006) (see also Section 10.2).

The AIC is simply defined as:

$$AIC = -2(LL - K)$$

where LL is the log-likelihood of a given model, and K the number of parameters (here, the number of variables). According to Akaike (1974), among the candidate models considered, the model selected should be the one which yields the smallest value of AIC because this model is estimated to be "closest" to the unknown truth from which the data were sampled. An AIC-based stepwise regression thus selects the combination of variables producing the lowest AIC which is the closest to the unknown *truth*.

If the number of tested parameters is too high relative to the number of observations (i.e. sample size greater than a 1:10 ratio of parameters to observation, see Harrell's rule of thumb in Chapter 8), a second order bias adjustment of AIC should be used, which leads to the *AICc* criterion:

$$AICc = AIC + \frac{2K(K+1)}{n - K - 1}$$

We previously stated that the rationale for using multi-model approaches is to draw inference from more than one single "best" (or "optimal") model by extending the concept of parameter estimation from one model and one dataset to a concept of likelihood of a model given the data. Different likelihoods are therefore obtained for the different models, which can then be summarized among all candidate models by the Akaike weights w_i:

$$w_i = \frac{\exp\left(-\frac{1}{2}\Delta_i\right)}{\sum\limits_{r=1}^{R}\exp\left(-\frac{1}{2}\Delta_r\right)}$$

where

$$\Delta_i = AIC_i - \min AIC$$

and where AIC_i is the AIC of the model i, and $minAIC$ is the smallest AIC value in the set of models. The "best" model is the one for which the difference with the minimum AIC across the set of candidate models is $\Delta_i = 0$. The model with the largest Δ_i is considered to be the least plausible of the candidate models. Note that it is still a candidate model, but with a lower plausibility weight.

The weight w_i is thus considered as the weight of evidence in favor of model i, being closest to the unknown truth when Δ_i is close to 0, and farthest from truth when Δ_i is the largest among all models. Those weights of evidence are known as model probabilities and are subsequently used by ecologists for two main purposes which we will now discuss.

Indeed, one of the primary goals of ecology, population and community biology is to classify the potential drivers of the system studied in terms of their importance. By analogy to the weight of evidence in favor of a given model, the weight of evidence of each predictor (w_{pi}) can be simply estimated as the sum of the model AICs weights (w_{pi}) over all models in which the selected predictor appears. The predictor with the highest w_{pi} (the closest to 1) is given the highest weight of evidence for explaining the response variable (the highest relative importance).

A second, very appealing, analytical option for MMI and model averaging is building predictions based on all (or a subset of models), weighted by their weight of evidence, and thus by their statistical performance. These "ensemble" predictions can be obtained very simply by calculating a weighted average of the predictions from all models. Models with a low weight of evidence basically have no predictive power, whereas models with similar AIC weights will contribute similarly, allowing concurrent predictors to contribute equally. The weighted average prediction is thus expressed by the formula:

$$\bar{P} = \sum_{i=1}^{R} w_i P_i$$

where $w_i P_i$ is the prediction (probability) from model i, weighted by its weight of evidence.

Associated with the idea of weighted means, a weighted variance or standard deviation, and associated confidence intervals, can also be easily estimated, providing a useful estimation of the uncertainty associated with the different candidate models.

As we have seen, MMI was initially conceived to be applied to one single type of regression model (e.g. GLM), but not to multiple model types (e.g. mixing GLM and GAM in a same framework). Along the same lines, researchers have proposed that a similar strategy could be applied when predicting species distributions from multiple types of models (Thuiller, 2004; Thuiller et al., 2004a; Araújo and New, 2007; Marmion et al., 2009). The idea, first championed by these authors for forecasting future distributions under climate change (Thuiller, 2004), was to build on MMI by compiling a set of competing models within and between diverse types of modeling techniques rather than only within a given modeling technique alone, and by extension using different input data (Araújo and New, 2007). The main advantage of such multi-technique approaches is that they account for the variability both within a given model type, as well as between model types (Thuiller, 2004). The theory behind this approach is currently much less advanced than for MMI because some techniques (e.g. NNs, random forest) used to fit HSMs are not based on maximum likelihood estimations and therefore do not allow the extraction of AIC or BIC criteria. Instead, researchers have proposed using the metrics of predictive performance conventionally used in habitat suitability modeling (see Part IV), such as the area under the curve (AUC) of a ROC plot (AUC, Swets, 1988) or the true skill statistic (Allouche et al., 2006) to weight the different models (see Part IV for details of these metrics), and have proposed weighted sum of probability values as follows:

$$\bar{P} = \sum_{i=1}^{R} w_i P_i$$

where $w_i P_i$ is the prediction from model i (fitted using any modeling technique), weighted by a weight of evidence in favor of this model, but this time based on a chosen predictive accuracy metric (e.g. AUC or TSS), ideally (but not necessarily) calculated on a left-out partition of the data, obtained for instance through cross-validation. As we will see in Part IV, cross-validation is one of the most widely accepted approaches for testing the predictive accuracy of habitat suitability modeling. A random part of the data is kept for calibration (i.e. training data) while the remainder is used to test the prediction of the model, and the whole approach is then repeated several times for a single model (e.g. GLM) and the average predictive accuracy is finally reported (Araújo et al., 2005a; Thuiller et al., 2009). The main advantage of an ensemble approach is that the predictions by all individual models can be retained and documented for each modeling unit (Figure 14.1), and additional summary statistics can be extracted within the same procedure, such as the weighted variance, the mean, the variance or the median and even the confidence interval. See section 17.4 for a complete example of ensemble modeling and ensemble forecasting under future conditions.

It should be noted, however, that a model that performs better than another with a given calibration dataset may not necessarily be better for a different region or for projections under future climates. Weighting by predictive performance may not therefore be the best choice. In climatology, model averaging has often been done with equal weights for all models, or even by down-weighting models that make similar predictions (Tebaldi and Knutti, 2007; Knutti, 2010; Knutti et al., 2010). In ecology, the question of how to best build multi-model ensembles from different approaches is still wide open. Accordingly, the bagging and boosting modeling approaches described in Chapter 14 can also be considered as ensembles of models (here within a same technique, i.e. a form of MMI) and their relatively good predictive performance further suggests that adding an element of stochasticity to an ensemble modeling process can improve predictive performance.

In a similar way, as with random forests, the biomod2 package builds on this cross-validation approach to model the habitat suitability of species using a set of cross-validated models. Instead of using the cross-validation for testing purposes only, the models calibrated on the training dataset are kept to predict the species' habitat suitability. A single algorithm such as GLM or GAM thus provides, for instance, ten different predictions from ten different sets of training data (and evaluated on ten test datasets). If five different algorithms are used, it then leads to 50 different habitat suitability predictions for a single species. As suggested before, those predictions can then be averaged and the total variance extracted. It also means that for each given pixel in the landscape, 50 different probabilities of occurrence are predicted and can be analysed using a probability density function (Araújo and New, 2007; Thuiller, 2007; Thuiller et al., 2009).

We present here one example of ensemble modeling and model averaging across the five different modeling techniques that we have already described, namely: GLM, GAM, MARS, FDA, and RF. To show the variation between different cross-validation runs, we will run a repeated ($N = 20$) split-sample cross-validation (while keeping the prevalence constant). Each single model is run on the training partition and evaluated on the test partition using the area under the ROC curve (AUC; see Part IV). For each run, the five single modeling techniques are used to project the species distribution. This is repeated 20 times with different partitioning of the original data into two separate calibration (training) and evaluation (test) sets (split samples; see Part IV).

Since running the five models 20 times is rather time consuming, we will use the same dataset as before but at a lower spatial resolution. The data used here are available in the `biomod2` package.

```
> library(biomod2)
> ### Load species and environmental data at lower resolution
> ### (100x100km)
> DataSpecies <- read.csv(system.file("external/species/mammals_
table.csv", package = "biomod2"))
> require(raster)
> myExpl <- stack(system.file("external/bioclim/current/bio3.grd",
package = "biomod2"), system.file("external/bioclim/current/
bio4.grd",
package = "biomod2"), system.file("external/bioclim/current/
bio7.grd",
package = "biomod2"), system.file("external/bioclim/current/
bio11.grd",
package = "biomod2"), system.file("external/bioclim/current/
bio12.grd",
package = "biomod2"))
```

Extract the environmental layers for the presence and absence points:

```
> Env <- extract(myExpl, DataSpecies[, c(2, 3)])
```

Combine the presence–absence data and the extracted environmental data:

```
> DataSpecies <- cbind(DataSpecies, Env)
```

Load the required packages:

```
> library(MASS)
> library(mgcv)
> library(earth)
> library(rpart)
> library(mda)
```

```
> library(Hmisc)
> nCV <- 20   # Number of cross-validations
> nRow <- nrow(DataSpecies)
```

Create a dataframe to store the evaluation results for each model for each cross-validation:

```
> Test_results <- as.data.frame(matrix(0, ncol = nCV, nrow = 5,
dimnames = list(c("GLM", "GAM", "MARS", "FDA", "RF"), NULL)))
```

Create an array to store the predicted habitat suitability for each single model × cross-validation combination:

```
> Pred_results <- array(0, c(nRow, 5, nCV),
dimnames = list(seq(1:nRow),
c("GLM", "GAM", "MARS", "FDA", "RF"), seq(1:nCV)))
```

Loop through the cross-validation runs:

```
> for (i in 1:nCV) {
# separate the original data in one sub set for calibration
# and the other for evaluation.
a <- SampleMat2(ref = DataSpecies$VulpesVulpes,
ratio = 0.7)   # function from the biomod2 package
calib <- DataSpecies[a$calibration, ]
eval <- DataSpecies[a$evaluation, ]
### GLM ###
glmStart <- glm(VulpesVulpes ~ 1, data = calib,
family = binomial)
glm.formula <- makeFormula("VulpesVulpes", DataSpecies[,
c("bio3", "bio7", "bio11", "bio12")], "quadratic",
interaction.level = 1)
glmModAIC <- stepAIC(glmStart, glm.formula, data = calib,
direction = "both", trace = FALSE, k = 2,
control = glm.control(maxit = 100))
# prediction on the evaluation data and evaluation using the
# AUC approach
Pred_test <- predict(glmModAIC, eval, type = "response")
Test_results["GLM", i] <- somers2(Pred_test,
eval$VulpesVulpes)["C"]
# prediction on the total dataset
Pred_results[, "GLM", i] <- predict(glmModAIC, DataSpecies,
type = "response")
### GAM ###
gam_mgcv <- gam(VulpesVulpes ~ s(bio3) + s(bio7) + s(bio11) +
s(bio12), data = calib, family = "binomial")
# prediction on the evaluation data and evaluation using the
# AUC approach
Pred_test <- predict(gam_mgcv, eval, type = "response")
Test_results["GAM", i] <- somers2(Pred_test,
eval$VulpesVulpes)["C"]
# prediction on the total dataset
Pred_results[, "GAM", i] <- predict(gam_mgcv, DataSpecies,
```

```
type = "response")
### MARS ###
Mars_int2 <- earth(VulpesVulpes ~ 1 + bio3 + bio7 + bio11 +
bio12, data = calib, degree = 2,
glm = list(family = binomial))
# prediction on the evaluation data and evaluation using the
# AUC approach
Pred_test <- predict(Mars_int2, eval, type = "response")
Test_results["MARS", i] <- somers2(Pred_test,
eval$VulpesVulpes)["C"]
# prediction on the total dataset
Pred_results[, "MARS", i] <- predict(Mars_int2,
DataSpecies, type = "response")
### FDA ###
fda_mod <- fda(VulpesVulpes ~ 1 + bio3 + bio7 + bio11 + bio12,
data = calib, method = mars)
# prediction on the evaluation data and evaluation using the
# AUC approach
Pred_test <- predict(fda_mod, eval, type = "posterior")[, 2]
Test_results["FDA", i] <- somers2(Pred_test,
eval$VulpesVulpes)["C"]
# prediction on the total dataset
Pred_results[, "FDA", i] <- predict(fda_mod, DataSpecies[,
c("bio3", "bio7", "bio11", "bio12")],
type = "posterior")[, 2]
### Random Forest ###
RF_mod <- randomForest(x = calib[, c("bio3", "bio7", "bio11",
"bio12")], y = as.factor(calib$VulpesVulpes),
ntree = 1000, importance = TRUE)
# prediction on the evaluation data and evaluation using the
# AUC approach
Pred_test <- predict(RF_mod, eval, type = "prob")[, 2]
Test_results["RF", i] <- somers2(Pred_test,
eval$VulpesVulpes)["C"]
# prediction on the total dataset
Pred_results[, "RF", i] <- predict(RF_mod, DataSpecies,
type = "prob")[, 2]
}
```

Once the cross-validation runs are computed, we can analyse the variation between the different runs and across the models. Here we will use the ggplot2 package as an example.

```
> library(ggplot2)
> AUC <- unlist(Test_results)
> AUC <- as.data.frame(AUC)
> Test_results_ggplot <- cbind(AUC,
model = rep(rownames(Test_results), times = 20))
> p <- ggplot(Test_results_ggplot, aes(model, AUC))
> p + geom_boxplot()
```

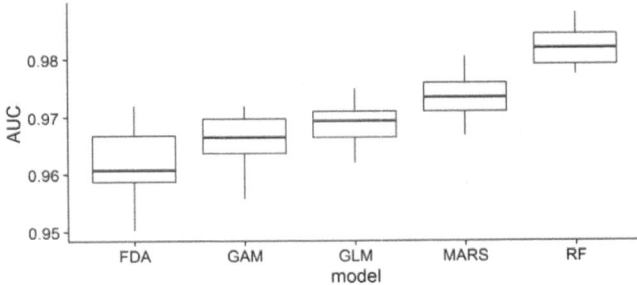

Figure 14.1 Variation in the area under the receiver-operating characteristic curve (AUC) among the cross-validation runs and between the different models.

Figure 14.1 highlights the variability among the cross-validation runs for a single model. The AUC varies by about 0.03 for FDA, which in turn reveals the largest variance compared to the other models, and a relatively lower predictive accuracy in this particular case. GLM and GAM show very similar mean AUC and similar variances, while MARS and RF show the best AUC values and generally a smaller variance between runs.

Let's now look at the variance in predicted habitat suitability for a given pixel (Figure 14.2). We chose a pixel where all the models predicted moderate habitat suitability. Using the gcplot2 package we can display the probability density function of the habitat suitability for that particular pixel.

```
> Pred_results[143, , ]   # select pixel 143 for which the species
# is predicted present by all models.

> HSM <- unlist(as.data.frame(Pred_results[143, , ]))
> HSM <- as.data.frame(HSM)
> Prob_density_ggplot <- cbind(HSM,
model = rep(rownames(Pred_results[143, , ]), times = 20))
> ggplot(Prob_density_ggplot, aes(x = HSM, fill = model)) +
geom_density(alpha = 0.5)
```

Interestingly, in Figure 14.2 there is a high variance in predicted habitat suitability for that particular pixel with MARS, GAM, and GLM predicting relatively high habitat suitability consistently across all cross-validation runs, while FDA and RF predict on average much lower (c. 0.2–0.3) suitability and with higher variance than the other three models. This variance is certainly linked to the classification algorithms behind both FDA and RF, which sometimes show sharp shifts in responses and are generally more sensitive to the underlying distribution of the data.

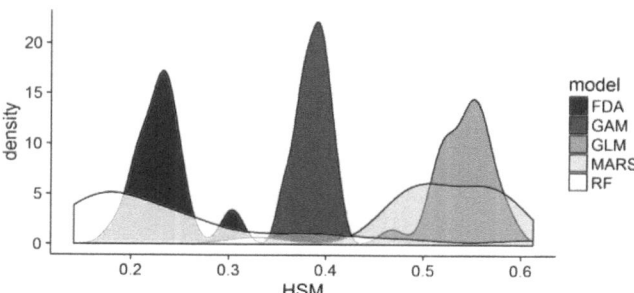

Figure 14.2 Probability density functions of the habitat suitability obtained for a given pixel across the different cross-validation runs, for each of the five modeling techniques. (*A black and white version of this figure will appear in some formats. For the color version, please refer to the plate section.*)

An ensemble model can now be used to map the habitat suitability of the species. We illustrate two different options, namely the mean and the median, together with the associated standard deviations that demonstrate the uncertainty of the ensemble predictions. Here the ensemble is generated in an unweighted form, and is therefore not weighted by predictive accuracy, as values for the latter are very high in all cases.

```
> ### Average prediction (mean and median) and variance
> Pred_total_mean <- apply(Pred_results, 1, mean)
> Pred_total_median <- apply(Pred_results, 1, median)
> Pred_total_sd <- apply(Pred_results, 1, sd)
```

Since the data have been modeled at a coarse (100 km) resolution, we will transform them back to a raster stack object to facilitate the representation.

```
> Obs <- rasterFromXYZ(DataSpecies[, c("X_WGS84", "Y_WGS84",
"VulpesVulpes")])
> Pred_total_mean_r <- rasterFromXYZ(cbind(DataSpecies[,
c("X_WGS84", "Y_WGS84")], Pred_total_mean))
> Pred_total_median_r <- rasterFromXYZ(cbind(DataSpecies[,
c("X_WGS84", "Y_WGS84")], Pred_total_median))
> Pred_total_sd_r <- rasterFromXYZ(cbind(DataSpecies[,
c("X_WGS84", "Y_WGS84")], Pred_total_sd))
> Out <- stack(Obs, Pred_total_mean_r, Pred_total_median_r,
Pred_total_sd_r)
> names(Out) <- c("Observed_Vulpes_vulpes",
"Ensemble_modeling_mean", "Ensemble_modeling_median",
"Ensemble_modeling_sd")
> plot(Out)
```

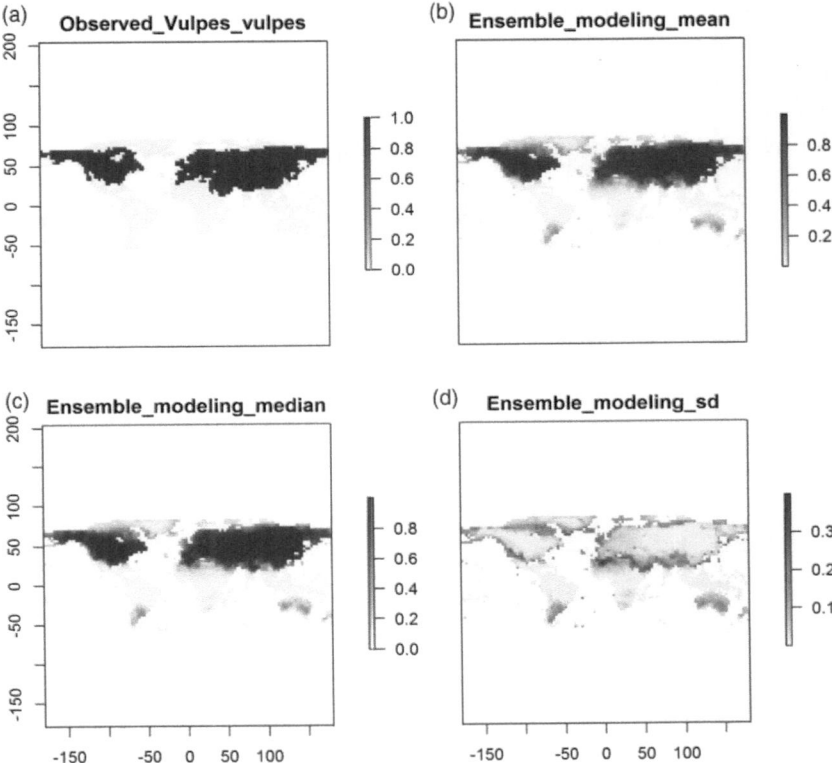

Figure 14.3 Observed presence and absence of *Vulpes vulpes* at the global scale (a), together with the two model averaging predictions (mean and median; b and c) and the ensemble modeling uncertainty (sd; d). (*A black and white version of this figure will appear in some formats. For the color version, please refer to the plate section.*)

As expected, the averaged habitat suitability values obtained using the "mean" and "median" approach are relatively similar since there are no extreme cases in the model (Figure 14.3). The ensemble modeling uncertainty map shows interesting patterns and highlights where the models and cross-validations show divergent results. Notably, North Africa, South America, Australia, or Greenland are places where the predictions from the ensemble approach diverge most significantly across models and cross-validation runs, and should be considered with care. They represent the edges of the observed distribution, where the model algorithms differ most in their predictions. We will see in Chapter 15 that there is greater divergence in the results across cross-validation runs and models when the models are used for forecasting future distributions (Thuiller, 2004; Pearson et al., 2006; Diniz et al., 2009; Buisson et al., 2010), demonstrating the need to account for more than one single modeling technique and more than one run when applying the HSM framework to predict future (or past) conditions (i.e. projections; see Part V).

PART IV · Evaluating Models: Errors and Uncertainty

In Part IV, we review and detail aspects of evaluating HSMs after their calibration, including the definition of the different types of errors and the types of metrics used to compare predictions with observations (Chapter 15), and the type of data needed to assess – as independently as possible – the predictive power of a model and the associated uncertainty estimates around the final predictions (Chapter 16). The data resampling approaches described in Chapter 16 can also be used to run sensitivity analyses and deliver uncertainty estimates, under the present or future conditions to which the model is applied (e.g. Buisson et al., 2010; Carvalho et al., 2011; Thibaud et al., 2014). These assessments are the ones commonly applied in the literature. A third type of assessment, less commonly used, is to assess the ecological realism of the models (e.g. shape of response curves, Elith et al., 2005; Merow et al., 2014) and associated predictions (Guisan et al., 2006a; Mateo et al., 2012; Thuiller et al., 2014b).

Model evaluation is a crucial step in any modeling exercise (Hastie et al., 2009), as it evaluates the capacity of a given model to reflect "truth," its inherent uncertainty in the parameter estimations, and whether it can be applied under other conditions. Evaluating HSMs is crucial if those models are to be used for conservation planning and biodiversity management (Vaughan and Ormerod, 2005; Guisan et al., 2013). Consequently, a sound evaluation primarily depends on the intended use of a model, and therefore on the aims of the underlying modeling study. For instance, estimating parameter uncertainty might be more relevant for making inferences about a given predictor variable, while prediction uncertainty might be more closely scrutinized when the model is used for purely predictive purposes (Hastie et al., 2009).

There is no universal definition of model evaluation, because a model can only be evaluated for an intended use (Rykiel, 1996). If a model can be used for different purposes, and if each of these requires a different evaluation criterion it may require as many different evaluation procedures as there are purposes for the model. The minimum requirement for a model that aims to explain patterns is that it is robust in its ecological assumptions, whereas a model that aims to predict distributions needs to be robust from a predictive capacity perspective (Mac Nally, 2000; Guisan et al., 2002). In the latter case, different evaluation criteria may also need to be taken into consideration, because – as we will see later in the chapter – different types of error (e.g. commission versus omission) may require different weights depending on the intended use of the model (e.g. for optimizing sampling or to define a reserve; Fielding and Bell, 1997; Fielding, 2002), as well as other quantitative aspects of the models (bias, parameter uncertainty). Evaluating the predictive capacity is the most common assessment used in HSM studies and we will focus much of this part on this type of evaluation. Evaluating bias and parameter uncertainty is not limited to HSMs, and we refer interested readers to the main textbooks on this topic (Claeskens and Hjort, 2008; Hastie et al., 2009). There are a number of interesting HSM papers on the questions surrounding uncertainty in parameter estimations and bias from the effects of spatial autocorrelation (Dormann, 2007b), sampling bias in observation records (Dennis et al., 1999; Phillips et al., 2009; Fithian et al., 2015), uncertainty in the gridded data used for deriving the environmental layers (McInerny and Purves, 2011), and collinearity in the set of predictors (Beale et al., 2010).

We have largely based our terminology on Hastie et al. (2009). We consider evaluation to be the overall procedure aiming to assess model strength, both in terms of the application of the model to predict independent data (predictive ability; "testing") and to the ecological meaningfulness of the underlying model (ecological realism). We therefore often use the term "evaluation" to refer to the entire procedure (i.e. including all the aspects evoked in Chapters 15 and 16), while we use the terms "validation" and "testing" to refer to more specific methodological procedures. In general, we tend to use the term "validation" for procedures that are used to optimize individual model building (e.g. calculating thresholds, pruning trees, etc.), or to compare different sets of models (using resampling procedures, such as cross-validation, repeated split sample, etc.; i.e. validation subsets or

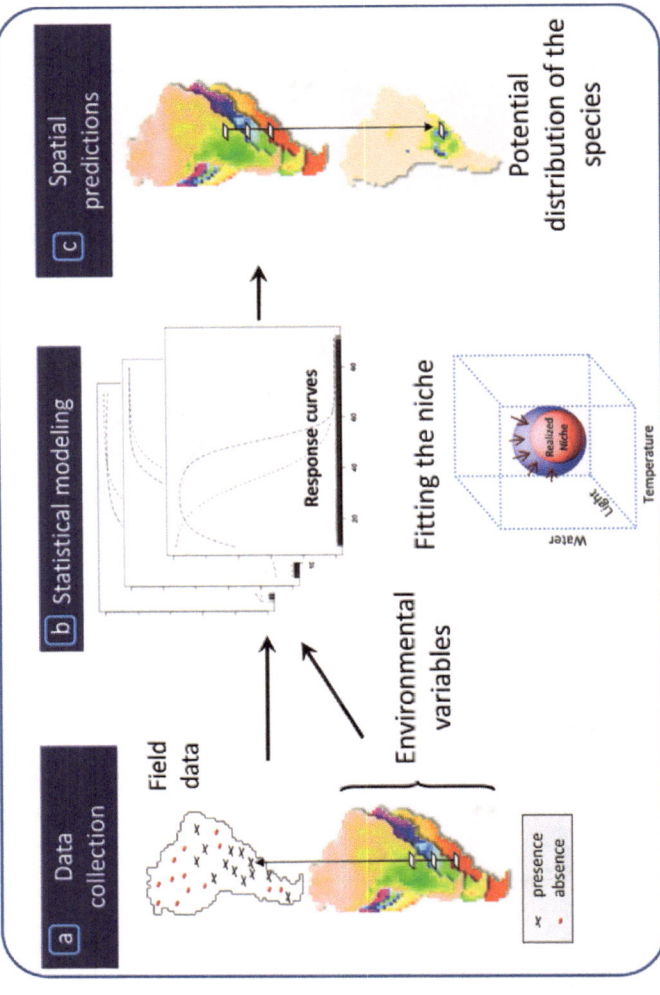

Figure 4.1 Principle of habitat suitability modeling, illustrated on a study area representing South America. (a) Field observations are collected and geo-referenced and the attributes of a set of environmental maps covering the area are extracted for each of them (see Part II). (b) Species observations are related to the environmental values using statistical approaches (Part III) to fit species response curves to each environmental predictor, in order to quantify the envelope of suitable habitat conditions, which represents the realized environmental niche of the species or of part of it if the study area does not encompass all the environmental conditions suitable for the species (see Chapter 3; the arrows suggest competitive exclusion of part of the fundamental niche from other species). (c) The fitted habitat suitability model is then used to combine the initial environmental maps and come up with a spatial projection of the model in geographic space (Part V), corresponding to the species' potential distribution. (*A black and white version of this figure will appear in some formats.*)

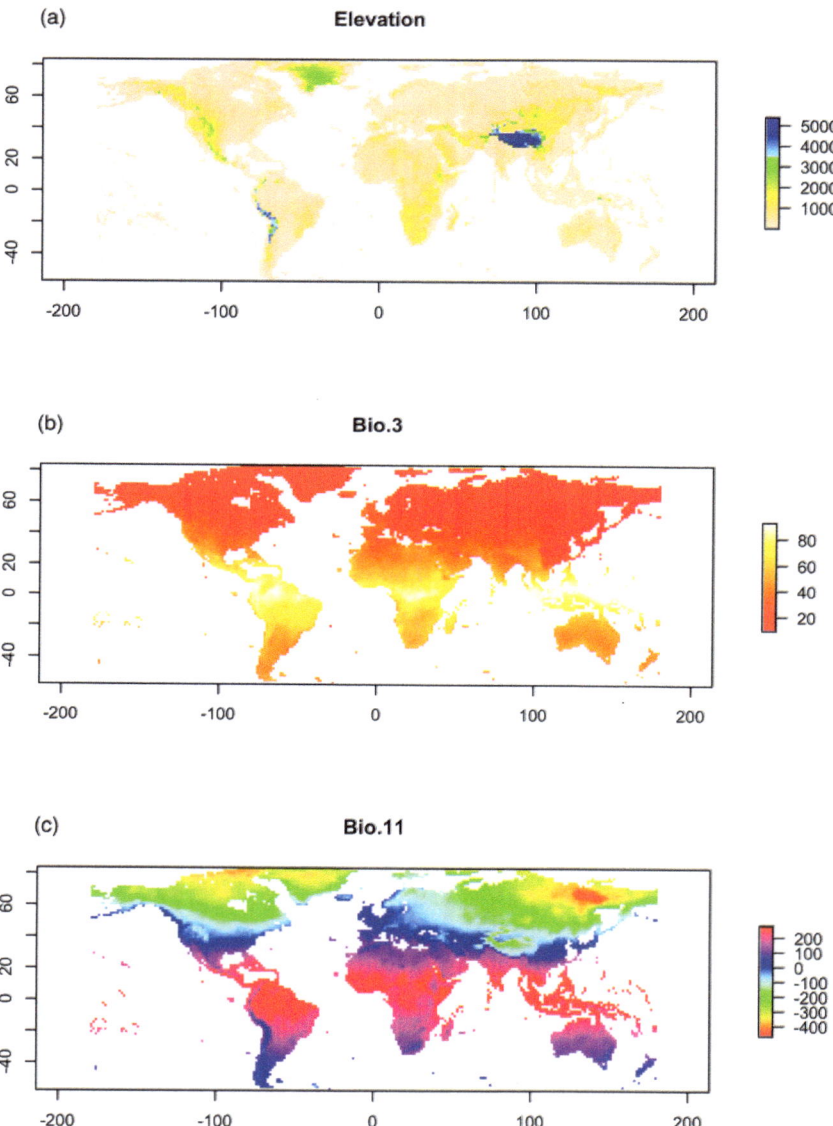

Figure 6.1 Illustration of three maps using the raster package: (a) elev1 = altitude, (b) bio3 = isothermality (×100), and (c) bio11 = mean temperature of the coldest quarter (×10). (*A black and white version of this figure will appear in some formats.*)

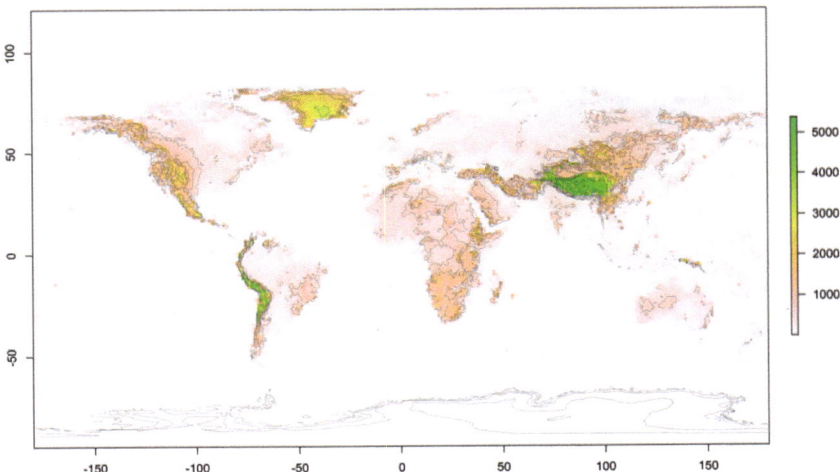

Figure 6.3 Elevation contours displayed over digital elevation model. (*A black and white version of this figure will appear in some formats.*)

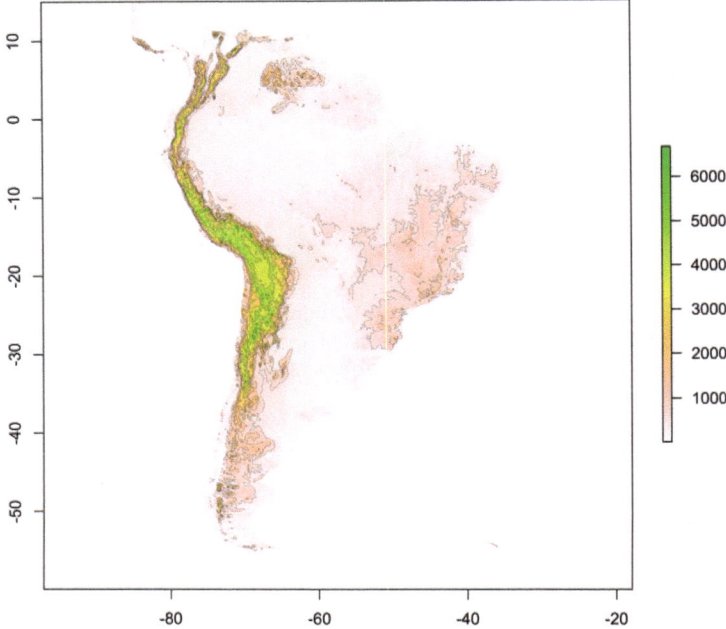

Figure 6.4 Contours displayed over a digital elevation model of South America. (*A black and white version of this figure will appear in some formats.*)

(a)   Modeled mean temperature of the coldest quarter

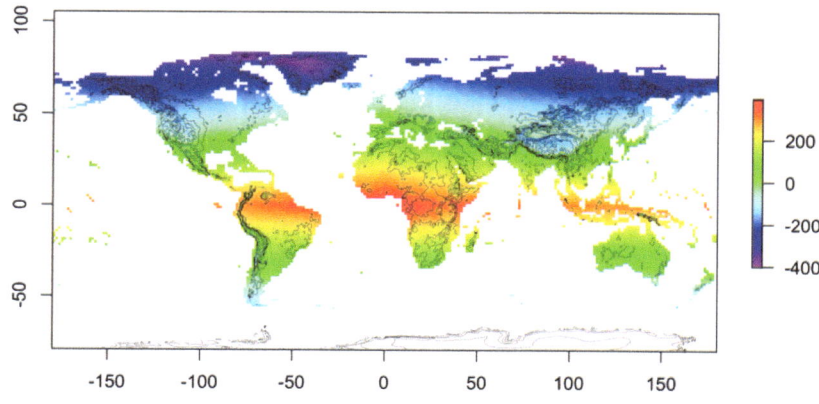

(b)

Difference modeled from observed mean temperature of the coldest quarter

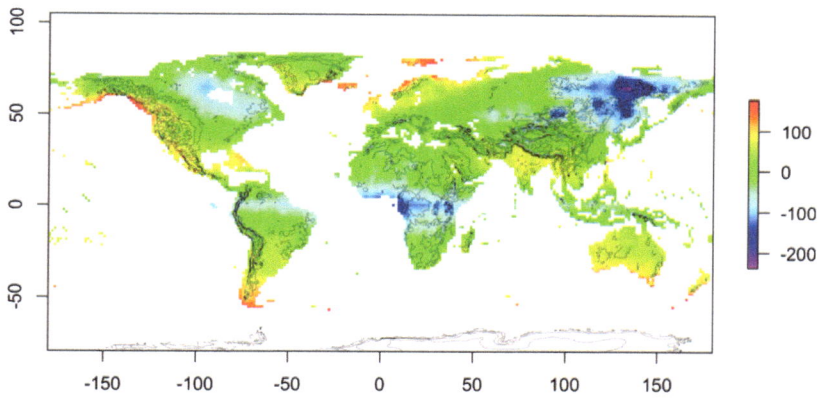

Figure 6.5 Elevation contours displayed over (a) modeled mean temperature of the coldest quarter and (b) difference between modeled and observed temperatures. (*A black and white version of this figure will appear in some formats.*)

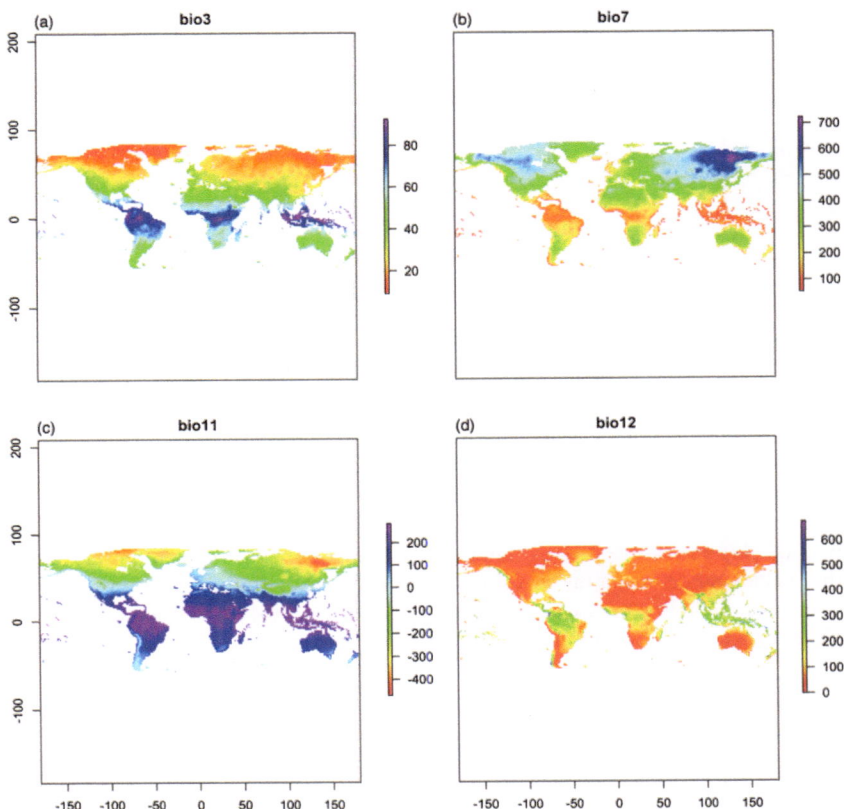

Figure 6.8 Visualization of all elements in a stacked grid. (*A black and white version of this figure will appear in some formats.*)

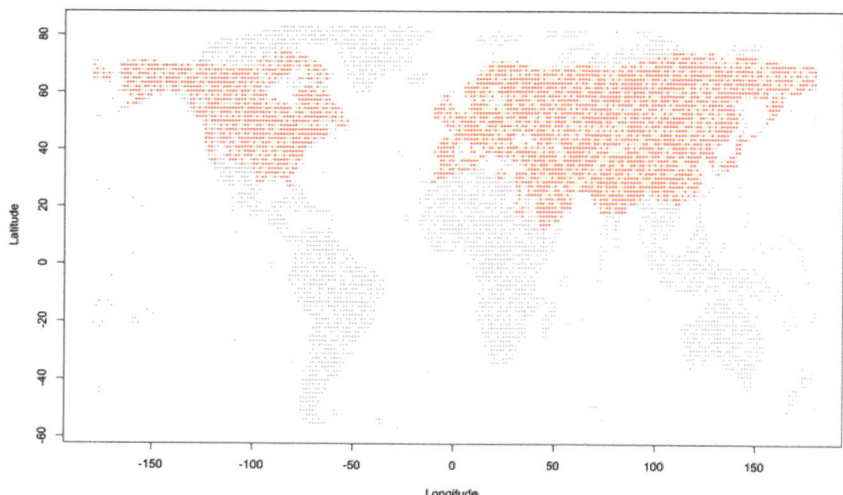

Figure 6.10 Distribution of presence and absence points sampled from a global range map of *Vulpes vulpes*. Presence calibration (red) and evaluation (green) points reflect the species' range, while the absence points of the calibration (darker gray) and the evaluation (lighter gray) datasets reflect areas of absence for the species. (*A black and white version of this figure will appear in some formats.*)

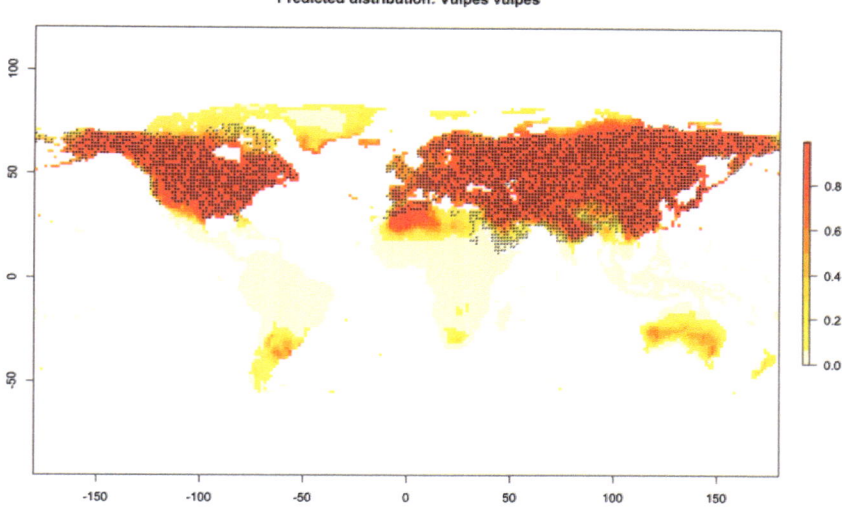

Figure 6.11 Map of the projected global distribution of *Vulpes vulpes* contrasted with the observed distribution points used to calibrate the simple four-parameter GLM model. (*A black and white version of this figure will appear in some formats.*)

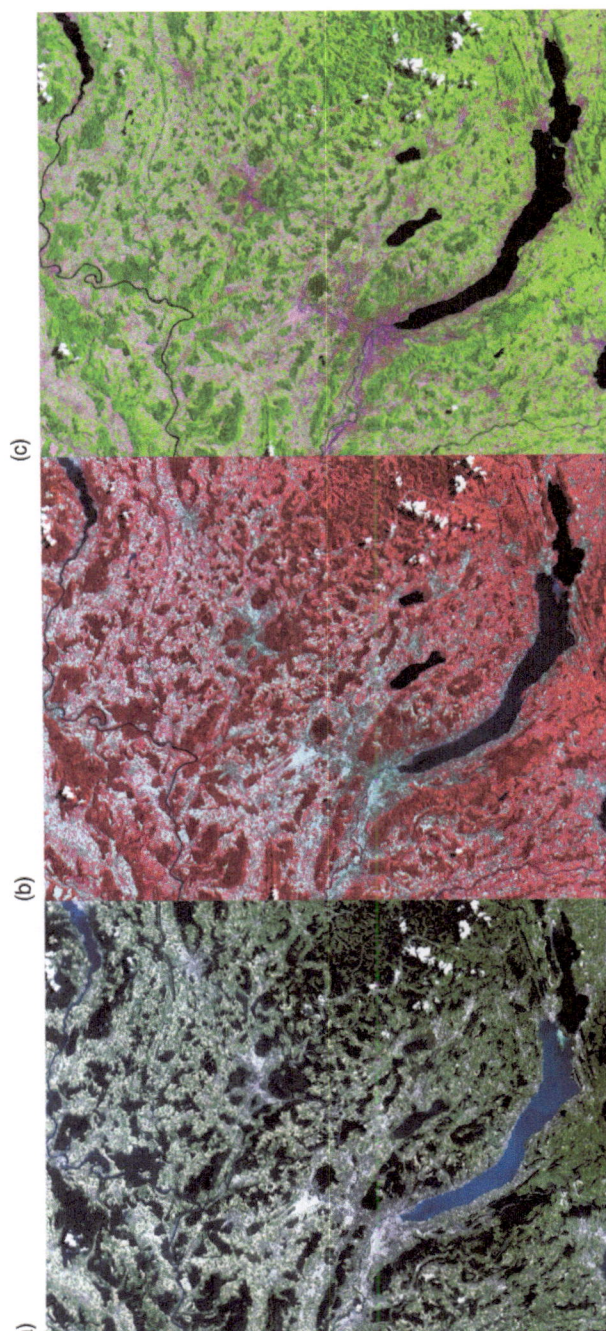

(a) (b) (c)

Figure 6.12 Illustration of three-color composites from Landsat bands of the area of Zurich in Switzerland. (*A black and white version of this figure will appear in some formats.*)

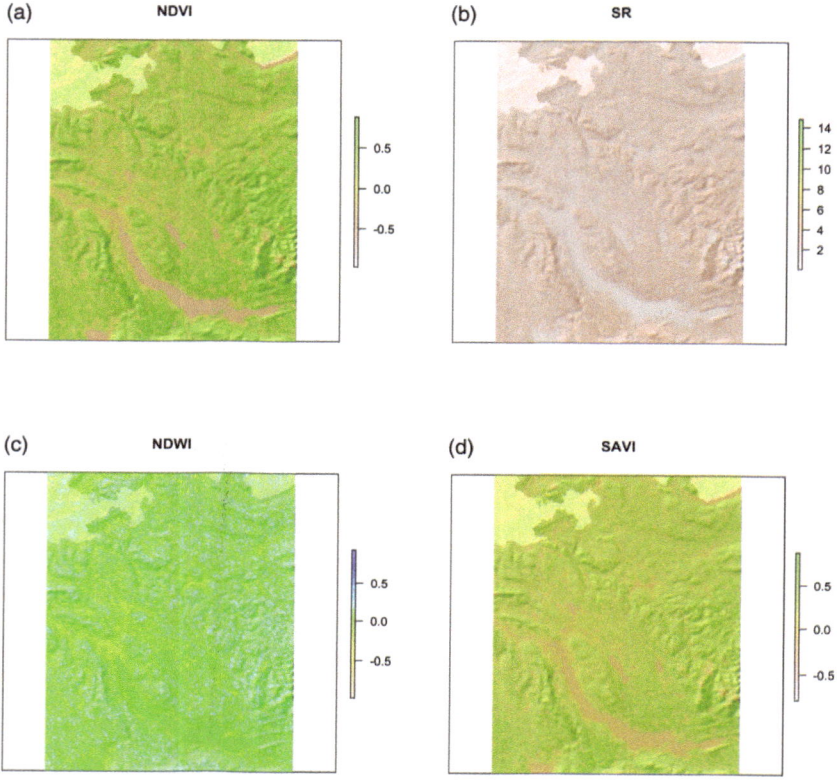

Figure 6.13 Visualization of four band–derived indices: (a) NDVI, (b) SR, (c) NDWI, and (d) SAVI. (*A black and white version of this figure will appear in some formats.*)

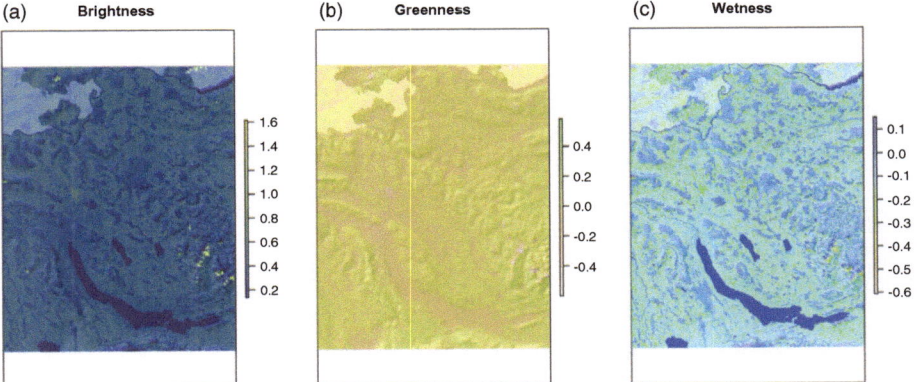

Figure 6.14 Visualization of the three tasseled cap indices: (a) brightness, (b) greenness, and (c) wetness. (*A black and white version of this figure will appear in some formats.*)

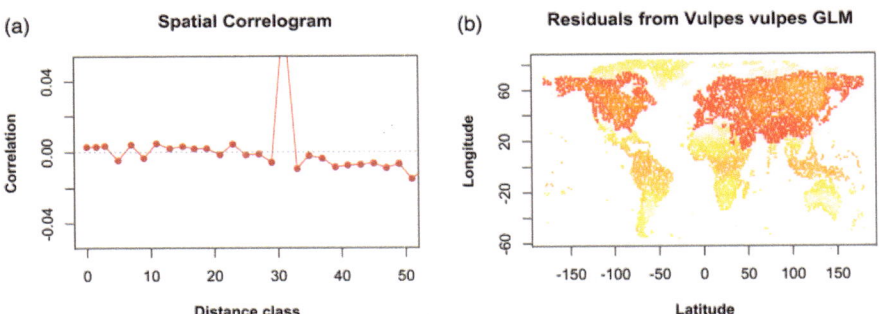

Figure 7.1 Spatial correlation (a) and spatial patterns (b) of model residuals for the *Vulpes vulpes* stepwise-optimized GLM model. The correlogram reveals a low correlation at a short distance. (*A black and white version of this figure will appear in some formats.*)

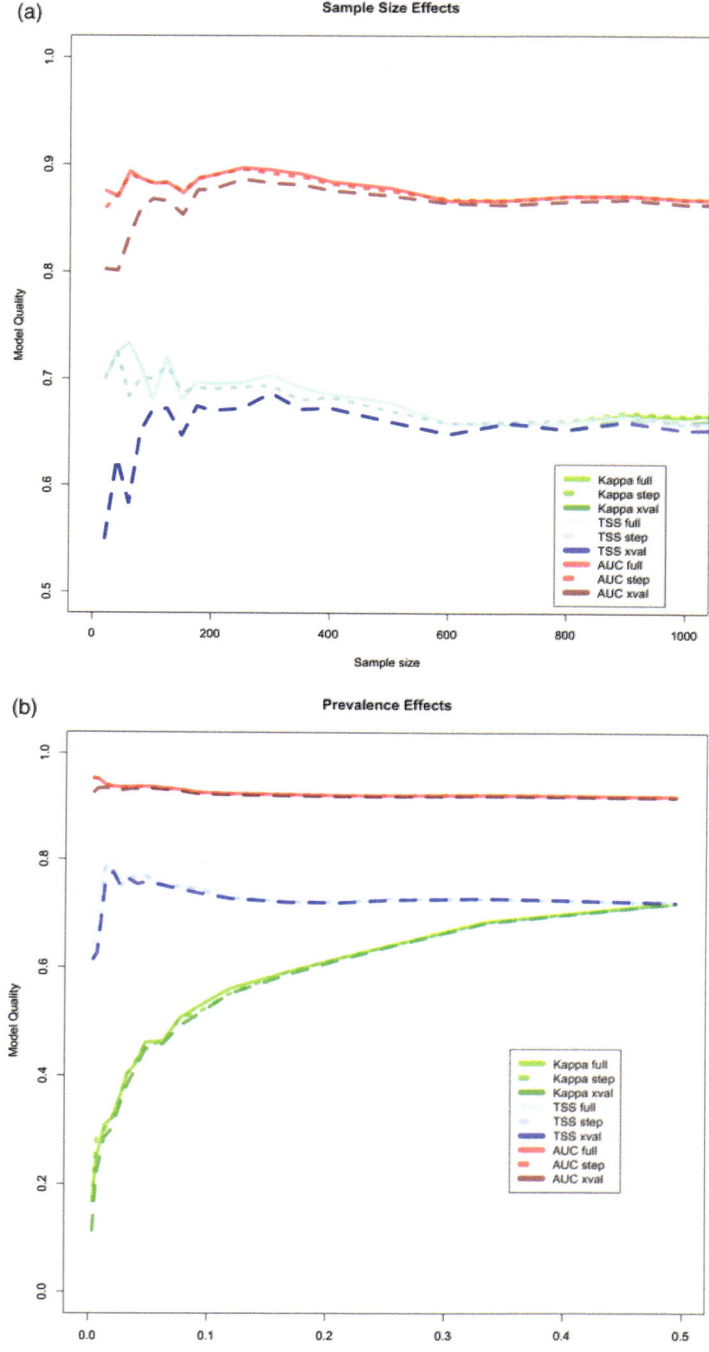

Figure 7.2 (a) Sample size and (b) prevalence effects on model accuracy in the global *Vulpes vulpes* dataset. Low prevalence has a strong effect on cross-validated Kappa and TSS accuracies. (*A black and white version of this figure will appear in some formats.*)

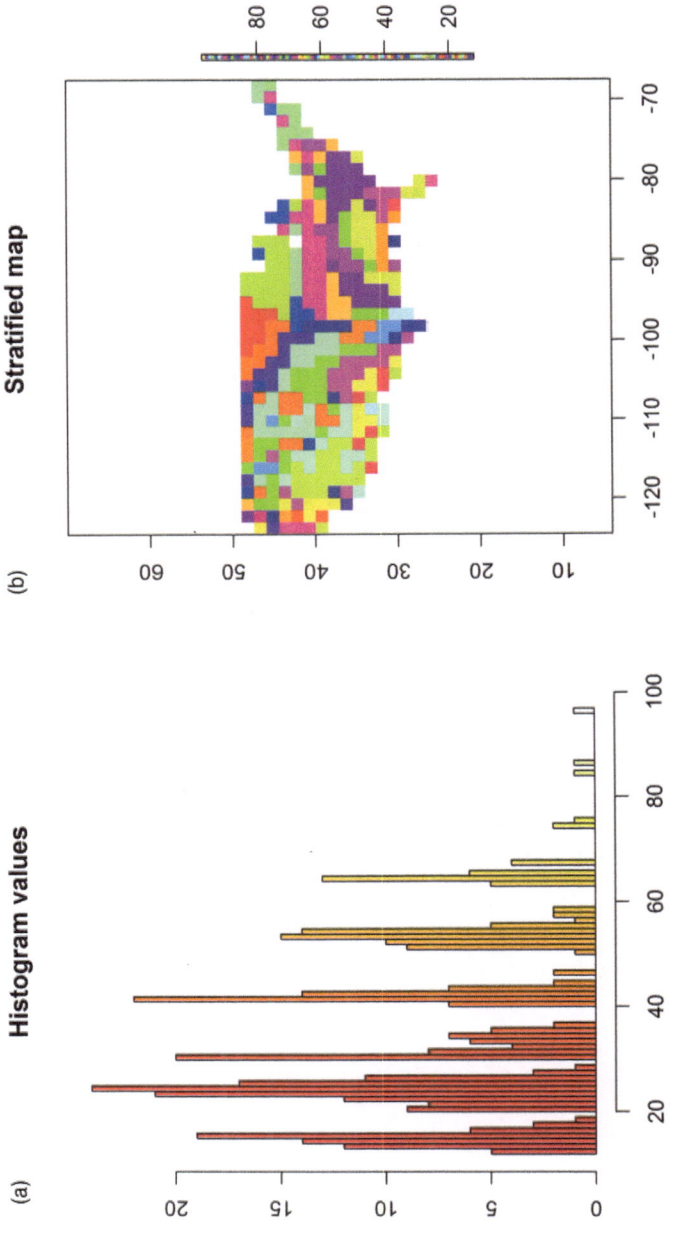

Figure 7.3 (a) Histogram of the frequencies of the classes, and (b) stratification map with the rainbow color scheme used to identify the strata from the environmental stratification of the study region. (*A black and white version of this figure will appear in some formats.*)

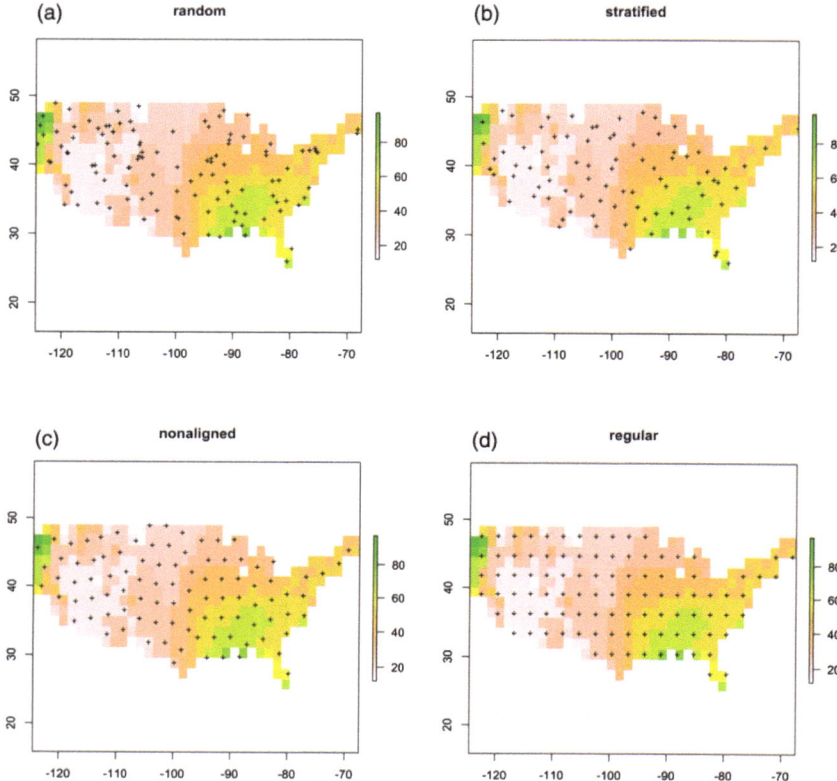

Figure 7.4 Spatial sampling designs: (a) random, (b) stratified, (c) non-aligned, and (d) regular. The color shades of the background indicate the identifier of the strata from the environmental stratification of the study region. (*A black and white version of this figure will appear in some formats.*)

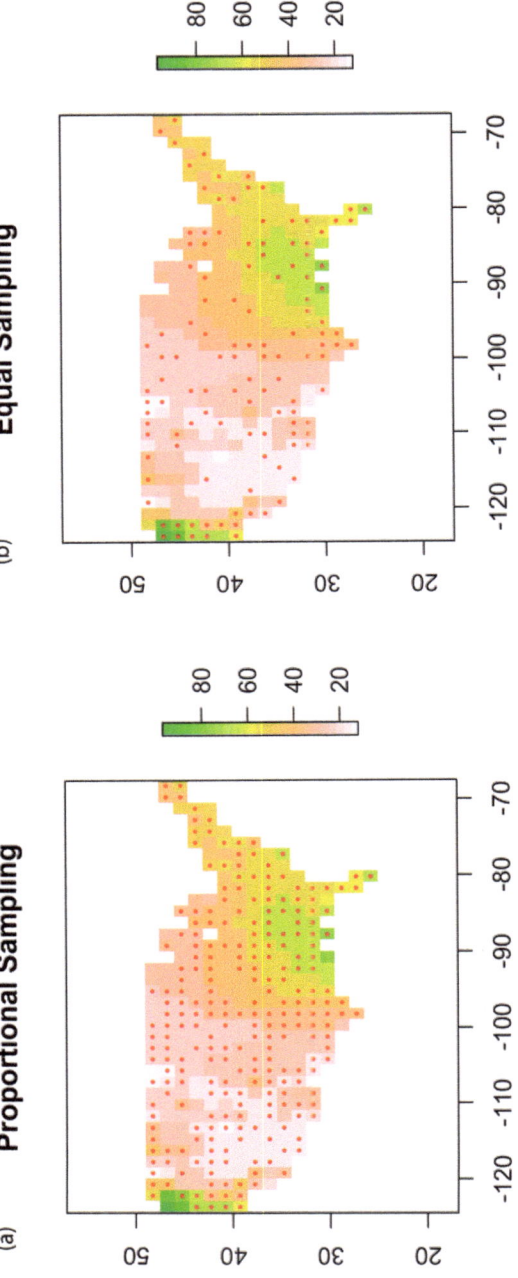

Figure 7.5 Sample points according to our random, environmentally stratified design. The color shades of the background indicate the identifiers of the strata from the environmental stratification of the study region. (*A black and white version of this figure will appear in some formats.*)

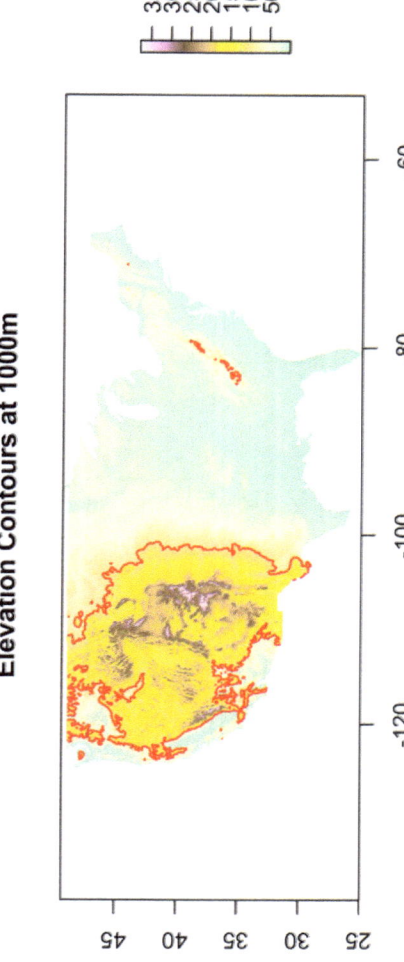

Figure 7.7 Elevation contour lines (1000 m) plotted in the study area. (*A black and white version of this figure will appear in some formats.*)

(a) Occurrence overview

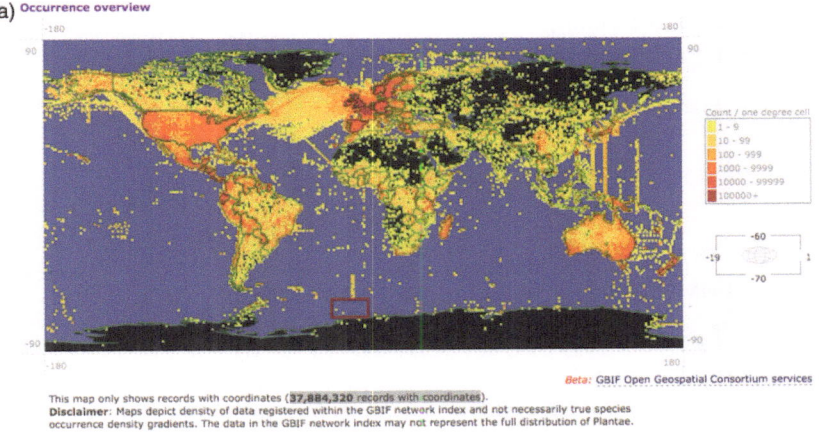

This map only shows records with coordinates (**37,884,320 records with coordinates**).
Disclaimer: Maps depict density of data registered within the GBIF network index and not necessarily true species occurrence density gradients. The data in the GBIF network index may not represent the full distribution of Plantae.

Beta: GBIF Open Geospatial Consortium services

(b) Map of results

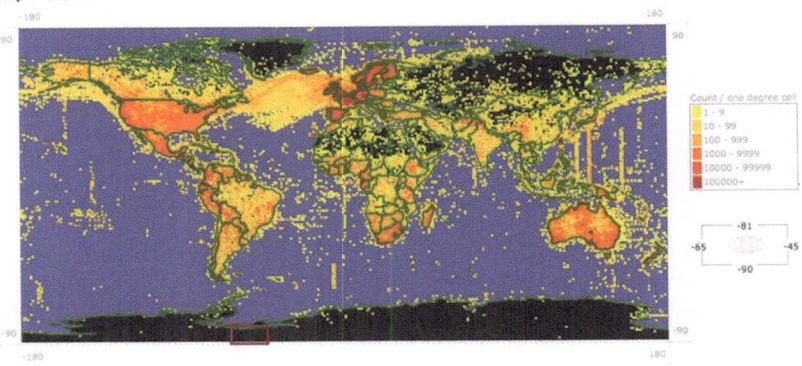

▸ Only records with no geospatial issues are plotted
▸ Your search returned **77,855,436** occurrences with coordinates.

Figure 7.9 Sampling intensity of plant samples in GBIF: (a) 37.8 Mio records as of October 26, 2009; (b) 77.8 Mio records as of February 12, 2013. It becomes evident that the collection intensity varies considerably in space as it is measured by density per 1 × 1 degree cells, which will cause bias in the predictions when the data is used without resampling. Over the period 2009 to 2013 the spatial sampling bias has not decreased visibly. (*A black and white version of this figure will appear in some formats.*)

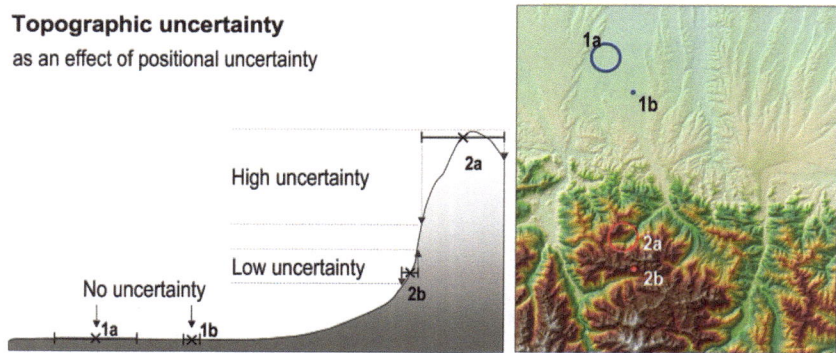

Figure 8.2 Relationship between positional and topographic uncertainty. Positional uncertainty matters especially where there are strong topographic and/or environmental gradients (2a/2b), and do not significantly influence our modeling where there are no such steep gradients visible (1a/b). (*A black and white version of this figure will appear in some formats.*)

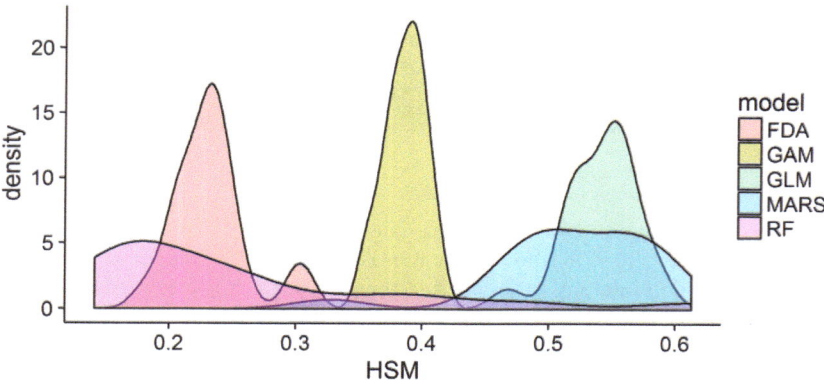

Figure 14.2 Probability density functions of the habitat suitability obtained for a given pixel across the different cross-validation runs, for each of the five modeling techniques. (*A black and white version of this figure will appear in some formats.*)

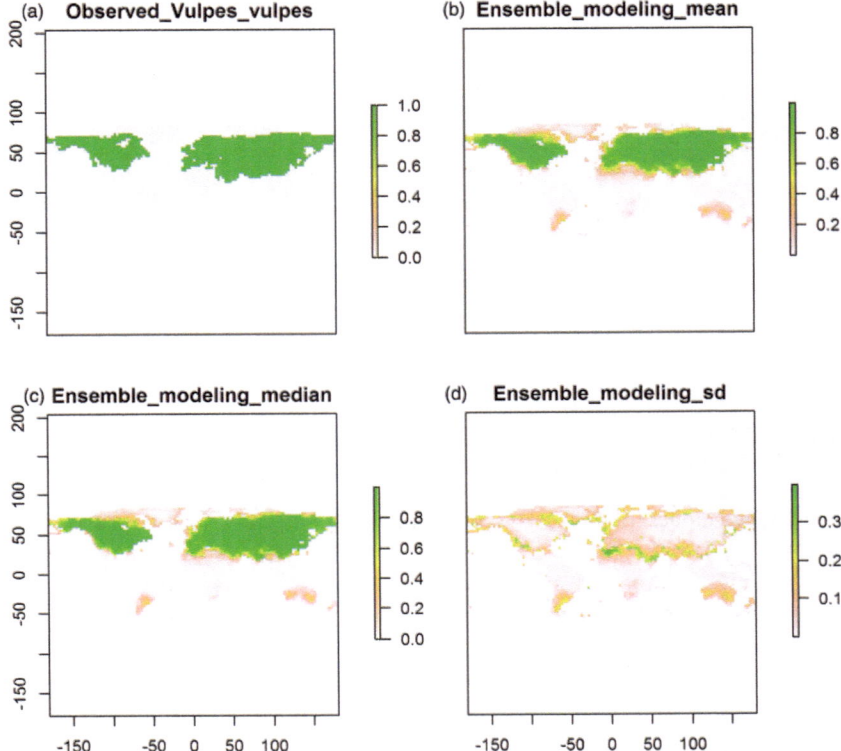

Figure 14.3 Observed presence and absence of *Vulpes vulpes* at the global scale (a), together with the two model averaging predictions (mean and median; b and c) and the ensemble modeling uncertainty (sd; d). (*A black and white version of this figure will appear in some formats.*)

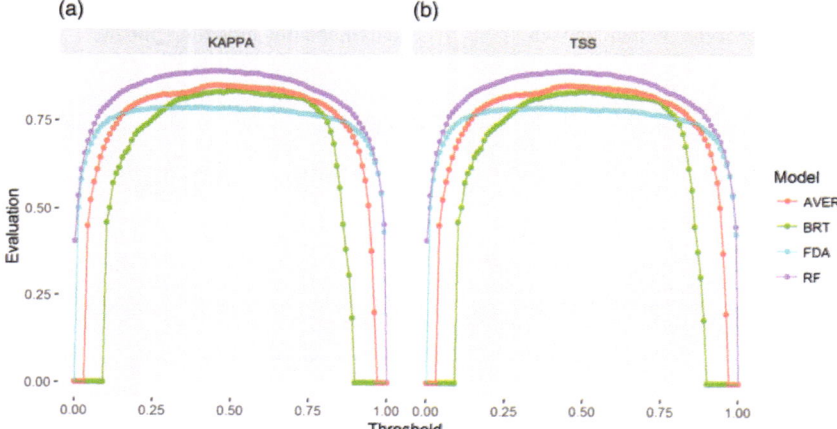

Figure 15.7 (a) Kappa and (b) TSS plots for species *Vulpes vulpes* modeled and predicted using three modeling techniques and their averaged ensemble: random forest (RF), flexible discriminant analysis (FDA) and boosting regression trees (BRT) and average model (AVER). (*A black and white version of this figure will appear in some formats.*)

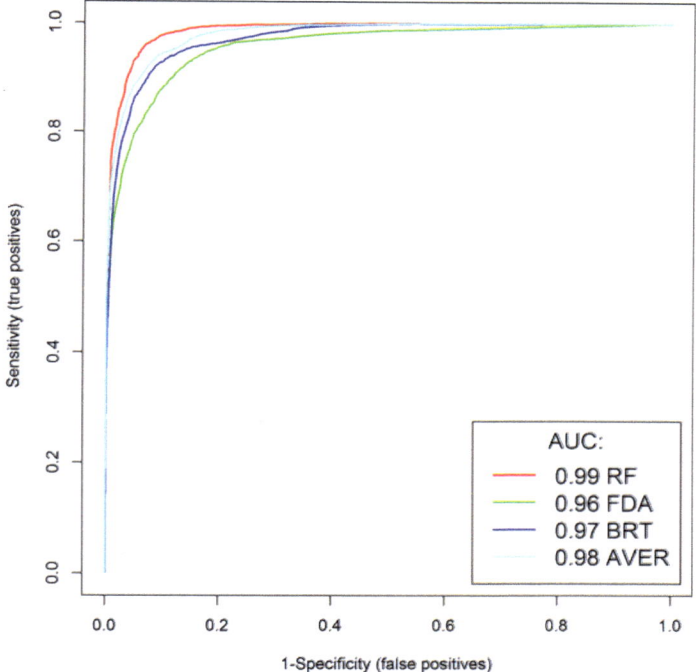

Figure 15.9 AUC ROC plots for the species *Vulpes vulpes* modeled and predicted using three modeling techniques and their averaged ensemble: random forest (RF), flexible discriminant analysis (FDA) and boosting regression trees (BRT) and average model (AVER). (*A black and white version of this figure will appear in some formats.*)

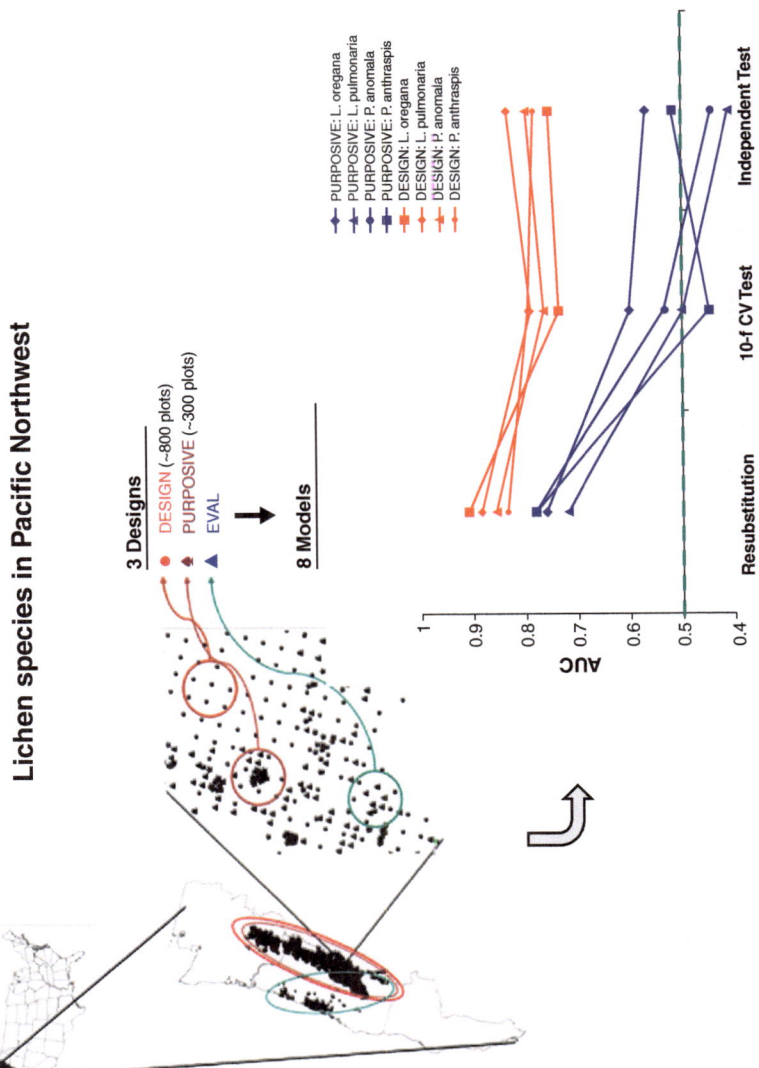

Figure 16.4 The use of three different types of samples to illustrate the importance of using partially independent (cross-validation, Section 16.1) and independent data (Section 16.2) in addition to internal resubstitution (Section 16.3) to evaluate model predictive power. The example is a modified version of that used by Edwards et al. (2006), with permission. (*A black and white version of this figure will appear in some formats.*)

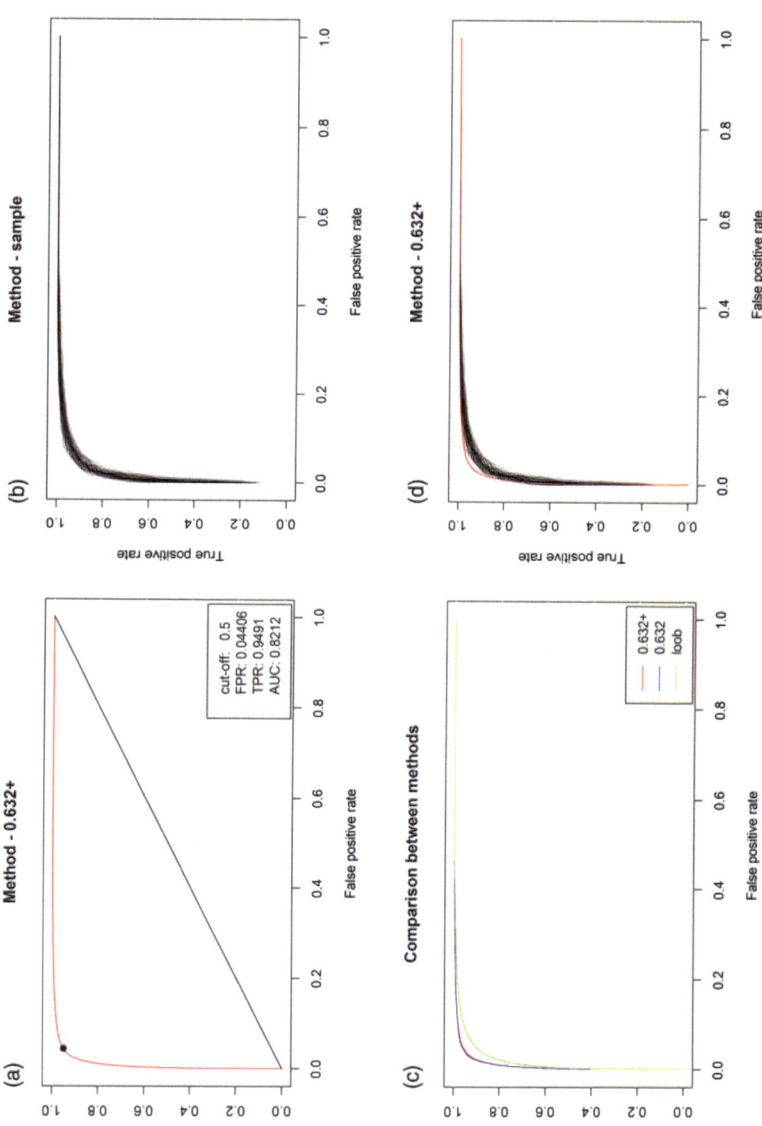

Figure 16.9 Plot of the Daim object generated by the Daim() function corresponding to ROC curves for various bootstrap evaluation methods. (a) The method "0.632+" discussed in the main text, (b) all the bootstrap samples, (c) a comparison of methods bottom left, and (d) all the bootstrap samples plotted together with the one of the"0.632+" method bottom right. FPR: false positive (presence) rate (1 – specificity), TPR: true positive (presence) rate (sensitivity), loob: leave-one-out bootstrap. See Efron and Tibshirani (1993), Efron and Gong (1983), and Efron and Tibshirani (1997). (*A black and white version of this figure will appear in some formats.*)

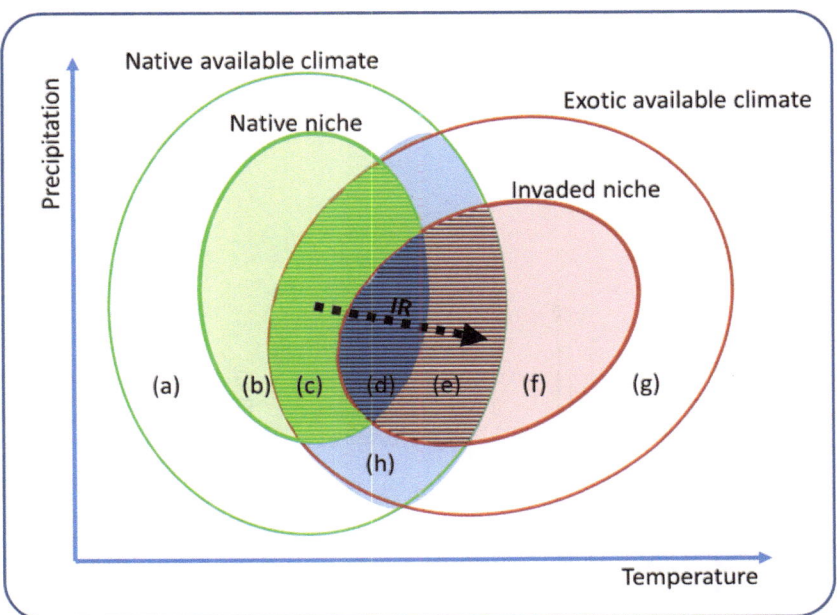

Figure 17.1 Schematic two-dimensional representation of the indices of niche change (unfilling, stability and expansion) presented in Broennimann et al. (2012) (see definitions in Guisan et al., 2014, Box 3). Solid thin lines show the density of available environments (see Box 4 in Guisan et al., 2014) in the native range (in green) and in the invaded range (in red). The gray area shows the most frequent environments common to both ranges (i.e. analog environments). The green and red thick lines show respectively the native and the invaded niches. Niche unfilling (U), stability (Se) and expansion (E) are shown respectively with green, blue and red hatched surfaces inside analog environments. The definition of a niche shift using the change of niche centroid only (inertia ratio, IR) is shown with a thick dotted arrow. In this context, the lower-case letters represent similar features in both graphs. (a) Available conditions in the native range, outside of the native niche and non-analog to the invaded range. (b) Conditions inside of the native niche but non-analog to the invaded range. (c) Unfilling, i.e. conditions inside of the native niche but outside the invaded niche, possibly due to recent introduction combined with ongoing dispersal of the exotic species, which should ultimately fill these conditions. (d) Niche stability, i.e. conditions filled in both native and invaded range. (e) Niche expansion, i.e. conditions inside the invaded niche but outside the native one, due to ecological or evolutionary change in the invaded range. (f) Conditions inside of the invasive niche but non-analog to the native range. (g) Available conditions in the invaded range but outside of the invasive niche and non-analog to the native range. (h) Analog conditions between the native and invaded ranges. Figure from Guisan et al. (2014), with permission. (*A black and white version of this figure will appear in some formats.*)

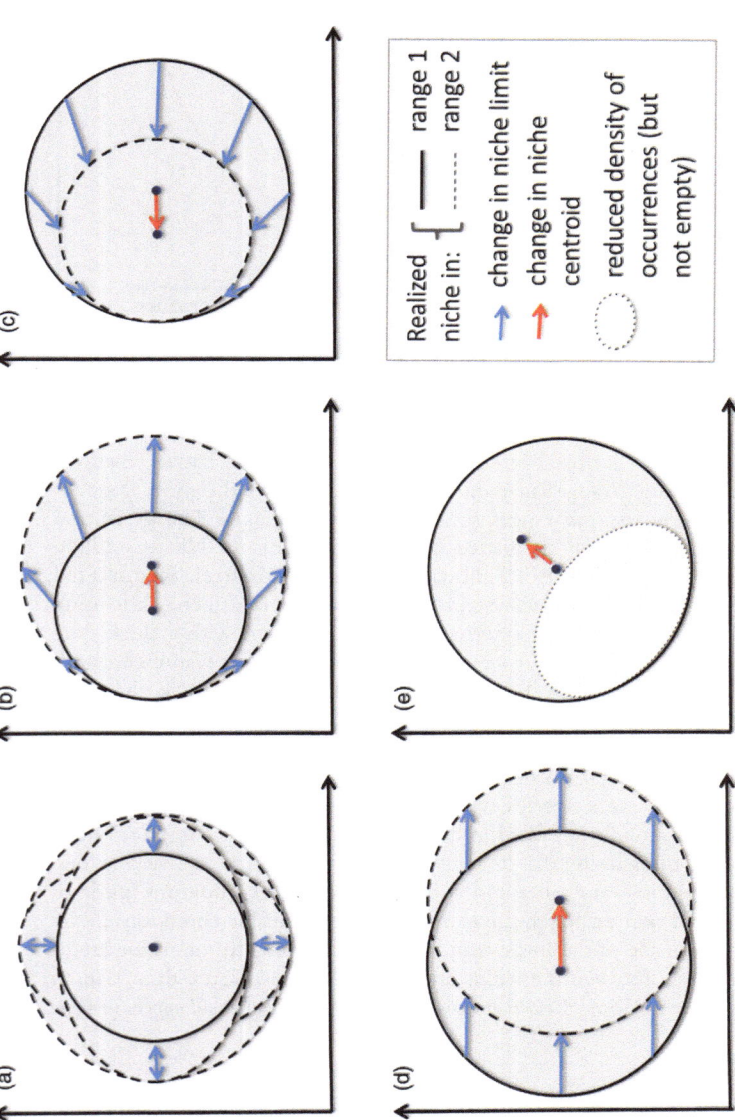

Figure 17.4 Theoretical scenarios of realized niche changes in space (e.g. following invasions) or time (e.g. under climate change). Change in: (i) the niche envelope (expansion or contraction) without change of the niche centroid, due to symmetric niche change, i.e. in two opposite or all directions in climatic space (a); (ii) the niche centroid with directional expansion (b) shrinkage (c) or displacement (d) or part of or the whole niche envelope, or (iii) the niche centroid only, due to a change of the density of occurrences within the same niche envelope in climatic space (e). The latter case would result in stability (no change) in Figure 17.1. Observed changes are likely to be combinations of these cases. Figure from Guisan et al. (2014), with permission. (*A black and white version of this figure will appear in some formats.*)

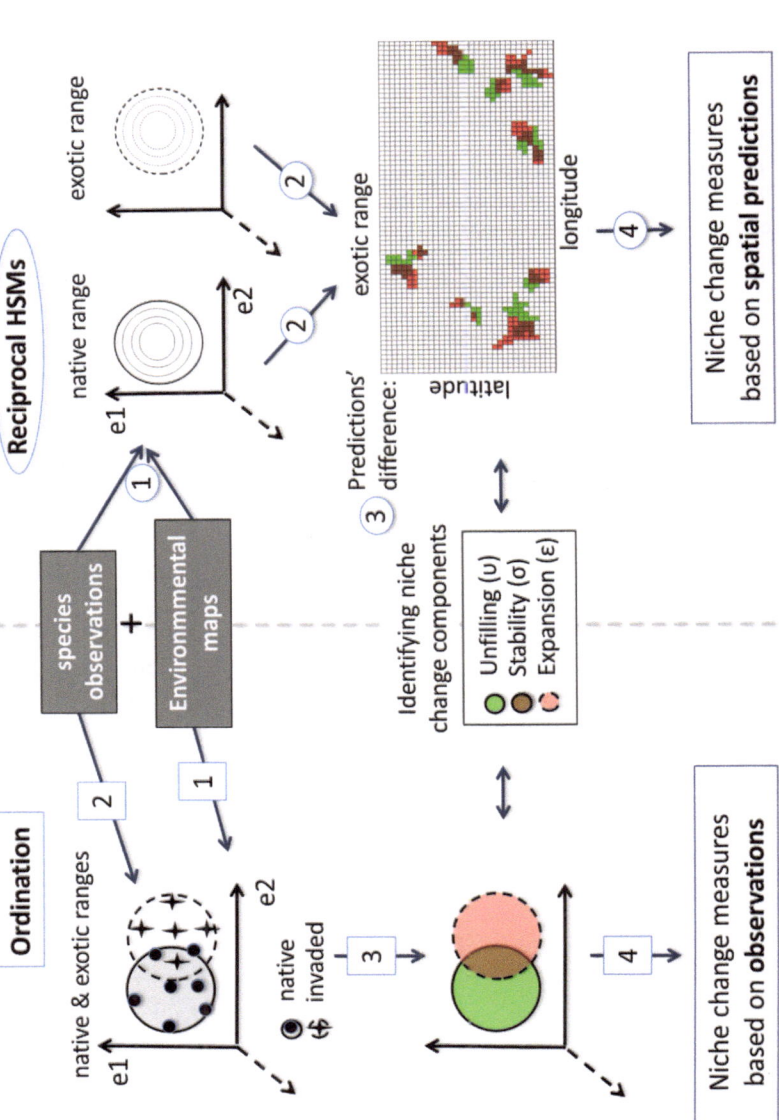

Figure 17.6 The two approaches commonly used to quantify niche changes between ranges). Ordination is based only on the observations, whereas HSM is based only on the predictions (see reference 22 and Box 1 in Guisan et al. 2014). The steps for ordination are (square numbers): 1. Definition of the reduced multidimensional environmental space; 2. Plotting the observations from each range in this space; 3. Comparing the niche defined from observations in each range; 4. Calculating the niche change metrics (see Box 3 in Guisan et al. 2014). The steps for HSMs are: 1. Fitting HSMs by relating field observations to environmental variables; 2. Projecting the HSMs in geographic space; 3. Computing differences in the projections; 4. Calculating the niche change metrics. See Guisan et al. (2014) for discussion of the respective strengths and weaknesses of the two approaches. Figure from Guisan et al. (2014), with permission. (*A black and white version of this figure will appear in some formats.*)

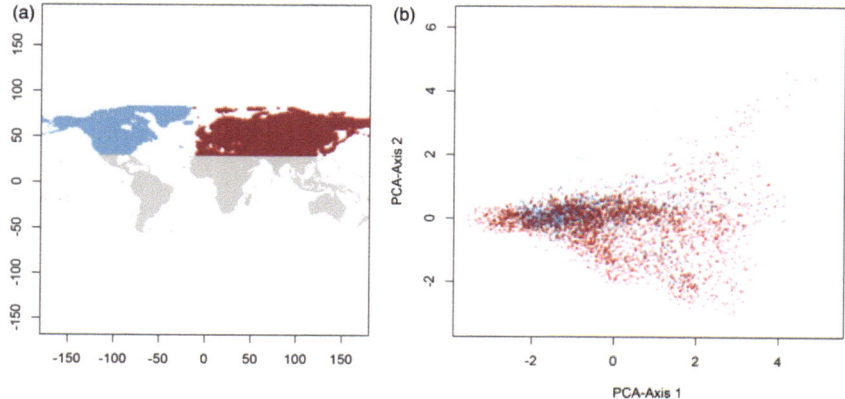

Figure 17.10 Differences in the ecological space of Vulpes vulpes between Old (red) and New (blue) World climates as mapped in (a) the geographic space and (b) the PCA space based on four bioclim variables. (*A black and white version of this figure will appear in some formats.*)

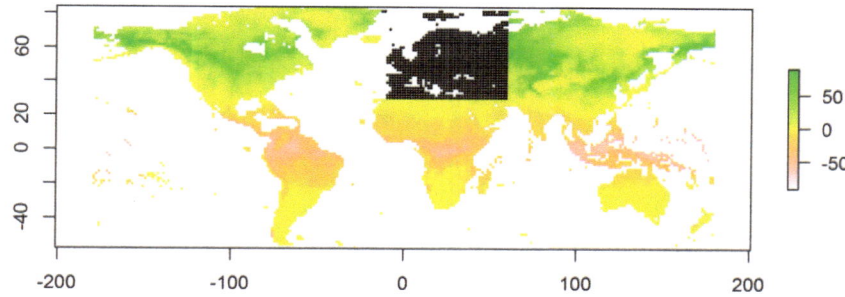

Figure 17.11 Multivariate environmental similarity surface in respect to European calibration data points for *Vulpes vulpes* in Europe. (*A black and white version of this figure will appear in some formats.*)

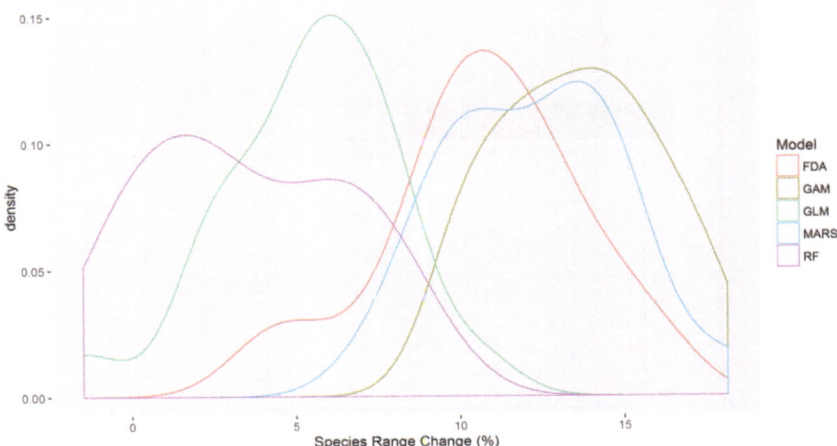

Figure 17.21 Density plot representing the variation in modeled species range for *Vulpes vulpes* for each technique due to the different repeated split sampling runs. (*A black and white version of this figure will appear in some formats.*)

Figure 19.1 Protea laurifolia flower and leaves. (Photo from www.flickr.com/photos/flowcomm/.) (*A black and white version of this figure will appear in some formats.*)

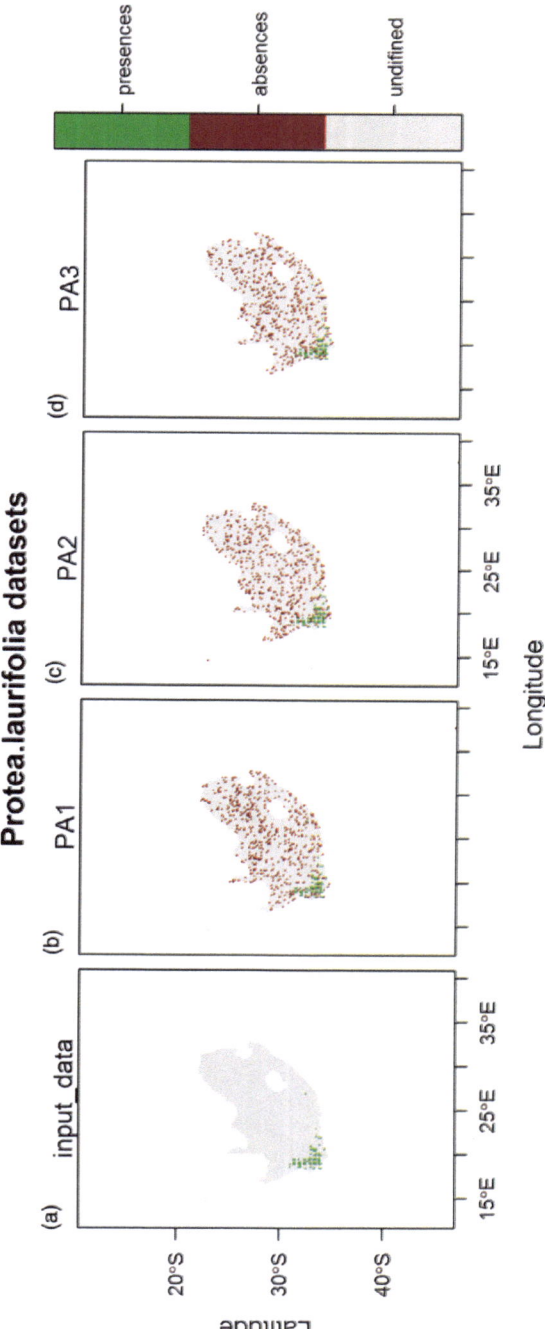

Figure 19.4 Plot of the species distribution (occurrences) and three selected sets of pseudo–absences. (*A black and white version of this figure will appear in some formats.*)

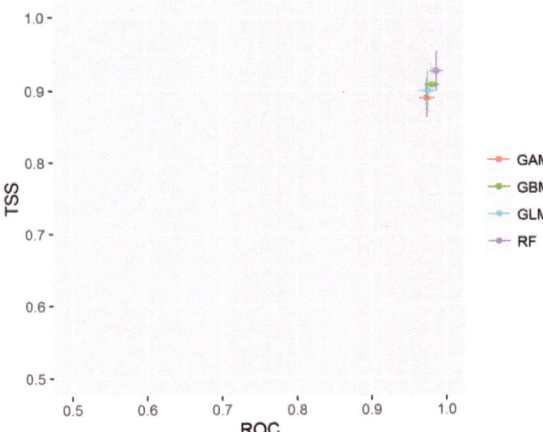

Figure 19.5 Plot of the mean of the model evaluation scores (by algorithms) according to two different evaluation metrics, ROC (AUC) and TSS. (*A black and white version of this figure will appear in some formats.*)

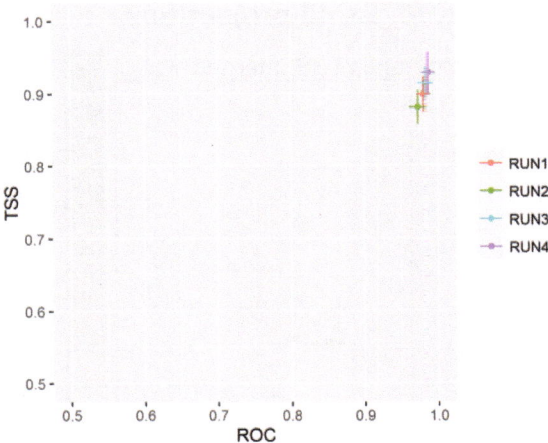

Figure 19.6 Plot of the mean of the model evaluation scores (by cross-validation) according to two different evaluation metrics, ROC (AUC) and TSS. (*A black and white version of this figure will appear in some formats.*)

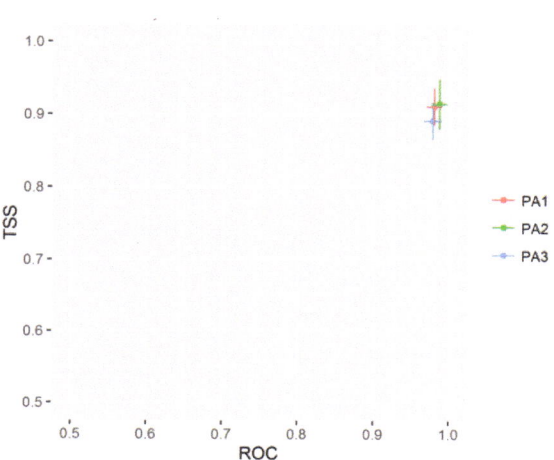

Figure 19.7 Plot of the mean of the model evaluation scores (by dataset) according to two different evaluation metrics, ROC (AUC) and TSS. (*A black and white version of this figure will appear in some formats.*)

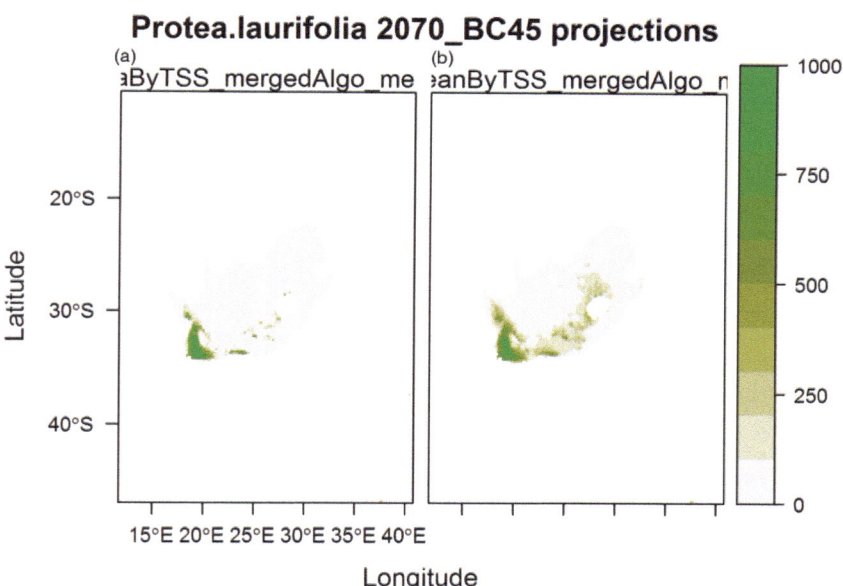

Figure 19.12 Plot showing the geographic projections using the weighted average ensemble model for *Protea laurifolia* under (a) current and (b) future conditions. (*A black and white version of this figure will appear in some formats.*)

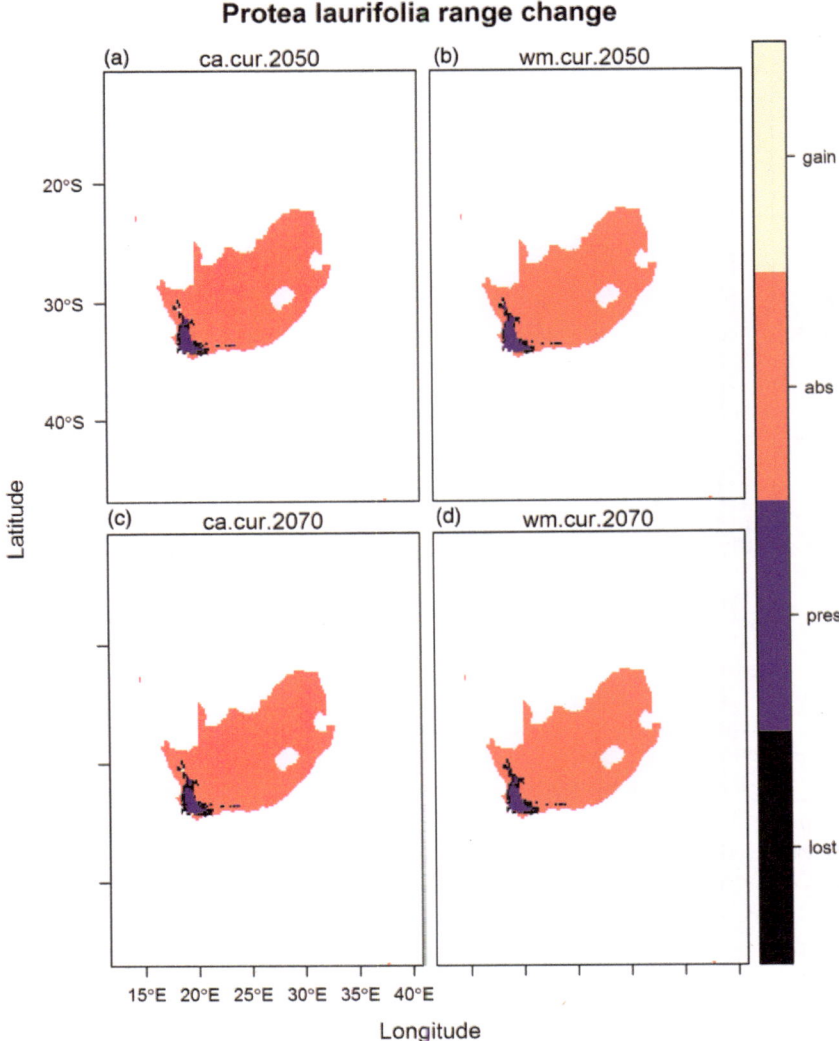

Figure 19.13 Plot of the predicted range changes for *Protea laurifolia* between present and future conditions. (*A black and white version of this figure will appear in some formats.*)

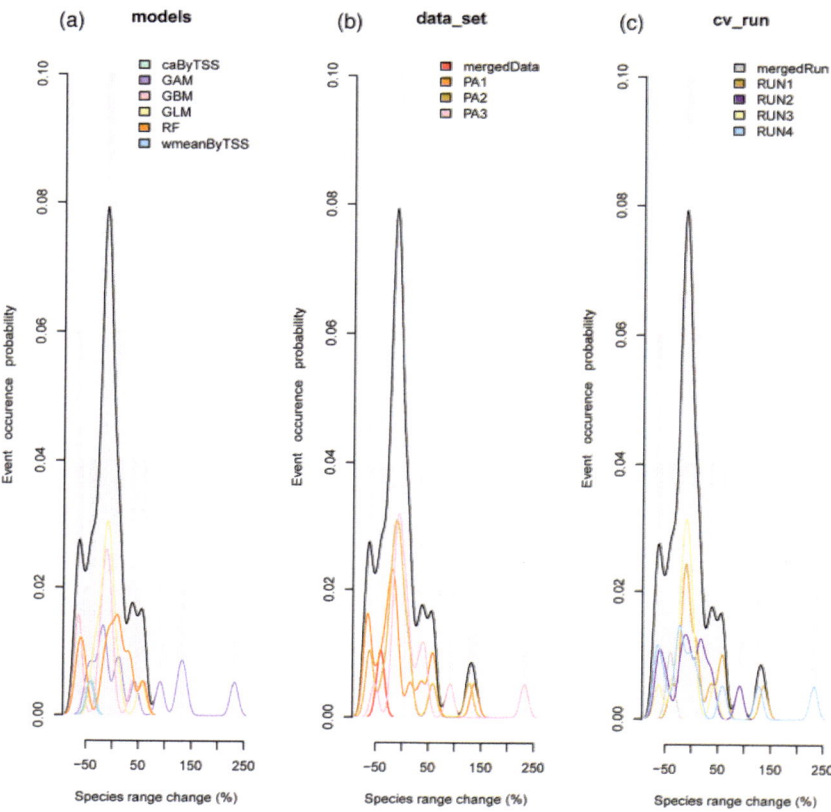

Figure 19.14 Density plot of the predicted species range changes according to the facets selected in the ensemble model. (*A black and white version of this figure will appear in some formats.*)

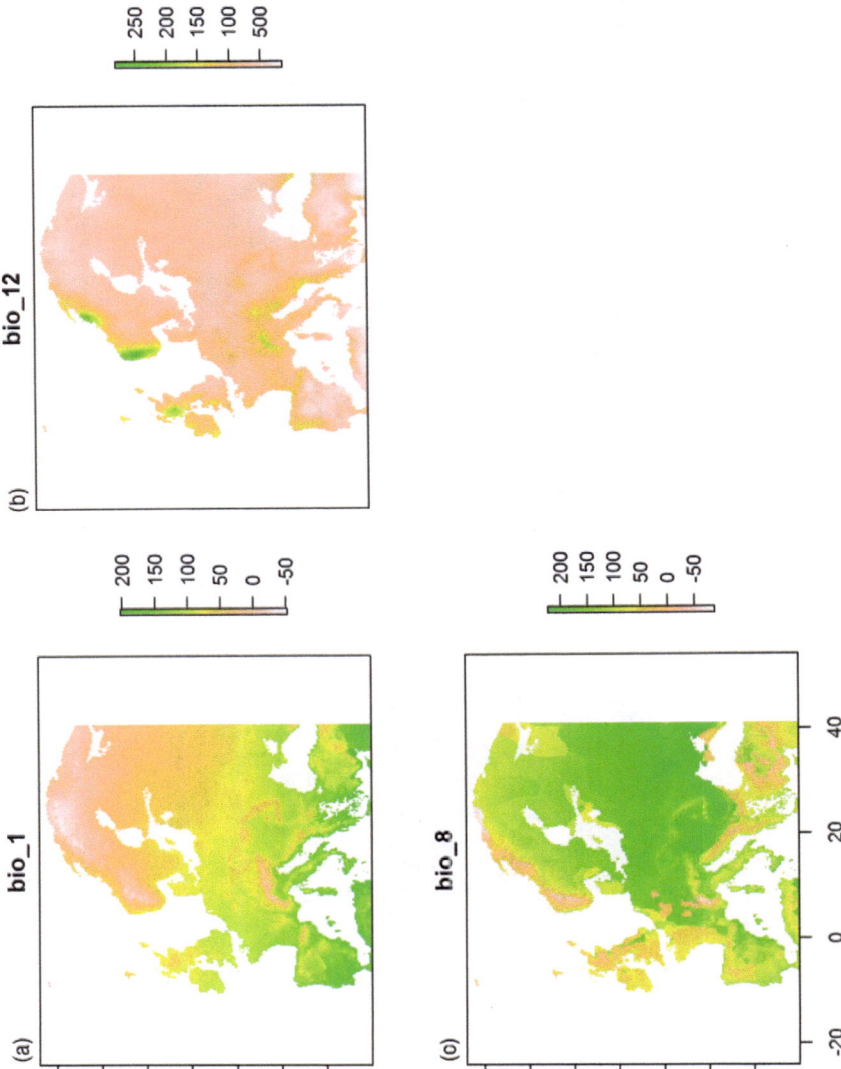

Figure 19.16 Maps of the three selected variables in Europe. (*A black and white version of this figure will appear in some formats.*)

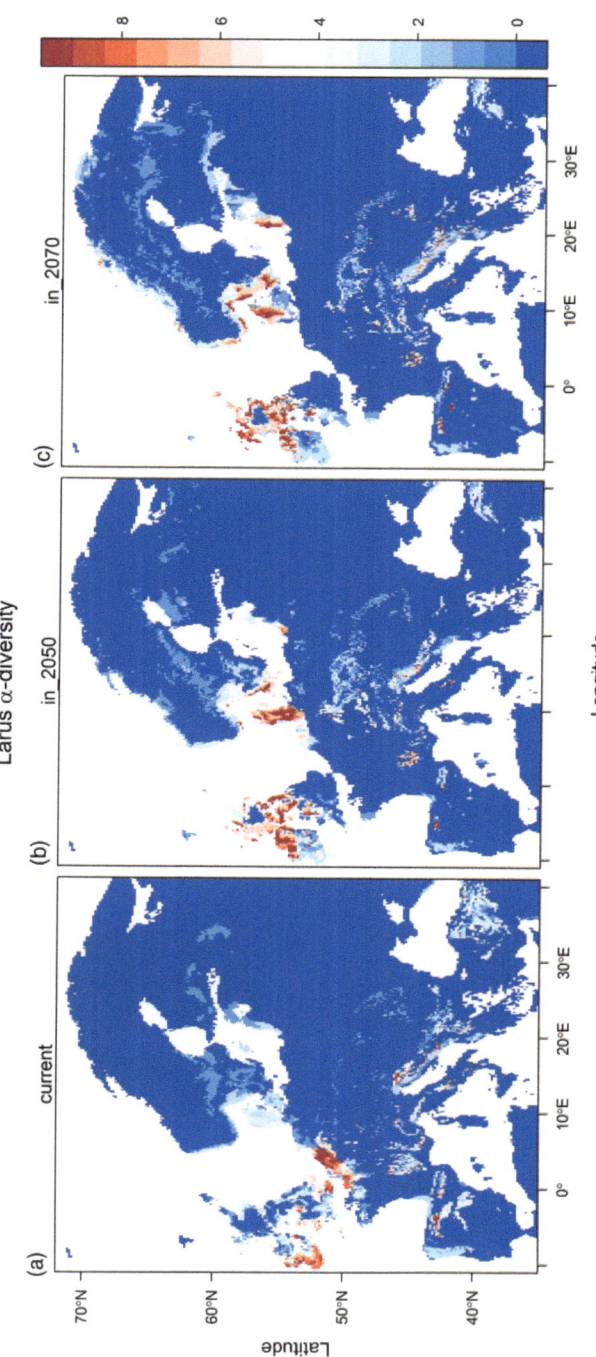

Figure 19.17 Species richness (alpha diversity) maps for the three time steps. (*A black and white version of this figure will appear in some formats.*)

partitions). Conversely, we tend to use the term "testing" for procedures that calculate final performance metrics on more "independent"
data (i.e. test set), such as accuracy, sensitivity, or specificity (Chapter
15). One advantage of the term "evaluation" as a general term is that it
does not call into question the veracity of the model (i.e. a model can
be partially valid, e.g. only in a region or time period). To summarize,
the training set is used to train the candidate algorithms, the validation
set is used to optimize them, compare their performance and eventually decide which algorithm or parameterization to use, and finally the
test set is used to obtain the final performance characteristics (Hastie
et al., 2009).

In the ideal case, we see the evaluation of the *predictive capacity* of a
model as a six-step procedure:

(i) Identifying the type of predictions that will be generated from the
    model, according to the defined objective;
(ii) Choosing an appropriate measure of agreement between the actual
     observations and the chosen type of predictions;
(iii) Choosing an appropriate validation procedure to optimize model
      building and/or select the final model(s) to be used for predictions
      (generating validation subsets or partitions);
(iv) Choosing an appropriate set of independent (or semi-independent)
     data to evaluate the model predictions, the "test set";
(v) Predicting to the test set; and
(vi) Applying the selected measure to the test set and delivering an
     agreement metric expressing model performance.

The series of steps (iv) to (vi) can also be repeated through resampling procedures. The first two steps are addressed in Chapter 15, the
third and fourth in Chapter 16, the fifth in Part V, and the sixth again
in Chapter 15. Step 1 also belongs to the conceptual model phase in
Part I, and is then mainly partly discussed in Parts III and V. Note that
while we consider model evaluation as explicitly implying a certain
degree of independence of the test dataset, the same metrics used for the
model evaluation can also be used to measure a model's goodness-of-fit
if measured on the training dataset (hereafter simply "fit"; i.e. how well
the model predictions compare to the same observations used to calibrate the model), which is also called a "resubstitution" evaluation by
some authors (e.g. Fielding and Bell, 1997). This is an important point
to keep in mind and will be discussed in Chapter 15 when presenting
the evaluation metrics.

We see an *uncertainty assessment* as a more complex procedure that can incorporate many different and complementary analyses (Elith et al., 2002; Barry and Elith, 2006), such as calculating uncertainty from the data, predictive algorithms, model parameters, or scenarios (e.g. Buisson et al., 2010; Carvalho et al., 2011; Wenger et al., 2013), or running sensitivity analyses by simulation (e.g. Dormann et al., 2008; Carvalho et al., 2011; Thibaud et al., 2014).

15 · *Measuring Model Accuracy: Which Metrics to Use?*

In this chapter we address the different types of evaluation metrics that can be used for different types of predictions. To do this, we assume that there is an independent evaluation dataset available, and that there are both predictions and observations for this dataset. Chapter 16 discusses which independent dataset to use for evaluation. How predictions are generated has already been partly addressed in Part III, and will be addressed again in Part V, as regards making predictions in time or space.

The type of metrics that can be used to evaluate a model primarily depends on the type of response variable that is being modeled (observed data; Figure 15.1) and the type of predictions that are produced by the model (predicted values).

In general, the observed and predicted values are on the same scale, except in two frequent cases in ecology: when a probability of presence or a continuous index of habitat suitability is to be compared to (i) presence–absence or (ii) presence-only data. In these two particular cases, dedicated metrics need to be used. Examples of quantitative (continuous or discrete), semi-quantitative (ordinal or ordered) and nominal (qualitative with >2 classes) models are rarer in the habitat suitability modeling literature than models of presence-only or presence–absence data. In this book, we therefore mainly focus on the two most common cases: observed presence–absence (binary) compared to probabilities of presence (Section 15.1), and single occurrences (presence-only data) compared to relative measures of probabilities or habitat suitability indices (Section 15.2). Metrics for the other situations are briefly described below and can be found in more standard statistical literature. The main metrics are summarized in Figure 15.1, with related references.

In all of the evaluation cases presented in Figure 15.1, it is usually good practice to consider more than one evaluation metric for comparing model predictions with observations in order to provide

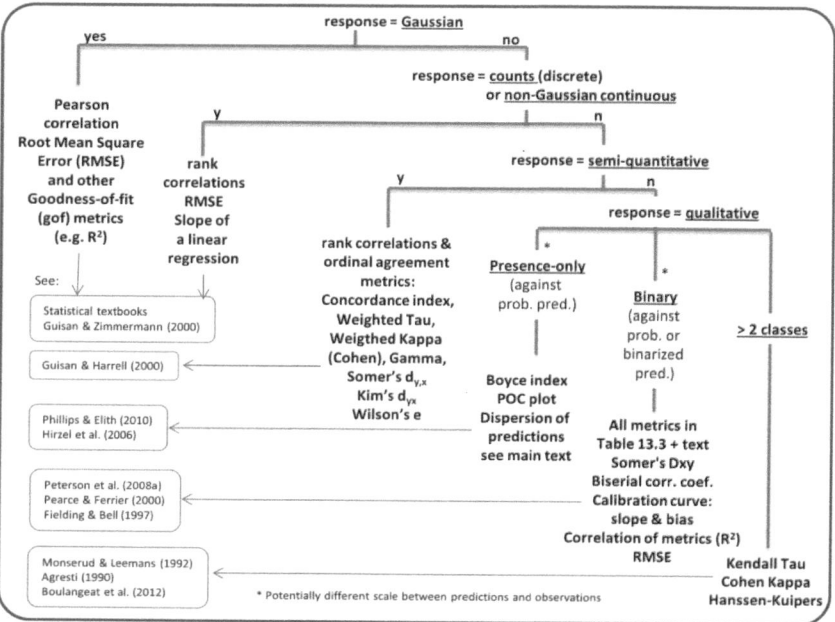

Figure 15.1 Non-exhaustive choice of adequate evaluation metrics based on the type of response variable. The most common cases, mainly discussed in this section of the book, are presence-only and presence–absence response variables. Typical variables for the different response types are: *Gaussian*: biomass or body size; *Counts (Poisson)*: abundance or species richness; *Semi-quantitative*: Braun–Blanquet abundance–dominance data (from phytosociological surveys); *Qualitative presence-only*: occurrences data from global databases (e.g. GBIF); *Qualitative binary*: presence–absence data; *Qualitative > 2 classes*: vegetation units or land-use/land-cover classes (e.g. from RS).

information on different aspects of the model's predictive power. For instance, in the case of presence–absence models, it may prove important to separately assess aspects of reliability, discrimination, and refinement (Murphy and Winkler, 1987; Pearce and Ferrier, 2000), or situations where different or equal weights can be assigned to different types of errors (Fielding and Bell, 1997). This will be developed below in the relevant section.

In the most classical statistical case where a quantitative continuous or discrete response is compared to observations on the same scale, such as comparing two variables with Gaussian (e.g. biomass) or Poisson (e.g. counts of individuals) distributions, there are a plethora of metrics available. Because the observed and predicted values are on the same scale, conventional statistics based on the sum of errors (residuals) and standard pairwise scatterplots can be used to assess the agreement. Conventional metrics include the Pearson correlation coefficient (r), the root mean square error (RMSE) or the coefficient of determination of a regression between observations and predictions (R^2). These are described in all statistical textbooks and we will therefore not develop these any further here. The comparison of two semi-quantitative variables (observed versus predicted) is much less frequent in the HSM literature, although there is plenty of data (e.g. from abundance–dominance classes in phytosociological surveys or visual estimates of ordered abundance classes for insects) that could be used to fit ordinal HSMs. Accordingly, there are fewer evaluation metrics available for semi-quantitative variables. This case will not be developed further here and we refer the readers to Guisan and Harrell (2000) and the references therein (e.g. Agresti, 1999) for a review of the main metrics available. The last case of comparing two qualitative variables with >2 classes is also less frequent in the HSM literature, where it is mostly used to predict vegetation classes (e.g. Brzeziecki et al., 1995). We refer readers here to specific papers or books (e.g. Agresti, 1990; Monserud and Leemans, 1992; see also Boulangeat et al., 2012a for a case where semi-quantitative data are simply considered as multi-classes qualitative data).

15.1 Comparing Predicted Probabilities of Presence to Presence–Absence Observations

The case of comparing probabilities of presence from the model (the predictions) to binary presence–absence data (the observations) is more complex than comparing two variables on the same scale because, depending on the level of precision of the predictions, a large number of predicted values can correspond to a single observed value (presence or absence). This implies that one can consider either the binary values from the perspective of the continuous probabilities, or conversely the continuous predictions from the perspective of the binary values. This results in two classes of evaluation measures, known as calibration and discrimination respectively. Calibration can be defined as "the extent to which a model correctly predicts conditional probability of presence," and discrimination as "the ability to distinguish between occupied and unoccupied sites" (Phillips and

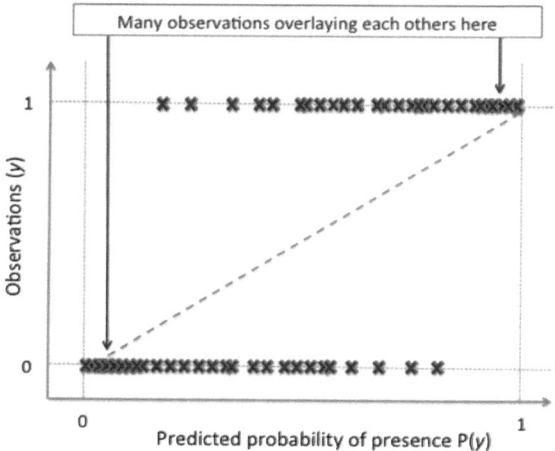

Figure 15.2 Representation of probabilistic predictions as a function of binary observations. In this plot, the relationship between the two variables can be difficult to visualize because numerous predictions may have the same value in each observation class.

Elith, 2010). For instance, the commonly used area under the ROC curve (AUC) metric only measures discrimination, while the point-biserial correlation (see Elith et al., 2006; r_{pb}, Linacre, 2008) measures both calibration and discrimination (Phillips and Elith, 2010). As the AUC is the predominant choice in published HSM studies, most models were only evaluated from this perspective. These two perspectives on the evaluation of presence–absence models should be used conjointly (as e.g. in Elith et al., 2006) when reporting on model evaluation, and the procedures for their use are developed in the next two sections.

15.1.1 Measuring Calibration

Checking the match between continuous predictions and binary observations in a plot visually is no easy matter because continuous predictions on the x-axis can only be compared to values of 0 and 1 on the y-axis, meaning also that many predictions close to 0 or 1 can stack on each other when they correspond to observed values of 0 or 1 respectively (Figure 15.2).

It may therefore be better to consider probability values binned in *k* equal width classes and to relate these to the observations. This

Table 15.1 *Table of contingency of observations (x_i) and predicted probabilities of presence binned in k classes, from Pearce and Ferrier (2000).*

	π_1	...	π_k	
$x = 0$	n_{01}	...	n_{0k}	$n_{0.}$
$x = 1$	n_{11}	...	n_{1k}	$n_{1.}$
	$n_{.1}$	...	$n_{.k}$	$n_{n..}$

can be displayed as a contingency table with two lines and k columns (Table 15.1).

The intersections of the observation classes (2) with the probability classes (k) represent the frequency of cases where a probability value in the corresponding class was predicted for the given observed value (0 or 1), or similarly the proportion of presences and absences in each prediction class. With a high performance model, one would expect high frequencies in low probability classes for absences (0) and high frequencies in high probability classes for presences (1). Random distribution of frequencies within the table cells would show that the relationship between the observed and predicted values does not differ from random.

These frequencies can also be represented graphically by plotting the ratio of presence to the total number of cases per class (i.e. the observed proportion of presences, or prevalence in the class) as a function of the probability classes (Figure 15.3b, with $k = 10$). With this type of graph, one would expect the points to be aligned diagonally, representing perfect calibration. The alignment is easy to interpret visually if points are properly aligned, but when the alignment deteriorates, it becomes more difficult to evaluate how close it is to the diagonal line. It can also be very sensitive to the number of bins chosen and to the number of observations in each bin. Therefore, a more convenient way of using the graph is to draw a curve through the points using a model-fitting algorithm (Figure 15.3c). Two approaches are commonly used here: (i) parametric regression, such as fitting a logistic regression with a generalized linear model (GLM; McCullagh and Nelder, 1989b); or (ii) semi-parametric regression, such as using a smoothing algorithm in a GAM; Hastie and Tibishirani, 1986), as a loess or spline function. The parametric approach can be particularly informative as it provides an intercept value, which

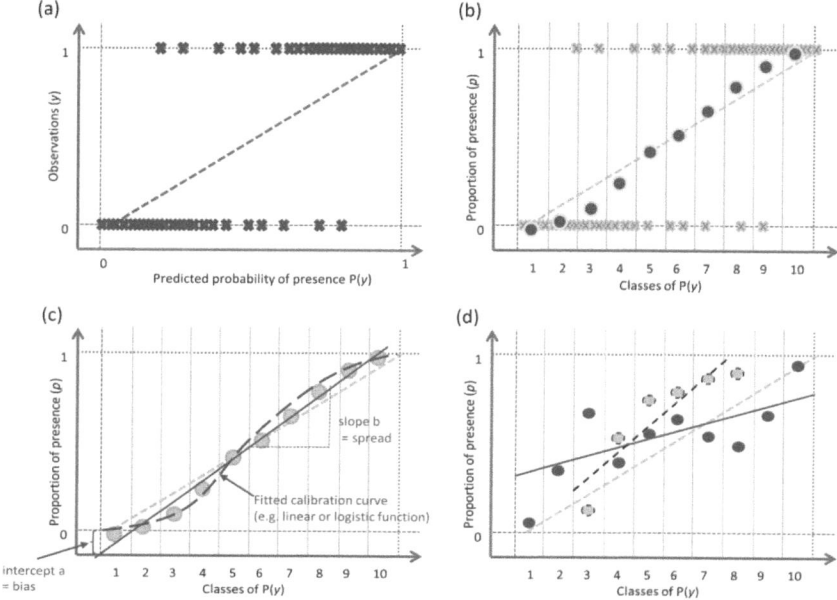

Figure 15.3 Principles of a calibration plot, the graphical representation of a calibration contingency table between observed presence–absence and classes of probabilities of presence, seen from the perspective of the observations. (a) Probabilistic predictions as a function of binary observations (same as 15.2). In this plot, the relationship between the two variables is difficult to visualize because many predictions can have the same value for each observation class. (b) Representation of the proportion of presences per class of predictions (here 10). (c) Fitting of a model through the points of graph b. The slope represents the spread of the predictions and the intercept the bias. (d) Different models will yield calibration curves with a different spread (high: plain red, low: dotted blue). See Pearce and Ferrier (2000) and Phillips and Elith (2010).

can be interpreted as an estimate of the bias in the predictions (Pearce and Ferrier, 2000).

Although an important feature of model predictions, evaluating model calibration using fully independent data has been largely ignored in the HSM literature (Pearce and Ferrier, 2000; Phillips and Elith, 2010) and there are few examples in real-life studies (examples are Ferrier and Watson, 1997; Edwards et al., 2005; Reineking and Schroder, 2006; Calvete et al., 2008; Heinanen and von Numers, 2009; Petitpierre et al., 2012). As an example, the calibration plots for the red fox models are shown in Figure 15.4.

The reason calibration curves are not frequently used in the literature to evaluate HSM predictions might be the absence of a single index, which can be used to compare studies. Different curves can be fitted to the points displayed in a plot (see Figure 15.3d) and it is therefore more difficult to extract one single value.

There are different ways of drawing these calibration plots in R. A first example is developed below using the same species (*V. vulpes*, the red fox) as in Part III. The `calibration.plot()` function in the `PresenceAbsence` package makes it possible to draw a calibration plot (Figure 15.4), although here only the points are drawn, and no trend is fitted through them.

```
> library(PresenceAbsence)
> mammals_data <- read.csv("tabular/species/mammals_and_bioclim_
table.csv", row.names=1)
## Create the RF model
> library(randomForest)
> RF <- randomForest(x = mammals_data[,c("bio3", "bio7",
"bio11", "bio12")],y = as.factor(mammals_data$VulpesVulpes),
ntree = 1000)
> RF.pred = predict(RF, type="prob")[,2]
## Create the FDA model
> library(mda)
> fda_mod = fda(VulpesVulpes ~ 1+bio3+bio7+bio11+bio12,
data=mammals_data,method=mars)
> FDA.pred = predict(fda_mod, mammals_data[,c("bio3", "bio7",
"bio11", "bio12")], type = "posterior")[,2]
## Create the BRT model
> library(gbm)
> BRT.mod <- gbm(VulpesVulpes~ bio3+bio7+bio11+bio12,
data=mammals_data, distribution = "bernoulli", n.trees = 2000,
interaction.depth = 7, shrinkage = 0.001, bag.fraction = 0.5,
cv.folds=5)
brt.mod.perf = gbm.perf(BRT.mod, method = "cv", plot.it = F)
> BRT.pred <- predict(BRT.mod, newdata=mammals_data[,c("bio3",
"bio7", "bio11", "bio12")], type="response",
n.trees=brt.mod.perf)

## Create an average prediction from the three single predictions
(RF, FDA, BRT)
> AVER.pred<-((RF.pred+FDA.pred+BRT.pred)/3)

## Create the final dataset containing all predictions
> ObsNum <- mammals_data[,8]
> plotID <- 1:nrow(mammals_data)

> EvalData <- data.frame(cbind(plotID, ObsNum, AVER.pred,
RF.pred, FDA.pred, BRT.pred))
> colnames(EvalData) <- c("plotID", "ObsNum", "AVER", "RF",
"FDA", "BRT")
> head(EvalData)
```

	plotID	ObsNum	AVER	RF	FDA	BRT
1	1	0	0.09601478	0.032000000	0.05846938	0.1975750
2	2	0	0.08302935	0.016216216	0.04233393	0.1905379
3	3	0	0.06324810	0.002793296	0.03732554	0.1496255
4	4	0	0.06923040	0.028735632	0.03761349	0.1413421
5	5	0	0.05382970	0.016438356	0.02384759	0.1212032
6	6	0	0.05121586	0.007978723	0.01947709	0.1261918

```
# Calibration plots for three single predictions and the
averaged model

> par(oma = c(0, 5, 0, 0), mar = c(4, 4, 4, 1), mfrow = c(2, 2),
cex = 0.7, cex.lab = 1.4, mgp = c(2, 0.5, 0))

> for (mod in 1:4) {
calibration.plot(EvalData, which.model = mod,
color = mod + 1, xlab = "", ylab = "")
}

> mtext("Predicted Probability of Occurrence", side = 1,
line = -1, cex = 1.4, outer = TRUE)
> mtext("Observed Occurrence as Proportion of Sites Surveyed",
side = 2, line = -1, cex = 1.4, outer = TRUE)
```

As we can see in Figure 15.4, the calibration plots are relatively similar for the three modeling techniques (each technique tending to slightly over- or under-predict at different places along the probability of occurrence gradient). Interestingly, predictions from the averaged model show a better calibration plot (close to the 1:1 line) compared to single algorithms, which further supports Laplace's idea that the average of multiple models predicts better than individual models (Araújo and New, 2007; see Part III and Part V).

In a next step, trend lines (i.e. models) can be added through the points with confidence intervals (CIs). Prediction bins which CI contains the diagonal represent the bins where the predictions and observations can be considered statistically identical. Such graph can be drawn using the scripts in Phillips and Elith (2010). The resulting plot is displayed in Figure 15.5.

```
> calibplot <- function(pred, negrug, posrug, ideal, ylim=c(0,1),
xlim=c(0,1), capuci=TRUE, xlabel = "Predicted probability of
presence", filename=NULL, title="Calibration plot", ...) {
if (!is.null(filename)) png(filename)
ylow <- pred$y - 2 * pred$se
ylow[ylow<0] <- 0
yhigh <- pred$y + 2 * pred$se
if (capuci) yhigh[yhigh>1] <- 1
plot(pred$x, ylow, type="l", col="orange", ylim=ylim,
xlim=xlim, xlab=xlabel, lwd=2, ...)
```

```
lines(pred$x, yhigh, lwd=2, col="orange")
lines(pred$x, sapply(pred$x, ideal), lty="dashed")
points(pred$x, pred$y, col="deepskyblue")
rug(negrug)
rug(posrug, col = "orange")
title(title)
if (!is.null(filename)) dev.off()
}

> smoothingdf <- 6
> smoothdist <- function(pred, res) {
require(splines)
gam1 <- glm(res ~ ns(pred, df=smoothingdf), weights=rep(1,
length(pred)), family=binomial)
x <- seq(min(pred), max(pred), length = 512)
y <- predict(gam1, newdata = data.frame(pred = x),
se.fit = TRUE, type = "response")
data.frame(x=x, y=y$fit, se=y$se.fit)
}
```

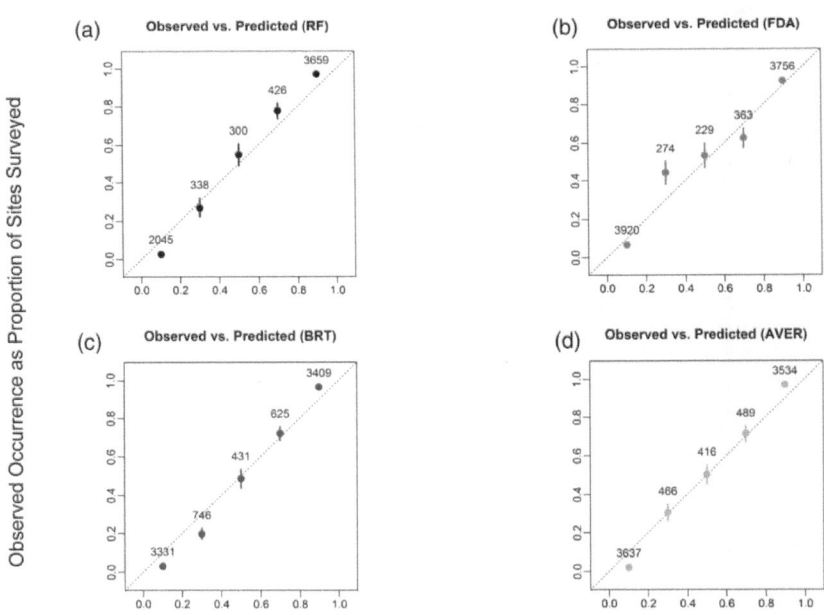

Figure 15.4 Example of calibration plots for species *Vulpes vulpes* modeled and predicted using three modeling techniques and their averaged ensemble: (a) random forest (RF), (b) flexible discriminant analysis (FDA), and (c) boosting regression trees (BRT) and (d) average model (AVER). Different models will yield calibration curves with different spreads.

```
> pacplot <- function(pred, pa, ...) {
predd <- smoothdist(pred, pa)
calibplot(predd, negrug=pred[pa==0], posrug=pred[pa==1],
ideal=function(x) x, ylab="Probability of presence", ...)
}


# binned calibration plot with equal width bins
> ecalp <- function(preds, acts, bins=10, do.plot=TRUE,
do.clear=TRUE, filename=NULL, title="Binned calibration
plot", ...){
g <- floor(preds*bins)
b <- 0:(bins-1)
p <- sapply(b, function(x) if (length(acts[g==x])==0) -1 else
sum(acts[g==x]) / length(acts[g==x]))
mx <- sapply(b, function(x,g) mean(preds[g==x]), g)
if(do.plot) {
if (!is.null(filename)) png(filename)
if (do.clear) {
plot(mx, p, xlim=c(0,1), ylim=c(0,1), ...)
} else {
points(mx, p, xlim=c(0,1), ylim=c(0,1), ...)
}
rug(preds[acts==0])
rug(preds[acts==1], col = "orange")
abline(0,1,lty="dashed")
title(title)
if (!is.null(filename)) dev.off()
}
return(p)
}


> Data<-EvalData[1:2000,]
#true probability of presence
> RF<-Data$RF
> FDA<-Data$FDA
> BRT<-Data$BRT
> AVER<-Data$AVER

# number of samples in the datasets
> ns <- 2000

# observed presence / absence, randomly drawn according to pt
> oRF <- rbinom(ns, 1, RF)
> oFDA <- rbinom(ns, 1, FDA)
> oBRT <- rbinom(ns, 1, BRT)
> oAVER <- rbinom(ns, 1, AVER)

> par(oma = c(0, 5, 0, 0), mar = c(4, 4, 4, 1), mfrow = c(2, 4),
cex = 0.7, cex.lab = 1.4, mgp = c(2, 0.5, 0))
> for (mod in 1:4) {
# binned calibration plot with equal width bins
ecalp(RF, oRF, title="(a) RF")
```

```
ecalp(FDA, oFDA, title="(b) FDA")
ecalp(BRT, oBRT, title="(c) BRT")
ecalp(AVER, oAVER, title="(d) AVER")
# presence-absence smoothed calibration plot
pacplot(RF, oRF, title="(e) RF")
pacplot(FDA, oFDA, title="(f) FDA")
pacplot(BRT, oBRT, title="(g) BRT")
pacplot(AVER, oAVER, title="(h) AVER")
}
```

15.1.2 Measuring Discrimination and Selecting a Prediction Threshold

Discrimination is a different view of the same comparison of probabilistic predictions with binary observations already seen, but from a "predictions" perspective (i.e. "calibration"), in the previous section (15.1.1). Let's start again from the graph shown in Figure 15.2, but this time look at it from the perspective of the observations. The simplest way of comparing our continuous predictions to the binary observations is to convert the probabilistic predictions to the binary scale [0, 1] of the observations. This is done by choosing a threshold value above which presences are predicted and below which absences are

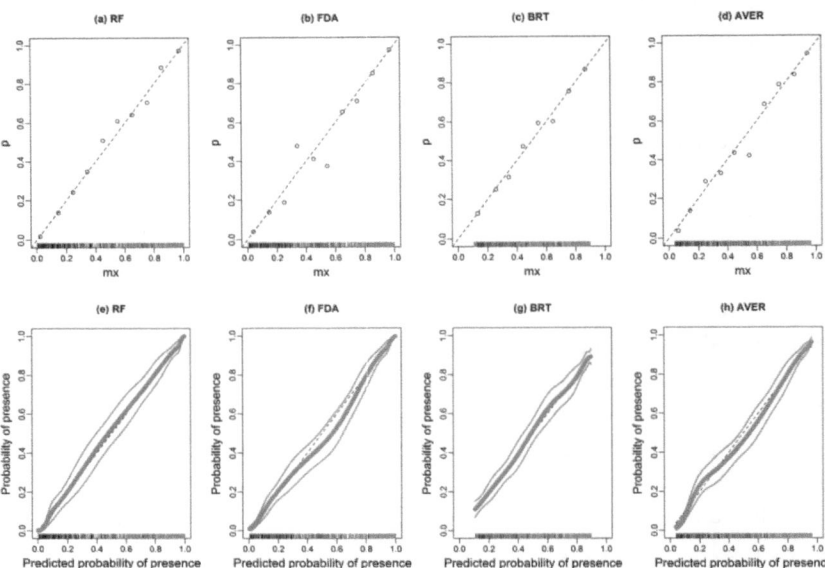

Figure 15.5 Example of calibration curves calibrated with simulation data reproduced from codes provided in Phillips and Elith (2010). The top row (a–d) shows binned calibration plot with equal width bins, the bottom row shows (e–h) presence–absence smoothed calibration plot.

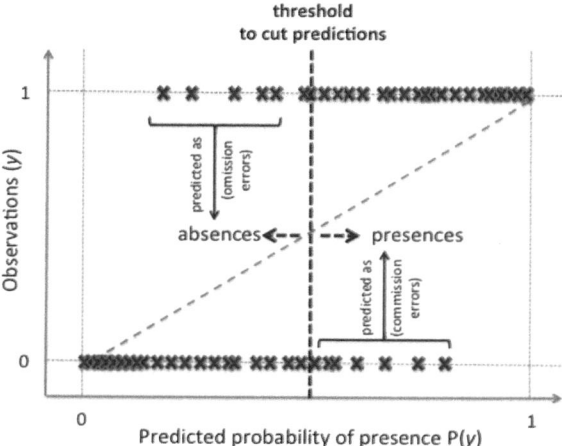

Figure 15.6 The same representation of probabilistic predictions as a function of binary observations, shown in Figure 15.2 and 15.3, but this time viewed from the perspective of the observations.

predicted (Figure 15.6; Guisan and Zimmermann, 2000; Freeman and Moisen, 2008). Using a selected threshold results in any observed presence found in the prediction values below the threshold to be incorrectly predicted as an absence (false absence), a type of error known as "omission". Similarly, any observed absence above the threshold is incorrectly predicted as a presence (false presence), a type of error known as "commission" (Figure 15.6).

Using a given threshold will result in four possible outcomes, best represented in a two-way confusion matrix (Table 15.2; also called contingency table): (i) correctly predicted presences (true presences, TP), (ii) falsely predicted presences (false presences, FP; also called commission errors), (iii) falsely predicted absences (false absences, FA; also called omission errors), and (iv) correctly predicted absences (true absences, TA). From this table, the agreement between observations and predictions can be quantified in numerous ways using the values obtained for TP, FP, FA, and TA.

In R, such table can be easily obtained using the command:

```
# Contingency table for one model (AVER) and one threshold (0.5)
> table(EvalData$AVER>0.5,EvalData$ObsNum)

          0    1
FALSE 4020  321
TRUE   319 3882
```

Table 15.2 *Two-way confusion matrix for comparing presence–absence observations to "binarized" (i.e. transformed into binary values) predictions (0/1). Several metrics can be derived from this table that are given in Table 15.3.*

			Observed		
			present **1**	**absent** **0**	**sum**
predicted	present	1	TP true presence	FP false presence commission error	*TP+FP* *total predicted presences*
	absent	0	FA False absence omission error	TA True absence	*FA+TA* *total predicted absences*
	sum		*TP=FA* *total presences*	*FP+TA* *total absences*	*N = TP + FP + FA + TA* *Total number of observations*

A large number of metrics have been proposed for discrimination purposes, of which the main ones used in habitat distribution modeling are given in Table 15.3.

Because threshold-dependent metrics are based on a confusion matrix, any of the measures of discrimination in Table 15.3 will necessarily depend on the *threshold* selected and used to build the confusion matrix in Table 15.2. So, how should this threshold be chosen?

Ultimately, the choice of the appropriate threshold depends on the goal of the study (Guisan and Zimmermann, 2000; Freeman and Moisen, 2008). In a conservation decision-making context, the threshold can be selected by evaluating the consequences of each type of error and, when necessary, by using different thresholds for different decisions (e.g. for invasive species, to decide when to monitor, when to eradicate, when to change categorization of threat; see Guisan et al., 2013). Prevalence-oriented thresholds (i.e. selecting a threshold that yields a predicted prevalence equal to the observed prevalence) might be preferable when the predictions need to reflect the observed prevalence (i.e. out of a restricted set of sampled locations), for instance when using models to prospect for rare species in the landscape (Guisan et al., 2006a). When both commission and omission errors need to be minimized, it might be more appropriate to choose a threshold that balances sensitivity and specificity (Liu et al., 2005).

Table 15.3 *The most commonly used metrics that can be derived from a two-way contingency table comparing presence–absence observations to binary predictions. These metrics are therefore threshold-dependent. TP = true presence, FP = false presence, FA = false absence, TA = true absence, N = TP + FP + FA + TA; see Table 15.2. See also Liu et al. (2011) for additional measures.*

Type	Metric	Abbreviation	Description	Range	Formula
Data properties	Sample size	N	Total number of observations	[1 : inf]	TP + FP + FA + TA
	Prevalence	PRE	Proportion of presences in the dataset	[0 : 1]	(TP + FA)/N
	Overall diagnostic power	ODP	Proportion of absences in the dataset	[0 : 1]	(FP + TA)/N = 1 − PREV
Optimist's view (no difference between types of errors)	Correct classification rate	CCR	Percentage of correct predictions (presences and absences)	[0 : 1]	(TP + TA)/N
	Misclassification rate	MR	Percentage of false predictions (presences and absences)	[0 : 1]	(FP + FA)/N
Observer's view (by column in Table 15.2)	Sensitivity (=true positive rate)	SE	Percentage of presences correctly predicted	[0 : 1]	TP/(TP + FA)
	False absence rate (=false negative rate)	FAR	Percent of presences falsely predicted	[0 : 1]	FA/(TP + FA) = 1 − SE
	Specificity (=true negative rate)	SP	Percentage of absences correctly predicted	[0 : 1]	TA/(TA + FP)
	False presence rate (=false positive rate)	FPR	Percentage of absences falsely predicted	[0 : 1]	FP/(FP + TA) = 1 − SP
Modeler's view (by row in Table 15.2)	Presence predictive power (=positive predictive power)	PPP	Percentage of all positive predictions being presences	[0 : 1]	TP/(TP + FP)
	Absence predictive power (=negative predictive power)	APP	Percentage of all negative predictions being absences	[0 : 1]	TA/(FA + TA)

Balanced view (full use of the confusion matrix, i.e. Table 15.2)				
Normalized mutual information	NMI	See Forbes (1995); non-monotonic when excessive error rates		$[-TP \star \ln(TP) - FP \star \ln(FP) - FN \star \ln(FA)-TN \star \ln(TA) + (TP+FP) \star \ln(TP + FP) + (FA + TA)\star \ln(FA + TA)]/[N \star \ln N - ((TP + FA) \star \ln(TP + FA) + (FP + TA) \star \ln(FP + TA))]$
Kappa	K	See Cohen (1960); sensitive to sample size and prevalence	[-1: 1]	$[(TP + TA)-(((TP + FA) \star (TP + FP) + (FP + TA) \star (FA + TA))/N)]/[N - (((TP + FA) \star (TP + FP) + (FP + TA) \star (FA + TA))/N)]$
Weighted Kappa	WK	See Cohen (1968); same as K but with weights assigned to TP, FP, FA and TA	[-1: 1]	Above formula weighted for TP, FP, FA and TA; see Cohen (1968)
Odds Ratio	OR	Infinite when either b or c are 0; i.e. same value when the algorithm is perfect or lacks one type of error	[0: inf]	$(TP \star TA)/(FP \star FA)$
True skill statistic (or Hanssen-Kuiper skill score)	TSS (or HKSS)	See Hanssen and Kuipers (1965); tends to converge to the prevalence for rare events (i.e. when TA is very large)	[-1: 1]	$[(TP \star TA) - (FP \star FA)]/ [(TP + FA) \star (FP + TA)] = SE + SP -1$

Recommended practice for generating threshold–dependent metrics is to (Hastie et al., 2009):

(i) Calibrate the model on a training dataset;
(ii) Use a second validation dataset to select a threshold (e.g. using a maximization approach; see text below in the same section);
(iii) Evaluate the model on a third, independent test dataset using the previously selected threshold.

A few studies have compared the use of different methods to select the optimized threshold (Liu et al., 2005; Jimenez-Valverde and Lobo, 2007; Freeman and Moisen, 2008; Liu et al., 2011; Nenzén and Araújo, 2011; Liu et al., 2013) with a view to building binary (presence–absence) prediction maps. Freeman and Moisen (2008) used data and models for 13 tree species in Utah to compare 11 different threshold criteria, based on their effect on the model's predictive performance and predicted prevalence (i.e. proportion of presence predicted out of the total number of pixels in the study area). They found that models for species with low observed prevalence or with the lowest predictive power were most sensitive to the choice of threshold, and that different thresholds result in variations in model performance. In another comprehensive study, Nenzén and Araújo (2011) compared the use of 14 different thresholds for projecting range shifts under climate change, and showed that the choice of threshold can strongly affect the projections, which corroborated earlier findings by Thuiller (2004).

The examples below illustrate different threshold selection strategies and metrics. They are developed with thresholds of 0.1 increments for simplicity, resulting in only 11 rows, but in practice, and in the graphs 0.01 increments are used, thereby resulting in tables with 101 rows. From the 11 thresholds (including 0 and 1) in our simplified example, one obtains values for PCC (percent correctly classified), sensitivity, specificity and Kappa (Table 15.3) for a model (here AVER) as follows:

```
## presence.absence.accuracy from package PresenceAbsence
# Example - Showing one model (AVER), eleven thresholds
> accu <- presence.absence.accuracy(EvalData, which.model = 4,
threshold = 11, st.dev = FALSE)
> accu[, -c(1, 2)] <- signif(accu[, -c(1, 2)], digits = 2)
> accu [c("threshold", "PCC", "sensitivity", "specificity",
"Kappa")]
```

	threshold	PCC	sensitivity	specificity	Kappa
1	0.0	0.49	1.00	0.00	0.00
2	0.1	0.84	0.99	0.70	0.69
3	0.2	0.90	0.98	0.82	0.80
4	0.3	0.91	0.97	0.86	0.82
5	0.4	0.92	0.95	0.90	0.84
6	0.5	0.93	0.92	0.93	0.85
7	0.6	0.92	0.90	0.94	0.84
8	0.7	0.91	0.87	0.96	0.83
9	0.8	0.90	0.81	0.98	0.79
10	0.9	0.84	0.69	0.99	0.68
11	1.0	0.51	0.00	1.00	0.00

One can also examine how the choice of threshold can change the predicted prevalence, i.e. the proportion of presences and absences across the prediction map, across the different models (RF=random forest, FDA=flexible discriminant analysis, BRT=boosted regression trees, AVER=average model of the three techniques; see Part III).

```
# Effect of threshold choice (11 thresholds) on predicted
prevalence
> pred.prev <- predicted.prevalence(EvalData, threshold = 11)
> pred.prev[, 2:6] <- round(pred.prev[, 2:6], digits = 2)
> pred.prev
```

	threshold	Obs.Prevalence	RF	FDA	BRT	AVER
1	0.0	0.48	1.00	1.00	1.00	1.00
2	0.1	0.48	0.60	0.58	0.73	0.64
3	0.2	0.48	0.54	0.54	0.59	0.56
4	0.3	0.48	0.51	0.51	0.53	0.52
5	0.4	0.48	0.49	0.49	0.49	0.49
6	0.5	0.48	0.47	0.47	0.47	0.47
7	0.6	0.48	0.46	0.45	0.45	0.44
8	0.7	0.48	0.44	0.43	0.43	0.43
9	0.8	0.48	0.41	0.41	0.39	0.40
10	0.9	0.48	0.35	0.38	0.00	0.33
11	1.0	0.48	0.00	0.00	0.00	0.00

The function ecospat.meva.table() in the ecospat package can be used to directly obtain the values of a series of different metrics for a model (see Table 15.3) and for a given threshold value. Let's again take the example of the AVER model, and use a threshold of 0.6.

```
> meva <- ecospat.meva.table (EvalData$AVER,
EvalData$ObsNum, 0.6)
> meva
```

```
$CONTINGENCY_TABLE
                 Observed values
Predicted values    0    1
           FALSE 4092  430
           TRUE   247 3773


$EVALUATION_METRICS
      Metric                              Value
1   "Prevalence"                         "0.492"
2   "Correct classification rate"        "0.9207"
3   "Misclassification rate"             "0.0793"
4   "Sensitivity"                        "0.8977"
5   "Specificity"                        "0.9431"
6   "Positive predictive power"          "0.0569"
7   "Negative predictive power"          "0.1023"
8   "False positive rate"                "0.9386"
9   "False negative rate"                "0.9049"
10  "Odds Ratio"                         "145.3641"
11  "Kappa"                              "0.8413"
12  "Normalized mutual information"      "0.3968"
13  "True skill statistic"               "0.8408"
```

The previous step provided values for different metrics for a given threshold. Let's now see in detail how to obtain values for one metric across different thresholds, taking Cohen's Kappa as the evaluation metric, and considering this time 0.01 increments (i.e. 99 thresholds).

```
> kappa100 <- ecospat.max.kappa(EvalData$AVER, EvalData$ObsNum)
> kappa100 [[2]]
      [,1]                          [,2]
[1,] "Maximum K"                   "0.8507"
[2,] "Correspondent threshold"     "0.44"
```

As we will see later, the same type of analysis can be run in `biomod2` and in other R packages (e.g. `PresenceAbsence`).

From these types of "across threshold" analyses, **a first type of** *threshold-independent evaluation measures of discrimination* can be derived. Here, the "optimized" threshold (on any dataset) is simply found by calculating the chosen evaluation metric for a range of possible thresholds (e.g. from 0 to 1, with an increment of 0.01), and by then selecting the one that maximizes the metric (assuming the response of the evaluation metric to the threshold is unimodal). This results in a "max" value for the chosen metric (e.g. max-Kappa or max-TSS; Table 15.3; see Liu et al., 2005). The underlying hypothesis is that the best possible value for the evaluation metric will reveal the predictive potential of the related model. Indeed, a model with poor predictive capacity will obtain a low

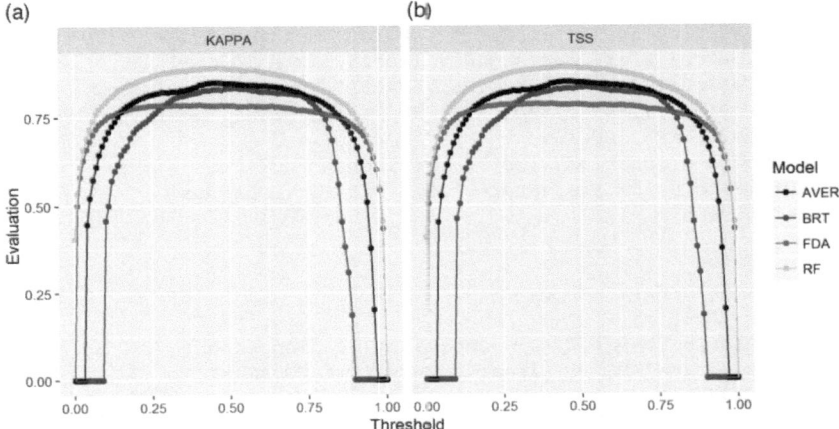

Figure 15.7 (a) Kappa and (b) TSS plots for species *Vulpes vulpes* modeled and predicted using three modeling techniques and their averaged ensemble: random forest (RF), flexible discriminant analysis (FDA) and boosting regression trees (BRT) and average model (AVER). (*A black and white version of this figure will appear in some formats. For the color version, please refer to the plate section.*)

score for the maximized evaluation metric, supporting the use of this approach. For instance, Landis and Koch (1977) suggested the following scale of judgment for Kappa: excellent K > 0.75; good 0.40 > K > 0.75; and poor K < 0.40. The advantage of this strategy is that it applies to any discriminant evaluation metric that can be calculated between binary observations and binarized predictions (see Table 15.3). In this regard, Liu et al. (2013) showed that the thresholding approach maximizing the true skill statistics (max-TSS), which is equivalent to maximizing the sum of sensitivity and specificity (max SSS), is particularly well suited as it produces the same threshold using either presence–absence data or presence-only data.

In `biomod2`, the following steps allow us to obtain all the threshold-dependent metrics simultaneously, along with graphs showing the variation in the values along the thresholds and the maximized statistics (Figure 15.7):

```
# Plotting the Kappa and TSS for each model using the function
Find.Optim.Stat() from the package biomod2
> library(biomod2)
> library(ggplot2)
> n=100
```

```
> dataToPlot <- as.data.frame(matrix(0, ncol=4, nrow=n*8, dimname
s=list(NULL,c("Evaluation","Threshold","Metric","Model"))))
> dataToPlot[,2] <- rep(seq(0,1,length.out = 100),8)
> dataToPlot[,3] <- rep(c("TSS","KAPPA"),each=100, times=4)
> dataToPlot[,4] <- c(rep("RF", 200), rep("FDA", 200),rep("BRT",
200),rep("AVER", 200))
> wrapper <- function(x, stat, Fit, Obs){
return(Find.Optim.Stat(Stat=stat,  Fit=Fit, Obs=Obs,
Fixed.thresh = x)[1])
}
> b=1
> for(i in 3:6){
a <- EvalData[,i]
dataToPlot[b:(b+99),1] <- sapply(seq(0,1,length.out = 100),
wrapper, stat='TSS', Fit=a, Obs=EvalData$ObsNum)
b <- b+100
dataToPlot[b:(b+99),1] <- sapply(seq(0,1,length.out = 100),
wrapper, stat='KAPPA', Fit=a, Obs=EvalData$ObsNum)
b <- b+100
}
> qplot(Threshold, Evaluation, data=dataToPlot, color=Model,
facets=~Metric, geom = c("point","line"))
```

Both Kappa and TSS combine information on the omission and commission error rates (see Tables 15.2 and 15.3) with the correctly predicted presences and absences. It can thus be informative to plot the variation of these metrics together with the variation in sensitivity and specificity across all thresholds, to see how variations in these overall metrics relate to the variation in the rate of correctly predicted presences and absences. This can be done using the `error.threshold.plot()` function in the `PresenceAbsence` package (Figure 15.8).

```
# Plotting the error statistics as a function of threshold in
four models
> data <- EvalData[1:6]
> N.models <- ncol(data) - 2
> par(oma=c(0,5,0,0), mar=c(4,4,4,1), mfrow=c(2,2), cex=0.7,
cex.lab=1.4, mgp=c(2, 0.5,0))
> for (mod in 1:N.models){
error.threshold.plot(data, which.model = mod, color = TRUE,
add.legend = TRUE, legend.cex = 0.7)
}
```

These plots can then be used to select a subjective threshold depending on the type of question being addressed. It should be noted that maximizing Kappa or TSS does not necessarily imply that the sensitivity and specificity will be balanced. For instance, in random forest the optimal threshold (i.e. which maximizes Kappa) is lower than a threshold that jointly maximizes sensitivity and specificity (the intersection of the two

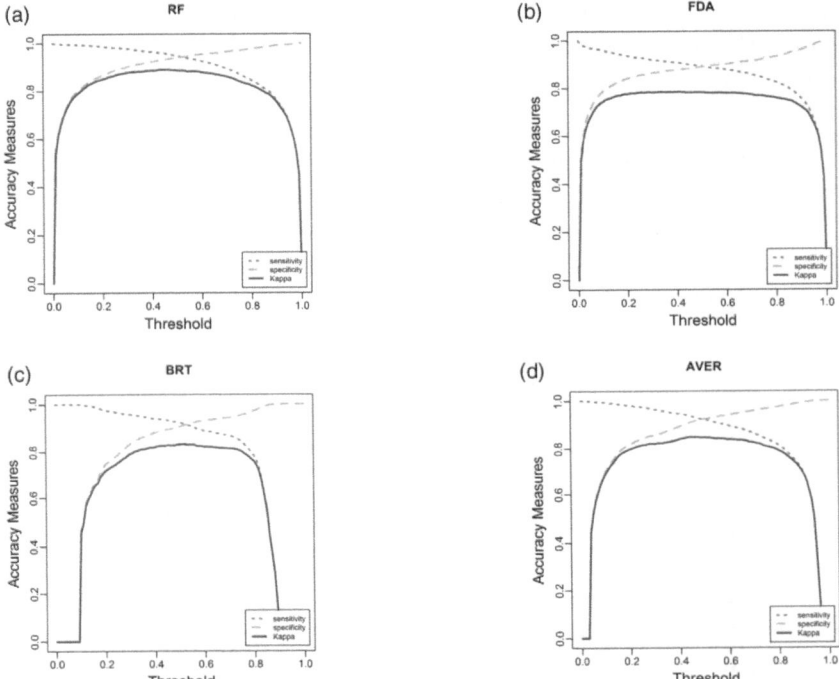

Figure 15.8 Error threshold plots for species *Vulpes vulpes* modeled and predicted using three modeling techniques and their averaged ensemble: (a) random forest (RF), (b) flexible discriminant analysis (FDA), and (c) boosting regression trees (BRT), and (d) average model (AVER).

curves along an axis of threshold values), the latter being often attributed to the threshold from the curve of a ROC plot (see below) since it is also equal to the threshold defining the inflection point of the curve.

A second type of threshold–independent discrimination metric, and an alternative to the previous maximization metrics, is to use an integrative approach that does not require the association of a metric with a given subjective or optimized threshold, but rather calculates it by integrating evaluation values across the whole range of possible thresholds. The AUC of a ROC[1] (see Swets, 1988; Fielding and Bell, 1997), originally developed during the World War II for signal detection and later used in medicine, is currently the most commonly used integrated discrimination metric in habitat distribution modeling. Instead of looking for the

---

[1] Receiver-Operating Characteristic

maximal value of a metric across all thresholds, it integrates all values at all thresholds into a single final metric, using a graphical approach. It is calculated in three steps:

(i) Calculate sensitivity (percentage of presences correctly predicted) and specificity (percentage of absences correctly predicted) (see Table 15.3) for all thresholds between 0 and 1.

(ii) Plot all values of sensitivity against the corresponding value of [1 − specificity], i.e. against the percentage of absences wrongly predicted (commission or false absence rate). This produces the ROC plot, with both axes ranging between 0 and 1, thus resulting in the total surface of the plot being equal to 1. When the threshold is zero, both sensitivity and [1 − specificity] take a value of one (upper-right corner of the plot), but when the threshold progressively increases, sensitivity decreases very slowly toward zero when the threshold is very close to one, while at the same time [1 − specificity] decreases far more rapidly, and both finally take a value of zero when the threshold is one. This results in the points in the plot being aligned on a curve that starts from the top-right corner (1,1 point) and steps-back logarithmically toward the lower-left corner (0,0 origin), i.e. curving more or less in close vicinity to the upper-left corner (Figure 15.9, below). With a perfect prediction, the curve is expected to pass through the upper-left corner (thus following the plot axes), whereas a prediction not differing from random would follow the 1:1 line where sensitivity = [1 − specificity]. An overall evaluation can thus be made by looking "visually" at how far the curve departs from the 1:1 line and how close it is to the upper-left corner.

(iii) A single quantitative value to characterize the model performance is however more convenient than a visual evaluation of the plot. A simple way of deriving such single value from the curve in the ROC plot is to calculate the surface below the curve (i.e. the area under the curve), where a maximum value is obtained when the curves goes through the upper-left corner, corresponding to a value of 1 for the whole plot area (i.e. 1 × 1), and the lowest value will be obtained when the curve precisely follows the 1:1 line, giving a value of 0.5 corresponding to the half-plot area.

A curve that goes below the 1:1 line means that the model yields predictions that are worse than random, i.e. counter-predictions (similar to negative correlation coefficients), with values between 0 and 0.5. Various judgment scales have been proposed to interpret AUC values. Swets (1988) defined AUC values between 0.5 and 0.7 as translating

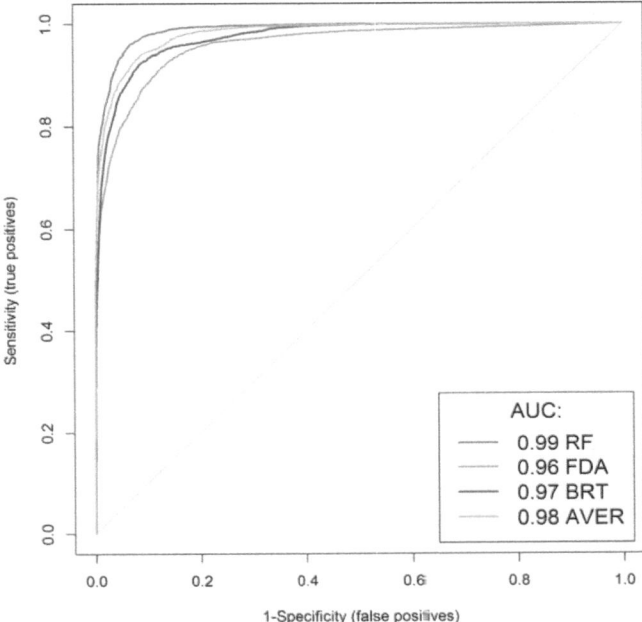

Figure 15.9 AUC ROC plots for the species *Vulpes vulpes* modeled and predicted using three modeling techniques and their averaged ensemble: random forest (RF), flexible discriminant analysis (FDA) and boosting regression trees (BRT) and average model (AVER). (*A black and white version of this figure will appear in some formats. For the color version, please refer to the plate section.*)

"poor" predictions, values between 0.7 and 0.9 as "useful" predictions, and values above 0.9 as good predictions. Araújo et al. (2005a) proposed a refined AUC scale, with: AUC > 0.90 being "excellent"; 0.80 < AUC < 0.90 being "good"; 0.70 < AUC < 0.80 "fair"; 0.60 < AUC < 0.70 "poor"; 0.50 < AUC < 0.60 "fail", and AUC < 0.5 being "counter-predictions" (i.e. similar to negative correlation between observations and predictions).

Producing an ROC plot and calculating the AUC can be done using the previous example and codes. In the latter, sensitivity and specificity were simply calculated for all thresholds between 0 and 1, and then the values of 1-specificity were plotted against sensitivity (Figure 15.9). A ROC plot and AUC value can also be obtained with the `auc.roc.plot()` function in the `PresenceAbsence` package, as illustrated here, again for *V. vulpes*.

```
#AUC ROC plot for all models
> auc.roc.plot(data, color=T, legend.cex=1.4, main="")
```

The four models for *V. vulpes* deliver here rather high AUC values, all at >0.95 (see the interpretation scale in the text above).

There is a *third type of threshold-independent discrimination metric*, mainly including the point-biserial correlation coefficient r_{pb} (Linacre, 2008), which does not require calculating intermediate indices at incremented threshold values (i.e. series of sensitivity and specificity values for calculating AUC), and in this sense can be considered more convenient. However, as we will see, this method also has limitations. The point-biserial correlation coefficient measures the strength of association between a binary (dichotomous) variable and a continuous variable, i.e. in our case between the observed presence–absence (binary) variable and the predicted probabilities. It is mathematically equivalent to the Pearson (product moment) correlation r and thus accordingly also ranges between –1 and +1, from a perfect negative (counter) association to a perfect positive association, with a value of 0 corresponding to an absence of association (Linacre, 2008). Two limitations of r_{pb} are that: (i) as for the Pearson correlation coefficient, the continuous variable must be distributed normally and (ii) the range of values becomes more constrained as the ratio of the binary variable moves away from 50/50, with implications for the interpretation of values (which are often very low; see for example Elith et al. 2006) and for significance testing. Alternatively, the same information can also be retrieved from formal dependence tests (e.g. *t*-test, Mann–Whitney, Kruskall–Wallis, chi-square).

Because it is equivalent to the Pearson correlation, it can be simply calculated using the `cor()` function in R.

```
# Measuring calibration and discrimination with Point-biserial
correlation (COR)
#Calculation of COR (point-biserial correlation) for AVER model
and P/A
> cor(AVER, ObsNum)
 [1] 0.8838176

#Calculation of COR (point-biserial correlation) for BRT model
and P/A
> cor(BRT, ObsNum)
 [1] 0.8626925
```

15.2 Comparing Probabilistic Predictions to Presence-Only Observations

As we saw in Chapter 7 (Part III), comparing simple occurrence (also called presence-only) data to predicted probability of presence (or some habitat suitability index) adds a level of complexity, because the usual discrimination metrics cannot be used here for model evaluation. Therefore, when building a model, the signal can theoretically only come from the distribution of occurrences (leading to the denomination "profile" techniques). We have already seen (Part III) that an alternative process is to still use discriminant techniques to build models using available presences and a subset of so-called "pseudo-absences" (or "background data"; see Phillips et al., 2009) as "absences". Most models built with pseudo-absences are also evaluated with standard statistics (e.g. AUC; see Section 15.1; Phillips et al., 2006; Roura-Pascual et al., 2006; Broennimann et al., 2007; Wolmarans et al., 2010; Yates et al., 2010; Elith et al., 2011; Gallien et al., 2012). However, the use of pseudo-absences renders model evaluation even more difficult than model fitting. Model evaluation metrics must be considered with great care when applied to presence–pseudo-absences (Hirzel et al., 2006; Phillips and Elith, 2010), because probabilities predicted with pseudo-absences have no absolute value per se. As a main effect, many of these evaluation metrics only provide relative evaluation values, because pseudo-absences can be generated in different ways, in varying numbers, and with various weights. This means that comparisons can only be made between models across species within a same study area or across models within a same species. Furthermore, as we will see, only occurrence data can be used in this case to evaluate calibration. If it is used to assess discrimination, the evaluation can only be partial because it can only properly assess the model's ability to predict presences. Presence-only evaluation metrics that consider how the presences (i.e. occurrences) are distributed along the range of predictions were developed to address this issue. The three simplest evaluation metrics for presence-only predictions are the absolute validation index (AVI; Hirzel and Arlettaz, 2003), the minimal predicted area (MPA; Engler et al., 2004), and the contrast validation index (CVI; Hirzel et al., 2004; see Hirzel et al., 2006). The AVI simply computes the proportion of presence points found above a fixed threshold in the predictions (see the previous discussion about thresholds) and has a value of between 0 and 1. The MPA is the smallest area containing all observation points (or another percentage, say 95%; i.e. MPA100, MPA95, etc.; Engler et al., 2004) and is thus similar to the AVI, but is expressed in units of geographic area, whereas the CVI is the AVI corrected by its theoretical value for a model predicting the species everywhere (i.e. chance performance) and takes a value of between 0 and 0.5 (Hirzel et al., 2006). As all three rely on a predefined threshold (i.e. threshold-dependent metrics), they all have the same problem with finding an optimal threshold as the threshold-dependent metrics applied to presence–absence models (Hirzel et al., 2006; see previous section).

Based on the previous model, AVI can be computed very simply in R as:

```
## Example for BRT model
# AVI
> obs <- (EvalData$BRT * EvalData$ObsNum)
> avi <- sum(obs > 0.5)/length(obs)
> avi
  [1] 0.4555139
# CVI
> avi0 <- sum(EvalData$ObsNum)/length(obs)
> cvi <- avi0 - avi
> cvi
  [1] 0.0365254
```

In this case, the threshold was set to 0.5. In the same way as for presence–absence, a cross-validation or split-sampling procedure can be used to find the threshold that optimizes the associated metrics.

Several other approaches have been proposed for evaluating presence-only predictions that do not require the selection of a single threshold. The first is the Boyce index (Boyce et al., 2002; Hirzel et al., 2006). As initially defined, this methods splits the model predictions into b regular bins (or classes, typically 10) and then assesses the proportion of presences actually found within each bin i compared to the proportion of modeling cells (i.e. pixels) in the same bin, i.e. the expected proportion if the presences were distributed randomly (called the predicted-to-expected (P/E) ratio F_i in Hirzel et al. (2006). A model that adequately predicts the distribution of a given species should predict large numbers of presences in the high prediction bins (i.e. high proportion of presences with high values of habitat suitability) and fewer and fewer presences as one moves toward the lower prediction bins (i.e. toward low habitat suitability for the species).

In this, it is similar to drawing a calibration plot with presence–absence data (see Section 15.1.1), but with background data instead of absences (Phillips and Elith, 2010). Accordingly, one would expect a monotonic relationship between the mean (or median) bin value and the predicted-to-expected (P/E) ratio F_i. The Boyce index can therefore be calculated as the Spearman correlation between the mean/median bin value and F_i (Boyce et al., 2002; Hirzel et al., 2006). It takes a value between -1 and +1, with a value tending toward +1 indicating good to perfect predictions, values around 0 indicating predictions no different from those obtained by chance, and values toward -1 indicating counter-predictions, i.e. observing presences in low suitability classes and observing absences in high suitability classes (Hirzel et al., 2006). This approach has been used, for example, to compare the performance of different techniques for modeling the distribution of the Caribou (*Rangifer tarandus*) in British Columbia (Canada), using independent test samples (Johnson and Gillingham, 2005). It proved to be particularly well suited to this task, where habitat suitability had to be expressed as a six-class index. However, in many cases, defining fixed classes is a very subjective step, revealing two important and early identified (Boyce et al., 2002) shortcomings of this metric that may affect the results: the number of bins *b* chosen and how their limits are defined along the predictions axis.

In order to work around these problems, and avoid the subjective selection of bins, one proposed refinement is to use a "window" of fixed width (the smallest possible) and move it iteratively along the model predictions (or habitat suitability) axis instead of defining fixed bins (Hirzel et al., 2006). F_i is then calculated at each iteration (i.e. after each window's displacement) and plotted against the mean predictions value in the window, and finally all F_i and mean predictions are plotted against each other to display a much smoother curve than is obtained with 5 or 10 fixed bins (as e.g. in Boyce et al., 2002; Hirzel et al., 2006). The smoothing of this presence-only calibration curve is very similar to certain smoothing procedures used to fit response functions in GAM, but it also makes the successive points partially dependent. At this point, a correlation can again be computed as when working with fixed bins, to yield a refined value (Hirzel et al., 2006). This refined Boyce evaluation approach was, for instance, used in a study assessing our ability to project niche models into new territories to predict species invasions (Petitpierre et al., 2012), because absences are not reliable when investigating colonizing species. Following on from this approach, a graphical solution generalizing calibration plots for presence-only situations (i.e. the POC-plot approach) was proposed that can also account for the different properties of the data (e.g. bias in occurrences, properties of the background data, etc.; Phillips and Elith, 2010), although the latter currently does not propose a single metric for evaluating the model.

The Boyce index B can be computed and plotted using the function `ecospat.boyce()` in the `ecospat` package (Broennimann et al., 2014b; Di Cola et al., 2017), as illustrated in the previous AVER model for the red fox, but here the predictions are only compared to the presence observations (the absences are not considered), to yield a high B value of 0.97 (Figure 15.10). The value is probably rather high here because the model tested was initially fitted with presence–absence data. Values for B can be lower for models fitted with presence data only (i.e. with profile techniques) or with presences and pseudo-absences (see Part III).

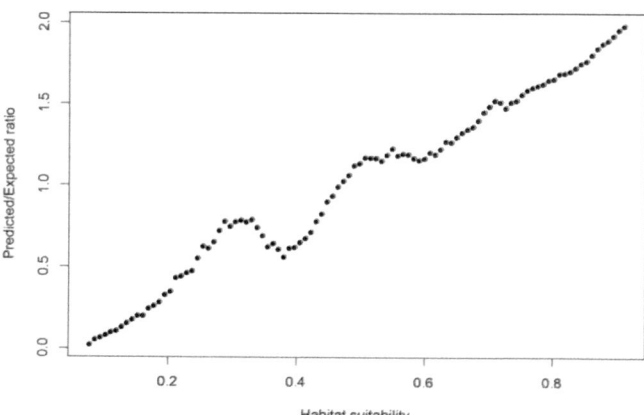

Figure 15.10 Boyce index plot of the *Vulpes vulpes* model fitted using an average of three modeling techniques (AVER) and predicted worldwide.

```
# Boyce index for one model (AVER)

> obs <- (EvalData$AVER [which(EvalData$ObsNum==1)])
> boyce <- ecospat.boyce (fit = EvalData$AVER , obs, nclass=0,
window.w="default", res=100, PEplot=T)
> boyce$Spearman.cor
$Spearman.cor
 [1] 0.986
```

The same type of plot, but with a curve that is also fitted to the points, can be obtained, using the `pocplot()` function provided in Phillips and Elith (2010), see Figure 15.11.

```
## POC function
# presence-only smoothed calibration plot

> pocplot <- function(pred, back, linearize=TRUE, ...) {
ispresence <- c(rep(1,length(pred)), rep(0, length(back)))
predd <- smoothdist(c(pred,back), ispresence)
c <- mean(back)*length(back)/length(pred)
if (linearize) {
fun <- function(x,y) c*y / (1-y)
predd$y <- mapply(fun, predd$x, predd$y)
predd$se <- mapply(fun, predd$x, predd$se)
ideal <- function(x) x
ylab <- "Relative probability of presence"
} }
else {
ideal <- function(x) x / (x + c)
ylab <- "Probability of presence"
}
```

```
calibplot(predd, negrug=back, posrug=pred, ideal=ideal,
ylab=ylab, capuci = FALSE, ...)
predd
}
# POC plot for AVER
> pocplot(AVER[ObsNum==1], AVER, title="AVER")
```

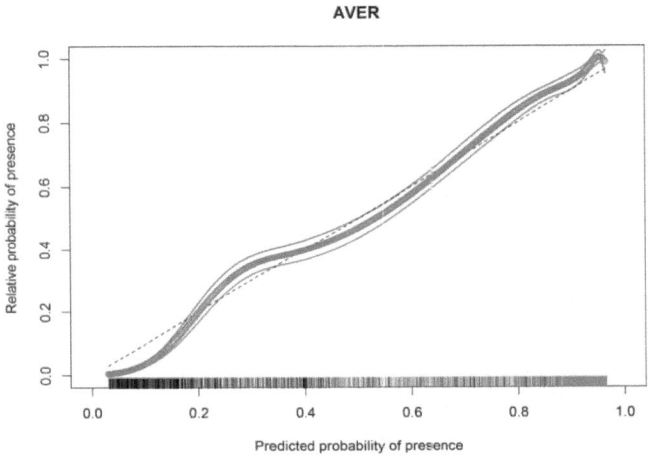

Figure 15.11 POC-plot of the *Vulpes vulpes* model fitted using an average of three modeling techniques (AVER) and predicted worldwide.

Alternative evaluation strategies for presence-only predictions, developed as refinements of the approaches presented until here, have also been proposed for particular situations (e.g. Anderson et al., 2003; Ottaviani et al., 2004; Peterson et al., 2008a). We will not go into any further detail about these special cases here, but refer interested readers to the original papers.

16 · *Assessing Model Performance: Which Data to Use?*

Once one or several evaluation metrics have been chosen, the next step is to determine which data to use for model evaluation. Using exactly the same data used to fit the model to calculate an agreement metric – a process often called resubstitution – is not considered a proper evaluation because the model is not tested on independent data (Section 16.1). Resubstitution procedures do, however, provide a baseline for comparing the same metrics measured on model predictions obtained on independent data (Sections 16.2 and 16.3). Randomization procedures can also be used here to complement the resubstitution procedure. The latter approaches additionally assesses the robustness of a model and its goodness-of-fit measures by randomizing the data (typically by permutation), and then testing in which proportion (across all models fitted with the randomized data) a similar model (e.g. with same coefficients and similar fit) can be obtained by chance (Section 16.1). Taking a honest evaluation perspective, involving some level of independent data, there are two basic strategies that can be followed depending on the degree of independence of the evaluation data (or test set) compared to the calibration data (or training set) (Guisan and Zimmermann, 2000; Araújo et al., 2005a):

(i) Using resampling procedures (e.g. jackknife, cross-validation, bootstrap) within the training set to assess the model's predictive power on partially independent data, known as "internal validation" by resampling (Section 16.2); and

(ii) Testing the model on fully independent data, kept separate from the beginning or ideally sampled a posteriori to test the model, known as "external evaluation" (Section 16.3).

With this in mind, "internal evaluation" can be considered as any assessment of a model within the dataset or region used to calibrate it (as

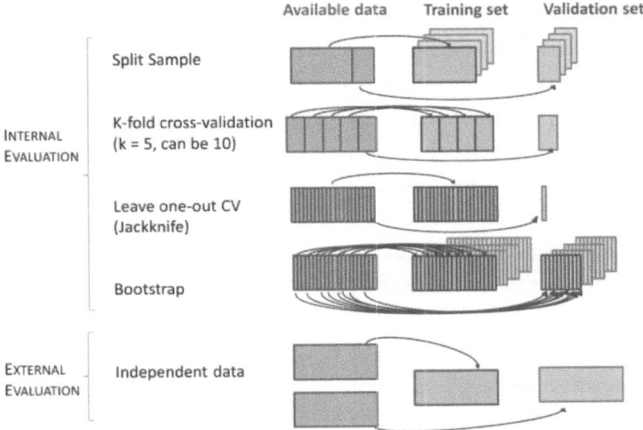

Figure 16.1 The different data partitioning strategies that can be used to evaluate a model. *k* = number of partitions. Upper arrows indicate model training; lower arrows indicate model evaluation. (Figure drawn with contributions by L. Maiorano.)

previously discussed) and "external evaluation" as any assessment on data with a demonstrated level of independence (e.g. different datasets with demonstrated spatial independence, a different region or time period; Randin et al. (2006).

The different possible evaluation strategies are summarized in Figure 16.1.

The choice of data partitioning strategy will depend on the initial sample size and what needs to be evaluated. Resubstitution coupled with randomization provides a baseline for assessing model fit and robustness, but is not a valid evaluation strategy, i.e. it cannot assess the predictive power of a model. Leave-one-out (Jackknife) cross-validation is appropriate for very small initial sample sizes that cannot be split into large partitions and provides additional "measures of influence" for individual observations. Classical *k*-fold (with *k* usually at 5 or 10) cross-validation provides a way of embedding exclusive stratification (e.g. geographic) into an evaluation procedure, but otherwise is similar to, but less powerful than, a repeated split sample. Repeated split sample is probably the most flexible and powerful of all evaluation strategies. Bootstrap with its ".632+" implementation (Efron and Tibshirani, 1997) can be seen as a special case of repeated split sampling (see e.g. Leathwick et al., 2006 for an application to HSMs). The principle is to apply a normal bootstrap to evaluate the model and, at each bootstrap iteration, to use the non-sampled sites (i.e. not in the bootstrap sample) as the independent dataset for evaluating the model fitted with the bootstrap sample. However, as any form of bootstrap offers the option of resampling the same observation multiple times (i.e. sample with replacement), the size of the left-out evaluation sets usually varies considerably across the bootstrap iteration loops, and each of these may further contain multiple duplications of the same observation site, potentially leading to issues of pseudo-replication. One advantage of bootstrap .632+ as an evaluation procedure for HSMs is that it also provides the typical bootstrap measures of bias and variance around model estimates (Efron and Tibshirani, 1993).

Ultimately, a fully independent evaluation with external data provides the most comprehensive strategy. Certain strategies complement each other or overlap, with internal and external validations being the most complementary (see e.g. Randin et al., 2006; Hastie et al., 2009; James et al., 2013). For instance, significant decreases between the results obtained using resubstitution and internal validation, and between internal and external validation, are often indicative of overfitting in a model (i.e. too many parameters in the model for the number of observations; Harrell et al., 1996). Comparing the answers obtained from these different approaches can thus provide additional information that is not delivered by any one of these strategies when considered alone.

16.1 Assessment of Model Fit Using Resubstitution and Randomization

Applying the evaluation metrics (e.g. AUC, TSS) to the same data used to build the model (training set) will generate other estimates of model's goodness-of-fit than the traditional ones (e.g. RMSE, R^2, adjusted-R^2), which can then be compared to the same metrics applied to predictions on partially or truly independent data. However, other aspects of the evaluation can also be assessed on the training set, such as testing the robustness of the model through randomization procedures, or assessing bias and variance of model parameters using bootstrap. This section addresses all measures of model evaluation that consider the full dataset without splitting it into training and validation (test) datasets.

16.1.1 Assessing Model Fit with Resubstitution
Resubstitution consists of applying the chosen evaluation metrics to the same dataset used to build the model. Strictly speaking, it is not a real

evaluation of a model. Instead, it provides additional measures of overall model goodness-of-fit (for any model type), or more specifically for presence–absence and presence-only models of model fitting, and for presence–absence models of model discrimination. This therefore provides insight into the properties of the fitted model, but this approach cannot provide an assessment of the model's power to predict to independent data (Edwards et al., 2006). Edwards et al. (2006) illustrated that resubstitution does not test the predictive ability of models by showing that a badly designed dataset still performs well in resubstitution testing, but not when tested against more independent data (Figure 16.2; i.e. using cross-validation or independent testing in their case; see Sections 16.2 and 16.3 below). Araújo et al. (2005a) also demonstrated that resubstitution metrics provided over-optimistic (and incorrect) estimates of predictive ability when SDMs were fitted on observed bird species distributions in the UK for the years 1967–1972 and tested against their observed distributions 20 years later.

By looking carefully at the values of the resubstitution metrics (evaluation or goodness-of-fit), one can, however, potentially detect model overfitting, for instance when the values obtained are suspiciously high. A better alternative is to compare them to values of evaluation metrics obtained through randomization, resampling or fully independent approaches (see Sections 16.1.2, 16.2 and 16.3).

16.1.2 Assessing Model Fit by Randomization (Permutation)
Another way of assessing model performance on the training data is to use randomization approaches (Figure 16.2). Although there are various ways of randomizing a dataset, here we will focus on the most common type of randomization: permutation. In the case of permutation, some variables (response or predictors) are permuted (i.e. through resampling without replacement), then a new model is fitted using the modified dataset (i.e. including one or more randomized variables), and the model parameters and fitted values are stored. By repeating this procedure a large number of times (say 999; see Figure 16.2), one can compare the value for each parameter in the model fitted with the original data to the range of values for the same parameters in all the models fitted with the randomized data. At this point, one can assess how frequently the original value (e.g. model fit, model coefficient, variable contribution) could be obtained by chance (out of all runs) and therefore calculate permutation-based p-values (Manly, 2006; see Figure 16.2). Permutation approaches can be particularly useful in cases where the data do not

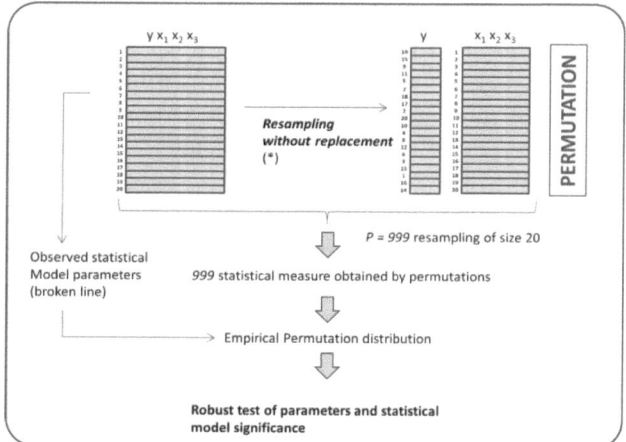

Figure 16.2 Permutation procedure applied to test the significance of parameters in predictive models. Here, the species presence–absence information is permuted, but the predictor variables can also be permuted in some applications (e.g. to assess the individual importance of each predictor in a model).

match the underlying assumptions of the models or statistical metrics used (Manly, 2006). For instance, permutation tests have been shown to be able to detect biological interactions in a wider range of models than using traditional F-tests in ANOVAs (Fraker and Peacor, 2008) or to discard insignificant predictors previously selected (using parametric significance tests) in HSMs (e.g. Pellet et al., 2004).

There are examples of the use of permutation approaches to randomize the response variable in species distribution model studies, but these are scarce (e.g. Jaberg and Guisan, 2001; Pellet et al., 2004). Often, randomization-based p-values are more restrictive than model-based p-values, potentially discarding variables that had significant p-values and associated interpretation obtained from parametric approaches, especially in the case of small datasets (e.g. Pellet et al., 2004). In this sense, adding randomization procedures can make the models more robust to small changes in the input data, for instance when very few outlying values render some tests significant (i.e. leverage points), possibly resulting in predictor variables being inappropriately selected in the models (Jaberg and Guisan, 2001).

The examples above used random permutations of the response variable (e.g. presence–absence or abundance of the species), but the same can be applied at the level of a predictor (or explanatory) variable, either to assess its importance in a model (Thuiller et al., 2009) or to assess how different modeling techniques select predictor variables (Elith, 2002; Lehmann et al., 2002). In machine-learning algorithms (i.e. random Forest, boosted regression trees), and also in some software packages (`biomod2`), variable importance in a model is estimated through the random permutation of one predictor variable at a time, either in the model or in a prediction dataset (so as to maintain the variable in the model structure) and assessment of the resulting drop in explained variance or in prediction accuracy, respectively (Strobl et al., 2009; Thuiller et al., 2009). The former shows whether a variable, kept in the model fitting or variable selection procedure, does or does not have an important effect on the model structure and explained variance. The latter is somewhat more complicated since it assesses the importance of a given variable in a new dataset used for prediction (in R wordings, only the `predict()` function is used, the model is not run again). This new dataset could be the original one in which one variable is permuted at a time, or an independent evaluation dataset, in a different geographic area or in a different time period. In this way, the importance of the variable can be evaluated according to the aim of the study. In the latter case, *in fine* it is not the change in explained variance or deviance that is estimated, but rather the correlation between the predictions made with the correct dataset and the predictions made with a dataset in which one given predictor is permuted. A similar approach to randomizing existing predictors is to add random predictor variables to a model and assess how frequently this or these random predictors are selected in the model when a variable selection approach is used (Elith, 2002; Lehmann et al., 2002). Such random procedures provide insight into the modeling technique's capacity to select meaningful predictors for the modeled species, as well as assess the risk of some techniques selecting spurious environmental predictors, and can thus prove useful for comparing the predictive capacity of different modeling techniques.

16.2 Internal Evaluation by Resampling

Approaches to internal evaluation by resampling include the traditional *k*-fold cross-validation, leave-one-out cross-validation (or jackknife), repeated split sample and bootstrap. As we will see, jackknife and repeated split sample are both special cases of cross-validation, and a special type of bootstrap (0.632+) can also be run to perform a form of repeated split sample validation. The main principle of this class of approaches is to start from a single, large enough dataset and to resample distinct partitions within it, one for model fitting (training set) and one for model evaluation (test set). The partitions can be defined one single time, as in the case of traditional k-fold or leave-one-out cross-validation, or reiteratively as in the case of repeated split sample and bootstrap. The rational is to fit the model(s) to one partition and use the second (left-out) partition to evaluate the model. In cases where the partitioning is repeated n times, the same data can thus be used once to fit the model and a second time to evaluate it, allowing for more reshuffling of the data and thus avoiding the risk of bias (by chance or caused by some methodological artifact) in the way partitions were defined. However, this also means that the data used for such evaluation are not really independent. They will notably come from a same area and time period, and so the evaluation may lack spatial or temporal independence (see Section 16.3).

All these resampling approaches generate two useful types of outputs (Verbyla and Litvaitis, 1989; Wiens et al., 2008; Arlot and Celisse, 2010): (i) the cross-validated or bootstrapped predictions for all observations we wanted to obtain in the first place, but also (ii) estimates of variation in model parameters (e.g. predictors selected, variance, or deviance explained), and therefore an indication of model stability, since in all cases several models are fitted (i.e. using the different partitions of the data; see Table 16.1).

Hereafter, we present and discuss the four main approaches: k-fold cross-validation, leave-one-out cross-validation, repeated split sample (two partitions, but no reciprocal fitting and testing here), and bootstrap (Table 16.1).

16.2.1 Evaluation Using k-Fold Cross-Validation

A k-fold cross-validation evaluation consists of splitting the original and unique dataset into k partitions (most commonly 5 or 10, as these values have been shown to best accommodate the trade-off to yield not too biased nor too high-variance error rates, James et al. (2013); Figure 16.1, Table 16.1). All except one partition are then used to fit the model, and the one left out is used to evaluate it (Figure 16.2). This procedure is run k times, each time leaving out a different partition. At the end, k models have been fitted and one set of predictions is available for each partition (since each partition was left out of the model fitting once), allowing the real observations to be compared to these semi-independent predictions using any of the metrics described in Chapter 15. K-fold cross-validation can also be applied by selecting

Table 16.1 *The four main resampling approaches and their characteristics.*

Resampling approach	Partitions	Partition rule	Repeats	Nr of models	Typical examples	Example reference (HSM)
k-fold cross-validation	k	Regular, random, stratified	0	k	k = 5 or 10	Zimmermann et al. (2009)
Leave-one-out cross-validation (jackknife)	n	Each observation left out at a time	0	n	k = n	Jaberg and Guisan (2001)
Repeated split sample cross-validation	2	Fixed ratio, random	R	R	R = (10 – 100)	Thuiller et al. (2011)
Bootstrap / Bootstrap .632+	2 (*)	Random with replacement	R	R	R = (50 – 200)	Leathwick et al. (2006); Moretti et al. (2006)

(*) except the very exceptional cases where all observations are resampled (highly improbable)

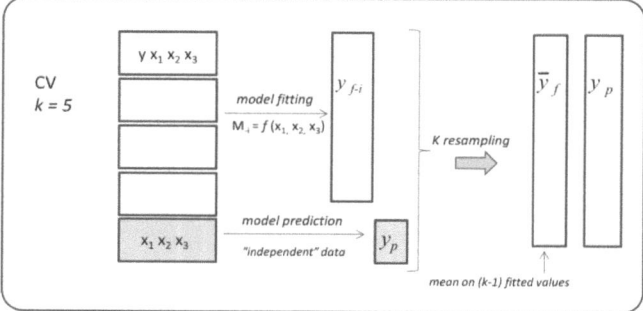

Figure 16.3 Procedure for *k*-fold cross-validation to evaluate a predictive model, illustrated for *k*=5. *k* is the fixed number of partitions. M_{-i} stands for the model fitted without partition *i* (i.e. *k*-1 partitions). $f()$ represents the model function used to fit the species–environment relationship, based on environmental variables x_1, x_2 and x_3, y_{f-i} is the vector of fitted values from the model fitted on all partitions except partition *i*, p_i is the vector of predictions made with M_{-i} on partition i. $\bar{y}_f$ is the vector of mean y_{f-i} across all M_{-i} models (for *i*=1 to *k*). y_p is the final vector of predictions made by appending the k y*pi* vectors, which can then be compared to the vector of initial observations y through the chosen evaluation metric. Because the initial partitions are fixed, this procedure can in some cases be repeated multiple times to randomize the partitions (but see Section 16.2.3).

partitions in geographic space, for instance to detect spatial non-stationary effects (Osborne and Suarez-Seoane, 2002). There are numerous examples of the use of *k*-fold cross-validation in HSM studies (e.g. Jensen et al., 2005; Edwards et al., 2006; Randin et al., 2006; Dormann et al., 2008; Zimmermann et al., 2009; Hanspach et al., 2010; Meier et al., 2011). Edwards et al. (2006) used 10-fold cross-validation of lichen species models in the Pacific Northwest (United States) to assess the robustness of two sampling schemes for building models. They showed that models built with data from a subjective sampling resulted in a large drop in predictive power when tested by cross-validation compared to resubstitution (Figure 16.3), but that models built with a well-designed sampling did not show such sharp reduction. In another study, Randin et al. (2006) used 10-fold cross-validation to show that models fitted with statistical techniques that tend to overfit models (i.e. closer to the training data; e.g. GAM; Chapter 8) show a more pronounced decrease in cross-validation evaluation (compared to resubstitution) than models fitted with techniques less likely to overfit (such as GLMs, Chapter 8).

K-fold cross-validation can therefore prove very useful, but also has limitations. In particular, it only delivers k estimates of each model parameter (unless the whole cross-validation itself is repeated numerous times; but then see repeated split sample in Section 16.2.3), which remains a small number (i.e. 5 or 10) for building proper empirical distributions and calculating robust uncertainty estimates. We will see in the following sections that repeating split-sample cross-validation a much larger number of times (see Section 16.2.3) or using the bootstrap (see Section 16.2.4) represent more interesting alternatives for estimating uncertainty around model parameters. Another limitation of k-fold cross-validation is specific to presence–absence models. In the latter case, running a k-fold CV requires that the species (or another binary feature being modeled) is sufficiently frequent across the dataset (i.e. with sufficient prevalence) that enough presences are still included in the training partitions to allow fitting the models. At the very least, in these cases, the presences should be stratified between the two partitions (e.g. Randin et al., 2006). Otherwise, if prevalence is very low in the training partition, model fitting might fail. In such cases, using leave-one-out cross-validation is preferable (Section 16.2.2; Jaberg and Guisan, 2001). Finally, the evaluation of HSMs using standard cross-validation was also shown to be sensitive to "spatial sorting bias", for some modeling techniques (i.e. Maxent and BIOCLIM; Hijmans, 2012). If the geographic distance between the training and the test data is affected by bias, stratifying the partitions used for cross-validation, with pairwise distance sampling for example, can reduce this bias (Hijmans, 2012).

This once again demonstrates the importance of embedding stratification procedures within a cross-validation exercise to ensure parity in presence–absences in the k-fold bins, and across time and space (see Figure 16.4). Instead of randomly selecting the partitions, they can also be geographically or temporally constrained. This might be done for example, to test whether the spatial or temporal structure in the data can be used to make relevant inferences or to test whether the models are good at extrapolating outside the training range by constraining the cross-validation procedure to ensure the partitions belong to different parts of the environmental space (e.g. to test for geographic non-stationarity; Osborne et al., 2007). In this regard, Wenger and Olden (2012) proposed a generalized cross-validation framework in the context of HSMs where data are "assigned non-randomly to groups that

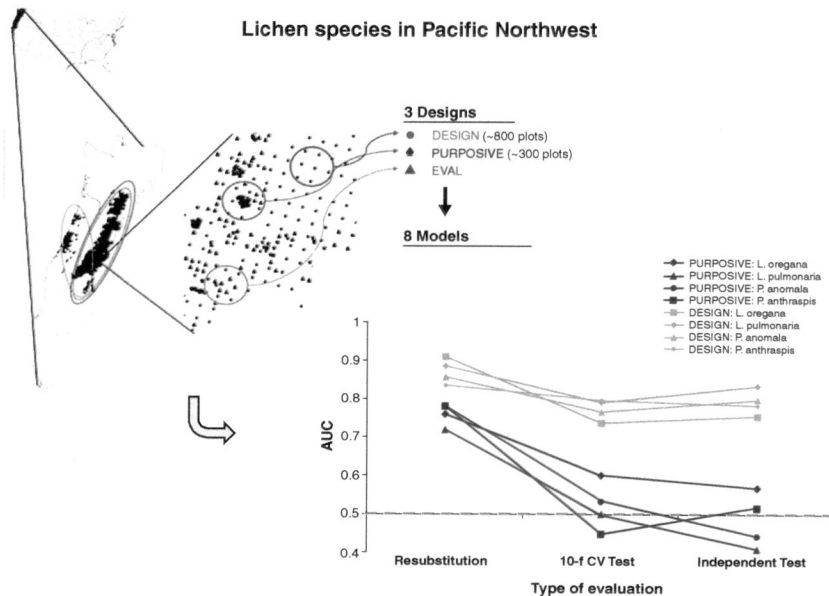

Figure 16.4 The use of three different types of samples to illustrate the importance of using partially independent (cross-validation, Section 16.2) and independent data (Section 16.3) in addition to internal resubstitution (Section 16.1) to evaluate model predictive power. The example is a modified version of that used by Edwards et al. (2006), with permission. (*A black and white version of this figure will appear in some formats. For the color version, please refer to the plate section.*)

are spatially, temporally or otherwise distinct, thus using heterogeneity in the data set as a surrogate for heterogeneity among data sets". As a recent example, Fithian et al. (2015) uses block cross-validation (randomly assigned contiguous spatial blocks) to limit spatial autocorrelation in the fitted models when testing the predictive accuracy of presence-only models.

To continue with the examples, we will use a simplified and smaller version of the dataset "`mammals_data.csv`", now called "`summary_mammals_data.csv`". Using it, we explore the use of k-fold cross-validation evaluation to estimate the test error rates that result from fitting various GLM models to the "`s_mammals_data`" set. All the following examples have been developed based on James et al. (2013) and their comparison of models allowing different levels of polynomials. The use of the `set.seed(555)` function is recommended to set a seed for R's random number generator, in order

Assessing Model Performance: Which Data to Use? · 281

to obtain precisely the same results as those shown below. Let's first attach the dataset.

```
> s_mammals_data <- read.csv("tabular/species/summary_mammals_
and_bioclim.csv", row.names=1)
> attach(s_mammals_data)
```

The poly() function can then be used within the glm() function to estimate the test error for the polynomial and higher-order (up to 10) regressions. We use glm() rather than lm() because it can be used together with cv.glm() that is part of the boot library.

```
> library(boot)
```

The cv.glm() function can be used to implement *k*-fold CV. In the following example we use k = 10, a common choice for *k* (see text above), in the mammal dataset. We set a random seed and initialize a vector in which the CV errors corresponding to the polynomial fits of orders 1–10 are stored.

```
> set.seed(555)
> cv.error.10=rep(0,10)
> for (i in 1:10){
glm.fit=glm(VulpesVulpes~poly(bio3+bio7+bio11+bio12,i),
family="binomial", data=mammals_data)
cv.error.10[i]=cv.glm(s_mammals_data,glm.fit,K=10)$delta[1]
}
> cv.error.10
[1] 0.1626184 0.1619637 0.1294147 0.1188130 0.1188086 0.1187090
[7] 0.1172326 0.1157738 0.1159589 0.1151248
```

Note that the computation time is much shorter than for a leave-one-out cross-validation (LOO-CV, see the next Section 16.2.2) due to the use of cv.glm() that cannot be used for LOO-CV. In this example, we see little evidence that using cubic or higher-order polynomial terms leads to lower test error than simply using quadratic fit.

The cv.glm() function produces a list with several components. The numbers in the delta vector contain the cross-validation results. On this dataset, the estimates are very similar to each other.

One other option is to perform an estimation of misclassification rate, sensitivity, specificity and AUC based on cross-validation (CV) using the Daim package.

```
> library(Daim)
> vulpes_data<-s_mammals_data[c(9:13,8)]
> vulpes_data$VulpesVulpes <- as.factor(vulpes_data$VulpesVulpes)
```

Note that the `Daim` function requires the species variable to be a factor. We will exemplify its use with a `randomForest` model, which should be predefined as follows:

```
> library(randomForest)
> myRF <- function(formula, train, test){
model <- randomForest(formula, train)
predict(model,test,type="prob")[,"pos"]
}
```

Then we calculate the optimal cut-point corresponding to the cv estimation of the sensitivity and the specificity using the `Daim()` function. The function `Daim.control()` can be used to control the parameters for the diagnostic accuracy of models (i.e. `Daim.control(method="cv"`, `k=10, k.runs=10)`, for de CV).

```
> set.seed(555)
> vulpes_RF_cv <- Daim(formula=VulpesVulpes~., model=myRF,
data=vulpes_data, labpos="1", control=Daim.control(method="cv",
k=10, k.runs=10), cutoff="cv")
> vulpes_RF_cv
Performance of the classification obtained by:

Call:
VulpesVulpes ~ bio3 + bio4 + bio7 + bio11 + bio12

Daim parameters:
  method = cv, k = 10, k.runs = 10, cutoff = cv, est.method =
obs, best.cutoff = 0.4.

Result:
-----------------------------------
Error:     | | cv     | | apparent  |
           ------------------------
           ------------------------
 | 0.0646 | | 0.0000    |
-----------------------------------


> summary(vulpes_RF_cv)
Performance of the classification obtained by:

Call:
VulpesVulpes ~ bio3 + bio4 + bio7 + bio11 + bio12

Daim parameters:
  method = cv, k = 10, k.runs = 10, cutoff = cv, est.method =
obs, best.cutoff = 0.4.
```

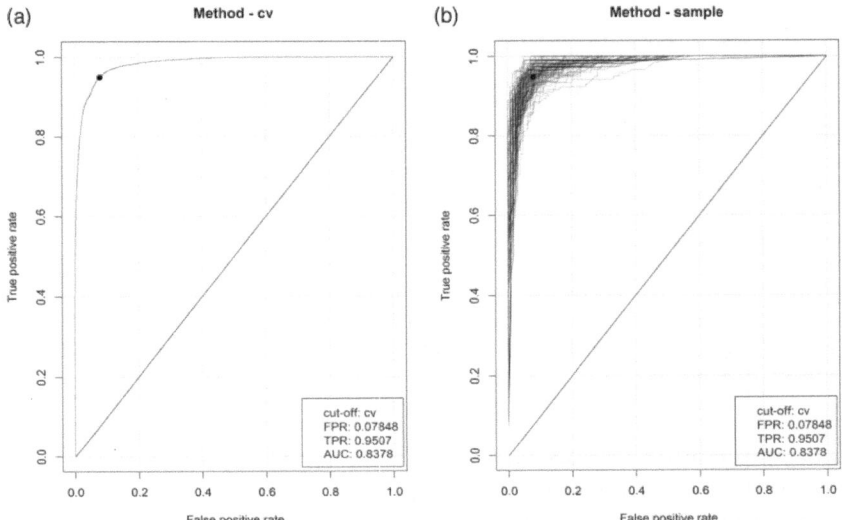

Figure 16.5 Plot of the Daim object generated using the Daim() function. (a) Cross-validation mean estimate of sensitivity and specificity. (b) All CV sample estimates of sensitivity and specificity.

```
Result:
-----------------------------------------
| Method:          |  | cv    |  | apparent  |
=========================================
| Error:           |  | 0.0646 |  | 0.0000   |
-----------------------------------------
| Sensitivity:     |  | 0.9507 |  | 1.0000   |
-----------------------------------------
| Specificity:     |  | 0.9215 |  | 1.0000   |
-----------------------------------------
| AUC              |  | 0.9830 |  | 1.0000   |
-----------------------------------------
```

Then we plot the Daim object generated by the Daim() function. Figure 16.5 shows the ROC plot of the randomForest model with an AUC of 0.8378.

```
> par(mfrow=c(1,2))
> plot(vulpes_RF_cv, method="cv")
> plot(vulpes_RF_cv, method="sample")
```

16.2.2 Evaluation Using Leave-One-Out Cross-Validation (Jackknife)

A LOO-CV evaluation − also called jackknife (Manly, 2006), see below and Table 16.1 − consists of fitting a model with all except

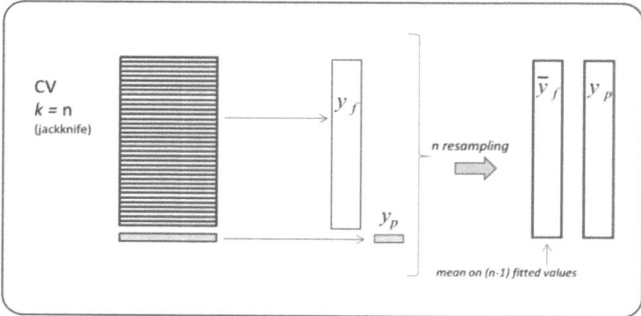

Figure 16.6 Procedure of a leave-one-out (jackknife) cross-validation for evaluating a predictive model.

one observation and using the model to predict to this single left-out observation. Then, the same operation is repeated n times, each time leaving out a different observation until all n observations have been left out once (Figure 16.6). This means that as many models are fitted as there are observations (n; Table 16.1). Each observation can therefore be associated with a prediction (i.e. from a model that was fitted without it), so that in the end, a vector of n predictions can be constructed. As for k-fold cross-validation, predicted values can then be compared to real observations using any of the evaluation metrics presented in Chapter 16.

If used to generate independent predictions, this approach is more appropriate with very small sample size, when too few species observations are available to conduct a k-fold cross-validation (Guisan and Zimmermann, 2000). Examples of uses of jackknife include HSMs for bats at a coarse resolution (and therefore small sample size) in Switzerland (Jaberg and Guisan, 2001), and habitat models of geckos in Madagascar (Pearson et al., 2007). However, unless the number of observations is very low, in most cases a repeated split sample cross-validation approach is a better option (16.2.3). In addition, jackknife does not thoroughly assess the stability of a model, because it only removes one observation at a time between models, and therefore the models and associated parameters do not differ drastically.

So, is this approach at all useful? There is indeed another, more important role for leave-one-out cross-validation, associated with the initial aim of the Jackknife: to calculate a measure of the influence of each single observation on the overall model or statistics (*influence measure*; Efron and Tibshirani, 1993). Because the models are fitted and each time another observation is left out, until they have all been removed once, it is easy to monitor changes in model parameters or evaluation metrics and identify observations causing greatest (or respectively the smallest) changes in these values. Jackknife can thus be turned into a *measure of influence* of single observations on a model or statistics, i.e. what it was initially designed for. In this respect, we can see that this approach plays an important role in the case of HSMs. Jackknife can help identify outlier observations that might have a major influence on the models (such as the Cook distance provided in the diagnostic plots for GLMs), either because inappropriate values are recorded for them, or because they truly play a specific biological role in the dataset (e.g. a fully different and isolated type of habitat, such as a cliff situation among grassland plots in a vegetation survey).

Next, we will explore the use of the LOO-CV to estimate the test error rates that result from fitting various GLM models on the "s_mam-mals_data" set. The cv.glm() function can also be used to implement LOO-CV. The LOO-CV estimate can be automatically computed for any GLM using the glm() and cv.glm() functions. Here, we must use the glm() function with family="binomial" to fit presence-absence data (numerical vector with 0s and 1s).

```
> glm.fit=glm(VulpesVulpes~bio3+bio7+bio11+bio12,
family="binomial", data=s_mammals_data)
> coef(glm.fit)
```

```
 (Intercept)           bio3           bio7           bio11           bio12
-1.0382810741 -0.1787174774   0.0216654829   0.0172735990   0.0009981371
```

As for the *k*-fold CV example, we call the boot library where the cv.glm() function is located.

```
> cv.err=cv.glm(s_mammals_data,glm.fit)
> cv.err$delta
```

```
[1]  0.1092130 0.1092129
```

As previously explained, the numbers in the delta vector contain the cross-validation results. In this case the numbers are identical (up to two decimal places) and correspond to the LOO-CV statistics. Our cross-validation estimate for the test error is approximately 0.1092.

This procedure is then repeated for increasingly complex polynomial fits. In order to automate the process, we use a for() loop to iteratively fit polynomial regressions from first order (simple linear) up to the fifth order, computes the associated cross-validation error, and store it in the *i*-th element of the vector cv.error. We begin by initializing the vector (note that this command takes a couple of minutes to run on a standard computer).

```
> cv.error=rep(0,5)
> for (i in 1:5){
glm.fit=glm(VulpesVulpes~poly(bio3,i),family="binomial",
data=s_mammals_data)
cv.error[i]=cv.glm(s_mammals_data,glm.fit)$delta[1]
}
> cv.error

[1] 0.1585545 0.1171832 0.1158225 0.1159716 0.1158523
```

As previously seen in the section on k-fold CV section, there is a sharp drop in the estimated test MSE (mean square error) between the linear and quadratic fits, and then no clear improvement when using higher-order polynomials.

```
> cv.error=rep(0,5)
> for (i in 1:5){
glm.fit=glm(VulpesVulpes~poly(bio3+bio7+bio11+bio12,i),
family="binomial",data=s_mammals_data)
cv.error[i]=cv.glm(s_mammals_data,glm.fit)$delta[1]
}
> cv.error

[1] 0.2144169 0.2011938 0.1978158 0.1904142 0.1902824
```

Note that the computation time is much longer than for k-fold cross-validation.

16.2.3 Evaluation Using Repeated Split Sample Cross-Validation

The split-sample cross-validation approach (also called "validation set approach" in James et al., 2013) consists of randomly splitting the initial dataset into two partitions (Table 16.1; typically 70%/30%), one for fitting the model and one for evaluating it. A first split-sample iteration is made and the model parameters and evaluation metrics (assuming that several are measured, but it could also be only one) are stored and a new split-sample iteration is made. The parameters and metrics from this second iteration are stored in the same place, and the iterations are repeated a total number of times R (not to be confused with the R software program; Table 16.1; Figure 16.7). Hence, single iterations are similar to the single iterations in the k-fold cross-validation, but the two approaches differ markedly in terms of the number of runs that are made: only k in total for k-fold (unless the whole procedure itself is repeated multiple times), but R iterations for repeated split sample cross-validation (see Table 16.1; with R>>k).

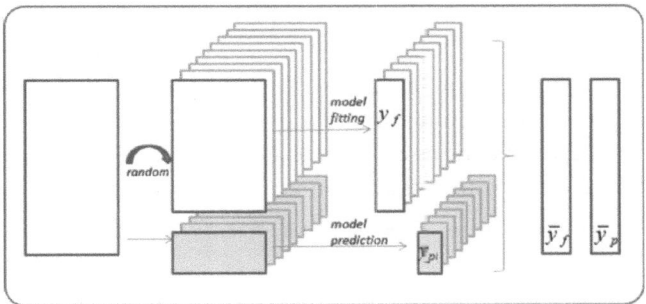

Figure 16.7 Procedure for the repeated split sample (i.e. repeated twofold) cross-validation for evaluating predictive models.

Repeated split sample is the standard option implemented in the `biomod2` package (Thuiller et al., 2009) and it is increasingly commonly used in habitat suitability modeling studies in general due to its simplicity and strength. Unlike the two previous approaches (*k*-fold and LOO cross-validation), it also generates estimates of model parameters and evaluation metrics, which can be used to assess model stability and assess uncertainty around model parameters and evaluation metrics. In this sense, it is a more informative approach. Its main limitation is the computing cost, as running repeated models (say 100) for a large number of species can be challenging. As with cross-validation, a more refined resampling procedure for split-sampling can be developed (e.g. using stratification) to reduce autocorrelation or to test the model's extrapolation ability. Furthermore, the original prevalence of the species in the dataset can be kept constant (or not) when partitioning into the training and evaluation dataset.

In the next example we explore the use of the repeated split sample cross-validation (or the validation set approach) in order to estimate the test error rates that result from fitting various GLM models on the "s_mammals_data" dataset. We begin by using the `sample()` function to split the set of observations into two halves, by selecting a random subset of 1244 observations out of the original 2488 observations. We refer to these observations as the training set.

```
> set.seed(555)
> train=sample(2488,1244)
```

Use the subset option in `glm()` to fit a glm using only observations in the training set.

```
> glm.fit = glm(VulpesVulpes~bio3+bio7+bio11+bio12,
family="binomial", data=s_mammals_data, subset=train)
```

Use the predict() function to estimate the response for all 2,488 observations, and the mean() function to calculate the MSE of the 1,244 observations in the validation set. Note that the -train argument below selects only the observations that are not in the training set.

```
> mean((VulpesVulpes-predict(glm.fit,s_mammals_data))[-train]^2)
[1] 7.349963
```

Therefore, the estimated test MSE for the GLM fit is 7.349963. We can use the poly() function to estimate the error for the second- and third-order (cubic) polynomial regressions.

```
> glm.fit2=glm(VulpesVulpes~poly(bio3+bio7+bio11+bio12,2),
family="binomial", data=s_mammals_data, subset=train)
> mean((VulpesVulpes-predict(glm.fit2,s_mammals_data))[-train]^2)

[1] 8.360697
```

```
> glm.fit3=glm(VulpesVulpes~poly(bio3+bio7+bio11+bio12,3),
family="binomial", data=s_mammals_data, subset=train)
> mean((VulpesVulpes-predict(glm.fit3,s_mammals_data))[-train]^2)
[1] 3.575069
```

The errors are 8.36 and 3.58, respectively. If we choose a different training set, then we will obtain somewhat different errors on the validation set.

```
> set.seed(555)
> train=sample(2488,1244)
> glm.fit=glm(VulpesVulpes~bio3+bio7+bio11+bio12,
family="binomial", subset=train)
> mean((VulpesVulpes-predict(glm.fit,s_mammals_data))[-train]^2)

[1] 10.1622
```

```
> glm.fit2=glm(VulpesVulpes~poly(bio3+bio7+bio11+bio12,2),
family="binomial", data=s_mammals_data, subset=train)
> mean((VulpesVulpes-predict(glm.fit2,s_mammals_data))[-train]^2)

[1] 9.018095
```

```
> glm.fit3=glm(VulpesVulpes~poly(bio3+bio7+bio11+bio12,3),
family="binomial", data=s_mammals_data, subset=train)
> mean((VulpesVulpes-predict(glm.fit3,s_mammals_data))[-train]^2)

[1] 3.920435
```

Using these observations split into a training set and a validation set, we find that the repeated split sample error rates for the models with linear, quadratic, and cubic terms are 10.16, 9.02, and 3.92, respectively.

A model that predicts VulpesVulpes using a cubic function of all variables performs better than a model that involves only a linear function of all variables, and there is little evidence here in favor of a model that uses a quadratic function of all variables. However, we will see later in Part VI that polynomial GLMs including cubic terms might better fit a present situation (e.g. to capture a bimodal distribution), but are likely to cause problems when used to project the species niche and distribution under future climate change scenarios.

16.2.4 Evaluation by Bootstrap

Bootstrap consists of resampling observations in a dataset (of size n), but, unlike the other approaches, with replacement. This means that an observation can be resampled several times within a same iteration. Another difference with previous approaches is that n observations are sampled at each iteration, i.e. the same number as the total sample size (though variants exist where a smaller number is resampled, but we will not address these here). The reason for resampling with replacement is that bootstrap was initially designed to provide empirical estimates of bias and variance in a dataset, which requires replacement (see Efron and Tibshirani, 1993; Manly, 2006). As a result, resampled subsets of the data can theoretically consist of a single observation repeated n times in one subset, or exactly the same initial dataset (i.e. all observations sampled once) in another subset, although the latter are extreme and highly unlikely cases. All subsets will fall in-between these two extremes, with various representations of each observation. On average across all resampled subsets, each will be composed of 63.2% of the observations (Efron and Tibshirani, 1997).

In bootstrap, the data are therefore not randomized, as in permutation, but an *"empirical distribution function"* (or sampling distribution) of the statistical measure of interest is built (See Figure 16.8). As well as providing a more robust estimate of the variance of an estimator, this approach can also be used to estimate the *bias* in model parameters (e.g. model coefficients, variance explained). By subtracting this bias, an unbiased estimate of the statistics of interest is obtained (Efron and Gong, 1983; Efron and Tibshirani, 1993). The bootstrap variance V is calculated as follows:

$$V = \frac{1}{R-1} \sum_{r=1}^{R} \left(t_r^* - \overline{t}^* \right)^2$$

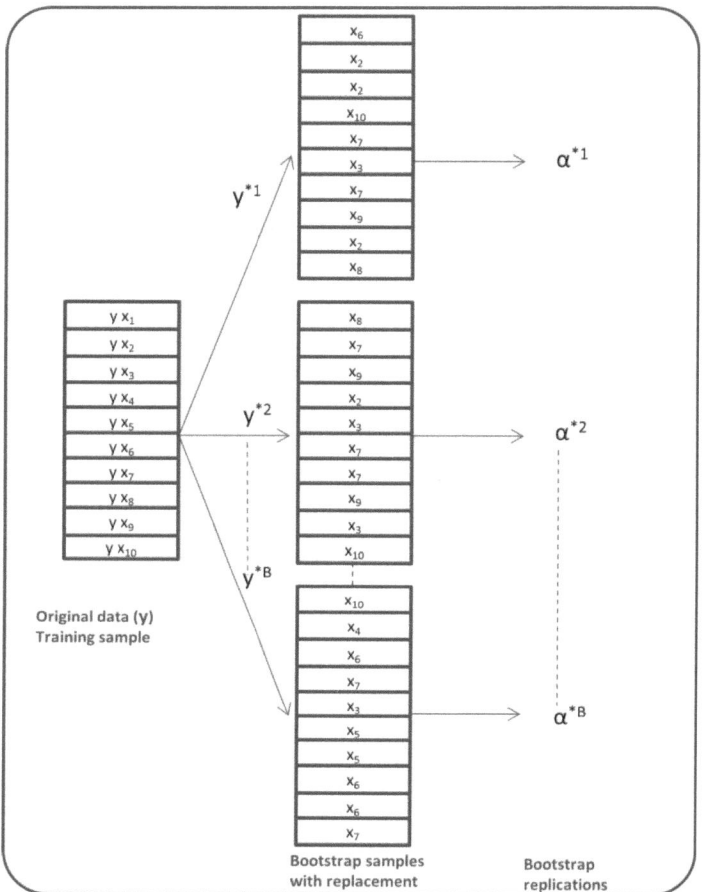

Figure 16.8 Procedure for the bootstrap approach exemplified on a small sample containing $n = 3$ observations. Each bootstrap dataset contains n observations, sampled with replacement from the original dataset. Each bootstrap dataset is used to obtain an estimate of α for evaluating predictive models. Adapted from Hastie et al. (2009) and James et al. (2013), with permission.

where R is the number of bootstrap samples, t_r^* is the value of the metric or model parameter of interest estimated in sample r, and $\bar{t}^*$ is the mean of the empirical bootstrap values.

The bias B is estimated as the difference between the model parameter calculated on all data (full model) and the mean of the estimates calculated on the empirical bootstrap values, as follows:

$$B = \frac{1}{R}\sum_{r=1}^{R}\left(t_r^* - t\right) = \overline{t}^* - t_0$$

With GLMs for instance, bias corrected values of the explained deviance of the predictor coefficients and other model parameters can be obtained in this way (Harrell et al., 1996). When the difference between the parameter of the full model and the mean of the empirical bootstrapped values is too high – what is called "optimisms from overfitting" (Harrell et al., 1996) – then the predictive ability of the model can be questioned (see SDM examples in Moretti et al., 2006; Marcelli et al., 2012).

Measuring bias and variance was the initial aim of bootstrap (Efron and Tibshirani, 1993), as measuring influence values was the initial aim of jackknife (see Section 16.2.2). However, bootstrap can also be used as an alternative to cross-validation to obtain data for evaluation (Efron and Tibshirani, 1997). We previously saw that on average in each subset, bootstrap selects 63.2% of all observations in the full dataset (Efron and Tibshirani, 1993). This means conversely that, again on average, at each iteration 36.8% of the observations will not be resampled and therefore can be used as data for evaluation. Although this was not the original intent of bootstrap, it can therefore be used as a powerful alternative to k-fold cross-validation. Referred to as the .632+ bootstrap, it provides more robust estimates than k-fold CV because it is repeated a much larger number of cases (Efron and Tibshirani, 1997; Hastie et al., 2009). However, one drawback is that, unlike the repeated split sample, a different number of observations is left out at each bootstrap iteration. This means the proportion of independent data is uneven between iterations, and therefore the programming of the procedure and statistical calculations is more complex. Although the potential exists (as presented in Efron and Tibshirani, 1997; see Robinson et al., 2011 for an example with GAMs used in forestry), for using bootstrap .632+ in the context of habitat suitability modeling, we are only aware of two examples (Wintle et al., 2005; Leathwick et al., 2006).

The bootstrap approach can thus be used to assess both the variability of the coefficient estimates and predictions from a statistical predictive model (normal bootstrap) and the predictive power when used as a cross-validation method (.632+ bootstrap). This section shows examples of applying bootstrap for both uses. We first illustrate the use to assess the variability of the estimates for b_0 and b_1, the intercept and slope terms for the previous linear regression model using bio3 to predict *V. vulpes*. We will compare the estimates obtained using the bootstrap to those obtained using the formulae for $SE(b_0)$ and $SE(b_1)$ described in James et al. (2013).

We first create a simple function, boot.fn() that takes the dataset to be used and a subset of the observations (obtained by bootstrap sampling) as arguments and returns the intercept and the slope estimates for the linear regression model. We then apply this function to the full set of 2488 observations in order to compute the estimates of b_0 and b_1 on the entire dataset using the usual linear regression coefficient estimate formula.

```
> boot.fn=function(data,index)
return(coef(glm(VulpesVulpes~bio3+bio7+bio11+bio12,
family="binomial",data=data,subset=index)))
> boot.fn(s_mammals_data,1:2488)

  (Intercept)           bio3           bio7          bio11          bio12
-1.0382810741 -0.1787174774   0.0216654829   0.0172735990   0.0009981371
```

The boot.fn() function can then be used to create bootstrap estimates for the intercept and slope terms by randomly sampling with replacement from among the observations. Here are two sample examples:

```
> set.seed(555)
> boot.fn(s_mammals_data,sample(2488,2488,replace=T))

  (Intercept)          bio3          bio7          bio11          bio12
-1.824877269 -0.162926619   0.022341309   0.016283213   0.001015902

> boot.fn(s_mammals_data,sample(2488,2488,replace=T))

  (Intercept)           bio3          bio7          bio11          bio12
-0.6875006581 -0.1827196538   0.0213501718   0.0177117856   0.0009370499
```

Next, we use the boot() function to compute the standard errors of 1000 bootstrap estimates for the intercept and slope terms.

```
> boot(s_mammals_data,boot.fn,1000)
ORDINARY NONPARAMETRIC BOOTSTRAP


Call:
boot(data = s_mammals_data, statistic = boot.fn, R = 1000)


Bootstrap Statistics :
          original          bias      std. error
t1* -1.0382810741   8.206176e-03   0.4820269002
t2* -0.1787174774  -1.144172e-03   0.0131623288
t3*  0.0216654829   9.378300e-05   0.0010757700
```

```
t4*   0.0172735990   1.141323e-04  0.0012264571
t5*   0.0009981371   4.299183e-06  0.0001572054
```

This indicates that the bootstrap estimate for $SE(b_0)$ is 0.4820, and that the bootstrap estimate is 0.0132 for $SE(b_1)$, 0.0011 for $SE(b_2)$, etc. These can be compared to the analytical standard errors for the regression coefficients obtained by the summary() function applied to the GLM:

```
> summary(glm(VulpesVulpes~bio3+bio7+bio11+bio12,
family="binomial", data=s_mammals_data))$coef

                 Estimate    Std. Error    z value      Pr(>|z|)
(Intercept)  -1.0382810741 0.4957518950  -2.094356 3.622826e-02
bio3         -0.1787174774 0.0112674083 -15.861454 1.171502e-56
bio7          0.0216654829 0.0010980693  19.730525 1.179391e-86
bio11         0.0172735990 0.0010227886  16.888728 5.446478e-64
bio12         0.0009981371 0.0001357794   7.351166 1.964859e-13
```

The standard error estimates for b_0 and b_1 obtained using the formulae are 0.4958 for the intercept and 0.0113 for the slope of bio3, 0.0011 for the slope of bio7, 0.0010 for the slope of bio11, and 0.0001 for the slope of bio12. Interestingly, these are somewhat different from the bootstrap estimates. This indicates a potential problem with the analytical coefficients. Below, we compute the bootstrap standard error estimates and the standard glm estimates that result from fitting the quadratic model to the data (bio3). Since this model provides a good fit to the data, there is now a better correspondence between the bootstrap estimates and the standard estimates of $SE(b_0)$, $SE(b_1)$, and $SE(b_2)$.

```
> boot.fn=function(data,index)
coefficients(glm(VulpesVulpes~bio3+I(bio3^2),
family="binomial",data=data,subset=index))
> set.seed(555)
> boot(s_mammals_data,boot.fn,1000)

ORDINARY NONPARAMETRIC BOOTSTRAP

Call:
boot(data = s_mammals_data, statistic = boot.fn, R = 1000)

Bootstrap Statistics :
         original          bias      std. error
t1* -5.826502239 -4.134456e-02 0.4758100058
t2*  0.535307238  3.321060e-03 0.0365987705
t3* -0.009090535 -5.514673e-05 0.0005977474

> summary(glm(VulpesVulpes~bio3+I(bio3^2),family="binomial",
data=s_mammals_data))$coef
```

	Estimate	Std. Error	z value	Pr(>|z|)
(Intercept)	-5.826502239	0.436563948	-13.34627	1.245227e-40
bio3	0.535307238	0.032049745	16.70239	1.259249e-62
I(bio3^2)	-0.009090535	0.000505708	-17.97586	3.011651e-72

We again use the Daim package to perform an estimation of the misclassification rate, sensitivity, specificity and AUC based on various bootstrap techniques in a randomForest model for *V. vulpes*. We calculate the optimal cut-point corresponding to various bootstrap estimation techniques for the sensitivity (TPR) and (1–specificity, FPR) using the Daim() function (Figure 16.9). The function Daim.control() can be used to control the parameters affecting the models' diagnostic accuracy. In the case of bootstrap, it can be defined as: Daim.control(method="boot", number=100).

```
> vulpes_RF <- Daim(formula=VulpesVulpes~., model=myRF,
data=vulpes_data, labpos="1", control=Daim.control(number=50))

> summary(vulpes_RF)

Performance of the classification obtained by:

Call:
VulpesVulpes ~ bio3 + bio4 + bio7 + bio11 + bio12

Daim parameters:
  method = boot, nboot = 50, replace = TRUE, boot.size = 1,
cutoff = 0.5,

  est.method = obs.

Result:
-------------------------------------------------------------
| Method:      | | .632+ | | .632  | | loob   | | apparent |
=============================================================
| Error:       | | 0.0472 | | 0.0448 | | 0.0710 | | 0.0000  |
-------------------------------------------------------------
| Sensitivity: | | 0.9491 | | 0.9518 | | 0.9237 | | 1.0000  |
-------------------------------------------------------------
| Specificity: | | 0.9559 | | 0.9582 | | 0.9339 | | 1.0000  |
-------------------------------------------------------------
| AUC          | | 0.9889 | | 0.9897 | | 0.9800 | | 1.0000  |
-------------------------------------------------------------

> par(mfrow=c(2,2))
> plot(vulpes_RF, method="0.632+", legend=TRUE)
> plot(vulpes_RF, method="sample")
> plot(vulpes_RF, method="0.632+",
main="Comparison between methods")
> plot(vulpes_RF, method="0.632", col="blue", add=TRUE)
```

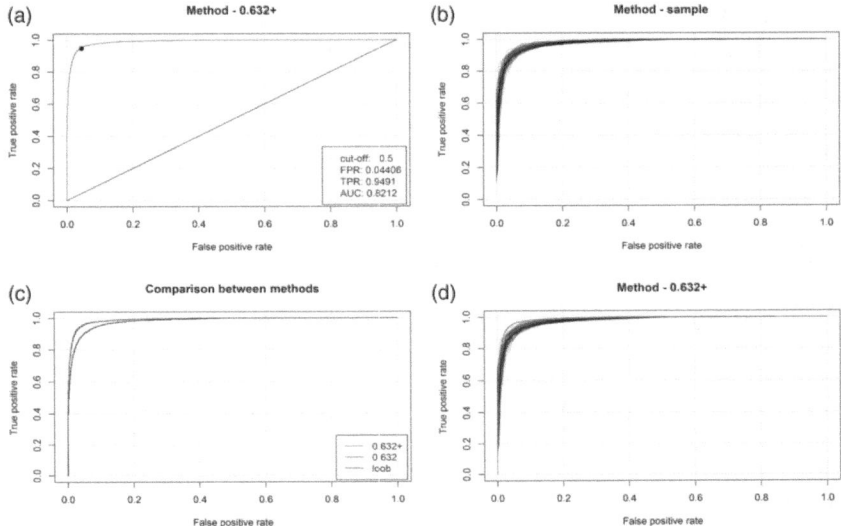

Figure 16.9 Plot of the `Daim` object generated by the `Daim()` function correspond-ing to ROC curves for various bootstrap evaluation methods. (a) The method "0.632+" discussed in the main text, (b) all the bootstrap samples, (c) a comparison of methods bottom left, and (d) all the bootstrap samples plotted together with the one of the "0.632+" method bottom right. FPR: false positive (presence) rate (1 − specificity), TPR: true positive (presence) rate (sensitivity), loob: leave-one-out bootstrap. See Efron and Tibshirani (1993), Efron and Gong (1983), and Efron and Tibshirani (1997). (*A black and white version of this figure will appear in some formats. For the color version, please refer to the plate section.*)

```
> plot(vulpes_RF, method="loob", col="green", add=TRUE)
> legend("bottomright", c("0.632+","0.632","loob"),
col=c("red","blue","green"), lty=1, inset=0.01)
> plot(vulpes_RF, all.roc=TRUE)
```

This function can also be used to obtain the optimal cut-point cor-responding to the "0.632+ bootstrap" estimation of the sensitivity and the specificity. In the following example, the best cut-point corresponds to 0.46.

```
> vulpes_RF2 <- Daim(formula=VulpesVulpes~., model=myRF,
data=vulpes_data, labpos="1", control=Daim.control(method="boot",
number=100), cutoff="0.632+")
> summary(vulpes_RF2)

Performance of the classification obtained by:

Call:
VulpesVulpes ~ bio3 + bio4 + bio7 + bio11 + bio12
```

```
Daim parameters:
  method = boot, nboot = 100, replace = TRUE, boot.size = 1,
cutoff = 0.632+,
  est.method = obs, best.cutoff = 0.46.
```

Result:

Method:	.632+	.632	loob	apparent
Error:	0.0462	0.0438	0.0694	0.0000
Sensitivity:	0.9570	0.9590	0.9351	1.0000
Specificity:	0.9508	0.9536	0.9266	1.0000
AUC	0.9891	0.9899	0.9803	1.0000

16.3 External Evaluation (Fully Independent Data)

An external evaluation consists of assessing a model's predictive power on a distinct, independent dataset to the one used to fit the model (Manel et al., 2001; Edwards et al., 2006), ideally in a distinct area or time period, fully independent of the area or time period used to fit the model (see Randin et al., 2006; Segurado et al., 2006; Bahn and McGill, 2007 for a discussion of what are independent data). A perfect example of independent data is when these are sampled after fitting the model, (i.e. post-modeling), stratified or not by the model predictions (Newbold et al., 2010; Figure 16.10).

Testing on an independent dataset has often been considered the most robust type of evaluation by ecological modelers (Fielding and Bell, 1997; Manel et al., 1999a; Araújo et al., 2005a). The assumption that fully independent evaluation is more robust initially appears to be an indefectible principle, but what are true independent data? This question is crucial since in most (if not all) cases, splitting an initial dataset or sampling two datasets differently in geographic or temporal space (as proposed by Wenger and Olden, 2012) will generate confounding problems, such that the separated samples may represent locally-adapted sub-taxa with divergent ecology, which may indeed require a separate model to be run for each sub-taxon. Several issues must thus be considered when running an evaluation on a dataset considered to be "independent", which in some circumstances might lead us to doubt the true independence of the data. These can be separated into four situations, depending on whether the independent dataset is: (i) within the same study area extent and same

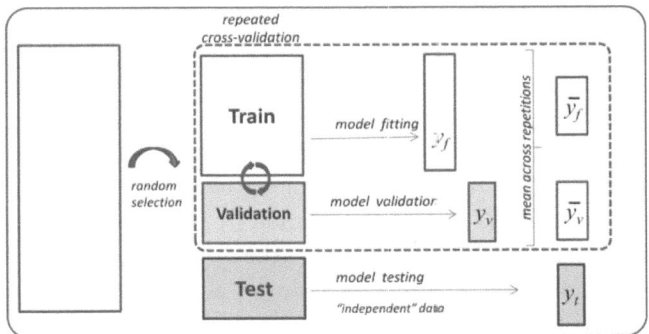

Figure 16.10 Procedure for internal (semi-independent) and external (with fully independent data) evaluation consisting of randomly dividing the dataset into three parts: a training set, a validation set, and a test set. The training set is used to fit the models; the validation set is used to estimate prediction error for model selection; the test set is used to assess the generalization error of the final chosen model. See Hastie et al. (2009).

time period, (ii) in a different geographic area but same time period, (iii) in a different time period but same geographic area, (iv) in a different area and time period (some cases of biological invasions under climate change typically fall into this category). However, since the considerations developed for point (ii) and (iii) should apply conjointly for point (iv), we will not discuss the latter any further herein.

When running an independent evaluation in the same area and time period, one of the potential problems is that the training and test datasets are not spatially independent, i.e. that their observations are spatially autocorrelated, therefore reducing the "independence" of the test observations, as regards the training observations. In order to assess this, the spatial independence between the test and training sets can be tested with spatial autocorrelation methods, as done for instance in Pottier et al. (2013) (see also: Bahn et al., 2006; Bahn and McGill, 2007; Beale et al., 2013; Fithian et al., 2015). In turn, in this evaluation in a same area and time period the spatial structure of the environmental predictors remains the same between the training and test sets, so that the model can be confidently transferred from one situation to the other. However, it will not guarantee that the model can be applied to another area (e.g. Randin et al., 2006) or time period (e.g. Araújo et al., 2005a) where the spatial co-variation between predictors is different or has changed (e.g. Wenger and Olden, 2012), as when attempting to anticipate biological invasions (e.g. Thuiller et al., 2005b) or the impact of climate change on biodiversity (e.g. Broennimann et al., 2006). In such cases, the model's transferability to these distinct conditions should ideally be tested, where this is feasible. Tests of geographical transferability have been conducted in geographic (Randin et al., 2006; Zanini et al., 2009; Petitpierre et al., 2017) and temporal (Araújo et al., 2005a; Pearman et al., 2008b; Tuanmu et al., 2011; Maiorano et al., 2013) spaces (see also Part V).

Transferability in geographic space is performed by fitting a model in one area and projecting it to another area where the same environmental predictor variables are available as maps (Randin et al., 2006; Segurado et al., 2006). However, and especially when transferring over large geographic distance where gene flow may not be guaranteed between the separated populations, one cannot exclude the possibility that genetic differentiation also occurs between them, potentially leading to niche divergences and therefore limiting model transferability, without invalidating the initial model. For instance, arctic–alpine species, present in the European Alps and in the European Arctic, seem to have retained the same cold tolerance limits, but not their warm tolerance limits (Pellissier et al., 2013a). In such cases, model transferability may not be an adequate test of a model's predictive power, and the models should preferably be fitted locally and projected to the same study area, e.g. to anticipate climate change impacts (as in Engler et al., 2011b).

Transferability in time is performed similarly to transferability in geographic space, by fitting models in one time period and projecting to a different time period where the same environmental predictor variables are available as maps (Araújo et al., 2005a; Araújo and Rahbek, 2006; Pearman et al., 2008b; Maiorano et al., 2013). Projecting from the present to the past is called hindcasting whereas projecting from the past to the present is called forecasting. Both are independent evaluations. Examples of hindcasting are projections of species models from the present time to several thousand years ago, where they are tested with independent pollen fossil data (Graham et al., 2004a; Martínez-Meyer et al., 2004; Pearman et al., 2008a; Maiorano et al., 2013), whereas examples of forecasting are projections of models from the recent past (using historical data) to the present time (Araújo et al., 2005a; Scherrer et al., 2017). The problem with both types of temporal evaluations is that factors not accounted for in the models but which influence species distribution may have changed between the time periods, or more importantly, unknown factors may have affected the population dynamics over time, such as disease outbreak, population declines due to stochastic events, human-based extirpation or biotic interactions. These issues hinder model performance in the test area without necessarily implying that the model is wrong or incorrectly parameterized. When conducting such experiments, it is difficult to conclude whether the calibrated HSMs show a good predictive performance or not. When they work well, there is a good chance that the habitat suitability model has good predictive performance. However, when they do not, sound conclusions are more difficult to reach. Another important issue is the availability of data over time for this type of evaluation, which remains scarce especially for past periods. The projection itself of models in time and space is treated in more depth in Part V.

PART V · Predictions in Space and Time

We have already seen in Parts I and IV that HSMs can be used to make predictions in time and/or space. For the purposes of convenience, in this section "projection" will be used to refer to any prediction made outside of the study area or time period used to train the model. We will also at times refer to this procedure as transferability in space and time. One can, for instance, project: (i) to a different area to anticipate biological invasions (e.g. Thuiller et al., 2005b; Gallien et al., 2010; Petitpierre et al., 2012); (ii) to future time periods to assess the possible impact of climate change on species ranges or diversity (e.g. Engler et al., 2011a; Thuiller et al., 2011); (iii) to both other areas and time periods, to assess the future state of invasions in a changed climate (e.g. Roura-Pascual et al., 2004; Broennimann and Guisan, 2008; Peterson et al., 2008b); (iv) to past periods (hindcasting; (e.g. Espíndola et al., 2012; Maiorano et al., 2013) or (v) to present distribution from past records (forecasting; Pearman et al., 2008b). However, additional assumptions have to be made to make these transfers.

This part is composed of a single chapter (Chapter 17), divided into four sections. The first section introduces the additional assumptions made when projecting models in space and time. The second and third sections then present approaches and examples of projections in space and time respectively. Finally, the fourth section presents the use of ensemble modeling for projections. This part is therefore based on, and complements, Parts III and IV, by showing how previously fitted and discussed models can be used to generalize projections in space and time. When predicting to different study areas or time periods (i.e. projecting), we will see that new issues arise, such as niche completeness, niche stability, and environmental analogy, and that these require careful consideration before making or interpreting any projections.

17 · *Projecting Models in Space and Time*

17.1 Additional Considerations and Assumptions When Projecting Models: Analog Environment, Niche Completeness, and Niche Stability

The first important consideration when projecting models in time or space is to assess whether the same environmental conditions prevail in both areas or periods, i.e. quantifying to what extent the environment (i.e. the envelope of environmental conditions found in an area in a given time period) is comparable between the area/period used for model fitting (training area/period) and the area/period used for projecting the model (projection area/period). There are two dimensions to this question, related to how *available* (Jackson and Overpeck, 2000; Ackerly, 2003) and how *analog* (Williams and Jackson, 2007; Fitzpatrick and Hargrove, 2009) the environment is between the two areas or time periods (Figures 17.1 and 17.2).

The available environment in each area/period is also known as the "realized environment" (Jackson and Overpeck, 2000) and comprises the particular combinations of the environmental variables (considered in the study) that exist in the area and/or at the time considered, ideally including the frequency with which these are encountered (i.e. their availability). Two areas or periods may have roughly the same envelope of realized environment, but may differ in terms of the frequency and dominance with which the different combinations (that make the envelope) exist in each area/period (Figure 17.2 shows and example for Eurasia and North America), with potential implications regarding the ecological and evolutionary processes that have taken place in each area (e.g. speciation/diversification, specific adaptations, etc.).

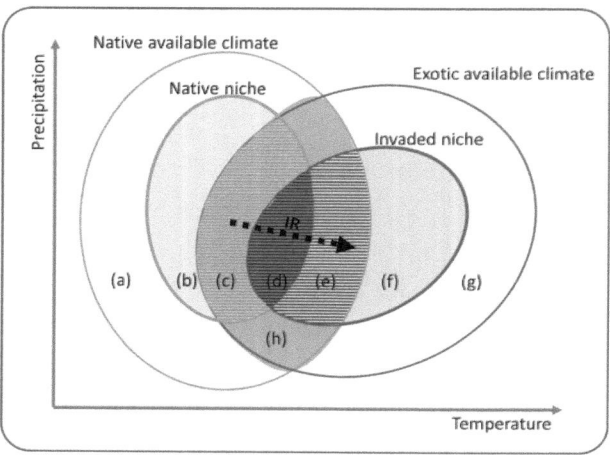

Figure 17.1 Schematic two-dimensional representation of the indices of niche change (unfilling, stability and expansion) presented in Broennimann et al. (2012) (see definitions in Guisan et al., 2014, Box 3). Solid thin lines show the density of available environments (see Box 4 in Guisan et al., 2014) in the native range (in green) and in the invaded range (in red). The gray area shows the most frequent environments common to both ranges (i.e. analog environments). The green and red thick lines show respectively the native and the invaded niches. Niche unfilling (U), stability (Se) and expansion (E) are shown respectively with green, blue and red hatched surfaces inside analog environments. The definition of a niche shift using the change of niche centroid only (inertia ratio, IR) is shown with a thick dotted arrow. In this context, the lower-case letters represent similar features in both graphs. (a) Available conditions in the native range, outside of the native niche and non-analog to the invaded range. (b) Conditions inside of the native niche but non-analog to the invaded range. (c) Unfilling, i.e. conditions inside of the native niche but outside the invaded niche, possibly due to recent introduction combined with ongoing dispersal of the exotic species, which should ultimately fill these conditions. (d) Niche stability, i.e. conditions filled in both native and invaded range. (e) Niche expansion, i.e. conditions inside the invaded niche but outside the native one, due to ecological or evolutionary change in the invaded range. (f) Conditions inside of the invasive niche but non-analog to the native range. (g) Available conditions in the invaded range but outside of the invasive niche and non-analog to the native range. (h) Analog conditions between the native and invaded ranges. Figure from Guisan et al. (2014), with permission. (*A black and white version of this figure will appear in some formats. For the color version, please refer to the plate section.*)

Figure 17.2 below can be generated with the following R code, using the niche quantification and comparison functions from the `ecospat` R package.

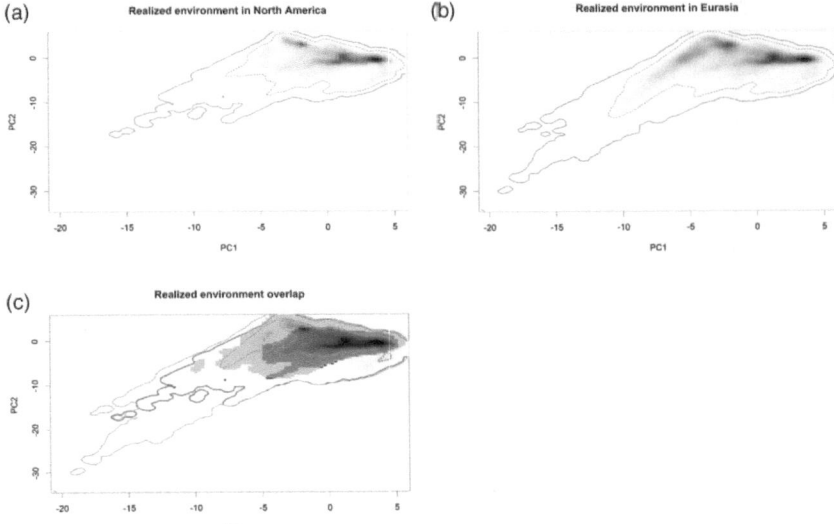

Figure 17.2 Comparison of the realized (available) environments in North America and Eurasia in a multivariate environmental space calculated from a principal component analyses (PCA) based on a large random sample of sites in both areas (Broennimann et al., 2012). Here, the overall envelope is displayed, with densities plotted. In many cases, only the simple contour of the envelope is provided. Note that this is not the niche of species, it is the full envelope of available environments in an area. (a) Realized environment in North America; (b) Realized environment in Eurasia; (c) Intersection of the realized environments between North America and Eurasia, showing analog (in dark, i.e. the intersection of the two envelopes), and non-analog (i.e. non-intersecting parts of each envelope; black shows conditions found only in North America, and grey shows conditions found only in Eurasia) situations.

```
> library(ecospat)
## Preparation of datasets
# load climate variable for all site of the Eurasian study area
(column names should be x,y,X1,X2,...,Xn)
> clim1<-read.table("tabular/bioclim/current/clim.vulpesNA_100.
txt",h=TRUE)

# load climate variable for all site of the North American study
area (column names should be x,y,X1,X2,...,Xn)
> clim2<-read.table("tabular/bioclim/current/clim.vulpesEU_100.
txt",h=TRUE)

# global climate for both ranges
> clim12<-rbind(clim1,clim2)

# loading occurrence sites for the species (column names should
be x,y)
> occ.sp1<-na.exclude(read.table("tabular/species/vulpes_
na.txt",h=TRUE)[c(1,2)])
```

```
> occ.sp2<-na.exclude(read.table("tabular/species/vulpes_
eu.txt",h=TRUE)[c(1,2)])

# create species occurrence dataset by adding climate variables
from the global climate datasets
# resolution should be the resolution of the climate data grid,
in this case at 100km

> occ.sp1 <- na.exclude(ecospat.sample.envar(dfsp=occ.sp1,colspx
y=1:2,colspkept=NULL,dfvar=clim1,colvarxy=1:2,colvar="all",
resolution=1))
> occ.sp2 <- na.exclude(ecospat.sample.envar(dfsp=occ.sp2,cols
pxy=1:2,colspkept=NULL,dfvar=clim2,colvarxy=1:2,colvar="all",
resolution=1))

## ANALYSIS - selection of parameters
# selection of variables to include in the analyses
> names(clim12)
> Xvar<-c(3:21)
> nvar<-length(Xvar)
# number of interation for the tests of equivalency and similarity
> iterations<-100
# resolution of the gridding of the climate space
> R=100

## row weighting and grouping factors for ade4 functions
> row.w.1.occ<-1-(nrow(occ.sp1)/nrow(rbind(occ.sp1,occ.sp2)))
# prevalence of occ1
> row.w.2.occ<-1-(nrow(occ.sp2)/nrow(rbind(occ.sp1,occ.sp2)))
# prevalence of occ2
> row.w.occ<-c(rep(0, nrow(clim1)),rep(0, nrow(clim2)),
rep(row.w.1.occ, nrow(occ.sp1)), rep(row.w.2.occ, nrow(occ.sp2)))

> row.w.1.env<-1-(nrow(clim1)/nrow(clim12))
# prevalence of clim1
> row.w.2.env<-1-(nrow(clim2)/nrow(clim12))
# prevalence of clim2
> row.w.env<-c(rep(row.w.1.env, nrow(clim1)),rep(row.w.2.env,
nrow(clim2)), rep(0, nrow(occ.sp1)), rep(0, nrow(occ.sp2)))

> fac<-as.factor(c(rep(1, nrow(clim1)),rep(2, nrow(clim2)),
rep(1, nrow(occ.sp1)),rep(2, nrow(occ.sp2))))

# global dataset for the analysis and rows for each sub dataset
> data.env.occ<-rbind(clim1,clim2,occ.sp1,occ.sp2)[Xvar]
> row.clim1<-1:nrow(clim1)
> row.clim2<-(nrow(clim1)+1):(nrow(clim1)+nrow(clim2))
> row.clim12<-1:(nrow(clim1)+nrow(clim2))
> row.sp1 <- (nrow(clim1)+nrow(clim2)+1):(nrow(clim1)+nrow(clim2
)+nrow(occ.sp1))
> row.sp2 <- (nrow(clim1)+nrow(clim2)+nrow(occ.sp1)+1):(nrow(cli
m1)+nrow(clim2)+nrow(occ.sp1)+nrow(occ.sp2))
## PCA-ENV
```

```
# measures niche overlap along the two first axes of a PCA
calibrated on all the pixels of the study areas
# Fit of the analyses using both ranges using the dudi.pca()
function from the ade4 package

> pca.cal <-dudi.pca(data.env.occ,row.w = row.w.env, center = T,
scale = T, scannf = F, nf = 2)
# predict the scores on the axes
> scores.clim12<- pca.cal$li[row.clim12,]
> scores.clim1<- pca.cal$li[row.clim1,]
> scores.clim2<- pca.cal$li[row.clim2,]
> scores.sp1<- pca.cal$li[row.sp1,]
> scores.sp2<- pca.cal$li[row.sp2,]

# calculation of environmental density using the ecospat.grid.
clim.dyn() function from the ecospat package

> z1 <- ecospat.grid.clim.dyn(scores.clim12,scores.clim1,
th.sp= 0,scores.sp1,R)
> z1$z.uncor <- z1$Z

> z2 <- ecospat.grid.clim.dyn(scores.clim12,scores.clim2,
th.sp= 0,scores.sp2,R)
> z2$z.uncor <- z2$Z

# plot realized environment
> ecospat.plot.niche(z1,title=" Realized environment in North
America ",name.axis1="PC1",name.axis2="PC2")
> ecospat.plot.niche(z2,title=" Realized environment in Eurasia
",name.axis1="PC1",name.axis2="PC2")
> ecospat.plot.niche.dyn (z1, z2, quant=0.8, title="Realized
environment overlap", name.axis1="PC1",name.axis2="PC2",
interest = 1, colz1 = "#00FF0050", colz2 = "#FF000050",
colinter = "#0000FF50", colZ1 = "green3", colZ2 = "red3")
```

This question of environmental availability is important because it can influence the way key assumptions (such as niche conservatism/stability; see below) are evaluated, and also how some models are parameterized (e.g. when selecting background data for presence-only models, see Part III). The question of environmental analogy is different, though complementary. It relates to the identification of environments that exist in one area/period but not in the other, i.e. *non-analog* environments (Figure 17.1 and 17.2). It therefore directly relates to the issue of interpolation (in analog environments) versus extrapolation (in non-analog environments). The latter is not straightforward to assess, as the environment can be analog for most variables except for one or a few, and these variables may be important in different ways for the species under investigation. In recent years, several tests have been developed to compare different environments, such as the multivariate environmental similarity surface (MESS) (Elith et al., 2010) or the ExDet (Mesgaran et al., 2014) analyses. Assessing environmental availability and analogy together is important because it can affect the quantification of the realized niche in each area or time period and thus affect the way models are built, compared, and projected, between areas and time periods (Guisan et al., 2014; see below). Projecting a model without comparing the environment in space and time implies that the environment is assumed to be identically available and analogous between the two time periods or areas.

A second implicit assumption when projecting models in time and/ or space is that the full realized niche is captured in the model. This assumption is implicit because a model built with data covering only a limited part of a species' geographic extent may result in truncated or biased response curves if the geographic truncation also results in environmental truncation (Thuiller et al., 2004a), which will introduce errors when projected to different areas or time periods (Thuiller et al., 2004a; Guisan and Thuiller, 2005; Barbet-Massin et al., 2010). A clear example is shown in Figure 17.3 (originating from Thuiller et al. (2004a), where the distribution of an endemic species in Southern Europe (France and Italy) is used as the starting point for building full-range GAM (see Part III), then the geographic extent is reduced and a GAM is fitted again and projected. This second GAM model based on the restricted geographic extent produces response curves that end abruptly and thus cannot properly inform on how to project the distribution in conditions outside the training range. As a result, an artificial trend is derived by the GAM that leads to this Mediterranean species being predicted in the north of Scandinavia (Figure 17.3), where the conditions are physiologically unsuitable for this species (Thuiller et al., 2004a).

The same type of problem was encountered for many bird species in the south of Europe by Barbet-Massin et al. (2010) when the north of Africa was not included to train the models before projecting them to warmer future climates. At least for climate, it is therefore extremely important to capture as much of the species' full realized niche as possible before projecting models to changed future environmental conditions. This is especially true for the climatic niche and in areas of limited extent where such climatic truncation may occur. One way to reduce the problem in the latter case is to build two models of species distributions at two extents and integrate them (e.g. using a Bayesian approach): a first extent large enough to include the full climatic niche of the species, and a second one which takes into account more local predictors (e.g. land use, topography, substrate; Pearson et al., 2004). This type of

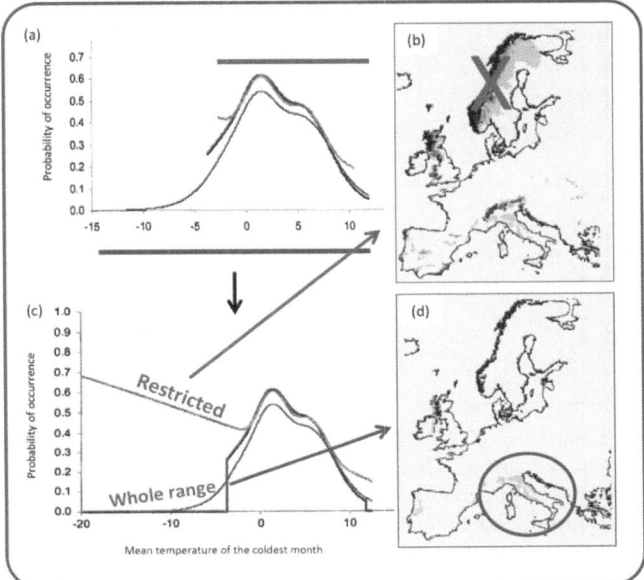

Figure 17.3 Illustration of the niche truncation problem for an oak species with restricted range in Europe, *Quercus crenata* Figure based on Thuiller et al. (2004b), with permission. (a) Response curves along the mean temperature of the coldest month from GAM (generalized additive models) fitted with restricted (two different levels of truncation) and full ranges (see the two thick lines above and below the plot), and (c) how GAM handles the extrapolation along the whole temperature gradient. Note that the curve is forced to zero in the second truncation case, whereas in the more severe truncation, the GAM forces the curve to increase again below temperatures of zero. (b) Spatial prediction based on the truncated model, showing incorrect predictions of the species in Scandinavia. (d) Spatial prediction with the full-range model, showing the correct prediction to the observed distribution range, in the South of France and Italy.

approach has recently been proposed for modeling invasion at regional scale, while using information at global scale and at a coarse resolution to avoid extrapolation out of the species' global niche (Gallien et al., 2012; Petitpierre et al., 2016). This ad-hoc hierarchical modeling represents one possible approach. It uses the prediction from the global model to weight the pseudo-absences in the regional model. The recent development of hierarchical Bayesian approaches in ecological modeling (e.g. Hooten et al., 2003; Carroll et al., 2010) should pave the way toward better ways of integrating models at the two extents. Keil et al. (2013) proposed a hierarchical Bayesian model for downscaling purposes, but their approach

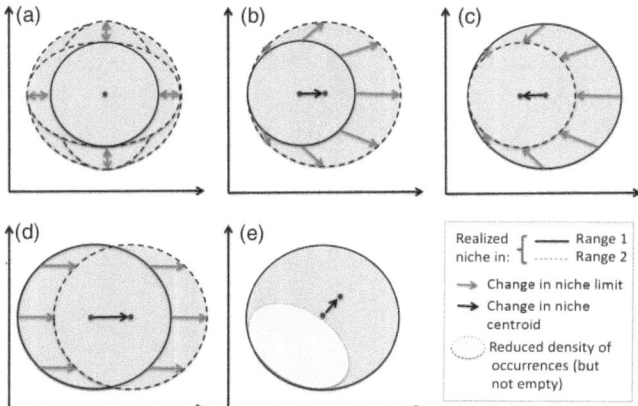

Figure 17.4 Theoretical scenarios of realized niche changes in space (e.g. following invasions) or time (e.g. under climate change). Change in: (i) the niche envelope (expansion or contraction) without change of the niche centroid, due to symmetric niche change, i.e. in two opposite or all directions in climatic space (a); (ii) the niche centroid with directional expansion (b) shrinkage (c) or displacement (d) of part of or the whole niche envelope, or (iii) the niche centroid only, due to a change of the density of occurrences within the same niche envelope in climatic space (e). The latter case would result in stability (no change) in Figure 17.1. Observed changes are likely to be combinations of these cases. Figure from Guisan et al. (2014), with permission. (*A black and white version of this figure will appear in some formats. For the color version, please refer to the plate section.*)

should also be applicable to the combination of models at different scales (see Chapter 20).

When projecting models in space and/or time, one projects the quantified realized niche (i.e. based on empirical field observations; Araújo and Guisan, 2006; Soberón, 2007). Another important assumption here is thus that this niche is implicitly considered to remain the same (i.e. stable) across different areas or time periods (i.e. to be conserved; Peterson, 2003; Pearman et al., 2008a; Guisan et al., 2014). If not, in which way does the niche change (e.g. expanding, shifting, shrinking; see Figure 17.4)? For example, if one wishes to project species distributions into a changed climatic future, then one may want to know how much and how the realized climatic niche of these species (on which the models are built) varied in the past and up to the present (Pearman et al., 2008a; Pearman et al., 2008b; Nogues-Bravo, 2009; Maiorano et al., 2013). When the niche was shown to fluctuate substantially in time or space, approaches to building the niche from different time periods (Maiorano et al., 2013) or areas (Broennimann and Guisan, 2008) may help circumvent (at least in part) the problem (see below).

Are species' realized niches actually conserved in space and time then? To answer this question, we should first recall that the niche fitted using HSMs is the expanded definition of the realized environmental niche, including constraints by biotic interactions and dispersal (Araújo and Guisan, 2006; Soberón, 2007; see Part I). The refined question is thus: does this extended species' realized environmental niche remain constant in time and space? And if not, how does it change? The answer then includes two components: one ecological and one evolutionary (Broennimann et al., 2007; Pearman et al., 2008a). The answer may be difficult to obtain in many cases because the different constraints can act as confounding factors on each other (e.g. limited dispersal may affect the niche in a way that is then interpreted as a change in biotic interactions or evolution in the new range; Guisan et al., 2012), and therefore the same change may result from both evolutionary and ecological processes. The need to check for niche stability applies for instance when projecting into the future from present observations in a single range and time period, but it is also worth questioning what kind of analyses can be used when there are observations available for different time periods or different geographic areas (e.g. biological invasions, comparisons of disjunct distributions, etc.). Findings across many studies show that both change and stability of the realized niche can be observed in the case of biological invasions, depending on the taxonomic groups considered and niche quantification methods used (see Guisan et al., 2014 for a review), but conservatism (stability) tends to predominate over evolutionary time (Peterson, 2011).

The ecological explanation of observed niche changes, for instance following invasions, may thus relate to changes in ecological interactions, mostly biotic but possibly also abiotic, and dispersal limitations in both the native and invaded ranges (Pearman et al., 2008a; Guisan et al., 2014). These factors can result in an expansion or a contraction of the niche (Figure 17.4), for instance as biotic interactions or dispersal barriers are increased or removed in the different time period or area (Pulliam, 2000; Pearman et al., 2008a; Soberón and Nakamura, 2009; Guisan et al., 2014).

As two different areas or time periods are likely to host different species assemblages and to have different patterns of geographic barriers and dispersal limitations, changes in the realized niche are also likely to

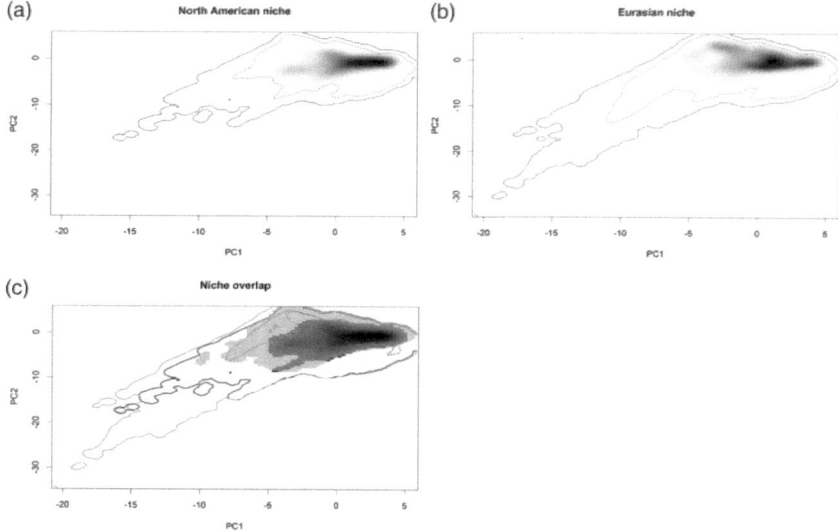

Figure 17.5 Comparison of the realized niches of the red fox (*Vulpes vulpes*) between its native distribution in Eurasia and both native and invaded distribution in North America. (a) Realized climatic niche in Eurasia; (b) realized climatic niche in North America; (c) overlap of the climatic niches between the two ranges, showing the stable (shared, overlapping) portion of the two niches in dark, and the differing niche conditions between the two ranges in black and grey, black showing conditions found only in North America and grey showing conditions found only in Eurasia.

be observed (as in Figure 17. 5). The question thus becomes how much of its fundamental niche a species occupies in the field at a given time (Maiorano et al., 2013), i.e. is it reduced to a smaller realized niche, and if so by how much? For instance, in the case of the red fox example used throughout this book, quantifying the niche in Eurasia and North America separately reveals slight niche differences, although the species is native in both ranges (Figure 17.5). This shows that each range only captures part of the full realized niche.

Figure 17.5 can be generated with the following R code. Following the previous calculation of environmental density, we can calculate the occurrence density for the species in each range.

```
# Calculation of occurrence density using the ecospat.grid.clim.
dyn() function from the ecospat package
> z1 <- ecospat.grid.clim.dyn(scores.clim12,scores.clim1,
th.sp= 0,scores.sp1,R)
> z2 <- ecospat.grid.clim.dyn(scores.clim12,scores.clim2,
th.sp= 0,scores.sp2,R)
```

```
# plot niche overlap
> ecospat.plot.niche(z1,title=" North American niche",
name.axis1="PC1",name.axis2="PC2",
> ecospat.plot.niche(z2,title="Eurasian niche",
name.axis1="PC1",name.axis2="PC2",
> ecospat.plot.niche.dyn (z1=z1, z2=z2, quant=0.8,
title="Niche overlap", name.axis1="PC1",name.axis2="PC2",
interest = 1, colz1 = "#00FF0050", colz2 = "#FF000050",
colinter = "#0000FF50", colZ1 = "green3", colZ2 = "red3")
```

When detecting changes in the realized niche, the next question is: are there any species properties that allow us to predict how much of its fundamental niche a species occupies? For instance, if a species is dominant across its full range of tolerances and has good dispersal ability, it is likely to occupy a larger part of its fundamental niche than subordinate species or species with limited dispersal abilities. This may influence whether the niche can be safely projected in space and time (Pearman et al., 2008b). However, very few studies have so far attempted to quantify the difference between the fundamental and realized niche (e.g. Malanson et al., 1992; Vetaas, 2002; Kearney and Porter, 2004; Wharton and Kriticos, 2004; Araújo et al., 2013). This is because this question is extremely difficult to assess from empirical data on species distributions, and experimental *in situ* and *ex situ* studies are also needed to explore this issue (but see Araújo et al., 2013).

The evolutionary explanation of niche change relates to a change of the fundamental niche of species, e.g. through evolution in the new range or in the new period (Dietz and Edwards, 2006). This could theoretically be caused by founder effects followed by genetic drift or natural selection in the case of biological invasions (Pearman et al., 2008b), as discussed by Lavergne and Molofsky (2007) for an invasive grass species.

Another crucial question here is to know how to *measure* such changes in the realized niche (Guisan et al., 2014)? Depending on the statistical approach and test used, there may be different answers to the same question (Pearman et al., 2008a; Warren et al., 2008; Guisan et al., 2014). For instance, Warren and colleagues (2008) reviewed two distinct tests of niche differences in geographical space, later generalized in environmental space by Broennimann et al. (2012). This highlights a first important dichotomy between existing tests in two approaches (Broennimann et al., 2012; Guisan et al., 2014; Figure 17.6): (i) tests in environmental space (i.e. ordination), using multivariate ordinations; (ii) tests in geographic space, using predictions of ecological niche models (Figure 17.6).

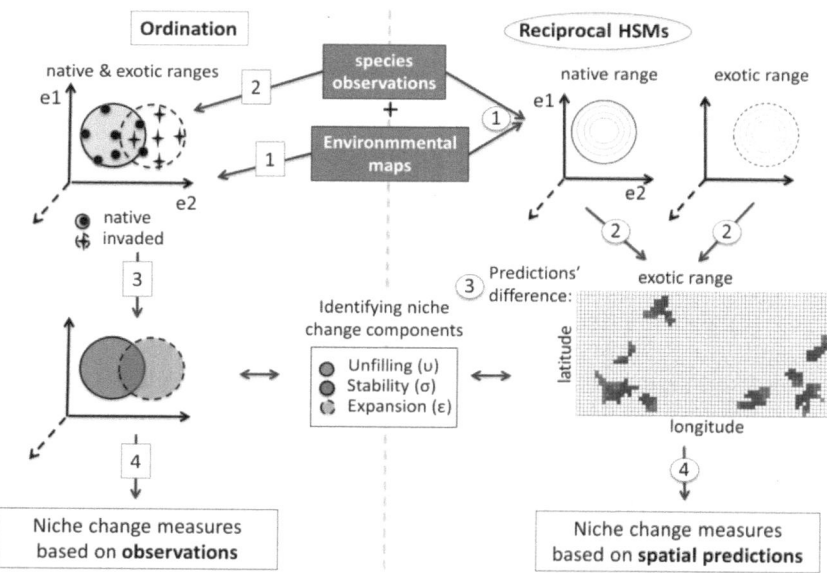

Figure 17.6 The two approaches commonly used to quantify niche changes between ranges). Ordination is based only on the observations, whereas HSM is based only on the predictions (see reference 22 and Box 1 in Guisan et al. 2014). The steps for ordination are (square numbers): 1. Definition of the reduced multidimensional environmental space; 2. Plotting the observations from each range in this space; 3. Comparing the niche defined from observations in each range; 4. Calculating the niche change metrics (see Box 3 in Guisan et al. 2014). The steps for HSMs are: 1. Fitting HSMs by relating field observations to environmental variables; 2. Projecting the HSMs in geographic space; 3. Computing differences in the projections; 4. Calculating the niche change metrics. See Guisan et al. (2014) for discussion of the respective strengths and weaknesses of the two approaches. Figure from Guisan et al. (2014), with permission. (*A black and white version of this figure will appear in some formats. For the color version, please refer to the plate section.*)

Using both approaches, niches can be further tested for being strictly equivalent (test of niche equivalency) or for being more similar to one another than to any random niche fitted in the same realized environment (test of niche similarity; Warren et al., 2008; Broennimann et al., 2012). For instance, in the case of biological invasions, the test of niche equivalency is usually so strict that it is rejected (often slightly, in both spaces) for most species between their native and invaded geographic ranges (Petitpierre et al., 2012), and so it would prevent projecting predictions to other areas for most species. On the other hand, niche similarity only tests if the two niches (in different time periods or areas) share a greater portion of their volume (i.e. niche overlap) than would be expected by chance in the same context, and so niche overlap may well be very small but significantly different from the overlap obtained at random. As a result, if the overlap between the two niches is small, projecting the model to a different range may fail even if the niche similarity is not rejected, (Broennimann et al., 2012; Petitpierre et al., 2012). This test of niche similarity is more useful for testing evolutionary hypotheses such as niche evolution among clades and especially between sister taxa (Broennimann et al., 2012), or niche conservatism among lineages (Graham et al., 2004b; Wiens and Graham, 2005), than it is for testing HSM projections. On the other hand, and as said above, the niche equivalency test is so strict that it is rejected in most cases when comparing niches between pairs of species (e.g. sister species in a phylogenetic tree), or for a same species in time or space.

How then can we test whether niches between areas or time periods are stable enough (i.e. whether the realized niche is sufficiently shared) to project the associated models in space and time? It has been shown in this regard that the degree of projection failure relates to the extent of niche change (Pearman et al., 2008b; Petitpierre et al., 2012). A pragmatic approach when species observations are available for different time periods or areas is to simply assess the relationship between niche overlap and the failure to project the related models to a range of different situations (e.g. different species, different areas or time periods being compared; in time, see Pearman et al., 2008b; in geographic space, see Petitpierre et al., 2017). When such assessment is possible, simply quantifying the extent of niche overlap then makes it possible to anticipate how well a model – and associated niche – can be projected to a different area (Petitpierre et al., 2012) or time period (Pearman et al., 2008b). More specifically, in the case of biological invasions, niche expansion into new climates (i.e. not found in the native range) or niche unfilling (i.e. native conditions not yet colonized) can be assessed as two specific components of partial niche overlap (Petitpierre et al., 2012; Guisan et al., 2014). A complete framework – the COUE scheme – for analysing niche changes in space and time is described for the case of biological invasions in Guisan et al. (2014). It can also be applied to changes in time, between the past and the present (and can also be adapted to compare the niche of distinct species, e.g. within phylogenetic trees). However, these assessments of observed niche change in time or space are not often possible due to the lack of available data, and therefore projections for most species have to be made by assuming that the realized niche remains stable in the new area or time period.

We will address some of these issues and questions again in the next sections, when illustrating different model projection contexts.

17.2 Projecting Models in Space

Any of the statistical models calibrated so far can be projected to the same or new areas, be it at the same or different spatial resolution or extent, or be it to points or raster objects. There are some important aspects and limitations that need to be considered; yet the basic methodology is always the same. Projecting species distributions requires applying a statistical model that represents the requirements of a species in environmental space back to a geographic space. Traditionally such projections have been done in a GIS, while model calibration was carried out in a statistical environment. Doing so directly in R instead of using a GIS is a huge advantage, in particular when projecting onto large grids. We can use the GIS functionality introduced in Part II to make best use of the projections.

In statistical terms, the `predict(model,data)` function is the key to carrying out any form of projection in R, where a statistical model object (`model`) is projected to new data (`data`). Below, we present some examples of how such projections can be made, and we discuss some important issues relevant to projections in space and time.

17.2.1 Predictions to the Training Area

In the simplest case, the model is spatially predicted in the area where it was fitted. This is the safest way of predicting habitat models of species, since we then make sure that the observations used to fit the model represent the requirements of the species in the target area of the projection. Below, we use the same dataset as that used in Section 10.3 to fit a GAM model.

```
> mammals_data<-read.table("tabular/species/
mammals_and_bioclim_table.csv", h=T, sep=",")
> library(mgcv)
> gam1 <- gam(VulpesVulpes~s(bio3)+s(bio7)+s(bio11)+s(bio12),
data=mammals_data, family="binomial")
```

This produces the red fox model using a point dataset. In this case, the points originate from a range map that has been sampled at regular spatial intervals, so the file does not differ much from a spatial raster file. In many other cases, however, individual point locations are available (as downloaded e.g. for *Pinus edulis* Engelm. in Section 6.2.9) as is usually the case when using or downloading museum-type data e.g. from

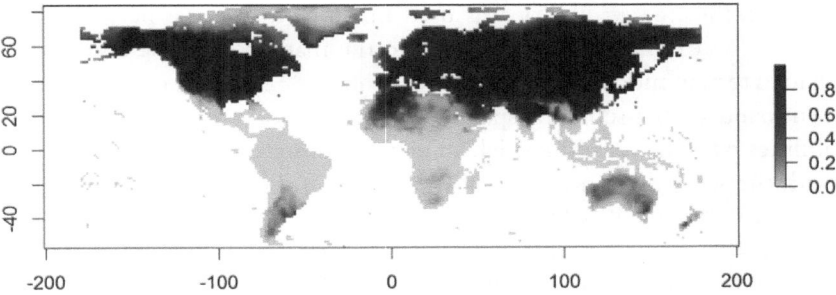

Figure 17.7 Simulated global habitat suitability of *Vuples vulpes* using a simple GAM model and five bioclim variables as predictors.

GBIF (see examples in Part VI for *Protea laurifolia*). In such cases, or when designing a sampling strategy (see Section 7.4), it is important to make sure that the whole area on which the model is later projected has been sampled. In fact, one usually aims to sample the whole niche space of a species (at least, the one available in the training area; see Section 17.1) in order to produce a reasonable projection of its spatial range. Raster layers are best used to represent the environment of the target area over which the model has to be projected. Here, four bioclim variables (bio3, bio7, bio11, and bio12) are used to develop the example. These variables need first to be loaded as spatial raster objects, and to be stacked into a raster stack object. After checking that the names of the raster layers match exactly the names used in the GAM model, and ideally after all previously discussed assumptions have been assessed and the model has been thoroughly checked (e.g. for biologically meaningful response curves; see Parts III and IV), the model "gam1" can finally be projected to the raster stack by using the predict() command, and the resulting map can be displayed (Figure 17.7).

```
> library(raster)
> bio3r.cu<-raster("raster/bioclim/current/grd/bio3.grd")
> bio7r.cu<-raster("raster/bioclim/current/grd/bio7.grd")
> bio11r.cu<-raster("raster/bioclim/current/grd/bio11.grd")
> bio12r.cu<-raster("raster/bioclim/current/grd/bio12.grd")
> biostack.curr<-stack(bio3r.cu,bio7r.cu,bio11r.cu,bio12r.cu)
> names(biostack.curr)
 [1] "bio3"  "bio7"  "bio11" "bio12"

> vulpes.curr <- predict(biostack.curr, gam1, type="response")
>library(fields)
> plot(vulpes.curr, col=two.colors(start="grey90",
end="firebrick4", middle="orange2"))
```

Here, the fields library was used in order to use the helpful two.color() command to assign colors to the map. In fact, one could use a shaded terrain model first, and then use the add=T option in the plot() command with a semi-transparent overlay to drape the projected probabilities over the terrain model. Semi-transparent colors can be obtained with the setting alpha=0.6, for example, in the two.color() command to obtain a 40% transparency.

The predict() command has several important options. Since the calibrated model was a GAM with a binomial family, the calibration of the model has been applied to logit-transformed *V. vulpes* distribution data. If type="response" in the predict() function is not set, then the default setting is used. For a GAM model this is type="link", which would result in a projection of the *V. vulpes* distribution in a logit-transformed data scale. This then requires applying the inverse logit transformation to get back to the scale of the response. The option type="response" applies this reverse transformation directly. Many different options exist for the different statistical models, and these can all be applied to the simple predict() command. The best way to obtain help with the different statistical models is to run the command ? predict.'model', where "model" stands for any statistical model family used (e.g. predict.gam).

One useful option is to not only project the fitted model, but to also store the model standard errors. This is available for some, but not for all, model types. One can then plot the spatial distribution of model standard errors. Currently, the option to map standard errors is not available for predictions to raster stacks. We have therefore illustrated the mapping of model standard errors in the example, using the mammals_data data frame to which we predict our GAM model (Figure 17.8).

```
> vulpes.se<-predict(gam1, mammals_data, type="response",
se.fit=TRUE)
> plot(mammals_data[,1:2],pch=15,cex=.25,col="grey70",
xlab="Longitude", ylab="Latitude")
> points(mammals_data[which(vulpes.se[[2]]>.05), 1:2],pch=15,
cex=.25,col="#FDD017")
> points(mammals_data[which(vulpes.se[[2]]>.10), 1:2],pch=15,
cex=.25,col="#E56717")
> points(mammals_data[which(vulpes.se[[2]]>.15), 1:2],pch=15,
cex=.25,col="#E42217")
> points(mammals_data[which(vulpes.se[[2]]>.20), 1:2],pch=15,
cex=.25,col="#9F000F")
> legend("bottomleft",legend=c("0.00 - 0.05","0.05 - 0.10",
"0.10 - 0.15", "0.15 - 0.20", "0.20 - 0.25"),
pch=c(15,15,15,15,15),col=c("grey70","#FDD017","#E56717",
"#E42217","#9F000F"),cex=.6,bg="white",title="GAM
Standard error")
```

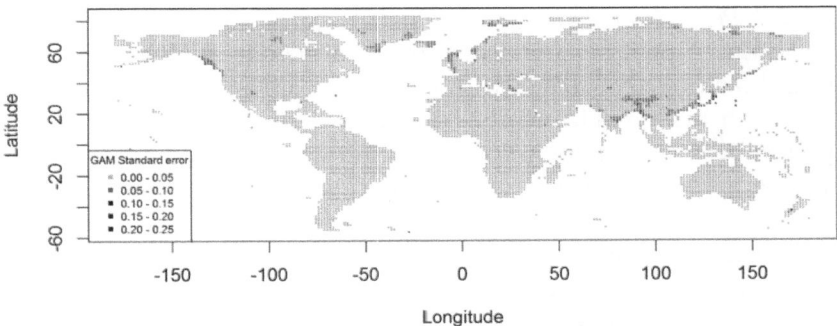

Figure 17.8 Spatial map of standard errors around the observation points for the GAM model of *Vulpes vulpes*.

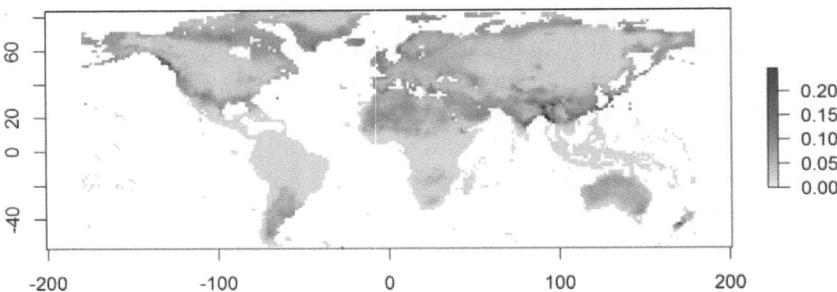

Figure 17.9 Spatial distribution of errors around predictions for the *Vulpes vulpes* GAM model.

Another alternative is to transform the dataframe from the `predict` function (`se.fit=T`) into a raster object using a bioclim raster as a mask (Figure 17.9).

```
>  vulpes.se_raster <- rasterize(cbind(mammals_data[,c(1:2)]),
y=bio3r.cu,field=vulpes.se[[2]])
> plot(vulpes.se_raster, col=two colors(start="grey90",
end="firebrick4", middle= "orange2"))
```

Usually, one finds an obvious pattern to such errors, and can clearly see that some, usually only a few, pixels have very high errors, while most pixels contain comparably low standard errors. In our example, such "error pixels" are mostly found along coasts and on the edge of the distribution that are rather marginal with regards to the species' distribution range.

17.2.2 Projections to New Areas

Projecting a model to a different area than the one used for model fitting should be approached with caution, as different combinations of environmental variables may be found between the training area and the projection area, i.e. non-analog situations (see Section 17.1). Such non-analog problems may arise specifically when projecting from one continent to another, a problem typically faced when projecting the niche (i.e. ensemble of suitable habitats) of an invasive species in the invaded range from a niche model fitted in the native range (Broennimann et al., 2007; Broennimann and Guisan, 2008; Gallien et al., 2012; Petitpierre et al., 2012). It is therefore strongly recommended to check how much novel (non-analog) environment (usually climate) exists between the calibration and the projection areas (Fitzpatrick and Hargrove, 2009; Elith et al., 2010). An easy way of doing this is to perform a PCA on the predictor variables used conjointly for both areas, and then to check their coverage and overlap in the PCA plot. The example below shows a PCA space built using `dudi.pca()` (from the `ade4` package) with the five bioclim variables used in the previous *V. vulpes* example. Differences between regions can be highlighted on a map by plotting the points in the Old World (> 20° west) in a different color to the points in the New World.

```
> library(ade4)
> vulpes_oldnew<-mammals_data[mammals_data$Y_WGS84>30.0,
c(1:2,8:13)]
> tmp1<-dudi.pca(vulpes_oldnew[,c(4,6:8)], nf=2, scannf=F)
> tmp2<-data.frame(cbind(vulpes_oldnew,tmp1$li))
> cols<-rep("#3090C733",nrow(vulpes_oldnew))
> cols[vulpes_oldnew$X_WGS84>-13]<-"#9F000F4D"
> par(mfrow=c(1,2))
> plot(bio3r.cu, legend=F,col="grey")
> points(vulpes_oldnew [,c(1:2)], col=cols, pch=16, cex=0.6)
> plot(jitter(tmp2$Axis1,amount=.3),jitter(tmp2$Axis2,
amount=.3),col=cols, pch=16,cex=.5,xlab="PCA-Axis 1",
ylab="PCA-Axis 2")
> par(mfrow=c(1,1))
```

From Figure 17.10, we can see that the differences between the two regions are rather minor, except in the upper left corner. This means that, in general, climates are similar in the two regions.

Another complementary approach is to apply a MESS method (Elith et al., 2010) as implemented in the `dismo` package in R. This approach measures the environmental similarity of a point (e.g. presence data) to the reference environment. In other words, it quantifies how far or close

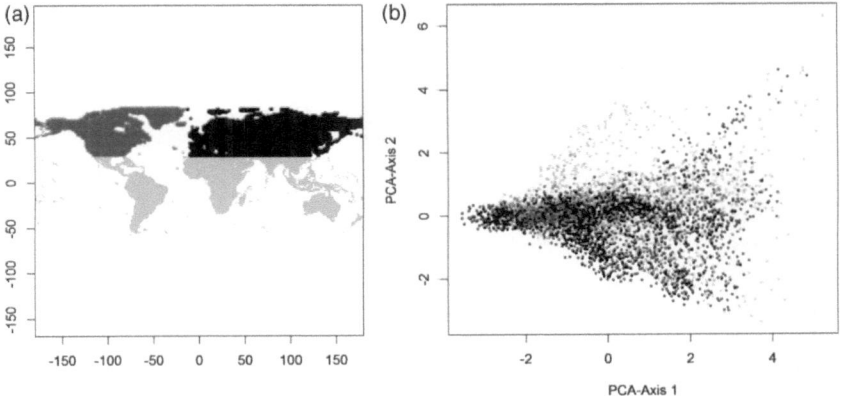

Figure 17.10 Differences in the ecological space of *Vulpes vulpes* between Old (red) and New (blue) World climates as mapped in (a) the geographic space and (b) the PCA space based on four bioclim variables. (*A black and white version of this figure will appear in some formats. For the color version, please refer to the plate section.*)

the projected area is to the training points. Negative values mean dissimilar points, the more negative these values are the more the points are dissimilar.

For instance, if we want to estimate the MESS between the training point of *V. vulpes* in Europe compared to North America:

```
> library(dismo)
> vulpes_east<-mammals_data[mammals_data$X_WGS84>-13.0,
c(1:2,8:13)]
> vulpes_ne<-vulpes_east[vulpes_east$Y_WGS84>30,]
> vulpes_europe<-vulpes_ne[vulpes_ne$X_WGS84<60,]
> Mess.Vulpes <- mess(biostack.curr, vulpes_europe[,c(4, 6:8)])
> plot(Mess.Vulpes)
> points(vulpes_oldnew[,1:2], col=cols, pch=16, cex=0.3)
```

Unsurprisingly, in Figure 17.11, we can see that the tropical belt has a rather different climate to Europe. The example is here trivial, but we strongly encourage researchers to take a look at the MESS analysis before conducting any extrapolations in space (it can also be applied in time). For instance, the MESS metric can be calculated with the ecospat package as follows:

```
> library(ecospat)
> mess.mammals <- ecospat.mess(mammals_data[,c(1:2, 8:13)],
mammals_europe])
> ecospat.plot.mess(mammals_data[,c(1:2)], mess.mammals)
```

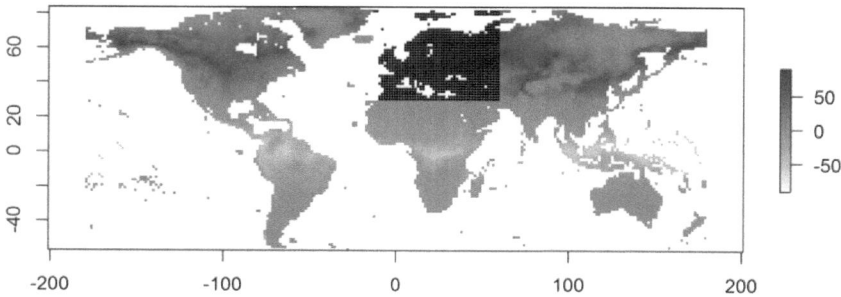

Figure 17.11 Multivariate environmental similarity surface in respect to European calibration data points for *Vulpes vulpes* in Europe. (*A black and white version of this figure will appear in some formats. For the color version, please refer to the plate section.*)

Projecting models to different areas (an issue of model "transferability") often changes their ranking in term of accuracy, i.e. the best models in the calibration area may not be the best in the new area, and vice versa. In the training range, more complex models often obtain a slightly better fit than simpler models (Elith et al., 2006). In a study comparing models fitted within and across two mountain regions in western Switzerland and eastern Austria, Randin and co-workers showed that while GAM was better than GLM for the respective within-region models, GLM models generally outperformed GAM models when applied between regions (Randin et al., 2006). GAM may be considered to be more complex than GLM in general, but see the review by Merow and co-workers (2014) for a discussion of simple versus complex models of species' habitat suitability. Any model type can be fitted in a complex or simple way, even boosted regression trees or Maxent, which usually are considered as being highly complex (e.g. Merow et al., 2014; Halvorsen et al., 2015).

As discussed in Section 17.1, and when projecting HSMs to another area, an important issue arises when the habitat suitability model is fitted from an incomplete species range, which is likely to (but does not necessarily) translate into incomplete niche quantification, known as niche truncation. Such incomplete sampling or coverage of the range or niche of a species may have unwanted effects when projecting to different regions, but also specifically when projecting to changed climates (Thuiller et al., 2004b; Barbet-Massin et al., 2010; Figure 17.3).

17.2.3 Changing Resolutions When Projecting Species Distributions
In most applications, the models are projected to the same spatial resolution at which they were trained. In Chapter 6, we discussed the

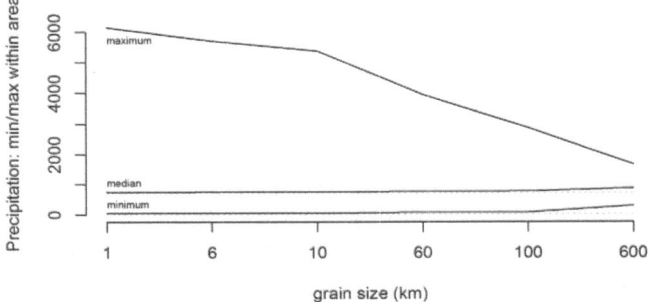

Figure 17.12 Effect of changing the grain size on a 1 km PRISM precipitation map for the extent of the United States. The three lines indicate the minimum, median and maximum values found in the study area at the grain size indicated by the x-axis. The dotted lines represent horizontal lines as references for median and minimum lines.

importance of considering the spatial grain of dependent and predictor variables. Changing the resolution (grain size) when projecting species distributions is feasible, but should be done with caution. At coarser grains, the range of values covered by climate and topographic maps decreases sharply. This means that maps at different spatial resolutions cover different extents and spans of environmental data (i.e. the coarser the grain, the smoother the mean values, and the smaller the span of possible values across the range). This section illustrates this grain effect when using the 1 km PRISM precipitation data already used in Section 6.2.8 for the United States. The 1 km grids were resampled at a 6, 10, 60, 100, and 600 km spatial resolution, and the minimum, median, maximum, and range of all observed precipitation values were then summarized in these six grids and plotted (Figure 17.12). This neatly illustrates that the high precipitation values in particular start to fade away beyond a resolution of 10 km. The reason for this decrease in environmental range and extremes toward coarser resolutions is discussed in Chapter 6. While the range of rainfall spans from 46 to 6148 mm at 1 km spatial resolution, the range decays sharply beyond 10 km. At 100 km, the range spans from 73 to 3940 mm, and at 600 km it spans from 288 to 1668 mm. Instead of spanning >6100 mm, it merely spans 1380 mm. This is mostly due to the averaging out of high-elevation pixels, which means that high-elevation species should still be associated with higher precipitation values in general. However, when projecting a model fitted at a coarse grain (say 100 km) to a finer grain (say 1 km), then the values around 1000 mm will be found at much lower elevations, and the species will be projected to incorrect locations. A similar effect is illustrated in a climate change context in Randin et al. (2009), where the authors show that models fitted and projected at coarse resolutions predict the extinction of alpine plant species that would persist according to finer grain models (see also Jimenez-Alfaro et al., 2012).

This means that when a model is fitted to a suite of environmental predictors at a large grain size, it cannot easily be projected to environmental data at a smaller grain size. This is especially true beyond a scale of 1:10 (Guisan et al., 2007a). Here, we lack the capacity to assign reasonable predictions to downscaled pixels (see Chapter 6 for a discussion and Bombi and D'Amen, 2012, for comparable analyses). However, there do not seem to be any major problems when scaling from 50 km to 20 km, as is often done across Europe for plants, for example (e.g. Thuiller et al., 2005a; Thuiller et al., 2011).

One way of reducing these grain issues when downscaling models is to add the scaling component directly to the model-fitting procedure in a Bayesian framework. This can be useful, for example if only coarse resolution biodiversity data is available (e.g. from atlas data such as the *Atlas Florae Europaeae*, Jalas and Suominen, 1972–1996). However, predictor variables are usually available at a much finer grain (e.g. 1 km, as in bioclim, Hijmans et al., 2005), which is also well suited to numerous model applications. With this in mind, Keil et al. (2013) proposed a hierarchical Bayesian framework that considers presence–absences at a fine resolution (the resolution at which we want to predict the species distribution, for instance 1 km) as latent variables, which are then modeled as a function of available fine resolution environmental variables (e.g. 1 km) and constrained by observed coarse resolution presences-absences (for instance 25 km) using logistic regression (Keil et al., 2013). This approach is very promising given the increasing availability of high-resolution data and the presence–absence data at various resolutions. It again shows the power of hierarchical Bayesian frameworks in such a context.

Projecting to a coarser resolution – i.e. upscaling – can also be tricky in some instances, and some of the same constraints apply, but in principle it is more straightforward than downscaling, as the former can be done statistically (i.e. aggregating cells with some statistics) whereas the latter often requires more dynamic (Berrocal et al., 2012) or hierarchical approaches (Keil et al., 2013). A simple solution in the case of HSMs is to first project the model at the same fine resolution at which the model was fitted, and to then upscale the projected habitat suitability or the classified presence–absence maps to a coarser resolution. This can be done using some simple aggregation statistics (e.g. average, or least common denominator). As an example, Thuiller et al. (2014a) calibrated HSMs for the French Alps for about 2750 species at very high resolution (250 m). Since they focused on diversity patterns and needed to stack their species projections, they looked for the optimal resolution at which observed species richness was best predicted by stacked HSMs (often called stacked species distribution models, S-SDMs). They showed that at a resolution of 2.5 km, the correlation between observed and predicted species richness was close to 0.9. At that resolution, the pervasive effects of biotic interactions and dispersal vanished, offering the best prediction of predicted diversity from the stacked HSMs (Thuiller et al., 2015).

In this way, it is possible to avoid projecting the fitted species–environment relationship to a different set of environmental predictors, potentially with a different meaning (Guisan and Thuiller, 2005).

17.3 Projecting Models in Time

As with projections in space, any of the statistical models fitted so far can be projected to the same or to new time periods, as long as we have comparable data on environmental predictors. Similar limitations as discussed for projections in space also need to be considered in time. Most projections are applied to future climates, often in order to assess the effect of climate change on biodiversity and ecosystem services (e.g. Thuiller et al., 2005a; Iverson et al., 2008; Lawler et al., 2009; Engler et al., 2011a; Civantos et al., 2012; Normand et al., 2013). Here, the risk assessment aspect and planning for mitigation measures is the primary goal of such studies (e.g. Araújo et al., 2004; Araújo et al., 2011). HSMs are now also increasingly projected to past climatic conditions (reviewed in Nogues-Bravo, 2009; Svenning et al., 2011). These studies aim to understand current distribution patterns (Espíndola et al., 2012; Schorr et al., 2012; Schorr et al., 2013; Patsiou et al., 2014); test past migration processes and its effect on current local or past global extinctions (Svenning and Skov, 2007; Araújo et al., 2008; Nogues-Bravo et al., 2008; Svenning et al., 2008; Lorenzen et al., 2011; Lima-Ribeiro et al., 2012); evaluate the capability and validity of HSMs for climate change projections (Pearman et al., 2008b; Davis et al., 2014); or improve niche calibrations for habitat suitability modeling (Maiorano et al., 2013).

One major aspect to consider for temporal projections is the link between the date of observation and the date or time window of predictors used (temporal matching, see Guisan and Thuiller, 2005). We usually assume that the general climate means can be linked to observations, irrespective of when the observation was made. This may be fine for many species but certainly not for all. Highly mobile and fast-responding species (e.g. migratory birds, phytoplankton or fishes in oceans) may respond more to seasonal climate and weather patterns than to long-term seasonal means (e.g. Reid et al., 1990). In such cases, temporal matching would need to be applied scrupulously, as recently used to predict the future distribution of arctic fishes, for example (Wisz et al., 2015). On the other hand, very slow migrating species may still show limited range filling, and possibly as a result, limited niche filling. The latter can occur if, for example, only part of a species' niche is actually colorized in the field, due to severe time lags in readjusting species' ranges (i.e. caused by limited dispersal) during the Holocene (Svenning et al., 2006). A study of European amphibians found better model fit when relating contemporary observations to past climate predictors than when relating them to current climate predictors (Araújo et al., 2008). This indicates that many amphibians currently do not seem to be at equilibrium with current climates, thus revealing limited niche filling. This issue of synchrony and equilibrium between observations and related environmental predictors is thus also particularly crucial when projecting HSMs across large time scales.

The following section illustrates how such projections in time can be performed, and which major limitations apply when projecting to past or future time frames.

17.3.1 Projecting to Future Environments

One of the main uses of habitat suitability modeling is to project the potential (realized) distribution of species to future climate conditions (forecasting), in order to study to what extent and where species would find new suitable habitats, how much suitable habitat would be lost, and how much biodiversity turnover is likely to result from these processes. Although the projection itself is a simple operation, the discussion of the resulting maps requires more careful consideration, especially regarding the potential pitfalls and shortcomings associated with this correlative approach. This is best illustrated with a simple example.

Let's start by loading into the workspace the data representing the future climate for the same bioclim variables used in Section 6.2, and checking the naming of the variables.

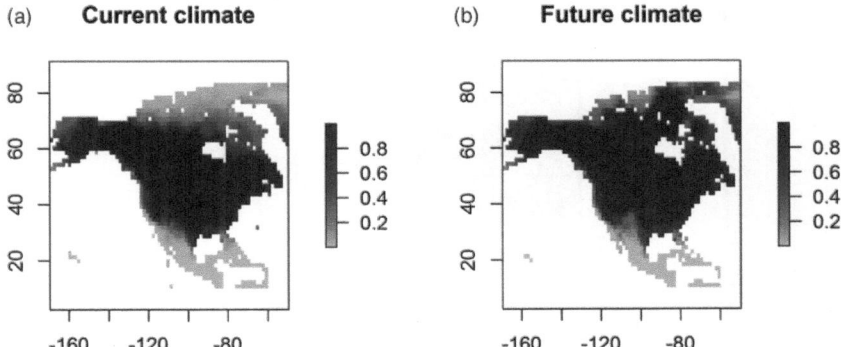

Figure 17.13 Projected habitat suitability of *Vulpes vulpes* under (a) current and (b) projected future climate, mapped over the extent of North America from a globally fitted GAM model.

```
> bio3r.fu<-raster("raster/bioclim/future/grd/bio3.grd")
> bio7r.fu<-raster("raster/bioclim/future/grd/bio7.grd")
> bio11r.fu<-raster("raster/bioclim/future/grd/bio11.grd")
> bio12r.fu<-raster("raster/bioclim/future/grd/bio12.grd")
> biostack.fut<-stack(bio3r.fu,bio7r.fu,bio11r.fu,bio12r.fu)
> names(biostack.fut)
 [1] "bio3"  "bio7"  "bio11"  "bio12"
```

It appears that due to the naming of the raster files in the "grd" folder on the hard drive, the names in the raster stack are exactly the same as those in the current climate stack (biostack.curr). This is important for the next step. Let's now project the *V. vulpes* GAM model to future climates and map the resulting predictions across North America in order to assess range changes. For this step, the names of the predictor variables have to precisely match those used to fit the model. The map shows that the predicted habitat suitability of *V. vulpes* is likely to expand toward more northern latitudes (Figure 17.13).

```
> vulpes.fut <- predict(biostack.fut, gam1, type="response")
> vulpes.na.cur<-crop(vulpes.curr, extent(-170,-50,10,90))
> vulpes.na.fut<-crop(vulpes.fut, extent(-170,-50,10,90))
> par(mfrow=c(1,2))
> plot(vulpes.na.cur, col=two.colors(start="grey90",
end="firebrick4", middle= "orange2"),main="Current climate")
> plot(vulpes.na.fut, col=two.colors(start="grey90",
end="firebrick4",middle= "orange2"),main="Future climate")
> par(mfrow=c(1,1))
```

At this stage, it is also interesting to map the standard errors to spot locations where the predictions are highly uncertain. Interestingly, these

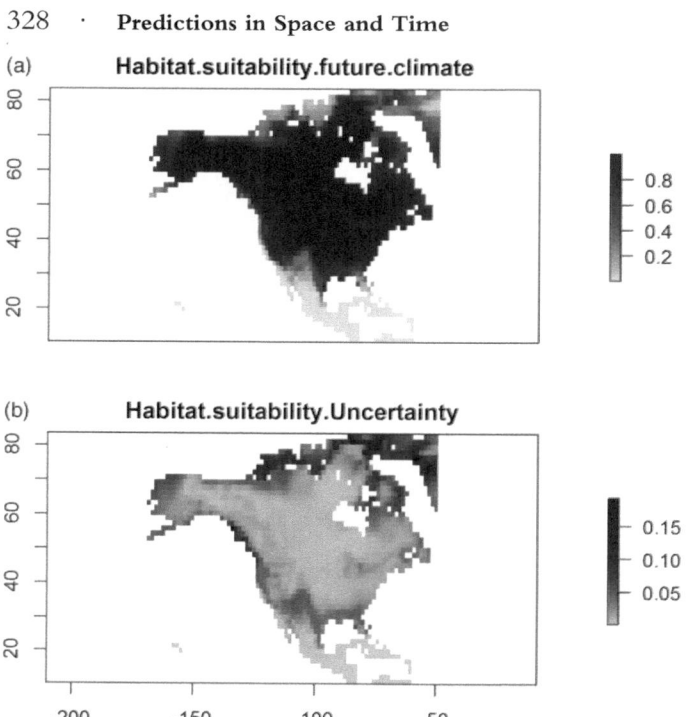

Figure 17.14 Future habitat suitability for *Vulpes vulpes* predicted by (a) a GAM model, and (b) its associated uncertainty.

areas also correspond to the most northern latitude the habitat suitability is projected to expand to (Figure 17.14).

```
> biostack.fut_df <- as.data.frame(rasterToPoints(biostack.fut))
> vulpes.fut_se <- predict(gam1, biostack.fut_df,
type="response", se.fit=T)
> vulpes.fut_se <- rasterFromXYZ(cbind(biostack.fut_df[,1:2],
vulpes.fut_se), biostack.fut)
> vulpes.fut_se<-crop(vulpes.fut_se, extent(-170,-50,10,90))
> names(vulpes.fut_se) <- c("Habitat suitability  future
climate", "Habitat suitability - Uncertainty")
> plot(vulpes.fut_se, col=two.colors(start="grey90",
end="firebrick4", middle= "orange2"))
```

For species that disperse rapidly, such as large birds or mammals, it should not be too difficult to track projected environmental change at the same pace as climate changes over time and space (i.e. the "velocity" of climate change in terms of species' exposure; Serra-Diaz et al., 2014). However, for small and slowly dispersing animals, or for most plants with slow migration rates, the projected future habitat suitability for the time considered may remain outside of a realistic dispersal domain (Bertrand et al., 2011). This issue has been much debated in the scientific literature, since such models often claim to project extinction and invasion/migration risks, and their impact on biodiversity turnover (e.g. Thomas et al., 2004; Thuiller et al., 2005a; Lawler et al., 2009; Engler et al., 2011a; Thuiller et al., 2011). Therefore, caution is recommended in the use of terminology, which many authors of such papers do, but not all, to strictly convey that such models project a "change in potential habitat suitability" at a given time and for a given place, with an associated "projected turnover". It is clear that, if species demography is not actively incorporated, simple HSMs cannot predict what percentage of the species will go extinct or what percentage will be able to migrate to keep up with changing conditions (Guisan and Thuiller, 2005). HSMs are therefore simply unable to tell us how fast the modeled species will respond to changing environmental conditions. Some attempts have therefore been made and tools were developed (e.g. Dullinger et al., 2012; Engler et al., 2012) to make such projections more realistic, either: by exploring the effect of assumed migration rates on future projections (Engler and Guisan, 2009; Engler et al., 2009), by inferring migration rates from expert data (e.g. Vittoz and Engler, 2007); from dynamic community process models, where land use and climate change with its local effect on landscape fragmentation and migration potential is included in projected re-adjustments of future species ranges (e.g. Meier et al., 2012); by explicitly combining habitat suitability modeling with a simple species population dynamics and sometime dispersal model (Keith et al., 2008; Anderson et al., 2009; Dullinger et al., 2012); or by using an integrated hierarchical model (Schurr et al., 2012). Numerous different approaches are currently being developed and this is a rapidly evolving field (Dormann et al., 2012). Some approaches use stand structure data to infer migration and dispersion processes (e.g. Saltre et al., 2013), while others directly infer demographic rates and migration rates from field measurements to fit a hierarchical Bayesian population model for modeling species' range dynamics (e.g. Pagel and Schurr, 2012; Schurr et al., 2012). These models suggest that for many species, the velocity of climate change (Loarie et al., 2009; Sandel et al., 2011; Dobrowski et al., 2013) will prevent them from tracking it successfully (Engler et al., 2009; Dullinger et al., 2012; Meier et al., 2012), meaning that by the end of the century many species may experience limited filling of their potential ranges.

Limited range filling (RF), a term promoted by Svenning and Skov (2004), refers to the fact that many species appear not to colonize all pixels that would appear (from the available climate maps) to contain suitable climates under current climate conditions. Under future conditions this inability to "fill" (i.e. colonize) suitable pixels is likely to increase. This affects our view of distribution equilibrium. Many species either do not appear to have reached equilibrium, or are constantly appearing and disappearing locally in a so-called pseudo-equilibrium (Franklin, 1995; Guisan and Zimmermann, 2000); an important assumption discussed in Part I. If only some pixels of the suitable environment remain un-colonized, this is unlikely to pose a problem in terms of model calibration and projection. However, if a species is clearly lagging behind its potential geographic distribution, it may also lead to a failure to colonize some portions of its environmental niche. Such situations can result in fitting an overly narrow niche, a problem known as the issue of *truncated niches* (see Section 17.1, and e.g. Thuiller et al., 2004b; Barbet-Massin et al., 2010). Such problems specifically occur when invasive species are modeled from the native range (Peterson, 2003). However, when combining data from the native and the invaded range (Broennimann and Guisan, 2008), we can discover to what extent invasive species have stabilized their migration or are still colonizing (Gallagher et al., 2010; Petitpierre et al., 2012) (see Gallien et al., 2010; Guisan et al., 2014 for reviews).

Not all the environmental conditions that constitute the environmental niche of a species are available in each region, i.e. the issue of the *available environment* (Jackson and Overpeck, 2000; Ackerly, 2003; Broennimann et al., 2014a; Guisan et al., 2014; Section 17.1). Therefore caution should be taken when projecting regionally fitted models to future conditions (Guisan et al., 2012). The future might hold novel, *non-analog* conditions (Williams and Jackson, 2007) that may not be novel when whole spatio-temporal ranges of species are used to train the models (Fitzpatrick and Hargrove, 2009; Barbet-Massin et al., 2010). The appearance of non-analog conditions under future climates is a severe problem, which statistical models cannot easily cope with since projections to such conditions represent an extrapolation to conditions outside of the training range (Fitzpatrick and Hargrove, 2009; Guisan et al., 2012). One way of avoiding such extrapolations is to clearly mark such regions on the prediction map (to visually distinguish predictions in analog from non-analog situations; e.g. Figure 10 in Guisan and Theurillat, 2000) or to apply a filter to the projections (Berteaux et al., 2006) to avoid projecting to regions that encompass non-analog climates in the future (Fitzpatrick and Hargrove, 2009). Non-analog climates seem to occur frequently through larger time periods, be it the Holocene past or the projected climate future (Jackson and Williams, 2004; Williams and Jackson, 2007; Williams et al., 2007), and it is recommended that such conditions are checked (Nogues-Bravo, 2009).

An alternative measure to reduce the impact of non-analog conditions is to avoid including too many environmental variables when defining the niche in the models if the latter are to be used to predict distributions under future climatic conditions, i.e. avoid making the models overly complex (Randin et al., 2006; Merow et al., 2014). When numerous climate variables are correlated with each other, these correlations, depending on their level, can have an impact on model calibration (Dormann et al., 2013) and thus on their transferability, especially when using more complex algorithms such as boosted regression trees or random forests that implicitly consider interactions among variables. The link between the number of variables and non-analog situations is that the more variables are used to define the niche, the easier it is to find combinations of environmental conditions that differ between two regions, and thus correspond to "non-analog" environments. Therefore, projecting complex (often over-fitted) models, including many predictors, to other regions is more likely to result in many cases in projecting to "novel", non-analog environments (usually climates). Avoiding predicting to these non-analog situations may then considerably restrict the projection domain and, accordingly, change the predicted rates of habitat loss compared to when simpler models including fewer predictor variables are used. See Merow and co-workers (2014) for a discussion on simple versus complex models and their use for predictions and projections.

Of course, numerous factors (discussed in Part I) can modulate climate change responses predicted by HSMs over time (Elith et al., 2010; Meier et al., 2012). The previously discussed capacity of species to migrate is an obvious one, but others include species' population dynamics (e.g. extinction debts; Dullinger et al., 2012), phenotypic plasticity (e.g. vegetative, through clonality for plants; de Witte and Stöcklin, 2010), interactions with other species (e.g. plants–pollinators; Araújo and Luoto, 2007), and species' evolutionary abilities (Thuiller et al., 2013). Species interaction is a crucial question that may be particularly decisive in determining a species' capacity to colonize newly suitable or remote areas (Thuiller et al., 2013; Wisz et al., 2013). Some simple dimensions of species interactions have been included in HSMs (e.g. Pellissier et al., 2010; Meier et al., 2011; Meier et al., 2012).

However, these approaches have remained limited to well-established interactions involving a small number of species at a time. A number of new perspectives for quantifying biotic interactions (Boulangeat et al., 2012a; Kissling et al., 2012) and new frameworks to incorporate them into HSMs and derive spatial projections (Mokany et al., 2012) have been proposed (Kissling et al., 2012; Wisz et al., 2013). Dynamic community models (Pagel and Schurr, 2012; Bocedi et al., 2014; Snell et al., 2014), such as forest gap models (e.g. TreeMig; Lischke et al., 2006), can also be used to account for interactions and sometimes for dispersal in extensively studied species groups such as trees (e.g. in TreeMig; Lischke et al., 2006), but cannot be applied to large numbers of species in many other, less well studied groups (Botkin et al., 2007). As an alternative, spatially and temporally explicit landscape models are now able to accommodate a reasonably large number of interacting functional groups (Boulangeat et al., 2014). In order to be of use for conservation and biodiversity management, these functional groups need to convey some sort of information on biodiversity. To this end, Boulangeat and colleagues have proposed a framework at the crossroads between functional and community ecology that can be used to construct meaningful functional groups for biodiversity models (Boulangeat et al., 2012b). This approach can be easily transcribed to other groups of organisms.

Perhaps the greatest obstacle to projecting well-fitted HSMs to future environmental conditions is the lack of data availability for what could be the most useful (because ecologically meaningful) predictors (Austin and Van Niel, 2011; Mod et al., 2016). Climate variables are more widely available for future conditions than soil variables (e.g. Bertrand et al., 2012; Dubuis et al., 2013), local land-use (e.g. Randin et al., 2009), or other socioeconomic data characterizing the landscape (Gellrich and Zimmermann, 2007; Quetier et al., 2010).

17.3.2 Projecting to Past Environments

Projecting models to past environmental conditions (hindcasting) is done in the same way as projecting to future conditions. As soon as past environmental data is available, the same procedure as for future conditions can be applied using the `predict()` function. For this reason, there are no examples of specific code herein. Instead, a number of additional issues are discussed that need to be considered when projecting species' habitat suitability to the past, or when fitting past distribution data to past climates in order to project them to current, or even future, climates.

In addition to the points discussed below, most, if not all of the points discussed under projections to future environments (see Section 17.3.1) apply here as well.

Another important issue that is often neglected in future projections of climate change is the fact that a species may have evolved a broader fundamental niche over its history than can be seen even when sampling the whole contemporary range of a species. Some conditions, which are within the niche of a target species, may simply no longer exist today, but may have existed in the past, during the species' history and the species may therefore have adapted to them (Nogues-Bravo, 2009). A study of tree species in Europe revealed that when combining training data from many millennia throughout the Holocene, the current (and therefore in all likelihood also the future) distribution of tree species is more accurately modeled than when only using current observations and current climates (Maiorano et al., 2013). This highlights the fact that past distributions may contain information on the niche that is not currently well visible or available, due to existence in the past of "non-analog" climates with no contemporary equivalent. Therefore, instead of simply projecting HSMs, fitted to current climate only, to future non-analog climates, the projections can be improved by calibrating these models across a larger range of conditions over time (thus "assembling" the niche over time) in order to reduce the amount of novel future conditions and increase the predictive power (Maiorano et al., 2013). The lack of suitable environmental data other than climate for the distant past is perhaps the most severe constraint to hindcasting species distributions over millennia.

Significantly more data are available for the recent past, typically the last 100 years, than for older periods, and thus we have a greater capacity to learn from this recent historical period, and can provide useful insights into the changes to be expected over the next 100 years. However, there are only relatively few examples of HSM projections into the past century (e.g. Araújo et al., 2005a; Dobrowski et al., 2011). Furthermore, the fact that recent changes in climate are occurring at faster rates than previous changes in climate (Dobrowski et al., 2013), together with the significant human impact on landscape during the same period, may have resulted in an apparent disequilibrium between species distribution and climate, which may limit our ability to observe all the biological effects of climate change predicted by the models, and may therefore also limit our capacity to understand all mechanisms of species' responses to climate change. Although numerous studies assessing the "biological fingerprints of climate change" have revealed that many species have responded to climate warming in recent decades (Root et al., 2003; Walther et al., 2005; Chen et al., 2011), the analysis of large regional datasets has also revealed that several species have not responded as expected (Lenoir et al., 2008; Moritz et al., 2008; Tingley et al., 2009; Crimmins et al., 2011). For instance, some species respond through downward shifts along elevation. Lenoir and co-workers (2010) discuss possible reasons for such divergent responses, including the predominant role of concomitant land-use changes restricted to some parts of environmental gradients.

Over longer time scales (>100 years to millennia), the lack of data becomes a more serious problem. There are very few global or continental datasets available that sufficiently reconstruct the climate of the past to the point where it can be included in habitat suitability modeling and biogeographic studies (e.g. Espíndola et al., 2012; Maiorano et al., 2013). Data for other variables such as soil or socioeconomic data are largely non-existent. Reconstructions are usually only available for very coarse land-cover classes (Ramankutty and Foley, 1999; Goldewijk, 2001; Pongratz et al., 2008; Kaplan et al., 2009; Hurtt et al., 2011), while for many other important variables (e.g. soils) there is either no information available for the Holocene or earlier climates at all, or none available at a reasonable thematic or spatial resolution.

Some past climate datasets used in HSMs are derived from GCM simulations, e.g. from PMIP2 simulations (Braconnot et al., 2007; Schorr et al., 2012; Schorr et al., 2013), while others are derived from climate reconstructions using hemisphere or continental climate proxy data such as tree-rings or sediment cores (Mauri et al., 2014). One major problem for most GCM-based reconstructions is that these GCMs usually only cover Holocene time periods up to the pre-industrial era. The gap between pre-industrial (*c.* 1750) and current (1950 and onward) climate is often neglected in studies that project species models into the past. This 200-year period between the pre-industrial era and the period when weather station data became readily available is most likely a time when there was significant climate change due to the emergence from the Little Ice Age in the northern hemisphere (~post 1850) and the warming from gas emissions due to industrialization. Also, the sea level at the last glacial maximum was approximately 125 m lower than it is today, meaning that large areas in the European North Sea or in the Gulf of Mexico were not inundated and could be colonized by plants and animals (Peltier, 2004). Failing to take into consideration the effects of climate and sea-level difference as discussed above may further lead to erroneous projections and biased conclusions regarding habitat suitability distribution.

17.4 Ensemble Projections

In Part III, we introduced the concept of model averaging and ensemble modeling. Several authors have raised the issue that although different techniques (e.g. GLM and random forest) are likely to provide similar species predictions under current (calibration) conditions, these predictions may drastically diverge when used to project species ranges in space or time (Thuiller, 2004; Lawler et al., 2006; Buisson et al., 2010). The same may happen when using different data sets (e.g. through split-sampling), different sets of background data (or pseudo-absences), different ways of parameterizing the same technique (e.g. linear versus polynomial terms in a GLM), different sets of environmental predictors or different environmental change scenarios. How to deal with variability in these different initial conditions, techniques, parameterizations, and bounding conditions is still an open question in ecology and other fields of science. When no one modeling option clearly stands out over several others, combining several plausible models into a final ensemble prediction is a good solution (Araújo and New, 2007). This is an approach inherited from the climate modelling community, where an ensemble of plausible realizations of future climate is usually preferred over one single climate change prediction.

As reviewed in Araújo and New (2007), an ensemble of forecasts is one of the most commonly accepted ways of accounting for projection variability since it relies on multiple projections across sets of initial conditions, algorithms (e.g. GLMs or boosted regression trees), parameters (e.g. quadratic vs. polynomial terms in a GLM, number of regression trees in a random forest), and bounding conditions (e.g. ensemble of climate predictions). Although ensemble forecasting was already a relatively well-accepted approach in other fields such as economics or climatology, it was relatively unknown in ecology in the early 2000s and did not emerge as a plausible alternative to single initial data algorithms until 2004 (Thuiller, 2004). The major advantage of combining a set of forecasts is that it provides a probability distribution per pixel as opposed to a single value. This makes it possible to extract average predictions as well as CIs given varying input data, algorithms, parameterization, and bounding conditions. Still, although such an approach has now become common practice (Diniz et al., 2009; Marmion et al., 2009), ensemble forecasts are often used as a single forecast by extracting an average or a weighted average based on different evaluation techniques, without considering the variability behind those averages, and without considering which metric to use for scoring the different projections. If forecasts have to be used in conservation planning or to be used as tools to guide decision-making (Guisan et al., 2013), they should present not only the main trend but also the variability around this trend (Meller et al., 2014).

We have already seen in Part III that predictions can vary for the red fox under current conditions across different sets of repeated split sampling and across models. This variation was however relatively small. Indeed, the red fox is certainly not the best example to demonstrate the potential use of ensemble modeling at that resolution, since its distribution is quite homogenous and relatively straightforward to model. We will now see that even for this species, model projections could differ quite substantially once projected into the future. Although the `biomod2` package can be used to automatically run an ensemble modeling procedure (see Part VI), here we will go through it sequentially.

We will first import the necessary data (distribution data, current and future climate layers) and then run a set of five techniques (GLM, GAM, MARS, FDA and random forest) with a 20-fold repeated split sampling procedure. Each calibrated model will be evaluated using the TSS statistics and then used to project the potential climatic suitability of the species at global scale under both current and future conditions. We will also transform the probability of occurrence into binary projections using the threshold that optimized the TSS statistics on the testing data.

Finally, we will build an ensemble forecast and analyse the uncertainty given the models, and the uncertainty given the data. We will also present an intuitive ensemble forecast that can represent agreement between models, uncertainty and prediction (i.e. committee averaging).

```
> library(MASS)
> library(earth)
> library(randomForest)
> library(mda)
> library(biomod2)
# Extract the future layers for the presence and absence
# points.
> FutureEnv <- as.data.frame(cbind(mammals_data [,c(2:9)],
extract(biostack.fut, mammals_data [,c(2,3)])))
> FutureEnv <- na.omit(FutureEnv)

# Create a dataframe to store the evaluation result for each
# model for each split-sampling
```

```
> nRow <- nrow(mammals_data)
> nCV <- 20 # the number of repeated split-sampling.
> Test_results <- as.data.frame(matrix(0,ncol=nCV,nrow=5,
dimnames=list(c("GLM","GAM","MARS","FDA","RF"), NULL)))
# Create an array to store the predicted habitat suitability
# for current conditions for each single model x single split-
# sampling
> Pred_results <- array(0,c(nRow, 5,nCV),
dimnames=list(seq(1:nRow), c("GLM","GAM","MARS","FDA",
"RF"), seq(1:nCV)))
# Create an array to store the predicted habitat suitability
# for future conditions for each single model x cross-
# validation combination
> ProjFuture_results <- array(0,c(nrow(FutureEnv), 5,nCV),
dimnames=list(seq(1:nrow(FutureEnv)),
c("GLM","GAM","MARS","FDA","RF"), seq(1:nCV)))
> ProjFuture_results_bin <- array(0,c(nrow(FutureEnv), 5,nCV),
dimnames=list(seq(1:nrow(FutureEnv)),
c("GLM","GAM","MARS","FDA","RF"), seq(1:nCV)))
# Build a function to create the calibration
and evaluation
# datasets
> SampMat <- function (ref, ratio. as.logi = FALSE) {
ntot <- length(ref)
npres <- sum(ref)
ncal <- ceiling(ntot * ratio)
pres <- sample(which(ref == 1), ceiling(npres * ratio))
absc <- sample(which(ref == 0), ncal - length(pres))
if (as.logi){
calib <- rep(FALSE, ntot)
calib[c(pres, absc)] <- TRUE
eval <- !calib
}
else {
calib <- c(pres, absc)
eval <- (1:ntot)[-c(pres, absc)]
}
return(list(calibration = calib, evaluation = eval))
}


# Loop across the 20-fold repeated split sampling
> for(i in 1:nCV) {
# separate the original data into one subset
for calibration
# and the other for evaluation.
a <- SampMat(ref=mammals_data$VulpesVulpes, ratio=0.7)
# function from the biomod2 package
calib <- mammals_data[a$calibration,]
eval <- mammals_data[a$evaluation.]
```

```
### GLM ###
glmStart <- glm(VulpesVulpes~1, data=calib, family=binomial)
glm.formula <-
makeFormula("VulpesVulpes",mammals_data[,c("bio3", "bio7",
"bio11", "bio12")],"quadratic",interaction.level=1)
glmModAIC <- stepAIC( glmStart, glm.formula, data = calib,
direction = "both", trace = FALSE, k = 2,
control=glm.control(maxit=100))
# prediction to the evaluation data and evaluation using the
# TSS approach
Pred_test <-  predict(glmModAIC, eval, type="response")
# The Find.Optim.Stat from biomod2 computes the TSS and will
# provide the cutoff that optimizes it. Within biomod2,
# probabilities of presence are transformed into integers after
# being multiplied by 1,000 (to save memory space). We will
# therefore multiply here the probability of occurrence by 1,000.
Test <- Find.Optim.Stat(Stat='TSS', Fit=Pred_test*1000,
Obs=eval$VulpesVulpes)
Test_results["GLM",i] <- Test[1,1]
# prediction on the total dataset for current and future.
Pred_results[,"GLM",i] <- predict(glmModAIC, mammals_data,
type="response")
ProjFuture_results[,"GLM",i] <- predict(glmModAIC, FutureEnv,
type="response")
# transform the probability of occurrence into binary
# predictions. Use the cutoff that optimizes the TSS
# statistics divided by 1,000.
ProjFuture_results_bin[ProjFuture_results[, "GLM",i]>=
(Test[1,2]/1000),"GLM",i] <- 1


### GAM ###
gam_mgcv <- gam(VulpesVulpes~ s(bio3)+s(bio7)+s(bio11)+s(bio12),
data=calib, family="binomial")
# prediction on the evaluation data and evaluation using the
# TSS approach
Pred_test <-  predict(gam_mgcv, eval, type="response")
Test <- Find.Optim.Stat(Stat='TSS', Fit=Pred_test*1000,
Obs=eval$VulpesVulpes)
Test_results["GAM",i] <- Test[1,1]
# prediction on the total dataset

Pred_results[,"GAM",i] <- predict(gam_mgcv, mammals_data,
type="response")
ProjFuture_results[,"GAM",i] <- predict(gam_mgcv, FutureEnv,
type="response")
ProjFuture_results_bin[ProjFuture_results[,"GAM",i] >=
(Test[1,2]/1000),"GAM",i] <- 1
### MARS ###
Mars_int2 = earth(VulpesVulpes ~ 1+bio3+bio7+bio11+bio12,
data=calib, degree = 2, glm=list(family=binomial))
# prediction on the evaluation data and evaluation using the
```

```
# TSS approach
Pred_test <- predict(Mars_int2, eval, type="response")
Test <- Find.Optim.Stat(Stat='TSS', Fit=Pred_test*1000,
Obs=eval$VulpesVulpes)
Test_results["MARS",i] <- Test[1,1]
# prediction on the total dataset
Pred_results[,"MARS",i] <- predict(Mars_int2, mammals_data,
type="response")
ProjFuture_results[,"MARS",i] <- predict(Mars_int2,
FutureEnv, type="response")
ProjFuture_results_bin[ProjFuture_results[,"MARS",i] >=
(Test[1,2]/1000),"MARS",i] <- 1


### FDA ###
fda_mod = fda(VulpesVulpes ~ 1+bio3+bio7+bio11+bio12,
data=calib,method=mars)
# prediction on the evaluation data and evaluation using the
# TSS approach
Pred_test <- predict(fda_mod, eval, type = "posterior")[,2]
Test <- Find.Optim.Stat(Stat='TSS', Fit=Pred_test*1000,
Obs=eval$VulpesVulpes)
Test_results["FDA",i] <- Test[1,1]
# prediction on the total dataset
Pred_results[,"FDA",i] = predict(fda_mod,
mammals_data[,c("bio3", "bio7", "bio11", "bio12")],
type="posterior")[,2]
ProjFuture_results[,"FDA",i] = predict(fda_mod,
FutureEnv[,c("bio3", "bio7", "bio11", "bio12")],
type="posterior")[,2]
ProjFuture_results_bin[ProjFuture_results[,"FDA",i] >=
(Test[1,2]/1000),"FDA",i] <- 1


### Random Forest ###
RF_mod = randomForest(x = calib[,c("bio3", "bio7", "bio11",
"bio12")],y = as.factor(calib$VulpesVulpes), ntree = 1000,
importance = TRUE)
# prediction on the evaluation data and evaluation using the
# TSS approach
Pred_test <- predict(RF_mod, eval, type="prob")[,2]
Test <- Find.Optim.Stat(Stat='TSS', Fit=Pred_test*1000,
Obs=eval$VulpesVulpes)
Test_results["RF",i] <- Test[1,1]
# prediction on the total dataset
Pred_results[,"RF",i] = predict(RF_mod, mammals_data,
type="prob")[,2]
ProjFuture_results[,"RF",i] = predict(RF_mod, FutureEnv,
type="prob")[,2]
ProjFuture_results_bin[ProjFuture_results[,"RF",i] >=
(Test[1,2]/1000),"RF",i] <- 1
} # End of the loop.
```

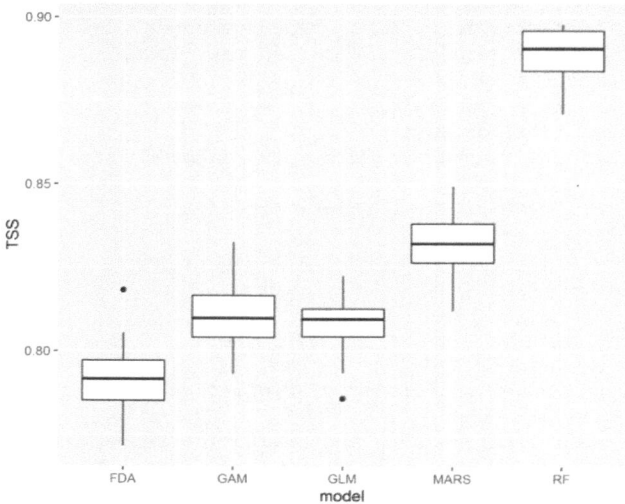

Figure 17.15 Variation in the true skill statistics in the 20-fold repeated split sampling procedure.

We have now run 20-fold repeated split sampling for five different techniques and used all the calibrated models to project the potential future climatic suitability for the species.

Let's first look at the quality of the model to judge whether some of the techniques or runs need to be discarded due to poor quality (Figure 17.15). We will use here the `ggplot` package.

```
# Variation in TSS between models and cross-validation runs
> library(ggplot2)
> TSS <- unlist(Test_results)
> TSS <- as.data.frame(TSS)
> Test_results_ggplot <- cbind(TSS,
model=rep(rownames(Test_results), times=20))
# Variability in predictive accuracy between cross-validation
# runs and models.
> p <- ggplot(Test_results_ggplot, aes(model, TSS))
> p + geom_boxplot()
```

The quality of all calibrated models is satisfactory since no model has a TSS of less than 0.775 (Figure 17.15). They can therefore all be kept to build the ensemble forecast.

Let's start by first building the most basic summary of the ensemble (i.e. mean, median and standard deviation).

```
# Average prediction (mean and median) and standard deviation
> Pred_total_mean <- apply(Pred_results,1,mean)
> Pred_total_median <- apply(Pred_results,1,median)
> Pred_total_sd <- apply(Pred_results,1,sd)


# Average projection into the future (mean and median) and
# variance
> ProjFuture_total_mean <- apply(ProjFuture_results,1,mean)
> ProjFuture_total_median <- apply(ProjFuture_results,1,median)
> ProjFuture_total_sd <- apply(ProjFuture_results,1,sd)


# Transformation in raster objects to facilitate the
# representation.
> Obs <- rasterFromXYZ(mammals_data[,c("X_WGS84", "Y_WGS84",
"VulpesVulpes")])
> Pred_total_mean_r <-
rasterFromXYZ(cbind(mammals_data[,c("X_WGS84",
"Y_WGS84")],Pred_total_mean))
> Pred_total_median_r <-
rasterFromXYZ(cbind(mammals_data[,c("X_WGS84",
"Y_WGS84")],Pred_total_median))
> Pred_total_sd_r <-
rasterFromXYZ(cbind(mammals_data[,c("X_WGS84",
"Y_WGS84")],Pred_total_sd))
> Out <- stack(Obs,Pred_total_mean_r,
Pred_total_median_r,Pred_total_sd_r)
> names(Out) <- c("Observed Vulpes vulpes","Ensemble
modeling_mean","Ensemble modeling_median", "Ensemble
modeling_sd")


# Habitat suitability maps for Vulpes vulpes predicted by the
# different model averaging methods and the associated
# uncertainty map.
> plot(Out)
```

As already seen in Part III, the observed presences and absences of the red fox are modeled relatively well under current climatic conditions (Figure 17.16). Both ensemble forecasts (mean and median) gave similar predictions. The uncertainty maps show areas where the models tend to differ across the different runs of repeated split sampling. These are mostly concentrated in North Africa where the models not only tend to over-predict southward, but also disagree with each other.

What would happen under future conditions? Here, we have used projections of future climate by 2080 under the A1FI scenario downloaded from the Worldclim dataset. Using the same strategy as for the current conditions, we first transform the point data into raster stacks:

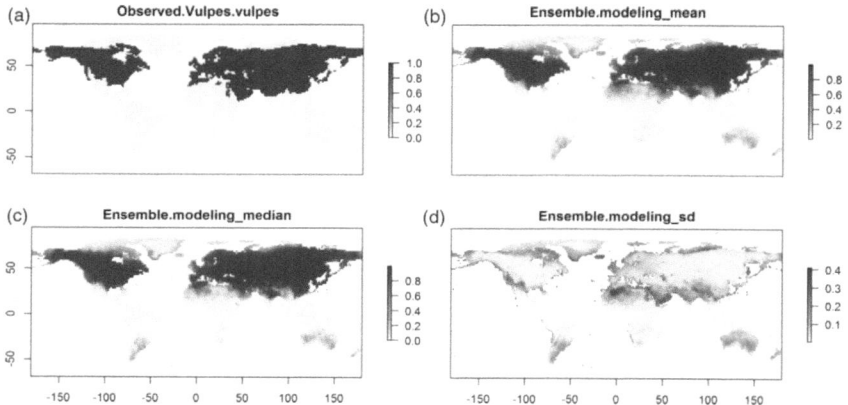

Figure 17.16 Observed presence and absence of *Vulpes vulpes* at (a) the global scale, together with (b and c) the two model averaging predictions (mean and median); and (d) the ensemble modeling uncertainty (sd).

```
# Transformation into raster objects to facilitate the
# representation.
> ObsF <- rasterFromXYZ(FutureEnv[,c("X_WGS84", "Y_WGS84",
"VulpesVulpes")])
> ProjFuture_total_mean_r <-
rasterFromXYZ(cbind(FutureEnv[,c("X_WGS84", "Y_WGS84")],
ProjFuture_total_mean))
> ProjFuture_total_median_r <-
rasterFromXYZ(cbind(FutureEnv[,c("X_WGS84", "Y_WGS84")],
ProjFuture_total_median))
> ProjFuture_total_sd_r <-
rasterFromXYZ(cbind(FutureEnv[,c("X_WGS84",
"Y_WGS84")],ProjFuture_total_sd))
> OutFut <- stack(ObsF, ProjFuture_total_mean_r,
ProjFuture_total_median_r,ProjFuture_total_sd_r)
> names(OutFut) <- c("Observed Vulpes vulpes","Ensemble
modeling_mean","Ensemble modeling_median", "Ensemble
modeling_sd")
# Future Habitat suitability maps for Vulpes vulpes predicted by
# the different model averaging methods and the associated
# uncertainty map.
# quartz()
> plot(OutFut)
```

Compared to Figure 17.17 we can see that the climatic suitability of the red fox is projected to expand northward in northern America and in Europe and Russia. This phenomenon has already been documented and is expected to be triggered by future climate change (Gallant et al., 2012). As we can see, the highest levels of uncertainty are found at the

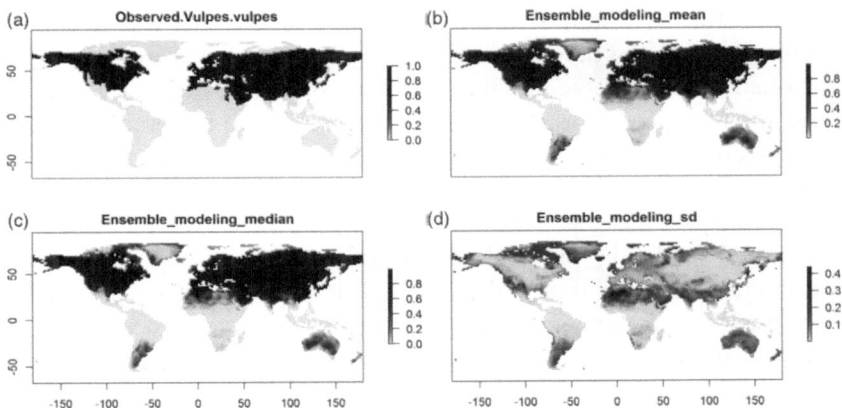

Figure 17.17 Observed presence and absence of *Vulpes vulpes* at (a) global scale, together with (b, c) the two model averaging projections for 2080 (mean and median) and (d) the ensemble modeling uncertainty (sd).

expanding margins of the ensemble forecast, especially in the north of Canada and south-west of Greenland. In general, this means that although there is a good chance that the species could find suitable climate in those regions in the future, the high variance between predictions requires us to interpret the results with caution. This is not the case for northern Europe.

A key issue when documenting future potential range shifts is to represent both the projection (either a single model or an ensemble forecast) and the associated uncertainty. A further non-negligible issue is that averaging the probability of occurrence from different models might not necessarily be the best approach since each model may respond differently to the original prevalence of the species, and the probability values may therefore not be comparable. In order to address these questions, one solution is to compute committee (consensus) averaging instead (Marmion et al., 2009; Meller et al., 2014).

Let's first address the issue of averaging probability values from different models. In order to achieve this successfully, the output of every model (i.e. probability values) can be transformed into binary data (i.e. presence and absence) using our preferred threshold (e.g. the one which maximizes TSS on the test data). This can be done for each technique and for each repeated split sampling run. In our previous example, we used five techniques and a 20-fold repeated split sampling procedure, obtaining 100 predictions. If these are all transformed into binary data and summed, they will sum to 100 if all techniques across all runs predict a presence, and to 0 if they all predict an absence. In other words, summing the binary data from the different techniques and repeated split sampling runs also addresses the issue of how to represent both the prediction and the uncertainty. Indeed, the committee averaging establishes the likelihood of being present given the data (i.e. repeated split sampling) and the techniques (i.e. the five techniques). When transformed in relative terms (i.e. divided by the total number of predictions), committee averaging close to either 0 or 1 shows a high level of agreement to predict the absence or the presence of a species, respectively, while a value of 0.5 shows that half the predictions suggest an absence and the other half a presence (maximum uncertainty given the data and techniques). More generally speaking, looking at the distribution of predicted values from ensemble models can reveal areas or parts of the environmental ranges with higher or lower consensus, which might, for instance, be associated with conditions where specialist species are found that can be easier to model than generalist species (Grenouillet et al., 2011).

The overall approach can also be expanded by removing models (i.e. by technique or split-sampling runs) that do not reach a minimum quality threshold (e.g. an AUC ≥ 0.7; Vicente et al., 2013) and also by weighting the models by the quality of their prediction on the test data (weighted average; Marmion et al., 2009).

We will show here the simplest example with no selection (all techniques for all runs have a high predictive power (TSS >0.75) and no weights attributed. In other words, the maximum sum equals to 100 (20 runs and 5 techniques).

```
# Committee averaging: Sum all binary projections
# from the 5 models and 20 repetitions.
> ProjFuture_CA<-as.data.frame(apply(ProjFuture_results_bin,1:2,
sum))
> ProjFuture_CA$CA <- rowSums(ProjFuture_CA)
```

We can first contrast the mean probability of the ensemble forecast and the committee averaging, which need to correlate to a certain extent.

```
# Link between committee averaging and mean probabilities across
# the models and repetitions.
> plot(ProjFuture_CA$CA,ProjFuture_total_mean,
xlab="Committee
averaging", ylab="Mean probability")
```

As we can see, there is a close relationship between mean probability and committee averaging (Figure 17.18). Importantly, it shows that

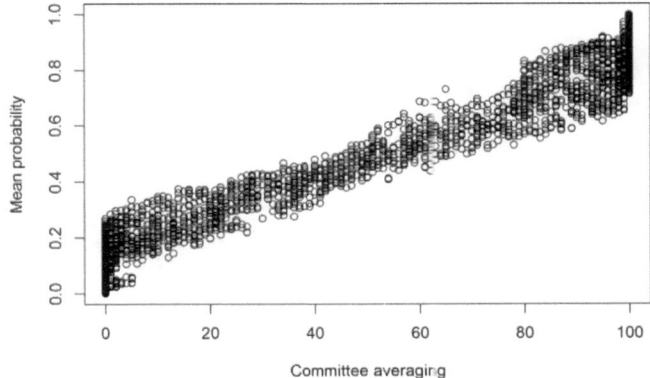

Figure 17.18 Relationship between *Vulpes vulpes* projections into the future from either committee averaging (*x*-axis) or mean probability across all techniques and repeated split sampling (*y*-axis).

when the committee averaging is close to 0 or 1, there is more variation in mean probability for those sizes. In other words, committee averaging provides a more informed view of the projections, including some degree of confidence (if all or none of the models agree).

Here, we have looked at committee averaging across all techniques and repeated split sampling runs, but the same approach could also be applied to each modeling technique or each run separately. We will here map the committee averaging for each technique (20 runs) and across all techniques and runs (Figure 17.19).

```
> ProjFuture_CAglm<-rasterFromXYZ(cbind(FutureEnv[,c("X_WGS84",
"Y_WGS84")],ProjFuture_CA$GLM))
> ProjFuture_CAgam<-rasterFromXYZ(cbind(FutureEnv[,c("X_WGS84",
"Y_WGS84")],ProjFuture_CA$GAM))
> ProjFuture_CAmars<-rasterFromXYZ(cbind(FutureEnv[,c("X_WGS84",
"Y_WGS84")],ProjFuture_CA$MARS))
> ProjFuture_CAfda<-rasterFromXYZ(cbind(FutureEnv[,c("X_WGS84",
"Y_WGS84")],ProjFuture_CA$FDA))
> ProjFuture_CArf<-rasterFromXYZ(cbind(FutureEnv[,c("X_WGS84",
"Y_WGS84")],ProjFuture_CA$RF))
> ProjFuture_CAall<-rasterFromXYZ(cbind(FutureEnv[,c("X_WGS84",
"Y_WGS84")],ProjFuture_CA$CA))
> OutFut_CA <- stack(ProjFuture_CAglm,
ProjFuture_CAgam,ProjFuture_CAmars, ProjFuture_CAfda,
ProjFuture_CArf,ProjFuture_CAall)
> names(OutFut_CA) <- c("CA_GLM","CA_GAM","CA_MARS","CA_FDA",
"CA_RF", "CA_ALL")
> plot(OutFut_CA, nc=2)
```

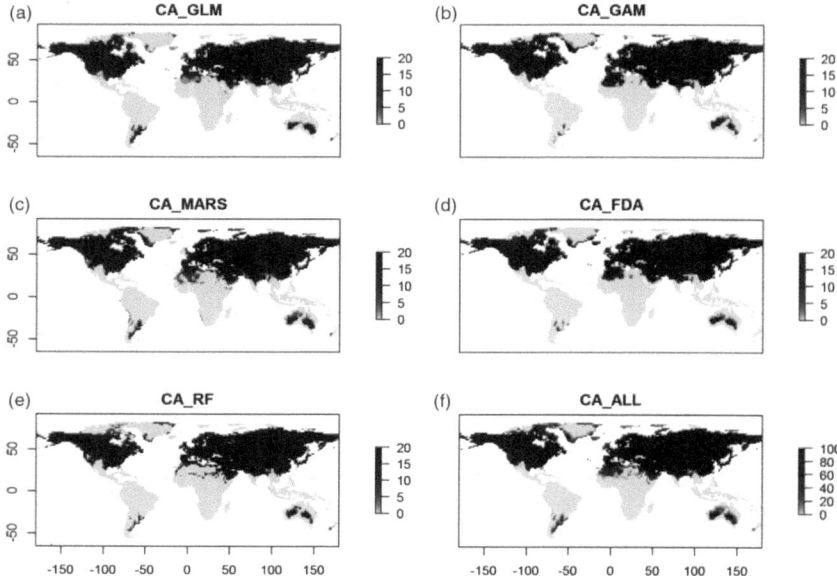

Figure 17.19 Future climatically suitable sites for *Vulpes vulpes* according to the different committee averaging procedures. (a) CA_GLM, (b) CA_GAM, (c) CA_MARS, (d) CA_FDA, (e) CA_RF represent the committee averaging for GLM, GAM, MARS, FDA and random forest across the 20 different repeated split sampling, while (f) CA_ALL represents the committee averaging across all techniques and repeated split sampling runs.

As we can see in Figure 17.19, some techniques yield more variations between runs than others. MARS appears to be more sensitive to the split-sampling procedure in the southern range of the red fox in North Africa, while FDA shows little variation. The committee averaging based on both the techniques and the runs reflects those variations. It shows that the likelihood of the species being present on the southern edge of the distribution in North Africa is close to 50 out of 100.

As previously reported, the distribution of the red fox is relatively well predicted and the variation between techniques and repeated split sampling is quite low compared to other studies (see many examples in Franklin, 2010a; Peterson et al., 2011). However, we will see that although moderate, this variation could have a significant impact on the species' measured sensitivity to climate change. As an example, let's measure the projected range change simply by comparing the current and future predictions.

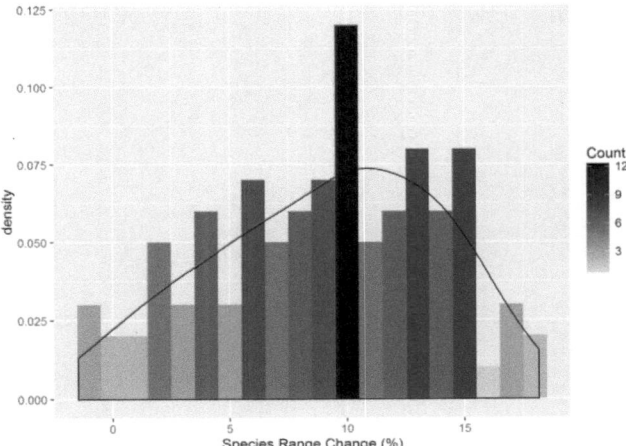

Figure 17.20 Histogram representing the variation in modeled species range for *Vulpes vulpes* across techniques and repeated split sampling runs.

```
# Species range change
> SRG <- 100*(colSums(ProjFuture_results_bin)-
sum(FutureEnv$VulpesVulpes))/sum(FutureEnv$VulpesVulpes)
> SRG_ToPlot <- as.data.frame(as.numeric(SRG))
> SRG_ToPlot$Model <- rep(c("GLM","GAM","MARS","FDA","RF"), 20)
> colnames(SRG_ToPlot)[1] <- "SRG"
> library(ggplot2)
> ggplot(SRG_ToPlot, aes(SRG)) + geom_histogram(aes(y =
..density.., fill = ..count..), binwidth=1) + geom_density()
+ scale_fill_gradient("Count", low = "lightgrey", high =
"black") + xlab("Species Range Change (%)")# Density
```

On average, we can see that the red fox is predicted to increase its total range by about 9–10% (Figure 17.20). However, we can also see that, depending on the technique and the split-sampling run, the expected species range could vary from a small reduction (-1%) to an almost 25% increase. Employing several techniques and several split-sampling runs means we can have more confidence in the projections, which in this case give an increase of around 10%.

Indeed, we can see that even for a given modeling technique, high levels variation can be found (Figure 17.21).

```
> p <- ggplot(SRG_ToPlot, aes(SRG, colour=Model))
> p + geom_density()+ xlab("Species Range Change (%)")
```

Figure 17.21 shows that, although the different techniques generally agree, some still have divergent outputs. For instance, GLM tends

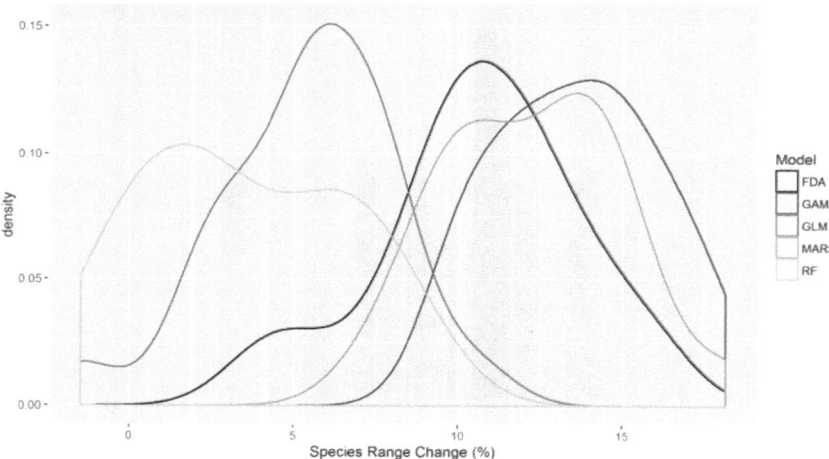

Figure 17.21 Density plot representing the variation in modeled species range for
Vulpes vulpes for each technique due to the different repeated split sampling runs. (*A
black and white version of this figure will appear in some formats. For the color version, please
refer to the plate section.*)

generally to predict a smaller range expansion than the other techniques,
especially in comparison to GAM.

In conclusion to this "ensemble projections" section, we have demonstrated the potential benefits of considering several repeated split sampling
runs and of using several modeling techniques, but other dimensions can
also be varied (e.g. different climate change predictions, different model
parameterization of a same technique, etc.; Araújo and New, 2007). We
have illustrated only one way to do such ensembling, but other approaches
exist that could also be explored, like using other consensus approaches
(e.g. using principal components of the predictions, i.e. PCA consensus;
Thuiller, 2004) than summing or averaging predictions, or using more
sophisticated approaches like Bayesian model averaging (Wintle et al.,
2003). In no way are we promoting the systematic and naïve averaging
of multiple models built with different options, but we do advise that at
least the variation between models is assessed and reported, as a way of
understanding the underlying uncertainty. Ensemble forecasts is certainly
not the only way of quantifying uncertainty, but it does help quantifying the confidence one can have in a set of predictions, and ensures the
conclusions of a given study are not overly-dependent on the choice of
techniques and/or data used.

PART VI · Data and Tools Used in this Book, with Developed Case Studies

18 · *Datasets and Tools Used for the Examples in this Book*

In this chapter, we describe the different datasets used in the different examples presented in this book. Many datasets can be accessed in directly the `biomod2` package available on CRAN and R-Forge [`install. packages("biomod2", repos="http://R-Forge.R-project. org")`].[1] Other datasets can be downloaded from the WorldClim website,[2] using the R code provided. Table 18.1 contains a summary of the datasets required to work through the examples and exercises. Some of these datasets are also available as text files on the book website,[3] for use in Part II.

```
> setwd("PATH/data/")
```

The mammal's dataset was obtained from the Global Mammal Assessment database (IUCN, 2001) for native terrestrial mammals. It includes presence records for the native range of the following species: *Connochaetes gnou* (black wildebeest), *Gulo gulo* (wolverine), *Panthera onca* (jaguar), *Pteropus giganteus* (Indian flying fox), *Tenrec ecaudatus* (tailess tenrec), and *Vulpes vulpes* (red fox). The environmental values associated to the dataset were obtained from WorldClim(Hijmans et al., 2005).[4] The environmental variables used are: bio3 (isothermality), bio4 (temperature seasonality), bio7 (temperature annual range), bio11 (mean temperature of coldest quarter) and bio12 (annual precipitation).

The main dataset used to perform the examples illustrated in this book is a set of native presence records in Eurasia and North America for the red fox, *V. vulpes*, and associated values for a set of environmental

---

[1] http://r-forge.r-project.org/projects/biomod/
[2] www.worldclim.org/download
[3] www.unil.ch/hsdm
[4] www.worldclim.org

Table 18.1 *A list of datasets used to work through the examples and exercises in this book. Unless otherwise indicated all datasets are available in the* biomod2 *package.*

Part	Chapter	Section	File name	Path: > setwd("PATH/data/")	Type	Source (Book website or code)
2	6	6.2.2	bio3.grd	~/raster/bioclim/current/grd/	grid	biomod2 package
2	6	6.2.2	bio7.grd	~/raster/bioclim/current/grd/	grid	biomod2 package
2	6	6.2.2	bio11.grd	~/raster/bioclim/current/grd/	grid	biomod2 package
2	6	6.2.2	bio12.grd	~/raster/bioclim/current/grd/	grid	biomod2 package
2	6	6.2.2	GTOPO30.tif	~/raster/topo/	Tiff	biomod2 package
2	6	6.2.4	isolines.shp	~/vector/globe/	Shape file	
2	6	6.2.5	latitude.tif	~/raster/other/	Tiff	
2	6	6.2.5	longitude.tif	~/raster/other/	Tiff	
2	6	6.2.6	hillshade.tif	~/raster/topo/	Tiff	
2	6	6.2.8	prec_30yr_normal_annual.asc	~/raster/prism/	ascii	PRISM project
2	6	6.2.8	tave_30yr_normal_annual.asc	~/raster/prism/	ascii	PRISM project
2	6	6.2.9	pinus_edulis_occ.csv.txt	~/tabular/species/	csv	GBIF July 2014
2	6	6.2.10	cal.txt	~/tabular/species/	txt	Calculated from the mammals_and_bioclim_table.csv dataset using the ecospat.caleval() function from ecospat package
2	6	6.2.10	eva.txt	~/tabular/species/	txt	Calculated from the mammals_and_bioclim_table.csv dataset using the ecospat.caleval() function from ecospat package

2	6	6.3.3	Swiss_Cantons.shp	~/vector/swiss/	Shape file	
2	6	6.3.3	L7_194027_2001_08_24_B10.TIF	~/raster/landsat/	Tiff	Landsat
2	6	6.3.3	L7_194027_2001_08_24_B20.TIF	~/raster/landsat/	Tiff	Landsat
2	6	6.3.3	L7_194027_2001_08_24_B30.TIF	~/raster/landsat/	Tiff	Landsat
2	6	6.3.3	L7_194027_2001_08_24_B40.TIF	~/raster/landsat/	Tiff	Landsat
2	6	6.3.3	L7_194027_2001_08_24_B50.TIF	~/raster/landsat/	Tiff	Landsat
2	6	6.3.3	L72194027_2001_08_24_B70.TIF	~/raster/landsat/	Tiff	Landsat
2	6	6.3.4	hill_250m_utm.tif	~/raster/topo	Tiff	biomod2 package
2	6	6.4.2	bioclim_table.csv	~/tabular/bioclim/current/	csv	biomod2 package
2	7	7.4.1	USA_states.shp	~/vector/usa/	Shape file	
2	7	7.4.1	GTOPO30.tif	~/raster/topo/	Tiff	https://lta.cr.usgs.gov/ GTOPO30
3	8	8.2	mammals_and_bioclim_table.csv	~/tabular/species/	csv	biomod2 package
3	14	14	mammals_table.csv	system.file("external/species/ mammals_table.csv", package="biomod2")	csv	biomod2 package
3	14	14	bio3.grd	system.file("external/ bioclim/current/bio3. grd",package="biomod2")	grd	biomod2 package
3	14	14	bio4.grd	system.file("external/ bioclim/current/bio4. grd",package="biomod2")	grd	biomod2 package

(continued)

Table 18.1 (*cont.*)

Part	Chapter	Section	File name	Path: > setwd("PATH/data/")	Type	Source (Book website or code)
3	14		bio7.grd	system.file("external/bioclim/current/bio7.grd",package="biomod2")	grd	biomod2 package
3	14		bio11.grd	system.file("external/bioclim/current/bio11.grd",package="biomod2")	grd	biomod2 package
3	14		bio12.grd	system.file("external/bioclim/current/bio12.grd",package="biomod2")	grd	biomod2 package
4	15	15.1.1	EvalData.txt	~/tabular/	txt file	models calculated from the mammals_and_bioclim_table.csv dataset
4	16	16.2.1	summary_mammals_and_bioclim.csv	~/tabular/species/	csv	
5	17	17.1	clim.vulpesEU_100.txt	~/tabular/bioclim/current	txt file	obtained from biomod2 package dataset
5	17	17.1	clim.vulpesNA_100.txt	~/tabular/bioclim/current	txt file	obtained from biomod2 package dataset
5	17	17.1	vulpes_eu.txt	~/tabular/species/	txt file	obtained from biomod2 package dataset
5	17	17.1	vulpes_na.txt	~/tabular/species/	txt file	obtained from biomod2 package dataset

5	17	17.3	bio3.grd	~/raster/bioclim/future/grd/	grd	
5	17	17.3	bio7.grd	~/raster/bioclim/future/grd/	grd	
5	17	17.3	bio11.grd	~/raster/bioclim/future/grd/	grd	
5	17	17.3	bio12.grd	~/raster/bioclim/future/grd/	grd	
6	19	19.1	protea laurifolia presence records		data frame	downloaded from GBIF with R code
6	19	19.1	worldclim data current	~/WorldClim_data	Tiff	downloaded from worldclim with R code
6	19	19.1	worldclim data future 50	~/WorldClim_data	Tiff	downloaded from worldclim with R code
6	19	19.1	worldclim data future 70	~/WorldClim_data	Tiff	downloaded from worldclim with R code
6	19	19.1	south africa shape file	download.file(url = "https://sourceforge.net/projects/biomod2/files/data_for_example/south_of_africa.zip", destfile = "south_of_africa.zip")	Shape file	biomod2 package
6	19	19.2	larus presence records		data frame	downloaded from GBIF with R code
6	19	19.1	worldclim data current	~/WorldClim_data	Tiff	downloaded from worldclim with R code
6	19	19.1	worldclim data future 50	~/WorldClim_data	Tiff	downloaded from worldclim with R code
6	19	19.1	worldclim data future 70	~/WorldClim_data	Tiff	downloaded from worldclim with R code

Table 18.2 *A list of the R packages used to perform the examples illustrated in this book.*

> library(ade4)	> library(ltm)
> library(adehabitatHS)	> library(maptools)
> library(ape)	> library(MASS)
> library(biomod2)	> library(mda)
> library(boot)	> library(mgcv)
> library(classInt)	> library(ncf)
> library(cowplot)	> library(nnet)
> library(Daim)	> library(PresenceAbsence)
> library(dismo)	> library(pROC)
> library(earth)	> library(randomForest)
> library(ecospat)	> library(raster)
> library(fields)	> library(rasterVis)
> library(gam)	> library(reshape2)
> library(gbm)	> library(rgbif)
> library(gbm)	> library(rgdal)
> library(ggplot2)	> library(rpart)
> library(gridExtra)	> library(snowfall)
> library(Hmisc)	> library(sp)
> library(landsat)	> library(usdm)

variables. It should be noted, however, that the presences of the ref fox in some other invaded areas, mainly Australia and New Zealand, are not included and therefore worldwide projections should be interpreted accordingly. This red fox dataset is well suited to the different examples because it has a global extent (world) which means everyone can easily understand the illustrations. A coarser resolution version of the dataset is also directly available in the biomod2 package, and the dataset used in this book, which has a higher resolution (166 km) is available through the book website.[5] In this type of dataset, the standard functions in R and several external libraries have been used to prepare the data and perform the modeling analyses. A list of the R packages required for the examples and analysis is also provided in Table 18.2.

Many of the resources used in this book are available on the book website at www.unil.ch/hsdm.

[5] www.unil.ch/hsdm

19 · *The Biomod2 Modeling Package Examples*

19.1 Example 1: Habitat Suitability Modeling of *Protea laurifolia* in South Africa

19.1.1 Brief Example Description

Objectives
Model the bioclimatic niche of a species endemic to South Africa and create a set of projections under a collection of climate change scenarios. A further objective is to measure the sensitivity of the species to climate change under the different scenarios.

Data
This example relies on presence-only data downloaded from GBIF, which will then require the creation of a set, or several sets, of pseudo-absence data. The explanatory variables are raster grid data downloaded from the WorldClim datacenter.

Modeling Steps

- Loading and formatting the presence-only data
- Loading and formatting the raster data
- Building a range of models and ensemble models using `biomod2`
- Decomposing the models' variability (predictive ability / predictions)
- Projections under current and future conditions
- Species' range change estimates.

19.1.2 Loading Required Packages
The `rgbif` package is required to obtain the species occurrences directly from the GBIF website.[1]

---

[1] www.gbif.org/

Figure 19.1 Protea laurifolia flower and leaves. (Photo from www.flickr.com/photos/ flowcomm/.) (*A black and white version of this figure will appear in some formats. For the color version, please refer to the plate section.*)

```
> if(!require(rgbif)){ # automatically install rgbif package
if needed
install.packages("rgbif")
library(rgbif)
}
```

Loading `biomod2`, `ggplot2`, `gridExtra`,

```
> library(biomod2)
> library(ggplot2)
> library(gridExtra)
```

19.1.3 Obtaining Species Data from a Datacenter

In this example, we will focus on *Protea laurifolia*, a plant species endemic to South Africa (Figure 19.1). It is a bearded protea with large, pale pink or cream flower heads during the winter months. It is also frost hardy, water-wise and tolerates a wider range of soil types than most protea. (See the detailed *Protea* description for further information on this plant and its ecology[2].)

Obtain the *Protea laurifolia* occurrences data from the GBIF website[3] using `rgbif` package.

---

[2] www.plantzafrica.com/plantnop/protealauri.htm
[3] www.gbif.org/species/5637308

The first step is to check that the species is indeed referenced in GBIF. The function name_suggest() searches all species that more or less fit the request. Only the species names from GBIF that exactly match the name of our species are kept.

```
> spp_Protea <- name_suggest(q = 'Protea laurifolia',
rank = 'species',
limit = 10000)
> (spp_Protea <- spp_Protea[grepl("^Protea laurifolia", spp_
Protea$canonicalName), ])
# A tibble: 3 × 3
       key         canonicalName     rank
     <int>                 <chr>    <chr>
1 7468831 Protea laurifolia SPECIES
2 8498444 Protea laurifolia SPECIES
3 5637308 Protea laurifolia SPECIES
```

The previous step identifies the taxonomic reference number of *Protea laurifolia*. It appears here that they are several matches for our species of interest. Using this reference is the safest way to request species occurrences in GBIF. We only want to keep the geo-referenced occurrences from the area in which it is endemic (i.e. South Africa).

```
> ## get species occurrences
> data <- occ_search(taxonKey = 5637308, country='ZA',
fields = c('name', 'key', 'country', 'decimalLatitude',
'decimalLongitude'),
hasCoordinate=T, limit=1000, return = 'data')
```

Now we print the summary of the extracted object data.

```
> data
$`7468831`
 [1] "no data found, try a different search"


$`8498444`
 [1] "no data found, try a different search"


$`5637308`
# A table: 290 × 5
                name       key decimalLongitude decimalLatitude      country
               <chr>     <int>            <dbl>           <dbl>        <chr>
1  Protea laurifolia 1212015147         19.10114       -33.62931 South Africa
2  Protea laurifolia 1212015345         19.04115       -31.37300 South Africa
3  Protea laurifolia 1212015363         19.04115       -31.37300 South Africa
4  Protea laurifolia 1212015372         19.04115       -31.37300 South Africa
5  Protea laurifolia  685238754         18.87500       -33.62500 South Africa
6  Protea laurifolia  699269103         19.15700       -32.63570 South Africa
7  Protea laurifolia  699269110         18.99160       -32.13538 South Africa
```

```
8  Protea laurifolia  462191010       19.62500       -32.87500 South Africa
9  Protea laurifolia  462191008       18.62500       -33.12500 South Africa
10 Protea laurifolia  462191007       19.37500       -32.37500 South Africa
# ... with 280 more rows


attr(,"args")

attr(,"args")$taxonKey

  [1] 7468831 8498444 5637308
attr(,"args")$country
  [1] "ZA"
attr(,"args")$hasCoordinate
  [1] TRUE
attr(,"args")$limit
  [1] 1000
attr(,"args")$offset
  [1] 0

attr(,"args")$fields
  [1] "name"              "key"              "country"
  [4] "decimalLatitude"   "decimalLongitude"
```

It appears that only the item "5637308" is data so we can remove the other ones.

```
> data <- data[['5637308']]
```

To prevent any problem with the pathway it is a good practice to remove blank spaces from species names.

```
# replace " " by "." in species names
> data$name <- sub(" ", ".", data$name)
> (spp_to_model <- unique(data$name))
  [1] "Protea.laurifolia"
# Total number of occurrences:
> sort(table(data$name), decreasing = T)
  Protea.laurifolia
                290
```

19.1.4 Obtaining the Environmental Data

For this example, we will use the WorldClim data from the worldclim website.[4] In order to avoid a time-consuming downloads and calculations, we will use 10-minute resolution raster grids. Since the objective

---

[4] www.worldclim.org/

is to look at both current and future potential distributions, we will not only download WorldClim data for current conditions, but also climatic projections for 2050 and 2070 from an arbitrary selected IPPC Fifth Assessment climate change scenario (here the BCC-CSM1-1 Regional Climatic Model combined with 4.5 Representative Concentration Pathways. Please refer to worldclim.org/CMIP5v1 for further explanations).

The first step is to download the Worlclim data archives from the data center:

```
# get WorldClim environmental variables
> dir.create("WorldClim_data", showWarnings = F)
# current bioclim
> download.file(url = "http://biogeo.ucdavis.edu/data/climate/
worldclim/1_4/grid/cur/bio_10m_esri.zip", destfile = "WorldClim_
data/current_bioclim_10min.zip", method = "auto")
>
# GCM -> BCC-CSM1-1, year -> 2050. RCP -> 4.5
> download.file(url = "http://biogeo.ucdavis.edu/data/climate/
cmip5/10m/bc45bi50.zip", destfile = "WorldClim_data/2050_BC_45_
bioclim_10min.zip", method = "auto")
>
# GCM -> BCC-CSM1-1, year -> 2070. RCP -> 4.5
> download.file(url = "http://biogeo.ucdavis.edu/data/climate/
cmip5/10m/bc45bi70.zip", destfile = "WorldClim_data/2070_BC_45_
bioclim_10min.zip", method = "auto")
```

Then we have to extract the files.

```
# unzip climatic files
> unzip(zipfile = "WorldClim_data/current_bioclim_10min.zip",
exdir = "WorldClim_data/current",
overwrite = T)
> list.files("WorldClim_data/current/bio/")
    [1] "bio_1"  "bio_10" "bio_11" "bio_12" "bio_13" "bio_14"
    [7] "bio_ 15" "bio_ 16" "bio_ 17" "bio_ 18" "bio_ 19" "bio_ 2"
   [13] "bio_ 3" "bio_ 4" "bio_ 5" "bio_ 6" "bio_ 7" "bio_ 8"
   [19] "bio_ 9" "info"
> unzip(zipfile = "WorldClim_data/2050_BC_45_bioclim_10min.
zip",
exdir = "WorldClim_data/2050/BC_45",
overwrite = T)
```

```
> list.files("WorldClim_data/2050/BC_45/")
  [1] "bc45bi5010.tif" "bc45bi5011.tif" "bc45bi5012.tif"
  [4] "bc45bi5013.tif" "bc45bi5014.tif" "bc45bi5015.tif"
  [7] "bc45bi5016.tif" "bc45bi5017.tif" "bc45bi5018.tif"
 [10] "bc45bi5019.tif" "bc45bi501.tif" "bc45bi502.tif"
 [13] "bc45bi503.tif" "bc45bi504.tif" "bc45bi505.tif"
 [16] "bc45bi506.tif" "bc45bi507.tif" "bc45bi508.tif"
 [19] "bc45bi509.tif"
> unzip(zipfile = "WorldClim_data/2070_BC_45_bioclim_10min.
zip",
exdir = "WorldClim_data/2070/BC_45", overwrite = T)
> list.files("WorldClim_data/2070/BC_45/")
  [1] "bc45bi7010.tif" "bc45bi7011.tif" "bc45bi7012.tif"
  [4] "bc45bi7013.tif" "bc45bi7014.tif" "bc45bi7015.tif"
  [7] "bc45bi7016.tif" "bc45bi7017.tif" "bc45bi7018.tif"
 [10] "bc45bi7019.tif" "bc45bi701.tif" "bc45bi702.tif"
 [13] "bc45bi703.tif" "bc45bi704.tif" "bc45bi705.tif"
 [16] "bc45bi706.tif" "bc45bi707.tif" "bc45bi708.tif"
 [19] "bc45bi709.tif"
```

At this point, we have all the species and climate data we need for this example.

19.1.5 Environmental Variables Selection

As previously discussed (see Chapter 6), in order to correctly select the variables, avoiding any potential collinearity problems is recommended. As we have seen there are several methods of variable selection. Here, we use a PCA to visualize the correlation between the variables and identify the main environmental gradients in the region to be used in the modeling process. The `ade4` package is used to perform this pre-analysis.

First, we need to extract species occurrences from our data table.

```
> ProLau_occ <- data[data$name == "Protea.laurifolia", ]
```

The bioclimatic variables are stored in grid format. We will first stack them, so they can all be found in one file.

```
> library (raster)
> bioclim_world <- stack(list.files("WorldClim_data/current/
bio",
pattern = "bio_", full.names = T), RAT = FALSE)
```

We will start with a shape file for the whole of Southern Africa from the `biomod2` package, stored in the `biomod2` repository. Then, since we are only interested in South Africa, we need to crop the larger rasters to only keep South Africa.

```
> download.file(url = "http://dfn.dl.sourceforge.net/project/
biomod2/data_for_example/south_of_africa.zip",
destfile = "south_of_africa.zip")
> unzip(zipfile = "south_of_africa.zip", exdir = ".",
overwrite = T)
> list.files("south_of_africa", recursive = T)
  [1] "protected_areas_south_africa.dbf" "protected_areas_south_
africa.prj"
  [3] "protected_areas_south_africa.sbn" "protected_areas_south_
africa.sbx"
  [5] "protected_areas_south_africa.shp" "protected_areas_south_
africa.shx"
  [7] "South_Africa.dbf"                 "South_Africa.prj"
  [9] "South_Africa.sbn"                 "South_Africa.sbx"
 [11] "South_Africa.shp"                 "South_Africa.shx"
> mask_south_of_africa <- shapefile("south_of_africa/South_Africa.
shp")
> bioclim_ZA <- mask(bioclim_world,
mask_south_of_africa[ mask_south_of_africa$CNTRY_NAME == "South
Africa", ])
> bioclim_ZA <- crop(bioclim_ZA, mask_south_of_africa)
```

In order to compare the *Protea laurifolia* niche (the conditions used by the species) to the environmental conditions in South Africa (the available conditions) we need to obtain the identifiers (ids) of the cells where *Protea laurifolia* occurs.

```
> points_laurifolia<-data.frame(ProLau_
occ[1:290,c("decimalLongitude", "decimalLatitude")])
> ProLau_cell_id <- cellFromXY(subset(bioclim_ZA,1), points_
laurifolia)
```

19.1.6 Principal Component Analysis

```
> library(ade4)
```

First, we convert the raster object into a data frame to run the PCA (but note that PCA can now deal with raster objects). We also need to remove the non-defined area from this dataset.

```
> bioclim_ZA_df <- na.omit(as.data.frame(bioclim_ZA))
> head(bioclim_ZA_df)
```

	bio_1	bio_10	bio_11	bio_12	bio_13	bio_14	bio_15	bio_16	bio_17	bio_18
12355	220	266	159	338	72	1	85	186	4	186
12356	220	266	159	340	73	1	85	187	5	187
12357	220	263	162	337	72	1	85	187	5	187
12358	219	260	163	340	72	1	85	189	5	189
12359	219	259	164	351	74	1	85	195	7	195
12360	225	265	170	334	70	1	86	185	7	185

	bio_19	bio_2	bio_3	bio_4	bio_5	bio_6	bio_7	bio_8	bio_9
12355	4	160	55	4279	338	49	289	266	159
12356	5	158	55	4290	337	50	287	266	159
12357	5	154	56	4054	331	57	274	263	162
12358	5	148	56	3894	326	63	263	260	163
12359	7	142	55	3822	323	69	254	259	164
12360	7	139	55	3827	328	78	250	265	170

The dudi.pca() function from the ade4 package makes it possible to conduct the PCA over the whole study area. Here, the decision has been made to keep two principal component axes to summarize the whole environmental space, but more could be used.

```
> pca_ZA <- dudi.pca(bioclim_ZA_df,scannf = F, nf = 2)
```

One preliminary test is to look for potential outliers in the environmental data.

```
# PCA scores on first two axes
> plot(pca_ZA$li[, 1:2])
```

Two points, located in the top-left corner of the graph (Figure 19.2), are far from all other points and can thus be considered as outliers. Because these outliers can seriously distort the analyses, and considering that in practice there are a number of good reasons for excluding them (e.g. if these are obviously numerical errors), they can easily be removed and the PCA performed again.

```
# tail of distributions
> sort(pca_ZA$li[, 1])[1:10]
   [1] -24.646 -24.463  -8.758  -8.462  -8.292  -8.290
-8.138  -8.136
   [9]  -8.131  -7.979
# IDs of points to remove
> (to_remove <- which(pca_ZA$li[, 1] < -10))
   [1] 4069 4070
# remove points and re-compute PCA
> if(length(to_remove)){ ## remove outliers
bioclim_ZA_df <- bioclim_ZA_df[ - to_remove,]
pca_ZA <- dudi.pca(bioclim_ZA_df, scannf = F, nf = 2)
  }
```

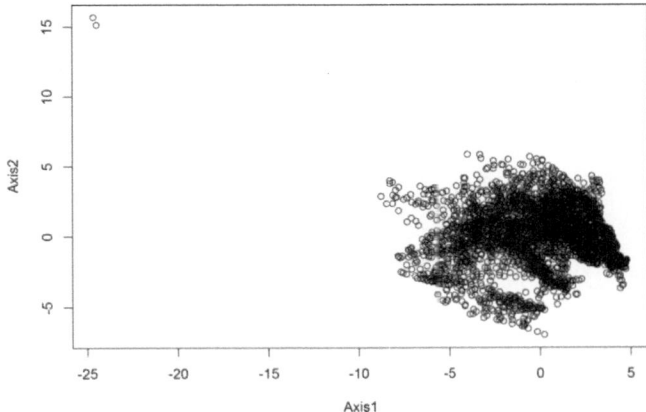

Figure 19.2 Plot of the principal component analysis scores of the first two axes of the South African environmental space.

The next step could be to investigate the distribution of our target species, *Protea laurifolia*, in the environmental space defined by the PCA. First, we investigate how *Protea laurifolia* is distributed along the first two PCA axes. Secondly, we illustrate the projection of the selected bioclimatic variables over the same two PCA axes.

```
> par(mfrow=c(1, 2))
# Discriminate Protea laurifolia presences from the entire
# South African environmental space.
> s.class(pca_ZA$li[, 1:2], fac= factor(rownames(bioclim_ZA_df)
%in% ProLau_cell_id, levels = c("FALSE", "TRUE" ),
labels = c("background", "ProLau")), col=c("red", "blue"),
csta = 0, cellipse = 2, cpoint = .3, pch = 16)
> mtext("(a)", side = 3, line = 3, adj = 0)
> s.corcircle(pca_ZA$co, clabel = .5 )
> mtext("(b)", side = 3, line = 3, adj = 0)
```

In the first graph we see that *Protea laurifolia* occupies a specific, although large, part of the South African environmental space, at least in the part defined by the first two PCA axes. Our target species thus has a clear range of tolerance, but is not highly specialized. The next analytical step is to highlight the relationship between the species occurrences and these specific environmental combinations.

The graph in panel (b) of Figure 19.3 is an important tool in terms of variable selection. We want to select a set of variables that are not too

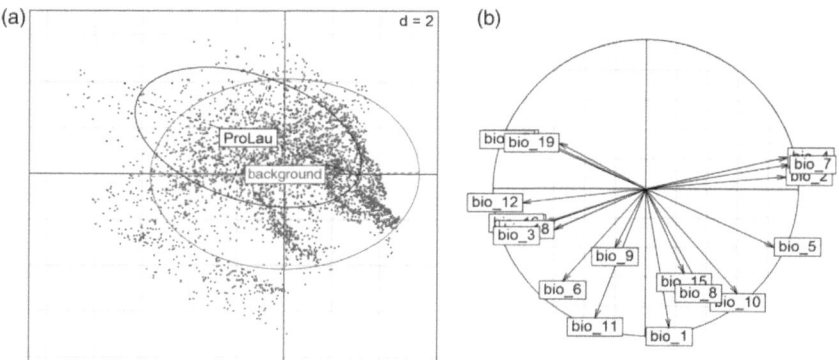

Figure 19.3 Distribution of the points of *Protea laurifolia* (ProLau) in the environmental space defined by the first two PCA axes (a) and correlation circle of the selected bioclimatic variables (see full names at worldclim website) as a function of the same first two PCA axes (b).

collinear (two variables pointing in orthogonal directions are independent, two variables pointing in the same or opposite directions are highly dependent, positively or negatively, respectively), and significantly contribute to the overall environmental variation (the longer the arrow, the more important the variable).

At first glance, bio_19 is a good variable for discriminating between our species occurrences and the rest of the environment. This is suggested by the fact that the longest axis of the species' ellipse (ProLau) in Figure 19.3a broadly follows the bio_19 variable axis.

Other choices, based on different criteria, could also be made when selecting the variables and ultimately it comes down to expert knowledge of the species' ecology and the study region to justify the final choice of predictor variables for the models. For this example, we will keep "bio_5", "bio_7", "bio_11", "bio_19".

These variables are (see also Worldclim website[5] for more information):

- bio_5 = max temperature of warmest month
- bio_7 = temperature annual range
- bio_11 = mean temperature of coldest quarter
- bio_19 = precipitation of coldest quarter.

[5] www.worldclim.org/bioclim

These include two temperature variables, a third variable related to variability in temperature and a fourth variable related to precipitation in winter.

To sub-select these four variables from the full environmental set:

```
> bioclim_ZA_sub <- stack(subset(bioclim_ZA, c("bio_5", "bio_7",
"bio_11", "bio_19")))
```

19.1.7 Biomod2 Modeling Procedure

We have now all the data required to quantify *Protea laurifolia*, niche, build a set of SDMs, and in this way investigate the link between the species and the environment. Several models, each with its own strengths and weaknesses (see Part III), are used to model the species–environment relationship and predict species distribution. In this regard, the `biomod2` package offers a range of suitable tools adapted to this objective.

Biomod2 Formatting

The first step is to put the data into the right format. This is done using the `BIOMOD_FormatingData()` function where we have to provide the species occurrences and associated coordinates, the environmental conditions, and the name of the species of interest. In this example we will start with presence-only data, but because most niche models need both presences and absences, we need to sample a set of pseudo-absences/background data from the South African landscape (see Part II). Since this process implies a stochastic procedure caused by the random selection (potentially stratified) of the pseudo-absences, it is recommended that several sets of pseudo-absences data are built to prevent sampling bias, especially for moderate or low numbers of pseudo-absences (say <1000). A suite of tests will be carried out to investigate the effect of each pseudo-absence selection on the predictive ability of the models. In `biomod2`, routines are built in to carry out the pseudo-absence selection under different procedures. Here we will use the simplest one, a random sampling, and repeat it three times with a selection of 500 pseudo-absences/background data.

```
> library(biomod2)
> ProLau_data <- BIOMOD_FormatingData(
resp.var = rep(1, nrow(ProLau_occ)), expl.var = bioclim_ZA_sub,
resp.xy = ProLau_occ[, c('decimalLongitude', 'decimalLatitude')],
resp.name = "Protea.laurifolia", PA.nb.rep = 3,
PA.nb.absences = 500, PA.strategy = 'random')
```

The data complies now with the `biomod2` formalism. We can print a summary and display a plot of the created object to check the data consistency.

```
# formatted object summary
> ProLau_data
  -=-=-=-=-=-=-=-=-=-=-=-=   'BIOMOD.formated.data.PA'
-=-=-=-=-=-=-=-=-=-=-=

   sp.name =  Protea.laurifolia

     282 presences,  0 true absences and  1315 undefined points in
dataset

      4 explanatory variables

        bio_5            bio_7            bio_11           bio_19
   Min.    :188    Min.     :142    Min.    : 35    Min.   : : 3.0
   1st Qu.:276    1st Qu.:241    1st Qu.: 96    1st Qu.: 21.0
   Median :300    Median :275    Median :111    Median : 37.0
   Mean    :299    Mean     :272    Mean    :112    Mean    : 72.2
   3rd Qu.:321    3rd Qu.:313    3rd Qu.:124    3rd Qu.: 80.0
   Max.    :383    Max.     :352    Max.    :189    Max.    :429.0

   3 Pseudo Absences dataset available (PA1 PA2 PA3) with  500
   absences in each (true abs + pseudo abs)


    -=-=-=-=-=-=-=-=-=-=-=-=-=-=-=-=-=-=-=-=-=-=-=-=-=-=-=-=-=
=-=-=-=-=-=-
# plot of selected pseudo-absences
> plot(ProLau_data)
```

Figure 19.4 displays the locations of the pseudo-absences in the three datasets (PA1, PA2, PA3) compared to the species occurrences. As expected, pseudo-absences are clearly randomly distributed over the whole study area.

Biomod2 Modeling

We now come to the main step where SDMs are parameterized and fitted.

Although the default parameters should reflect the settings most commonly used in published SDM studies, users still have the option to fine-tune the parameters for each algorithm separately. In the example below, we specify the use of quadratic terms and first-order interactions in GLMs, to limit the number of trees to 1000 in GBMs, and to use the "mgcv" package to fit the GAMs. As we decided to run the RFs with

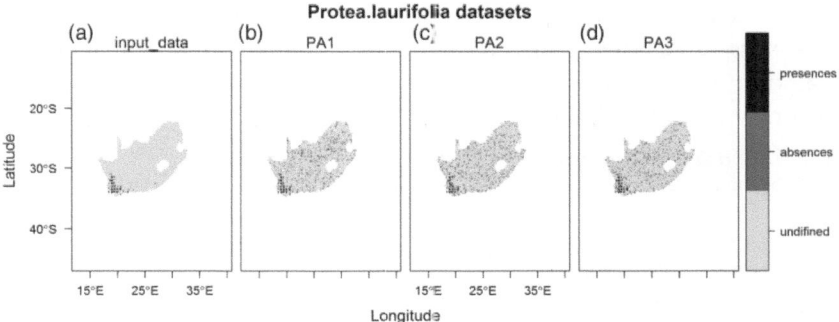

Figure 19.4 Plot of the species distribution (occurrences) and three selected sets of pseudo-absences. (*A black and white version of this figure will appear in some formats. For the color version, please refer to the plate section.*)

biomod2 default parameters they are not mentioned in the BIOMOD_ ModelingOptions() function. There are numerous other options that can be set up. Some of these have been reviewed in Part III. Please refer to the Biomod2 manual in CRAN for further details.

```
> ProLau_opt <- BIOMOD_ModelingOptions(
GLM = list(type = 'quadratic', interaction.level = 1),
GBM = list(n.trees = 1000), GAM = list(algo = 'GAM_mgcv'))
```

Four different algorithms have been selected here. A traditional algorithm, GLM, its semi-parametric extension (allowing smoothing functions in the predictors), GAM, and two bagging and boosting approaches (GBM and RF), all described in Part III. Other algorithms are also available in biomod2 (e.g. Maxent, classification trees, NNs, or MARS) and can be set up using the same syntax.

Since there is no independent dataset available in our example to evaluate the models, a repeated data-splitting procedure (cross-validation) is carried out. This procedure is described in detail in Part IV, but in short, the models are calibrated on 80% of the data (training set) and evaluated on the remaining 20% (validation set). This entire procedure is repeated four times (NbRunEval). By default, each model is evaluated by the TSS, ROC (=AUC) and KAPPA metrics. Other metrics are available in biomod2 and can be seen in the related documentation.

```
> ProLau_models <- BIOMOD_Modeling(
data = ProLau_data, models = c("GLM", "GBM", "RF", "GAM"),
models.options = ProLau_opt, NbRunEval = 4, DataSplit = 80,
VarImport = 3, do.full.models = F, modeling.id = "ex2")
```

In our example, these options lead to 48 models being built (4 algorithms × 4 cross-validations × 3 pseudo-absences samplings) and evaluated. The function `get_evaluations()` can be used to visualize the predictive accuracy of each individual model:

```
# get model evaluation scores
> ProLau_models_scores <- get_evaluations(ProLau_models)

# ProLau_models_scores is a 5 dimension array containing the
# scores for the models
> dim(ProLau_models_scores)
  [1] 3 4 4 4 3
> dimnames(ProLau_models_scores)
  [[1]]
  [1] "KAPPA" "TSS"     "ROC"

  [[2]]
  [1] "Testing.data" "Cutoff"          "Sensitivity"
"Specificity"

  [[3]]
  [1] "GLM" "GBM" "RF"   "GAM"

  [[4]]
  [1] "RUN1" "RUN2" "RUN3" "RUN4"

  [[5]]
  Protea.laurifolia_PA1 Protea.laurifolia_PA2 Protea.laurifolia_PA3
                  "PA1"                 "PA2"                 "PA3"
```

Graphical tools can also be used to assess the influence of the different choices made when parameterizing the models (e.g. the choice of algorithm (Figure 19.5), cross-validation run (Figure 19.6), pseudo-absences sampling (Figure 19.7)) according to the selected evaluation metrics. Here we focus on TSS and AUC (ROC scores) only. On these graphs, the points represent the mean of evaluation score for a given condition and the lines represent the associated standard deviations.

```
> models_scores_graph(ProLau_models, by = "models" ,
metrics = c("ROC","TSS"), xlim = c(0.5,1), ylim = c(0.5,1))

> (ProLau_models_var_import <-
get_variables_importance(ProLau_models))
```

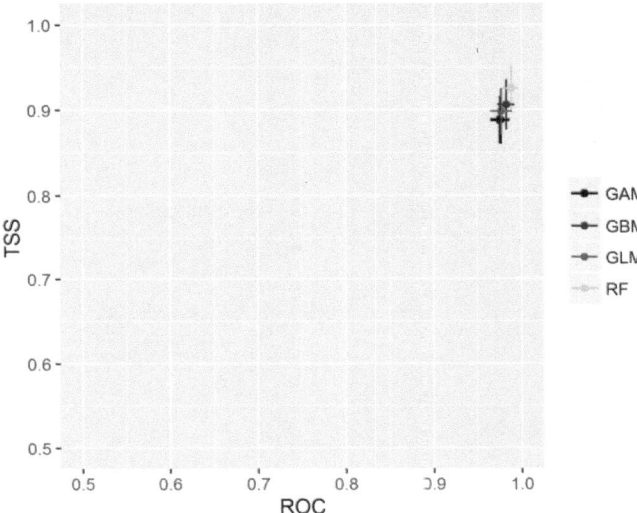

Figure 19.5 Plot of the mean of the model evaluation scores (by algorithms) according to two different evaluation metrics, ROC (AUC) and TSS. (*A black and white version of this figure will appear in some formats. For the color version, please refer to the plate section.*)

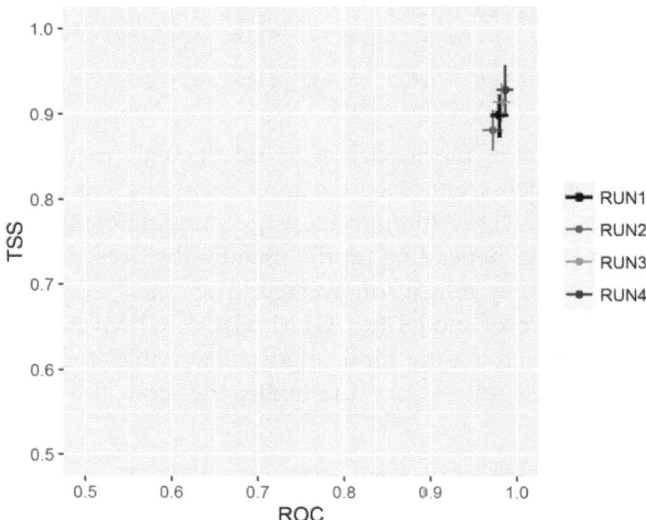

Figure 19.6 Plot of the mean of the model evaluation scores (by cross-validation) according to two different evaluation metrics, ROC (AUC) and TSS. (*A black and white version of this figure will appear in some formats. For the color version, please refer to the plate section.*)

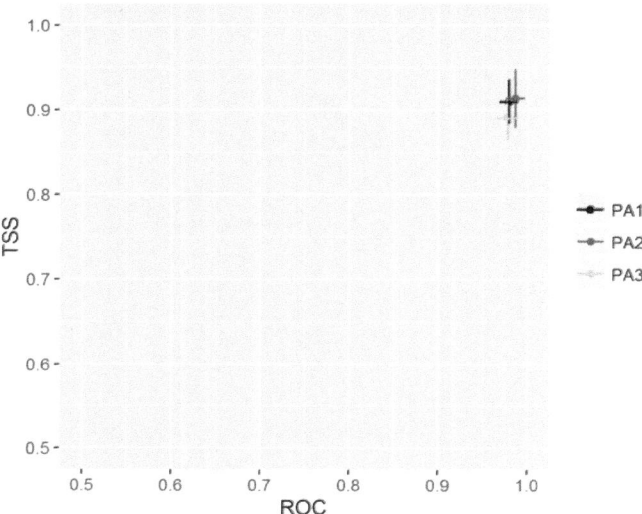

Figure 19.7 Plot of the mean of the model evaluation scores (by dataset) according to two different evaluation metrics, ROC (AUC) and TSS. (*A black and white version of this figure will appear in some formats. For the color version, please refer to the plate section.*)

```
> models_scores_graph(ProLau_models, by = "cv_run" ,
metrics = c("ROC","TSS"), xlim = c(0.5,1), ylim = c(0.5,1))

> models_scores_graph(ProLau_models, by = "data_set" ,
metrics = c("ROC", "TSS"), xlim = c(0.5, 1), ylim = c(0.5, 1))
```

The predictive accuracy of the models is excellent, with very high evaluation scores. RF models seem to be the most accurate on average, followed by GBM, GAM and GLM. Although there are some differences in the model evaluation scores depending on the pseudo-absences, dataset, and cross-validation run, they remain on average lower than score differences due to the choice of model (i.e. GLM, GAM, GBM or RF). We should also check which are the most important variables for each of the models. In Biomod2 for instance, the higher the score, the more important the variable.

```
# calculate the mean of variable importance by algorithm
> apply(ProLau_models_var_import, c(1,2), mean)
            GLM      GBM       RF      GAM
  bio_5   0.09942 0.002917 0.01975 0.03550
  bio_7   0.38792 0.053583 0.06750 0.21058
  bio_11  0.05400 0.042667 0.05367 0.05275
  bio_19  0.77558 0.941833 0.84175 0.84325
```

If we average variable importance across the different pseudo-absences sets and cross-validation runs, we see that variable importance is coherent across models. The variable bio_19 appears to be the most influential for all models. This high level of importance of bio_19 in the model prediction fits well with our first hypothesis put forward during variable selection. The variable bio_7 appears to be the second most influential. However, the next two variables seem to be less decisive in defining species distribution. Accordingly, the range of the modeled species — here *Protea laurifolia* — appears to be largely determined by the sum of winter precipitation (bio_19) and annual temperature range (bio_7; the difference between the maximal temperature of the warmest month and the minimal temperature of the coldest month).

It is therefore interesting to analyse more precisely how each environmental variable influences the species' probability of presence. This can be done using the evaluation strip procedure as proposed by Elith et al. (2005) and previously illustrated in Part III. These plots are simple graphical visualizations of the response curve of each variable in a model for one species, but including multiple variables. They are made by producing a prediction from a given model on a new dataset in which only one variable is allowed to vary (usually a sequence between the minimum and maximum of these variables) while the others are kept constant (usually by setting all values to their mean). A plot of these predictions makes it possible to visualize the species' modeled response to the given variable, conditional on all other variables (being held constant) in the model (Figures 19.8, 19.9, 19.10 and 19.11).

```
# To do this we first have to load the produced models.
> ProLau_glm <- BIOMOD_LoadModels(ProLau_models, models = 'GLM')
> ProLau_gbm <- BIOMOD_LoadModels(ProLau_models, models = 'GBM')
> ProLau_rf <- BIOMOD_LoadModels(ProLau_models, models = 'RF')
> ProLau_gam <- BIOMOD_LoadModels(ProLau_models, models = 'GAM')
> glm_eval_strip <- biomod2::response.plot2(
models  = ProLau_glm, Data = get_formal_data(ProLau_models,
'expl.var'), show.variables = get_formal_data(ProLau_models,
'expl.var.names'), do.bivariate = FALSE,
fixed.var.metric = 'median', legend = FALSE,
display_title = FALSE, data_species = get_formal_data(ProLau_
models, 'resp.var'))

> gbm_eval_strip <- biomod2::response.plot2(
models  = ProLau_gbm, Data = get_formal_data(ProLau_models,
'expl.var'), show.variables= get_formal_data(ProLau_models,
'expl.var.names'), do.bivariate = FALSE,
fixed.var.metric = 'median', legend = FALSE,
```

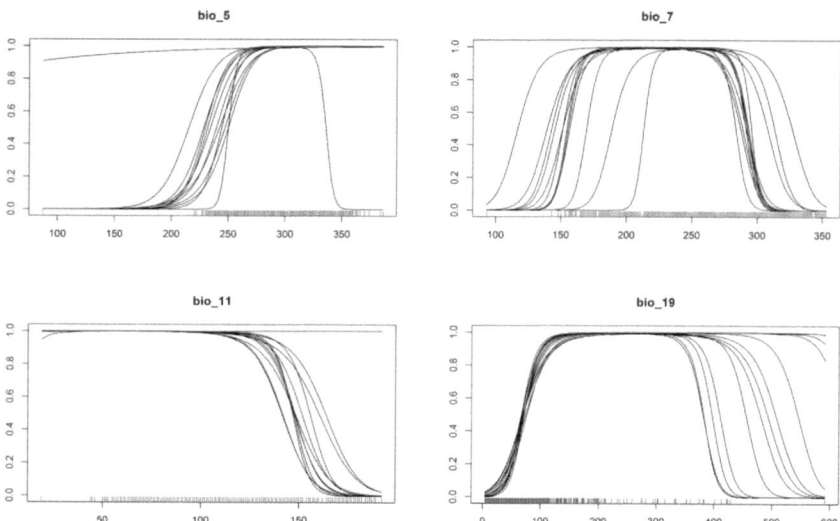

Figure 19.8 Plot of the response curves of a model (glm) to each variable.

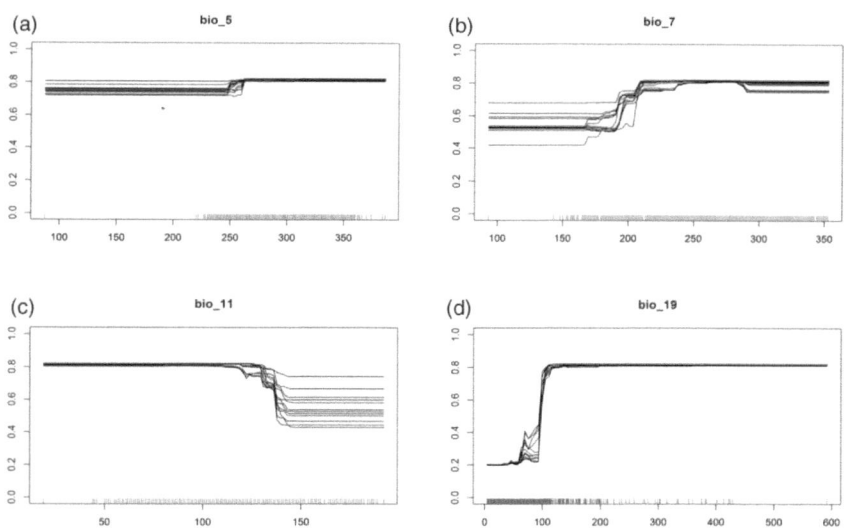

Figure 19.9 Plot of the response curves of four variables in a GBM (generalized boosting model) for *Protea laurifolia*.

```
display_title = FALSE, data_species = get_formal_data(ProLau_
models, 'resp.var'))

> rf_eval_strip <- biomod2::response.plot2(
models  = ProLau_rf, Data = get_formal_data(ProLau_models,
'expl.var'), show.variables= get_formal_data(ProLau_models,
'expl.var.names'),
do.bivariate = FALSE, fixed.var.metric = 'median',
legend = FALSE, display_title = FALSE,
data_species = get_formal_data(ProLau_models, 'resp.var'))

> gam_eval_strip <- biomod2::response.plot2(
models  = ProLau_gam, Data = get_formal_data(ProLau_models,
'expl.var'), show.variables = get_formal_data(ProLau_models,
'expl.var.names'), do.bivariate = FALSE,
fixed.var.metric = 'median', legend = FALSE,
display_title = FALSE, data_species = get_formal_data(ProLau_
models, 'resp.var'))
```

Each response line corresponds to a different model (involving a given pseudo-absence sampling and a given cross-validation sampling) (Figure 19.8, 19.9, 19.10 and 19.11).

From these response curves, we can conclude that *Protea laurifolia* appears unable to survive with less than a certain amount of winter precipitation (bio_19). Despite the variation between models, we can also deduce the optimum environmental conditions for the other variables.

Biomod2 Ensemble Modeling

It can also be seen from the TSS or AUC scores that fairly accurate models can be produced to describe the distribution of *Protea laurifolia*. However, different models can diverge and it can be of interest to summarizes and represent the information obtained across all models. This can be done by building "ensemble" models that combine the information from individual models fitted with different modeling techniques. We will use the BIOMOD_EnsembleModeling() function to build these ensemble models. Here, to reduce the number of outputs, we only consider two "ensembling" options: committee averaging and weighted mean. We also want to produce the coefficient of variation of the ensemble models that informs us of the extent to which predictions agree (or diverge) between models. This should help us to identify the areas where the model predictions are most divergent.

In the present case, we have made the decision to mix all models (i.e. all techniques, all pseudo-absences sampling, all cross-validation runs) to

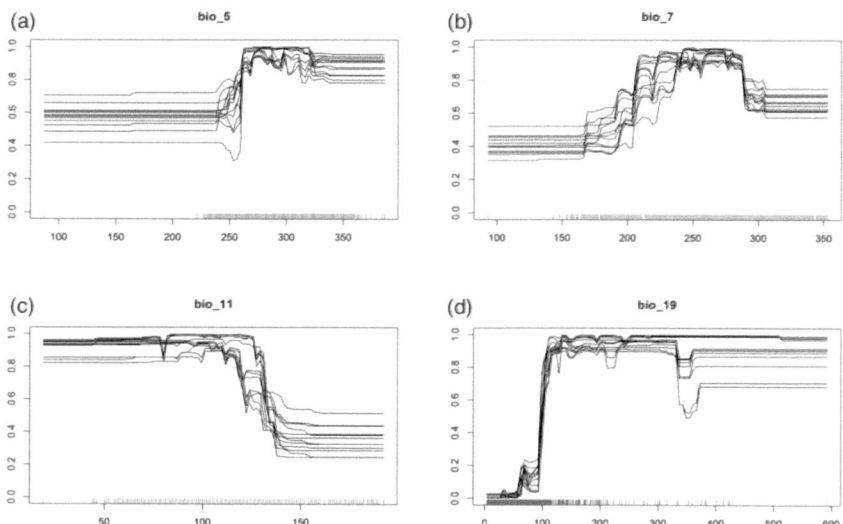

Figure 19.10 Plot of the response curves of four variables in a RF (random forest) model for *Protea laurifolia*.

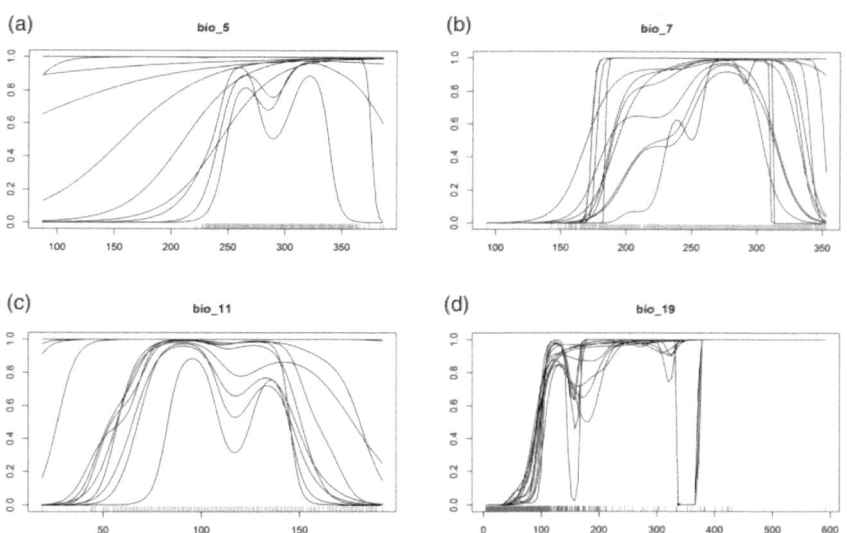

Figure 19.11 Plot of the response curves of four variables in a GAM (generalized additive model) for *Protea laurifolia*.

produce our ensembles of models. TSS is used as the evaluation reference (see Part IV) for committee building and defining weights. This means that only models with a TSS greater than or equal to 0.8 are kept to build the final ensemble. The final ensembles of models are then assessed using the same metrics as in the previous steps.

```
> ProLau_ensemble_models <- BIOMOD_EnsembleModeling(
modeling.output = ProLau_models, em.by = 'all',
eval.metric = 'TSS', eval.metric.quality.threshold = 0.8,
models.eval.meth = c('KAPPA', 'TSS', 'ROC'), prob.mean = FALSE,
prob.cv = TRUE, committee.averaging = TRUE,
prob.mean.weight = TRUE, VarImport = 0)
```

As for single algorithm models, we can check the scores for the ensembles of models.

```
> (ProLau_ensemble_models_scores <- get_evaluations(ProLau_
ensemble_models))
```

$Protea.laurifolia_EMcvByTSS_mergedAlgo_mergedRun_mergedData

	Testing.data	Cutoff	Sensitivity	Specificity
KAPPA	NA	NA	NA	NA
TSS	NA	NA	NA	NA
ROC	NA	NA	NA	NA

$Protea.laurifolia_EMcaByTSS_mergedAlgo_mergedRun_mergedData

	Testing.data	Cutoff	Sensitivity	Specificity
KAPPA	0.906	884.0	93.62	98.02
TSS	0.939	237.0	100.00	93.92
ROC	0.995	239.5	100.00	93.92

$Protea.laurifolia_EMwmeanByTSS_mergedAlgo_mergedRun_mergedData

	Testing.data	Cutoff	Sensitivity	Specificity
KAPPA	0.908	752.0	93.62	98.10
TSS	0.938	642.0	96.81	96.96
ROC	0.995	642.5	96.81	96.96

We can see that all our ensembles of models perform well (with evaluation scores of higher than 0.9 for all three evaluation metrics). Committee averaging seems to provide a slightly better evaluation than weighted mean, so we will keep the former to present the results hereafter.

Biomod2 Projections

Having built a range of SDMs and two ensemble models for *P. laurifolia* and shown how accurate these models were, we will now turn our attention to current and future spatial distributions of our focal species, using the ensemble of models built under the committee averaging rule. We will use the functions BIOMOD_Projection() and BIOMOD_EnsembleForecasting() to make single technique and ensemble projections, respectively.

Let's start with current conditions. We will use the same environmental raster maps as previously used to build the individual models. The decision was taken to also build presence–absence projections using the threshold that maximizes TSS evaluation scores. These binary maps will then be useful for estimating species range changes.

```
> ### Current projections ###
> ProLau_models_proj_current <- BIOMOD_Projection(
modeling.output = ProLau_models, new.env = bioclim_ZA_sub,
proj.name = "current", binary.meth = "TSS",
output.format = ".img", do.stack = FALSE)

> ProLau_ensemble_models_proj_current <-
BIOMOD_EnsembleForecasting( EM.output = ProLau_ensemble_models,
projection.output = ProLau_models_proj_current,
binary.meth = "TSS", output.format = ".img",
do.stack = FALSE)
```

The spatial projections for current conditions are stored in the "proj_current" directory.

The Worlclim bioclimatic scenarios downloaded from the Worlclim website were used to project future distributions. Although there are a wide range of scenarios available, we will only focus here on GCM BCC-CSM1-1 coupled with the RCP 45 bioclimatic scenario for year 2050 and 2070. The first step is to load this data and extract the areas we are interested in. Then, simply apply the same function with the same parameters as used for current conditions.

```
> ### Future projections ###
> ## load 2050 bioclim variables
> bioclim_world_2050_BC45 <-
stack(c(bio_5 = "WorldClim_data/2050/BC_45/bc45bi505.tif",
bio_7 = "WorldClim_data/2050/BC_45/bc45bi507.tif",
bio_11 = "WorldClim_data/2050/BC_45/bc45bi5011.tif",
bio_19 = "WorldClim_data/2050/BC_45/bc45bi5019.tif"))

> ## crop of our area
> bioclim_ZA_2050_BC45 <- crop(bioclim_world_2050_BC45,
mask_south_of_africa)
> bioclim_ZA_2050_BC45 <- mask(bioclim_ZA_2050_BC45,
mask_south_of_africa[ mask_south_of_africa$CNTRY_NAME == "South
Africa", ])
> bioclim_ZA_2050_BC45 <- stack(bioclim_ZA_2050_BC45)
```

```
> ## Save this rasterstack on the hard drive if needed.
> ProLau_models_proj_2050_BC45 <- BIOMOD_Projection(
modeling.output = ProLau_models,
new.env = bioclim_ZA_2050_BC45,
proj.name = "2050_BC45",
binary.meth = "TSS",
output.format = ".img",
do.stack = FALSE)
> ProLau_ensemble_models_proj_2050_BC45 <-
BIOMOD_EnsembleForecasting(
EM.output = ProLau_ensemble_models,
projection.output = ProLau_models_proj_2050_BC45,
binary.meth = "TSS",
output.format = ".img",
do.stack = FALSE)


> ## load 2070 bioclim variables
> bioclim_world_2070_BC45 <- stack(c(
bio_5 = "WorldClim_data/2070/BC_45/bc45bi705.tif",
bio_7 = "WorldClim_data/2070/BC_45/bc45bi707.tif",
bio_11 = "WorldClim_data/2070/BC_45/bc45bi7011.tif",
bio_19 = "WorldClim_data/2070/BC_45/bc45bi7019.tif"))
> ## crop of our area
> bioclim_ZA_2070_BC45 <- crop(bioclim_world_2070_BC45,
mask_south_of_africa)
> bioclim_ZA_2070_BC45 <- mask(bioclim_ZA_2070_BC45,
mask_south_of_africa[ mask_south_of_africa$CNTRY_NAME == "South
Africa", ])
> bioclim_ZA_2070_BC45 <- stack(bioclim_ZA_2070_BC45)

> ## You may save these rasters on the hard drive.
> ProLau_models_proj_2070_BC45 <- BIOMOD_Projection(
modeling.output = ProLau_models,
new.env = bioclim_ZA_2070_BC45,
proj.name = "2070_BC45", binary.meth = "TSS",
output.format = ".img", do.stack = FALSE)

> ProLau_ensemble_models_proj_2070_BC45 <-
BIOMOD_EnsembleForecasting(
EM.output = ProLau_ensemble_models,
projection.output = ProLau_models_proj_2070_BC45,
binary.meth = "TSS", output.format = ".img", do.stack = FALSE)
```

At this stage, we have built all our model predictions for current and future conditions. Although it is possible to graph maps for all different ensemble forecasting approaches, here we have only mapped the weighted mean ensemble model, for present and future conditions. Note that the units of projections are predicted habitat suitability multiplied by 1000 (thus on a 0–1000 scale).

```
> ## get the ensemble models projection stack
> stk_ProLau_ef_2070_BC45 <-
  get_predictions(ProLau_ensemble_models_proj_2070_BC45)
> ## keep committee averaging and weighted mean projections only
> ## ensemble models
> stk_ProLau_ef_2070_BC45 <- subset(stk_ProLau_ef_2070_BC45,
grep("EMca|EMwmean", names(stk_ProLau_ef_2070_BC45)))
> ## simplify the layer names for plotting conveniences
> names(stk_ProLau_ef_2070_BC45) <-
sapply(strsplit(names(stk_ProLau_ef_2070_BC45), "_"),
getElement, 2)
> ## plot the projections
> levelplot(stk_ProLau_ef_2070_BC45,
main = "Protea Laurifolia ensemble projections\nin 2070
with BC45",
col.regions = colorRampPalette(c("grey90", "yellow4", "green4"))
(100))
```

These maps in Figure 19.12 suggest that in 2070 the only remaining areas suitable for our species will be found in the south-west corner of South Africa.

19.1.8 Species Range Change

We have seen that suitable climates for the species are likely to shift geographically in the future. Next, we might want to quantify and represent this change over time. In order to do so, we can use the BIOMOD_ RangeSize() function. This function needs binary projection maps. We must therefore first load the binary files we produced during the previous projection step.

```
> ## load binary projections
> ProLau_bin_proj_current <- stack(c(
ca = "Protea.laurifolia/proj_current/individual_projections/
Protea.laurifolia_EMcaByTSS_mergedAlgo_mergedRun_mergedData_
TSSbin.img",
wm = "Protea.laurifolia/proj_current/individual_projections/
Protea.laurifolia_EMwmeanByTSS_mergedAlgo_mergedRun_mergedData_
TSSbin.img"))
> ProLau_bin_proj_2050_BC45 <- stack(c(
ca = "Protea.laurifolia/proj_2050_BC45/individual_projections/
Protea.laurifolia_EMcaByTSS_mergedAlgo_mergedRun_mergedData_
TSSbin.img",
wm = "Protea.laurifolia/proj_2050_BC45/individual_projections/
Protea.laurifolia_EMwmeanByTSS_mergedAlgo_mergedRun_mergedData_
TSSbin.img"))
> ProLau_bin_proj_2070_BC45 <- stack(c(
ca = "Protea.laurifolia/proj_2070_BC45/individual_projections/
Protea.laurifolia_EMcaByTSS_mergedAlgo_mergedRun_mergedData_
TSSbin.img",
```

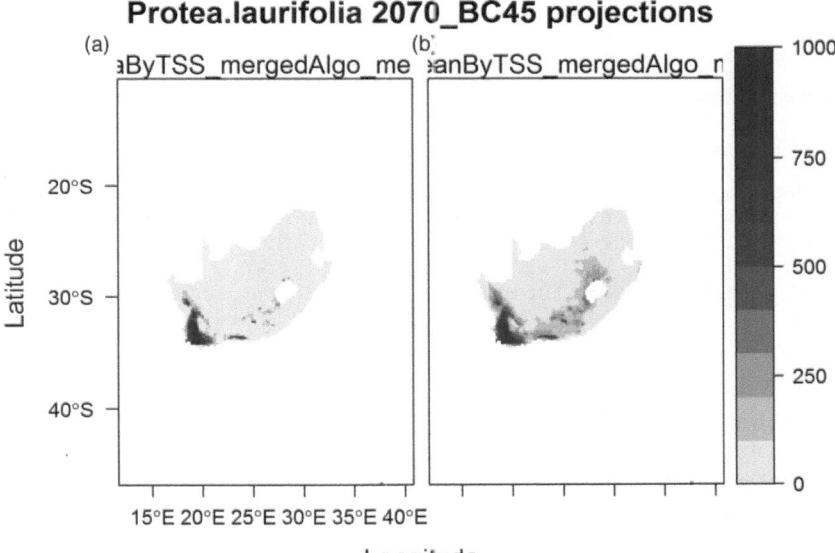

Figure 19.12 Plot showing the geographic projections using the weighted average ensemble model for *Protea laurifolia* under (a) current and (b) future conditions. (*A black and white version of this figure will appear in some formats. For the color version, please refer to the plate section.*)

```
wm = "Protea.laurifolia/proj_2070_BC45/individual_projections/
Protea.laurifolia_EMwmeanByTSS_mergedAlgo_mergedRun_mergedData_
TSSbin.img"))
```

We need then to calculate the change in species ranges (SRC: species range change) using the BIOMOD_RangeSize() function. This function produces two outputs: a table that contains a summary of species range change statistics and a spatial map that summarizes where species will gain or lose suitable conditions.

```
> ## SRC current -> 2050
> SRC_current_2050_BC45 <- BIOMOD_RangeSize(
ProLau_bin_proj_current,
ProLau_bin_proj_2050_BC45)

> SRC_current_2050_BC45$Compt.By.Models
```

	Loss	Stable0	Stable1	Gain	PercLoss	PercGain	SpeciesRangeChange
ca	97	37930	268	30	26.57	8.219	-18.36
wm	78	38107	140	0	35.78	0.000	-35.78

```
      CurrentRangeSize FutureRangeSize.NoDisp FutureRangeSize.
FullDisp
```

```
ca              365                   268                    298
wm              218                   140                    140
# SRC current -> 2070

> SRC_current_2070_BC45 <- BIOMOD_RangeSize(
ProLau_bin_proj_current,
    ProLau_bin_proj_2070_BC45)

> SRC_current_2070_BC45$Compt.By.Models
    Loss Stable0 Stable1 Gain PercLoss PercGain SpeciesRangeChange
ca   151   37950     214   10    41.37     2.74             -38.63
wm   103   38107     115    0    47.25     0.00             -47.25
    CurrentRangeSize FutureRangeSize.NoDisp FutureRangeSize.FullDisp
ca               365                    214                      224
wm               218                    115                      115
```

From the SRC output tables, we can see that our species will lose suit-able habitat in the future. According to the ensemble model, *P. laurifolia's* habitat could be reduced by 25% in 2050 and by nearly 40% in 2070.

These predicted changes in distributions can be plotted as follows (see Figure 19.13):

```
> ProLau_src_map <- stack(SRC_current_2050_BC45$Diff.By.Pixel,
SRC_current_2070_BC45$Diff.By.Pixel)
> names(ProLau_src_map) <- c("ca cur-2050", "wm cur-2050", "ca
cur-2070", "wm cur-2070")

> library(rasterVis)
> my.at <- seq(-2.5,1.5,1)
> myColorkey <- list(at = my.at, ## where the colors change
labels = list(labels = c("lost", "pres", "abs", "gain"),
## labels
at = my.at[-1]-0.5 ## where to print labels
))
> rasterVis::levelplot(ProLau_src_map,
main = "Protea laurifolia range change",
colorkey = myColorkey, layout = c(2,2))
```

As expected, the areas that are likely to become unsuitable in the future are mostly located at the borders of the species range. The final analytical step might be to try to understand how the different modeling tech-niques, pseudo-absences sampling, and cross-validation runs influence the predicted species range changes. In the previous example, we only used one climate change scenario, but in an extended exercise, the same func-tion could also be used to further compare the importance of different climate change scenarios. We will use the `ProbDensFunc()` function to try to disentangle the effects of these distinct modeling facets (techniques, pseudo-absences, cross-validations) on the species range changes.

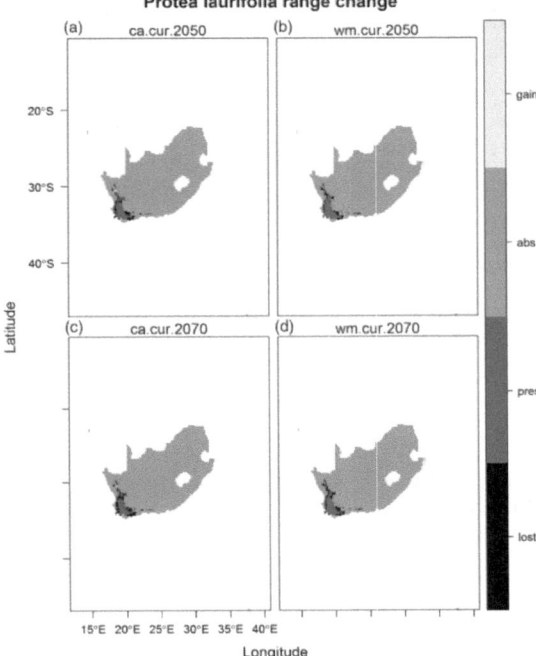

Figure 19.13 Plot of the predicted range changes for *Protea laurifolia* between present and future conditions. (*A black and white version of this figure will appear in some formats. For the color version, please refer to the plate section.*)

19.1.9 Impact of Model Scenario/Time Slice

We first have to choose a reference against which all further model predictions will be compared. The decision is to consider the projections using the committee averaging ensemble model as the reference. We will focus on projections for 2070.

```
> ref <- subset(ProLau_bin_proj_current, "ca")
```

Next, we have to define which facets we want to consider in the analysis, build a table with all possible combinations of these facets (called groups in this instance) and load the corresponding projections.

```
> ## define the facets we want to study
> mods <- c("GLM", "GBM", "RF", "GAM", "caByTSS",
"wmeanByTSS")
> data_set <- c("PA1", "PA2", "PA3", "mergedData")
> cv_run <- c("RUN1", "RUN2", "RUN3", "RUN4", "mergedRun")
```

```
> ## construct the combination of all facets
> groups <- as.matrix(expand.grid(models = mods,
data_set = data_set, cv_run = cv_run, stringsAsFactors = FALSE))

> ## load all projections we have produced
> all_bin_proj_files <- list.files(path = "Protea.laurifolia",
pattern = "_TSSbin.img$", full.names = TRUE,
recursive = TRUE)

> ## We want to focus on current versus 2070. We thus removed the
# projections by 2050.
> current_and_2070_proj_files <- grep(all_bin_proj_files,
pattern="2070", value=T)


> ## only keep projections that match our selected
facets groups
> selected_bin_proj_files <- apply(groups, 1,
function(x){
proj_file <- NA
match_tab <- sapply(x, grepl, current_and_2070_proj_files)
match_id <- which(apply(match_tab, 1, all))
if(length(match_id)) proj_file <- current_and_2070_proj_
files[match_id]
return(proj_file)
})


> ## remove non-matching groups
> to_remove <- which(is.na(selected_bin_proj_files))
> if(length(to_remove)){
groups <- groups[-to_remove, ]
selected_bin_proj_files <- selected_bin_proj_files[-to_remove]
  }


> ## build a stack of selected projections
> proj_groups <- stack(selected_bin_proj_files)
```

At this stage, we have a set of projections and corresponding groups defined from the selected facets. We can now apply the `ProbDensFunc()` function:

```
> ProbDensFunc(initial = ref, projections = proj_groups,
groups = t(groups), plothist = FALSE, cvsn = FALSE)
  $stats
     lower_limit upper_limit
  50%      -70.82      -48.93
  75%      -75.97      -36.05
  90%      -78.11      -11.16
  95%      -78.11       11.16
```

to get a density plot that shows the predicted species range change according to the selected facets in the ensemble model (Figure 19.14).

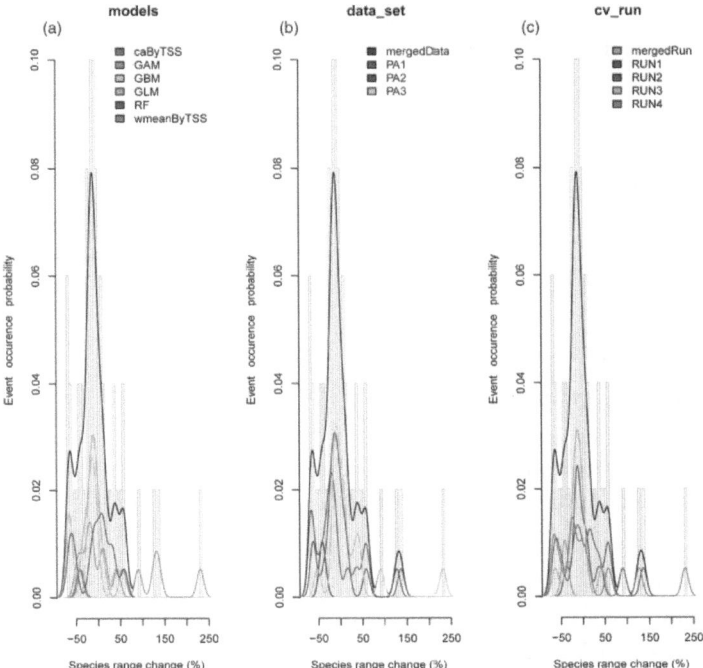

Figure 19.14 Density plot of the predicted species range changes according to the facets selected in the ensemble model. (*A black and white version of this figure will appear in some formats. For the color version, please refer to the plate section.*)

19.1.10 Conclusion

Many options and model parameterization options can be considered when building projections of species distribution, and these influence the predictions resulting from this type of analysis. Results and related conclusions thus depend strongly on the initial modeling choice. In this regard, a model ensembling approach provides more information on the variation and uncertainty in the predictions relating to these choices, and producing an ensemble of predictions that is based on the consensus across all choices.

19.2 Example 2: Creating Diversity Maps for the *Laurus* Species

19.2.1 Brief Description of the Example

Objectives

The objectives are to apply SDMs to a set of species (herein the *Larus* genus) and stack them to produce a resulting prediction of species diversity. The final aim of such analysis could be to high-light the diversity hotspots for a genus in an area (here the whole of Europe).

Data

All data for this example come from online data centers (IUCN, Worldclim). Most data is downloaded as raster grids. We will transform the initial species occurrences data to fit a XY + "species_name" formalism.

Methodological Steps

- Loading data from the web
- Formatting the data
- Building a range of models and ensemble models
- Building diversity indices and diversity maps.

19.2.2 Obtaining Species Data

For this example, we will extract species occurrences data directly from the GBIF data center. We aim to model species distribution for the whole *Larus* genus. This genus is mainly composed of coastal birds, mostly present in the northern hemisphere. The first step is to obtain the identifiers (IDs) for species matching the *Larus* genus.

```
> if(!require(rgbif)){
install.packages("rgbif")
require(rgbif)
  }


> ## get the species list belonging to the Larus genus
> spp_larus <- name_suggest(q = 'Larus ', rank = 'species',
limit = 1000)
> ## clean up the species list
> (spp_larus <- spp_larus[grepl("^Larus ",
spp_larus$canonicalName), ])
        key          canonicalName    rank
  1  6065742   Larus tridactylus SPECIES
  2  6065807        Larus roseum SPECIES
  3  6065808     Larus eburneus SPECIES
  4  2481127   Larus hyperboreus SPECIES
  5  2481139   Larus argentatus SPECIES
  6  2481145    Larus saundersi SPECIES
  7  2481172      Larus marinus SPECIES
  8  2481193    Larus audouinii SPECIES
```

```
 9 2481134     Larus delawarensis SPECIES
10 2481135   Larus brunnicephalus SPECIES
11 2481136        Larus scopulinus SPECIES
12 2481137        Larus ridibundus SPECIES
13 2481147         Larus pacificus SPECIES
14 2481148        Larus hartlaubii SPECIES
15 2481153          Larus eburnea SPECIES
16 2481155        Larus ichthyaetus SPECIES
17 2481162        Larus atlanticus SPECIES
18 2481164           Larus sabini SPECIES
19 2481170        Larus hemprichii SPECIES
20 2481171         Larus armenicus SPECIES
21 2481177          Larus pipixcan SPECIES
22 2481180          Larus modestus SPECIES
23 2481181        Larus cachinnans SPECIES
24 2481194    Larus melanocephalus SPECIES
25 2481200        Larus fuliginosus SPECIES
26 2481143           Larus minutus SPECIES
27 2481144    Larus leucophthalmus SPECIES
28 2481146       Larus californicus SPECIES
29 2481154            Larus livens SPECIES
30 2481156        Larus glaucoides SPECIES
31 2481159           Larus thayeri SPECIES
32 2481160       Larus brevirostris SPECIES
33 2481169       Larus philadelphia SPECIES
34 2481174            Larus fuscus SPECIES
35 2481190      Larus cirrocephalus SPECIES
36 2481191             Larus rosea SPECIES
37 2481192         Larus atricilla SPECIES
38 2481195          Larus serranus SPECIES
39 2481149       Larus occidentalis SPECIES
40 2481161   Larus novaehollandiae SPECIES
41 4408508       Larus parasiticus SPECIES
42 4408605          Larus heuglini SPECIES
43 4408604            Larus roseus SPECIES
44 5789287          Larus atlantis SPECIES
45 5789288        Larus mongolicus SPECIES
46 5789289             Larus vegae SPECIES
47 5789290         Larus graellsii SPECIES
48 5789291       Larus taimyrensis SPECIES
49 5789292        Larus barabensis SPECIES
50 2481130         Larus heermanni SPECIES
51 2481131      Larus maculipennis SPECIES
52 2481173       Larus dominicanus SPECIES
53 2481179          Larus belcheri SPECIES
54 2481182        Larus tridactyla SPECIES
55 2481187             Larus canus SPECIES
56 2481197      Larus crassirostris SPECIES
57 2481198           Larus bulleri SPECIES
58 2481199          Larus relictus SPECIES
59 2481132         Larus scoresbii SPECIES
```

```
60 2481163      Larus schistisagus SPECIES
61 2481178             Larus genei SPECIES
62 2481196       Larus glaucescens SPECIES
63 6486606        Larus icthyaetus SPECIES
64 4848476   Larus brachyrhynchus SPECIES
65 4848477         Larus kumlieni SPECIES
66 4848482        Larus argenteus SPECIES
67 4848483 Larus noraehollandiae SPECIES
68 4848478       Larus intermedius SPECIES
69 4848479       Larus hemiprichii SPECIES
70 4848480     Larus smithsonianus SPECIES
71 4966631              Larus vero SPECIES
72 4352366        Larus michahellis SPECIES
73 4352367          Larus furcatus SPECIES
74 4352368          Larus hartlaubi SPECIES
75 4352371             Larus rossii SPECIES
```

Having extracted the species IDs, we can then query the associated species' occurrences. We will then reformat the data.

```
> ## get species occurrences
> occ_larus <- occ_search( taxonKey = spp_larus$key,
continent='europe',
  fields = c('name', 'key', 'country', 'decimalLatitude',
'decimalLongitude'), hasCoordinate = TRUE, limit = 500,
return = 'data')
> ## remove null items
> occ_larus <- occ_larus[sapply(occ_larus,
function(x){!is.null(dim(x))})]
> ## combine all data in a single data.frame
> data <- do.call(rbind, occ_larus)
```

In order to avoid any problems with naming the file pathway (on the local computer), it is a good practice to remove spaces within species names.

```
> ## replace " " by "." in species names
> data$name <- sub(" ", ".", data$name)
```

We then decided to remove species with less than 20 occurrences to ensure the reliability of the estimates of species niches.

```
> ## only keep species with more than 20 occurrences
> table(data$name)
        Chroicocephalus.genei Chroicocephalus.philadelphia
                           34                            3
    34 3
Chroicocephalus.ridibundus         Hydrocoloeus.minutus
                          500                          500
            500 500
Larus.argentatus                   Larus.audouinii
                          500                           32
            500 32
```

```
Larus.cachinnans                    Larus.canus
                            500                                 500
            500 500
Larus.delawarensis                  Larus.fuscus
                            286                                 500
            286 500
Larus.glaucoides                  Larus.hyperboreus
                            500                                 500
            500 500
Larus.ichthyaetus                   Larus.marinus
                             26                                 500
             26 500
Larus.melanocephalus                Larus.michahellis
                            500                                 500
            500 500
Larus.schistisagus                  Larus.thayeri
                             37                                   1
             37 1
Leucophaeus.atricilla            Leucophaeus.pipixcan
                             54                                  19
                54 19
Xema.sabini
                            139
> (spp_to_model <- names(table(data$name))
[table(data$name) > 20])
    [1] "Chroicocephalus.genei"       "Chroicocephalus.ridibundus"
    [3] "Hydrocoloeus.minutus"        "Larus.argentatus"
    [5] "Larus.audouinii"             "Larus.cachinnans"
    [7] "Larus.canus"                 "Larus.delawarensis"
    [9] "Larus.fuscus"                "Larus.glaucoides"
   [11] "Larus.hyperboreus"           "Larus.ichthyaetus"
   [13] "Larus.marinus"               "Larus.melanocephalus"
   [15] "Larus.michahellis"           "Larus.schistisagus"
   [17] "Leucophaeus.atricilla"       "Xema.sabini"
```

This leaves us with 18 species for building models.

19.2.3 Obtaining the Environmental Data

For this example, we will use the Worldclim data.[6] In order to avoid time-consuming downloads and calculations, we will use 10-minute resolution raster grids. As the objective is to look at both current and future potential distributions, we will download Worldclim data for current conditions, as well as projections for 2050 and 2070 for an arbitrarily selected climate change scenario from the IPPC Fifth Assessment (here BCC-CSM1-1 Regional Climatic Model combined with 4.5 Representative Concentration Pathways; Please refer to worldclim.org/CMIP5v1 for further explanations).

---

[6] www.worldclim.org/

The first step is to download the Worlclim data archives from the datacenter.

```
> ## get Worldclim environmental variables
> dir.create("WorldClim_data", showWarnings = FALSE)
> ## curent bioclim
> download.file(
url = "biogeo.ucdavis.edu/data/climate/worldclim/1_4/grid/cur/
bio_10m_esri.zip",
destfile = "WorldClim_data/current_bioclim_10min.zip",
method = "auto")
> ## GCM -> BCC-CSM1-1, year -> 2050, RCP -> 4.5
> download.file(
url = "http://biogeo.ucdavis.edu/data/climate/cmip5/10m/bc45bi50.
zip",
destfile = "WorldClim_data/2050_BC_45_bioclim_10min.zip",
method = "auto")
>
> ## GCM -> BCC-CSM1-1, year -> 2070, RCP -> 4.5
> download.file(
url = "http://biogeo.ucdavis.edu/data/climate/cmip5/10m/bc45bi70.
zip",
destfile = "WorldClim_data/2070_BC_45_bioclim_10min.zip",
method = "auto")
```

Then, to extract the files:

```
> ## unzip climatic files
> unzip(zipfile = "WorldClim_data/current_bioclim_10min.zip",
exdir = "WorldClim_data/current",overwrite = TRUE)
> list.files("WorldClim_data/current/bio/")
    [1] "bio_1"  "bio_10" "bio_11" "bio_12" "bio_13" "bio_14"
    [7] "bio_ 15" "bio_ 16" "bio_ 17" "bio_ 18" "bio_ 19" "bio_ 2"
   [13] "bio_ 3" "bio_ 4"  "bio_ 5" "bio_ 6" "bio_ 7" "bio_ 8"
   [19] "bio_ 9" "info"
> unzip(zipfile = "WorldClim_data/2050_BC_45_bioclim_10min.
zip",
exdir = "WorldClim_data/2050/BC_45",overwrite = T)
> list.files("WorldClim_data/2050/BC_45/")
    [1] "bc45bi5010.tif" "bc45bi5011.tif" "bc45bi5012.tif"
    [4] "bc45bi5013.tif" "bc45bi5014.tif" "bc45bi5015.tif"
    [7] "bc45bi5016.tif" "bc45bi5017.tif" "bc45bi5018.tif"
   [10] "bc45bi5019.tif" "bc45bi501.tif" "bc45bi502.tif"
   [13] "bc45bi503.tif" "bc45bi504.tif" "bc45bi505.tif"
   [16] "bc45bi506.tif" "bc45bi507.tif" "bc45bi508.tif"
   [19] "bc45bi509.tif"
> unzip(zipfile = "WorldClim_data/2070_BC_45_bioclim_10min.
zip",
```

```
exdir = "WorldClim_data/2070/BC_45",overwrite = T)
> list.files("WorldClim_data/2070/BC_45/")
    [1] "bc45bi7010.tif" "bc45bi7011.tif" "bc45bi7012.tif"
    [4] "bc45bi7013.tif" "bc45bi7014.tif" "bc45bi7015.tif"
    [7] "bc45bi7016.tif" "bc45bi7017.tif" "bc45bi7018.tif"
   [10] "bc45bi7019.tif" "bc45bi7C1.tif" "bc45bi702.tif"
   [13] "bc45bi703.tif" "bc45bi704.tif" "bc45bi705.tif"
   [16] "bc45bi706.tif" "bc45bi707.tif" "bc45bi708.tif"
   [19] "bc45bi709.tif"
```

At this point, we have all the species (from GBIF) and climatic (from Worldclim) data we need to develop this biomod2 example.

19.2.4 Formatting and Selecting Climatic Data

In order to save on computation time, we will focus on Europe only. The first step in our analysis will be to reduce the geographic scope of our study area. Next, we will select a small number of variables. We will then use this subset of variables for the rest of the analyses.

```
> ## load libraries
> library(biomod2)
> library(gridExtra)
> library(rasterVis)


> ## define the extent as Europe
> europe_ext <- extent(-11, 41, 35, 72)
>
> ## load environmental variables within a 'RasterStack' object
> stk_current <- stack(list.files'
path = "WorldClim_data/current/bio/", pattern = "bio_",
full.names = T), RAT=F)
>
> ## Clip the environmental variables to the European extent
> stk_current <- crop(stk_current, europe_ext)
```

From this, we can plot the environmental rasters to ensure that they are correctly handled.

```
> # plot(stk_current)
```

A purely correlative approach based on Pearson correlations is used to select a subset of environmental variables. Here, we decided to use a graphical representation (color scale) of correlations between variables in order to easily identify a "good" subset of reasonably uncorrelated variables (Figure 19.15).

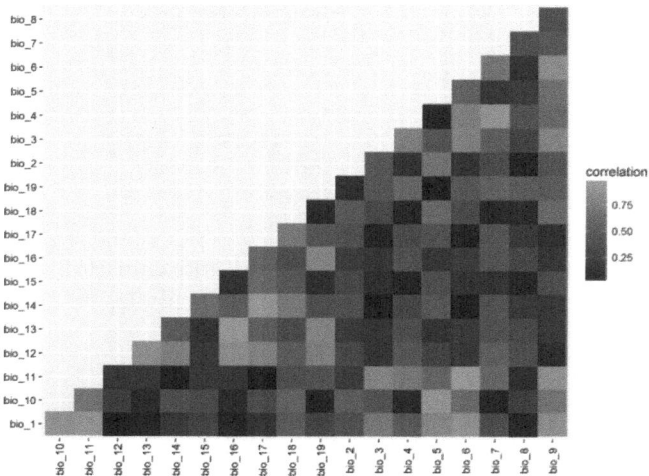

Figure 19.15 Plot of pairwise correlations between variables.

```
> ## convert our environmental stack into a data frame
> current_df <- as.data.frame(stk_current)
> current_df <- na.omit(current_df)
>
> ## calculate Pearson correlations between pairs of variables
> cor_current <- cor(current_df)
>
> ## reformat correlation table for the graphical analyses
> cor_current[upper.tri(cor_current, diag = TRUE)] <- NA
> cor_current_resh <- na.omit(melt(cor_current))
> colnames(cor_current_resh) <- c("var1", "var2", "correlation")
>
> ## only consider the absolute value of correlations
> cor_current_resh$correlation <- abs(cor_current_
resh$correlation)
>
> ## make a correlation plot
> gg_cor <- ggplot(cor_current_resh, aes(x = var1, y = var2 , fill
= correlation))
> gg_cor <- gg_cor + geom_tile() + xlab("") + ylab("") +
theme(axis.text.x  = element_text(angle = 90, vjust = 0.5))
> print(gg_cor)
```

The correlation graph produced can be useful for rapidly selecting variables that are not too closely correlated. In the graph, light colors indicate close correlations (Figure 19.15). Let's select the following three variables:

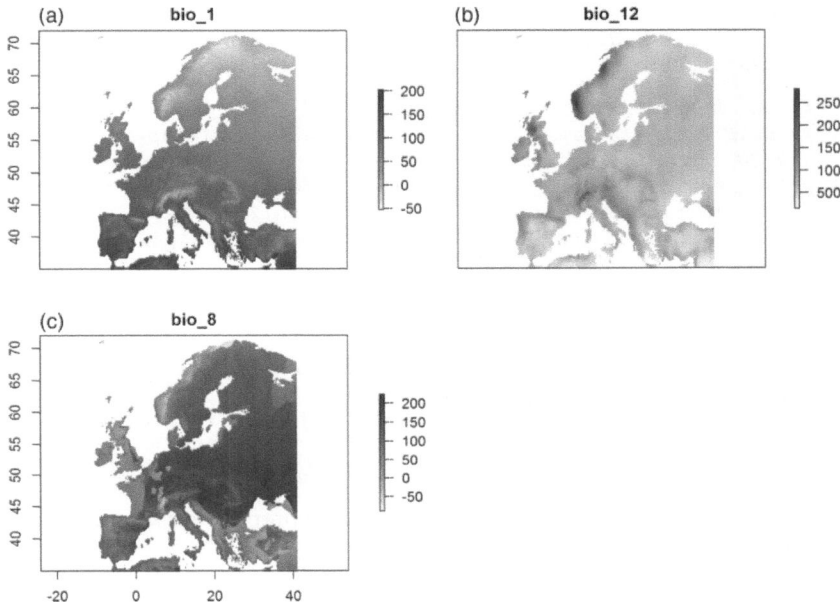

Figure 19.16 Maps of the three selected variables in Europe. (*A black and white version of this figure will appear in some formats. For the color version, please refer to the plate section.*)

- bio1 = annual mean temperature
- bio12 = annual precipitation
- bio8 = mean temperature of wettest quarter.

```
> selected_vars <- c("bio_1", "bio_12", "bio_8")
>
> ## check correlation between selected variables
> (cor_sel <- cor(current_df[, selected_vars]))
            bio_1    bio_12    bio_3
  bio_1    1.00000 -0.01478   0.0560
  bio_12  -0.01478  1.00000  -0.40665
  bio_8    0.05600 -0.40665   1.0000
```

We have seen that the maximum correlation between our three variables is 0.41 (between bio_8 and bio_12, a value well below the 0.7 figure usually considered acceptable (see Dormann et al., 2013). Let's extract these variables from the pool of bioclimatic variables available for current and future conditions (Figure 19.16)

```
> ## only keep the selected variables
> stk_current <- stack(subset(stk_current, selected_vars))
> plot(stk_current)

> ## do the same for future scenarios
> ## NOTE: respect layer names and order across stacks
> stk_2050_BC_45 <- stack(c(
bio_1 = "WorldClim_data/2050/BC_45/bc45bi501.tif",
bio_12 = "WorldClim_data/2050/BC_45/bc45bi5012.tif",
bio_8 = "WorldClim_data/2050/BC_45/bc45bi508.tif"), RAT
= FALSE)
> stk_2050_BC_45 <- stack(crop(stk_2050_BC_45, europe_ext))
>
> stk_2070_BC_45 <- stack(c(
bio_1 = "WorldClim_data/2070/BC_45/bc45bi701.tif",
bio_12 = "WorldClim_data/2070/BC_45/bc45bi7012.tif",
bio_8 = "WorldClim_data/2070/BC_45/bc45bi708.tif"), RAT
= FALSE)
> stk_2070_BC_45 <- stack(crop(stk_2070_BC_45, europe_ext))
```

The variable selection, data collection, and data preparation steps are now completed, which means we can fit the models.

19.2.5 Species Niche Modeling

Since the objective here is not to study a single species but a pool of species, we need to define a generic niche modeling procedure that will then be applied to each of our species of interest in the same way.

Building a Modeling Wrapper

The generic modeling procedure will be stored in the `biomod2_wrapper()` function that takes the species' name as the argument. The modeling procedure can be summarized into four main steps:

1. Formatting the data:
   - putting data into the required `biomod2` format
   - drawing three sets of 1000 random pseudo-absences for each species
2. Building the models:
   - selecting three model algorithms: GLM, FDA, and RF
   - selecting all default `biomod2` options for all three algorithms
   - running a repeated (3×) split-sample cross-validation for each model; models are calibrated on 70% of the data, and evaluated on the remaining 30%; the procedure is repeated three times

- evaluating models with TSS and the ROC, and producing graphical representations of evaluation scores to check the quality of the models
3. Building ensemble of models:
   - keeping only those models with a TSS score greater than 0.7
   - combining all models together to produce the ensemble model
   - computing four ways of ensembling the models fitted for each species with the three modeling techniques: the simple mean, the coefficient of variation, the committee averaging and the weighted mean of single model predictions; resulting in four ensemble models
   - evaluating the four ensemble models using TSS and ROC
4. Making projections:
   - producing projection maps for all single models and ensemble models for both current and future conditions
   - producing continuous (habitat suitability) and binary (presence–absence) projections based on the threshold optimizing each model TSS
   - saving all projections as .img files.

```
# build a biomod2 modeling wrapper
>
> biomod2_wrapper <- function(sp){
cat("\n> species : ", sp)
## get occurrence points
sp_dat <- data[ data$name == sp, ]
## formatting the data
sp_format <- BIOMOD_FormatingData(
resp.var = rep(1, nrow(sp_dat)),
expl.var = stk_current,
resp.xy = sp_dat[, c("decimalLongitude", "decimalLatitude")],
resp.name = sp, PA.strategy = "random",
PA.nb.rep = 3, PA.nb.absences = 1000)
## print formatting summary
sp_format
## save image of input data summary
if(!exists(sp)) dir.create(sp)
pdf(paste(sp, "/", sp ,"_data_formated.pdf", sep=""))
try(plot(sp_format))
dev.off()

## define the model options
sp_opt <- BIOMOD_ModelingOptions()
## model species
```

```
sp_model <- BIOMOD_Modeling(
sp_format,
models = c('GLM', 'FDA', 'RF'),
models.options = sp_opt,
NbRunEval = 3,
DataSplit = 70,
Yweights = NULL,
VarImport = 3,
models.eval.meth = c('TSS', 'ROC'),
SaveObj = TRUE,
rescal.all.models = FALSE,
do.full.models = FALSE,
modeling.id = "ex3")

## save some graphical outputs
#### model scores
pdf(paste(sp, "/", sp, "_models_scores.pdf", sep = ""))
try(gg1 <- models_scores_graph(sp_model, metrics = c("TSS",
"ROC"), by = 'models', plot = FALSE))
try(gg2 <- models_scores_graph(sp_model, metrics = c("TSS",
"ROC"), by = 'data_set', plot = FALSE))
try(gg3 <- models_scores_graph(sp_model, metrics = c("TSS",
"ROC"), by = 'cv_run', plot = FALSE))
try(grid.arrange(gg1, gg2, gg3))
dev.off()

## build ensemble models
sp_ens_model <- BIOMOD_EnsembleModeling(
modeling.output = sp_model, chosen.models = 'all',
em.by = 'all', eval.metric = c('TSS'),
eval.metric.quality.threshold = c(0.7),
models.eval.meth = c('TSS','ROC'), prob.mean = TRUE, prob.
cv = TRUE, prob.ci = FALSE, prob.ci.alpha = 0.05, prob.median
= FALSE, committee.averaging = TRUE,
prob.mean.weight = TRUE,
prob.mean.weight.decay = 'proportional')
## make the projections
proj_scen <- c("current", "2050_BC_45", "2070_BC_45")
for(scen in proj_scen){
cat("\n> projections of ", scen)
## Single model projections
sp_proj <- BIOMOD_Projection(
modeling.output = sp_model,
new.env = get(paste("stk_", scen, sep = "")),
proj.name = scen, selected.models = 'all', binary.meth = "TSS",
filtered.meth = NULL, compress = TRUE,
build.clamping.mask = TRUE,
do.stack = FALSE, output.format = ".img")
## Ensemble model projections
sp_ens_proj <- BIOMOD_EnsembleForecasting(
EM.output = sp_ens_model,
projection.output = sp_proj, binary.meth = "TSS",
```

```
compress = TRUE, do.stack = FALSE.
output.format = ".img")
}
return(paste(sp, " modelling completed !", sep = ""))

}
```

Multi-Species Niche Modeling

Now, our modeling wrapper is properly defined and only has to be applied to our list of species. Since we want to apply this function to each species, it needs to be embedded in a loop. Below, we describe two ways of building a loop, in sequential or parallel mode. Provided your machine has several processors (i.e. CPUs or cores), we strongly recommend using the parallel version (via the `parallel` package).

```
> if(require(snowfall)){ ## parallel computation
## start the cluster
sfInit(parallel = TRUE, cpus = 2)
## here we only require 2 cpus
sfExportAll()
sfLibrary(biomod2)
## launch our wrapper in parallel
sf_out <- sfLapply(spp_to_model, biomod2_wrapper)
## stop the cluster
sfStop()
} else { ## sequencial computation
for (sp in spp_to_model){
biomod2_wrapper(sp)
}
## or with a lapply function in sequential model
## all_species_bm <- lapply(spp_to_model, biomod2_wrapper)
}
```

For each species, a directory is created on the hard drive. This directory contains all `biomod2` modeling and projection outputs for that species (see the other example in Part III, or the biomod2 examples and vignettes (explanation regarding the specific functionalities of a package based on examples).

19.2.6 Producing Alpha-Diversity Maps

The final step in this example will be to extract binary (presence–absence) projections for each species, based on the weighted mean ensemble models projections for current and future (2050, 2080) environmental conditions, and then combine them to produce one current and two future alpha diversity (i.e. species richness) maps.

For each diversity map (present and future), the procedure is always the same: first load all presence–absence predictions from the selected models for a given date, then sum all projections to obtain the number of species predicted in each pixel on the map, and finally plot the map.

```
> ## we focus on a single ensemble model (e.g. weighted mean)
> ## we have to load the binary projections for each species
>
> ## current conditions
> ### load binary projections
> f_em_wmean_bin_current <- paste(spp_to_model,
"/proj_current/individual_projections/", spp_to_model,
"_EMwmeanByTSS_mergedAlgo_mergedRun_mergedData_TSSbin.img",
sep = "")
> ### sum all projections
> if(length(f_em_wmean_bin_current) >= 2){
## initialisation
taxo_alpha_div_current <- raster(f_em_wmean_bin_current[1])
for(f in f_em_wmean_bin_current){
taxo_alpha_div_current <- taxo_alpha_div_current + raster(f)
    }
  }
> ## 2050 conditions
> ### load binaries projections
> f_em_wmean_bin_2050 <- paste(spp_to_model,
"/proj_2050_BC_45/individual_projections/", spp_to_model, "_
EMwmeanByTSS_mergedAlgo_mergedRun_mergedData_TSSbin.img",
sep = "")

> ### sum all projections
> if(length(f_em_wmean_bin_2050) >= 2){
## initialisation
taxo_alpha_div_2050 <- raster(f_em_wmean_bin_2050[1])
for(f in f_em_wmean_bin_2050){
taxo_alpha_div_2050 <- taxo_alpha_div_2050 + raster(f)
}
}

> ## 2070 conditions
> ### load binaries projections
> f_em_wmean_bin_2070 <- paste(spp_to_model, "/proj_2070_BC_
45//individual_projections/", spp_to_model, "_EMwmeanByTSS_
mergedAlgo_mergedRun_mergedData_TSSbin.img", sep = "")
> ### sum all projections
> if(length(f_em_wmean_bin_2070) >= 2){
## Initialisation
taxo_alpha_div_2070 <- raster(f_em_wmean_bin_2070[1])
for(f in f_em_wmean_bin_2070){
taxo_alpha_div_2070 <- taxo_alpha_div_2070 + raster(f)
}
}
```

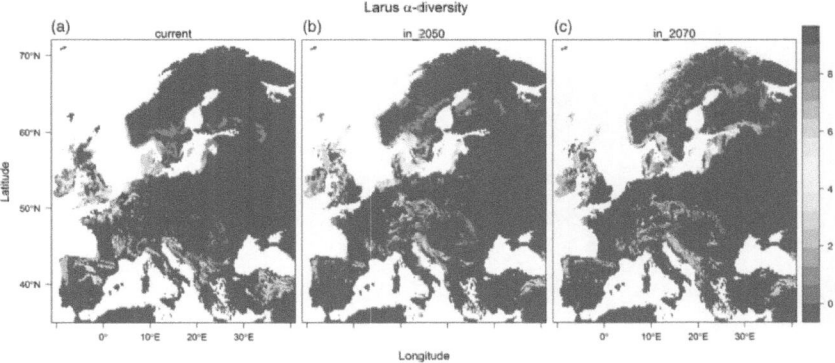

Figure 19.17 Species richness (alpha diversity) maps for the three time steps. (*A black and white version of this figure will appear in some formats. For the color version, please refer to the plate section.*)

We have now produced three species richness ("alpha diversity") maps for our three time steps (current, 2050, 2070). The last step is to display the maps (Figure 19.17).

```
> ## plot the results
> levelplot(stack(c(current = taxo_alpha_div_current,
in_2050 = taxo_alpha_div_2050, in_2070 = taxo_alpha_div_2070)),
main = expression(paste("Larus ", alpha, "-diversity")),
par.settings = BuRdTheme)
```

19.2.7 Conclusion

In this example, we have seen how to model a list of species in parallel mode using online databases, and how to stack them to produce some simple species richness (alpha diversity) maps. More advanced tuning of models, uncertainty analyses, and subsequent analysis may be included to address more complex questions. We refer interested users to the Biomod2 documentation for further details.

PART VII · Conclusions and Future Perspectives

In this last part, we briefly discuss the advances already made in HSMs and present the issues currently in development or under debate, which therefore constitute valuable topics for future HSM research.

20 · *Conclusions and Future Perspectives in Habitat Suitability Modeling*

The aim of this book was to present HSMs, and the associated theory and methods. As we have seen, this field has developed tremendously, but much still remains to be done to better formalize existing approaches in solid mathematical frameworks. Several aspects of the field are still under development or were making significant progress at the time of publication of this book. Here, we have identified some important topics which are currently developing rapidly and could not therefore be fully discussed in this book. We have mainly identified topics relating to: (i) further progress in HSMs through metagenomics and remote sensing; (ii) point-process models for presence-only HSM; (iii) hierarchical Bayesian approaches to integrate models at different scales; (iv) ensembles of small models for rarer species; (v) improving methods to build ensembles of models, e.g. using Bayesian approaches; (vi) modeling communities through multi-species modeling and joint-species distribution modeling; and (vii) use of artificial data to assess various methodological aspects of HSMs, such as which factors affect model building or model performance.

20.1 Further Progress in HSMs through Metagenomics and Remote Sensing

Our understanding of ecological niches and species' environmental requirements is evolving fast, and new findings are likely to emerge through major developments in two fields: (i) the very rapid progress in environmental metagenomics, with increasingly large sampling surveys in various ecosystems (marine, freshwater, and terrestrial), and

(ii) Remote Sensing (RS) of the environment, though the use of increasingly powerful sensors and remote devices (e.g. drones).

In recent years, metagenomics – and the related fields of metatranscriptomics and metabolomics – has yielded an impressive deluge of DNA (resp. RNA, proteins, and metabolites) data associated with natural environments from soils, marine water, sediment, freshwater, wastewater, or polluted mud samples (Howe and Chain, 2015; Aguiar-Pulido et al., 2016; Nesme et al., 2016). These data are now increasingly widely used to both identify and describe the biological communities found in these biota (mainly microbes, but also numerous eukaryotes, including protista, fungi, plants, invertebrates, etc.), as well as how these relate to environmental variations, and ultimately if specific functions are found in specific environments (Oulas et al., 2015; Bendall et al., 2016; Fondi et al., 2016). These new data are likely to further our understanding not only of how ecosystems function but also of biotic interactions in general and how these shape the geographic and environmental distributions of organisms. This will hopefully lead to the improved inclusion of these factors in HSMs (as proposed by Kissling et al., 2012; Wisz et al., 2013; Pellissier et al., 2013c).

In parallel, significant progress has also been made in the field of RS of the environment, driven by the ongoing development of sensor technologies (e.g. Ferreira et al., 2016; Rödder et al., 2016) and remote devices, such as drones and other small flying devices (Tang and Shao, 2015; Christie et al., 2016; Zhang et al., 2016). These have excellent potential for predicting species distributions (Pottier et al., 2014) and can also potentially be used in analyses and models of environmental genomics data (Larsen et al., 2015).

20.2 Point-Process Models for Presence-Only Data

As briefly mentioned when presenting Maxent, the last few years have seen the development of new approaches to modeling presence-only data. In this book, we have mainly considered approaches that deal with presence and pseudo-absence data, placing these cases of presence-only data in a binomial context. However, presence-only data cannot be easily described as a binomial process, but should be rather modeled as Poisson point processes (Renner et al., 2015). We have seen in Part III that generating pseudo-absence (or background) data and weighting them in models used for presence–absence data could nevertheless do the trick, and we have seen in Part IV that there are also evaluation metrics available to assess the predictive power in such cases. However, this ad-hoc way of fitting presence-only models is far from optimal, because selecting the right amount of pseudo-absence data in the right way is a challenging task (Barbet-Massin et al., 2012). Moreover, since there is no true absence data, the fitted value is actually a relative likelihood of presence, with an unknown scaling linking it to the true probability of presence (Renner et al., 2015). Instead, if we see presence-only data as a set of locations at which a species has been observed in a region, we can then focus on the intensity of presence records per unit of area or environment with respect to the available environment. The response variable is not the presence or absence of a given species, but rather a measure of the number of presence records per unit area or environment. That way, the intensity is a function of both the spatial distribution of the presence-only data in environmental space, and the spatial measurement units (i.e. resolution).

Point-process models (PPMs) are the tools of choice for modeling such a response variable. They were introduced a long time ago in spatial statistics (Cressie, 1993) and have recently been introduced in the field of ecology and HSMs (Renner and Warton, 2013; Renner et al., 2015). We refer interested readers to Renner et al. (2015) which offers the most comprehensive overview of PPMs for presence-only data. Renner et al. (2015) also describes how to fit Poisson PPMs that consider (or not) the spatial dependence between species records, and how to select the so-called quadrature points (related to background data). This makes it possible to run the models at the appropriate resolution and for a pre-determined number of locations. Interestingly, while Renner et al. (2015) presented PPMs in a log-likelihood GLM framework, most of the approaches presented in this book could be made available for modeling point processes. This perspective offers new avenues for further developments in this area.

20.3 Hierarchical Bayesian Approaches to Integrate Models at Different Scales

In Chapter 17, we proposed using hierarchical Bayesian approaches to integrating models at different scales to avoid climatic niche truncation effects in projections. However, these approaches are still in their infancy. Here we will look at these approaches again as future research perspectives. We suggest that the hierarchical Bayesian model of Keil et al. (2013), developed for downscaling purposes, could also be applied to integrating models at different scales. In their approach, high-resolution species occurrences are treated as latent variables that are fitted in a GLM-type model against a set of high-resolution environmental variables, which is, at the same time, constrained by observed coarse resolution presence and absence data using logistic regression. The authors demonstrated that their approach yields less-biased estimates of probabilities of occurrence than the traditional downscaling approach (e.g. Araújo et al., 2005b). This approach has recently been further improved to build meta-models that integrate predictions from models with different resolutions, different sources or even different types (Talluto et al., 2016). In this type of framework, the model of interest (at the highest resolution and perhaps for a smaller extent) can be fitted according to the environmental variables, while a prior probability could be defined from a model calibrated at global scale for instance (e.g. Gallien et al., 2012 in a non-integrated framework; Dorazio, 2014).

20.4 Ensemble of Small Models for Rarer Species

A major problem of fitting HSMs is when the number of occurrences or presences available to fit the models is too small, which directly limits the number of predictors that can be included in the models, and the total number of degrees of freedom used to fit the response curves for the predictors included. This has two immediate implications: (i) for modeling rare species, although these are also the most in need of models to support their conservation; this situation has been called the "rare species modeling paradox" by Lomba et al. (2010); and (ii) for modeling communities and ecosystems, as the least frequent species in the landscape can be difficult to model, and thus to include in attempts to predict communities from individual species (Guisan and Rahbek, 2011), although these often form an important part of the structure and functions of communities and ecosystems (Lyons and Schwartz, 2001; Lyons et al., 2005; Mouillot et al., 2013).

A solution to this rare species modeling problem has been proposed in the form of an ensemble of bivariate models (Lomba et al., 2010), later generalized to an ensemble of small models (ESM) by Breiner et al. (2015). The principle here is to fit small models, typically with 1–3 predictors at a time, so that the number of predictors in each small model is not overly large compared to the number of occurrences or presences available for the species (typically a maximum of ten occurrences (or presences in the case of presence–absence data) per degree of freedom, Harrell et al., 1996), thus avoiding overfitting any of the small models.

All or some selected combinations of the initial set of environmental predictors are then chosen to fit the series of small models, and all small models are finally combined through an ensemble modeling procedure (Araújo and New, 2007; Thuiller et al., 2009) that weights each model according to its cross-validated predictive performance. Spatial predictions can then be derived from the weighted ESM (Breiner et al., 2015). This approach has been shown, using >100 rare and uncommon plant species in Switzerland, to surpass standard HSMs fitted with all predictors at once, even when the latter were based on ensemble modeling procedures (Breiner et al., 2015). ESMs tend to outperform standard HSMs especially with low numbers of occurrences. At equivalent performance to standard HSMs, ESMs are costly as they require more computing power. Another interesting finding of Breiner et al. (2015) is that building ESMs with a single modeling technique – say GLMs, as these are easy to interpret from a biological point of view (Guisan et al., 2002) – yields the same results as building ESMs with multiple techniques (Breiner et al., 2015), which makes it possible to reduce computing power and improve model interpretability. The ESM method, however, would gain from further methodological development, and in particular would benefit from being rooted in a more formal mathematical framework (see the discussion on the online blog of the journal *Methods in Ecology and Evolution*[1]).

20.5 Improving the Modeling Techniques to Fit Simple and Ensemble HSMs

In the last 20 years, numerous modeling techniques have been developed that have improved predictive capacity (Elith et al., 2006), but there is still room for new improvements. One promising option is the use of Bayesian machine-learning approaches, e.g. by fitting Gaussian Processes HSMs through deterministic numerical approximations, which makes it possible to fit "smooth, but potentially complex response functions that can account for high-dimensional interactions between predictors" (Golding and Purse, 2016). Another alternative would be to combine the best options from existing approaches. For instance, Maxent, which has been recently shown to be a special case of GLM) for Poisson point-process data (Renner and Warton, 2013), includes complex functions (e.g. hinge features) not currently available in usual GLMs nor in most other

---

[1] https://methodsblog.wordpress.com/2016/05/24/esms-for-rare-species

approaches (Phillips et al., 2006), but these could be used more generally. Many new machine-learning techniques have also been recently proposed (Maher et al., 2014), which could include interesting options to use in other approaches. In this regard, what is still missing is a map of the mathematical correspondences between all modeling techniques, which would identify techniques that might be thought to be very different but would be shown – through such formalization – to be mathematically close or equivalent (with possible differences in their implementation; e.g. Maxent versus PPP GLM; Renner and Warton, 2013).

Similarly, the use of ensemble approaches to modeling and forecasting has seen a large increase in popularity in recent years (Araújo and New, 2007; Tebaldi and Knutti, 2007; Bombi et al., 2009; Coetzee et al., 2009; Diniz et al., 2009; Thuiller et al., 2009; Buisson et al., 2010; Meller et al., 2014). However, the methods used to ensemble the sub-models have not progressed accordingly, and the most common methods used are simple weighted averages or some slightly more complex approaches (e.g. based on multivariate ordinations) (see options in Thuiller et al., 2009). However, more advanced approaches could still be developed and tested. More crucially, it will be important to establish well-designed strategies for deciding which techniques should be combined in the final ensemble. Mixing techniques based on a single model (e.g. GLM, GAM, MAXENT; see Part III) with others that themselves are composed of an ensemble of sub-models (e.g. RF, BRT) may not be optimal in this regard, but these issues have not yet been addressed.

20.6 Multi-Species Modeling and Joint-Species Distribution Modeling

One of the strongest criticisms of HSMs is that species interactions are usually not explicitly modeled (Thuiller et al., 2013). This criticism has its theoretical origins in niche theory (see Part I). An HSM represents a realized niche, because it is based on the locations where a species is actually found. Ideally, it would also be interesting to know the species' fundamental niche, which would provide information on the full potential distribution of the species. One of the main factors thought to cause the realized niche to deviate from the fundamental niche is interference by other species (via numerous different processes including competition, facilitation or predation). From a more practical standpoint, ignoring species interactions in HSMs could lead to biased predictions (reviews in Kissling et al., 2012; Wisz et al., 2013). As community datasets become increasingly freely available, HSMs studies are starting to take advantage of these data. The most common approach is to add the occurrences of other species as predictors alongside environmental predictors. Results using this approach have shown that biotic predictors have a strong influence on predictions (Araújo and Luoto, 2007; Heikkinen et al., 2007; Meier et al., 2010; Pellissier et al., 2010) but there are clear limitations. The use of one species as a response and another as a predictor assumes a one-way relationship, one species is influencing the other, but not the other way round. However, in many types of ecological interactions, both species might be expected to have an influence on the other (even if one eventually outcompetes the other). Depending on the type of community to be modeled, the number of biotic interactions to include may also become overwhelming.

One alternative is to model all species simultaneously (Guisan and Zimmermann, 2000; Gelfand et al., 2006; Ovaskainen and Soininen, 2011; Pollock et al., 2012) rather than modeling each species individually. Simultaneous species modeling approaches exist and are known as multi-species, hierarchical, or mixed effect models. One advantage is that a hierarchical modeling framework offers flexibility in assigning submodels (Vesk, 2013). For example, functional traits (Pollock et al., 2012; Jamil et al., 2013; Vesk, 2013; Brown et al., 2014) or phylogenetic diversity (Ives and Helmus, 2011) can then be added to help explain species distributions. These models are thought to have advantages for community datasets, because they can produce more accurate predictions for rare species (Hui et al., 2015). Recently, multi-species models have been extended to explicitly consider correlations between species in an HSM (i.e. SDM; see Chapter 1 and Glossary) framework, which is known as a joint-species distribution model (JSDM; Clark et al., 2014; Pollock et al., 2014; Warton et al., 2015). JSDM searches for correlations between the occurrences of other species and the residuals of an HSM (i.e. the variance not explained by the abiotic predictors). The JSDM is thought to have the potential to go beyond conventional HSMs (SDMs) in two key ways: by offering the potential to detect possible species interactions (Ovaskainen and Meerson, 2010; Pollock et al., 2014) and the potential to improve prediction (Clark et al., 2014; Harris, 2015). Recent developments have included latent variable models (Warton et al., 2015) and a neural network version of JSDM, which to date had been based on GLM (Harris, 2015). These approaches are currently under development and are likely to become the tool of choice in the future as soon as their computational limits (no more than 70–80 species at the same time) are resolved.

20.7 Use of Artificial Data

Last but not least, we face a vital challenge related to our capacity to ensure that the analytical and modeling methods that modelers develop and use do indeed have the potential to predict species distributions and species assemblages in the way that we would expect. In this regard, new virtual simulation approaches using artificial data offer the novel perspective of better rooting modeling approaches in proper methodological and theoretical frameworks, where the limits of the analyses can be clearly identified and quantified (Hirzel et al., 2001; Hirzel and Guisan, 2002; Austin et al., 2006; Meynard and Quinn, 2007; Zurell et al., 2010; Bombi and D'Amen, 2012; Meynard and Kaplan, 2013; Thibaud et al., 2014). This "virtual ecologist" view (see Zurell et al., 2010) allows researchers to ask questions such as: Does my modeling approach have the power to answer my initial question? Can inferences about hypothesized underlying mechanisms be made with my data? How much variation and related uncertainty can I get by changing key parameters in my model? How does error propagate in my modeling process? (Austin et al., 2006; Zurell et al., 2010). In the case of predicting communities, how much error in my individual species models will yield excessively noisy community predictions? Virtual simulations can also be used to assess community assembly processes (Sokol et al., 2011; Münkemüller et al., 2012). However, much progress is still necessary here to identify the best ways to generate artificial data.

Glossary and Definitions of Terms and Concepts

Compiled from definitions in Guisan and Zimmermann (2000); Scott et al. (2002); Guisan and Thuiller (2005); Elith et al. (2006); Pearman et al. (2008a); Elith and Leathwick (2009); Franklin (2010a); Peterson et al. (2011); Guisan et al. (2014); Maher et al. (2014), and various online dictionaries (e.g. Wikipedia, National Institute of Standards and Technology, data-mining dictionary, etc.).

It is arranged by topic:

Methods, Approaches, Models, Techniques, Algorithms

Algorithm: "A computable set of steps to achieve a desired result" (from http://xlinux.nist.gov/dads//HTML/algorithm.html). Different algorithms can be used to build a model with a given statistical technique (e.g. GLM; least-square versus maximum likelihood).

Approach: A general term to refer to a group of procedures – e.g. modeling approaches, statistical approaches – sharing a common methodological root. Different modeling approaches can be statistical versus dynamical models, or niche/habitat (this book) versus dispersal/historical ones, explanatory versus predictive, etc.

Method: Systematic procedure (usually analytical, but possibly also theoretical or conceptual) to conduct a series of operations or processes intended to answer a question. Very general term used to refer to any methodological procedure.

Model: A general term to qualify any representation of a real process or pattern fitted, for a given feature (e.g. vegetation type, species, gene, etc.) with a specific algorithm (within one statistical approach and statistical technique).

Technique: "A procedure to complete a task" (from Wikipedia[1]), here used to refer to an analytical procedure, mainly statistical, but sometime dynamical. Examples of different techniques used to fit HSMs are GLMs, GAMs, RFs, etc. (see related parts in the Glossary). Different statistical approaches can be used to model habitat suitability (e.g. regression approaches, tree-based approaches; see Part III), which will include different possible techniques (e.g. GLMs, GAM, MARS, for regressions). Different algorithms can then be developed to implement a given technique.

ENM, SDM, HSM, etc.: Different Names and Acronyms for the Same Models!

Numerous different names and acronyms are used to refer to statistical, correlative models of species niches and distributions. We consider them all as equivalent, though stressing different aspects of a same modeling approach, because all of them are strictly based on the same empirical data and quantitative methods: species observations are related to a set of environmental predictors to quantify the realized environmental niche and project it in geographic space to predict species distributions (see next section and Part I; see Appendix S1 in Guisan et al., 2013). The model names refer to various components of the niche: the niche itself, the envelope that defines it, or the fact that the niche is defined by the envelope of suitable habitats (i.e. biotopes). All these models capture the realized environmental niche of species (Araújo and Guisan, 2006), but depending on the context (species, data, methods), it may represent a varying subset of the fundamental niche (i.e. only a tiny part of it or nearly the full fundamental niche). For some dominant species in productive conditions, the realized niche is likely to be close to the fundamental niche, whereas for species reaching extreme environments or for pioneer or opportunistic species, it may only represent a small and/or stochastic subset (Pearman et al., 2008a; Pearman et al., 2008b; Nogues-Bravo, 2009; Maiorano et al., 2013). Below, we present the main acronyms found in the literature.

CEM: Climate-envelope model; a subtype of EEM (see below) restricted to fit the climate niche only

CMM: Climate-matching model

EEM: Environmental envelope model

[1] http://en.wikipedia.org/wiki/technique

ENM: Ecological niche model
HDM: Habitat distribution model
HSM: Habitat suitability model
NBM: Niche-based model
RSF: Resource selection function (model)
SDM: Species distribution models; correlative SDMs are generally used, but a few examples of mechanistic approaches also exist.

Environment, Habitat, Niche, Niche–Biotope Duality, and Distribution

Species distributions are partly ruled by the physiologically based environmental requirements of species, what is called the fundamental environmental niche. Within it, species can only occupy those parts that are accessible and from where they are not biotically excluded (e.g. by a competitor, or by the lack of a facilitator), what is called the realized environmental niche sensu lato (Soberón, 2007). The fundamental niche is made of the envelope (i.e. ensemble) of abiotic conditions (i.e. habitats or biotopes) where the species can maintain populations, as defined from the species' fundamental needs The realized niche is made of the envelope of habitats/biotopes where the species occurs at a given time, as defined from its ecological requirements (including dispersal and competitive abilities). One unique habitat (i.e. a unique environmental combination) in environmental space can correspond to many locations in geographic space, which are not necessarily distributed as a gradient. The later is called the niche-biotope duality and was first raised by Hutchinson (see Colwell and Range, 2009; Guisan et al., 2014).

Accessible range: The geographic locations within a given area that are accessible to a species given its current distribution and the timescale considered in the study. It is thus conditional upon spatial configuration and the species' dispersal ability (see e.g. Soberón, 2007; Barve et al., 2011).

Analog climate: A combination of climate factors found in one area or time period that is within the envelope of climatic conditions found in a different area or time period used for comparison (see Williams and Jackson, 2007). Antonym: non-analog climate.

Available environment: The set of environmental conditions that exist in a given area (see Jackson and Overpeck, 2000, Box 3).

Synonyms: realized environment (whole range, not species specific), background environment.

Environment: The set of conditions, biotic or abiotic (i.e. physical and chemical), that characterize a site. The abiotic environment is typically climatic, geologic, or edaphic (soil). The biotic environment is defined by the various communities of organisms that co-exist.

Environmental niche: The environmental (usually abiotic) requirements of a species allowing the maintenance (infinitely if conditions do not change) of natural populations (sensu Hutchinson, 1957); it is either fundamental if defined by its physiological tolerances (see fundamental environmental niche) or realized if additionally constrained by biotic interactions (see realized environmental niche) and dispersal.

Fundamental niche: The envelope of environmental (abiotic) conditions defining the fundamental niche in an n-dimensional environmental space. It depicts the ecophysiological requirements of species (see e.g. Soberón, 2007). Synonym: physiological niche.

Habitat: A description of environmental conditions (abiotic or biotic) at a given locality, at a particular scale of space and time, where an organism either actually or potentially lives (adapted from Kearney, 2006).

Niche conservatism: The tendency for species to retain their niche in space and time. Synonym: niche stability (see e.g. Wiens and Graham, 2005).

Niche envelope: The envelope of conditions in multivariate environmental space defining a species' niche. The boundary of the envelope can be defined in many different ways (e.g. percentiles; see Broennimann et al., 2012).

Niche expansion: Proportion of the exotic niche non-overlapping with the native niche.

Niche overlap: The intersection of two niches in n-dimensional environmental space.

Niche shift: A change in the centroid (see above) or limits of the niche envelope in environmental space. Synonym: niche change.

Niche stability: Proportion of the exotic niche overlapping with the native niche

Niche unfilling: Proportion of the native niche non-overlapping with the exotic niche.

Niche-biotope duality: The reciprocal correspondence between the niche conditions in multidimensional environmental space and the physical locations that a species actually occupies in geographical space (derived from Colwell and Rangel, 2009).

Non-analog climate: See analog climate.

Potential niche: The intersection between the fundamental niche and the realized environment (see Jackson and Overpeck, 2000; Soberón and Nakamura, 2009).

Realized environment: The range of environmental conditions existing in a given area. Since not all possible combinations of two or more environmental factors exist on Earth, the realized environment defines those that are actually observed. As a corollary, in a given area, the niche of a species is necessarily nested within the realized environment (see niche) and thus the latter constrains the former.

Realized niche: The environmental (abiotic) niche of a species as quantified from field observations, i.e. the fundamental niche modulated by biotic exclusions, population dynamics (such as source–sink dynamics) and dispersal limitations (Soberón, 2007; Colwell and Rangel, 2009). Synonym: ecological niche.

Trophic niche: The trophic position a species occupies within a community or ecosystem, relative to the other species (Elton, 1927; Chase and Leibold, 2003).

Technical Acronyms for the Most Commonly Used Modeling Techniques

Here we list the acronyms of the most commonly used modeling techniques used to model habitat suitability and predict species distribution, with their meaning and one main reference as used in habitat suitability modeling and species predictions. Many are described, e.g. in Moisen and Frescino (2002), Elith et al. (2006) and Tsoar et al. (2007), else see the specific reference in the related description. Several others, introduced from machine learning, can be found in Maher et al. (2014). Some techniques also have a community-level implementation that accounts more for species interactions (see Elith et al., 2006; Baselga and Araújo, 2009).

ANN: Artificial neural network (see Manel et al., 1999b)

BIOCLIM: Bioclimatic rectilinear envelopes (see Busby, 1991)

BRT: Boosted regression trees (see Elith et al., 2008); one type of GBM

CART: Classification and regression trees (see De'Ath and Fabricius, 2000)

CCA: Canonical correspondence analysis (see Guisan et al., 1999)

ENFA: Ecological niche factor analysis (see Hirzel et al., 2002a)

GAM: Generalized additive model (see Guisan et al., 2002); similar to GLMs, but using smoother functions to adjust data-driven response curves.

GARP: Genetic algorithm for rule-set production (see Stockwell, 1999)

GBM: Generalized boosting model (or machine; see Elith et al., 2008); often implement as BRT

GDM: Generalized dissimilarity modeling (Ferrier et al., 2007)

GLM: Generalized linear model (see Guisan et al., 2002); similar to GAMs, but using polynomial functions to adjust parametric response curves

MARS: Multivariate adaptive regression splines (see Leathwick et al., 2005)

MAXENT: Maximum entropy (Phillips et al., 2006)

MDA: Multiple discriminant analyses (see Manel et al., 1999b)

PCA-SP: PCA-species. Predictive approach to predict species and habitats distribution based on a principal component analysis run on the presence-only species occurrence (Robertson et al., 2001)

RF: Random forest (see Prasad et al., 2006)

SRE: Species range envelope; terminology used in the BIOMOD package (Thuiller et al., 2009); see BIOCLIM

SVM: Support vector machine (see Drake et al., 2006)

References

Aarssen, L. W. and Schamp, B. S. 2002. Predicting distributions of species richness and species size in regional floras: applying the species pool hypothesis to the habitat template model. *Perspectives in Plant Ecology Evolution and Systematics*, 5, 3–12.

Ackerly, D. D. 2003. Community assembly, niche conservatism, and adaptive evolution in changing environments. *International Journal of Plant Sciences*, 164, S165–S184.

Agresti, A. 1990. *Categorical Data Analysis*. New York: Wiley.

Agresti, A. 1999. Modelling ordered categorical data: recent advances and future challenges. *Statistics in Medicine*, 18, 2191–2207.

Aguiar-Pulido, V., Huang, W., Suarez-Ulloa, V., et al. 2016. Metagenomics, metatranscriptomics, and metabolomics approaches for microbiome analysis. *Evolutionary Bioinformatics Online*, 12, 5.

Akaike, H. 1974. A new look at statistical model identification. *IEEE Transactions on Automatic Control*, AU-19, 716–722.

Alagador, D., Martins, M. J., Cerdeira, J. C., Cabeza, M. and Araújo, M. B. 2011. A probability-based approach to match species with reserves when data are at different resolutions. *Biological Conservation*, 144, 811–820.

Albert, C. H., Yoccoz, N. G., Edwards, T. C., et al. 2010. Sampling in ecology and evolution: bridging the gap between theory and practice. *Ecography*, 33, 1028–1037.

Albouy, C., Velez, L., Coll, M., et al. 2014. From projected species distribution to food-web structure under climate change. *Global Change Biology*, 20, 730–741.

Algar, A. C., Kharouba, H. M., Young, E. R. and Kerr, J. T. 2009. Predicting the future of species diversity: macroecological theory, climate change, and direct tests of alternative forecasting methods. *Ecography*, 32, 22–33.

Allen, T. F. H. and Hoekstra, T. W. 1992. *Towards a Unified Ecology*. New York: Columbia University Press.

Allouche, O., Tsoar, A. and Kadmon, R. 2006. Assessing the accuracy of species distribution models: prevalence, kappa and the true skill statistic (TSS). *Journal of Applied Ecology*, 43, 1223–1232.

Amante, C. and Eakins, B. W. 2009. ETOPO1 1 Arc-Minute Global Relief Model: Procedures, Data Sources and Analysis. *NOAA Technical Memorandum*.

Anderson, B. J., Akcakaya, H. R., Araujo, M. B., et al. 2009. Dynamics of range margins for metapopulations under climate change. *Proceedings of the Royal Society B. Biological Sciences*, 276, 1415–1420.

418 · **References**

Anderson, R. P. and Raza, A. 2010. The effect of the extent of the study region on GIS models of species geographic distributions and estimates of niche evolution: preliminary tests with montane rodents (genus *Nephelomys*) in Venezuela. *Journal of Biogeography*, 37, 1378–1393.

Anderson, R. P., Peterson, A. T. and Gomez-Laverde, M. 2002. Using niche-based GIS modeling to test geographic predictions of competitive exclusion and competitive release in South American pocket mice. *Oikos*, 98, 3–16.

Anderson, R. P., Lew, D. and Peterson, A. T. 2003. Evaluating predictive models of species' distributions: criteria for selecting optimal models. *Ecological Modelling*, 162, 211–232.

Araújo, M. B. and Guisan, A. 2006. Five (or so) challenges for species distribution modelling. *Journal of Biogeography*, 33, 1677–1688.

Araújo, M. B. and Luoto, M. 2007. The importance of biotic interactions for modelling species distributions under climate change. *Global Ecology and Biogeography*, 16, 743–753.

Araújo, M. B. and New, M. 2007. Ensemble forecasting of species distributions. *Trends in Ecology and Evolution,* 22, 42–47.

Araújo, M. B. and Pearson, R. G. 2005. Equilibrium of species' distributions with climate. *Ecography*, 28, 693–695.

Araújo, M. B. and Peterson, A. T. 2012. Uses and misuses of bioclimatic envelope modeling. *Ecology*, 93, 1527–1539.

Araújo, M. B. and Rahbek, C. 2006. How does climate change affect biodiversity? *Science,* 313, 1396–1397.

Araújo, M. B., Cabeza, M., Thuiller, W., Hannah, L. and Williams, P. H. 2004. Would climate change drive species out of reserves? An assessment of existing reserve-selection methods. *Global Change Biology*, 10, 1618–1626.

Araújo, M. B., Pearson, R. G., Thuiller, W. and Erhard, M. 2005a. Validation of species–climate impact models under climate change. *Global Change Biology*, 11, 1504–1513.

Araújo, M. B., Thuiller, W., Williams, P. H. and Reginster, I. 2005b. Downscaling European species atlas distributions to a finer resolution: implications for conservation planning. *Global Ecology and Biogeography*, 14, 17–30.

Araújo, M. B., Thuiller, W. and Pearson, R. G. 2006. Climate warming and the decline of amphibians and reptiles in Europe. *Journal of Biogeography*, 33, 1712–1728.

Araújo, M. B., Nogues-Bravo, D., Diniz-Filho, J. A. F., et al. 2008. Quaternary climate changes explain diversity among reptiles and amphibians. *Ecography*, 31, 8–15.

Araújo, M. B., Alagador, D., Cabeza, M., Nogues-Bravo, D. and Thuiller, W. 2011. Climate change threatens European conservation areas. *Ecology Letters*, 14, 484–492.

Araújo, M. B., Ferri-Yanez, F., Bozinovic, F., et al. 2013. Heat freezes niche evolution. *Ecology Letters*, 16, 1206–1219.

Arlot, S. and Celisse, A. 2010. A survey of cross-validation procedures for model selection. *Statistics Surveys*, 4, 40–79.

Austin, M. 1987. Models for the analysis of species' response to environmental gradients. *Vegetatio*, 69, 35–45.

Austin, M. 1990. Community theory and competition in vegetation. *Perspectives on Plant Competition*, 215, 238.

Austin, M. and Gaywood, M. 1994. Current problems of environmental gradients and species response curves in relation to continuum theory. *Journal of Vegetation Science*, 5, 473–482.

Austin, M., Nicholls, A., Doherty, M. and Meyers, J. 1994. Determining species response functions to an environmental gradient by means of a β-function. *Journal of Vegetation Science*, 5, 215–228.

Austin, M. and Smith, T. 1989. A new model for the continuum concept. *Plant Ecology*, 83, 35–47.

Austin, M., Belbin, L., Meyers, J., Doherty, M. and Luoto, M. 2006. Evaluation of statistical models used for predicting plant species distributions: role of artificial data and theory. *Ecological Modelling*, 199, 197–216.

Austin, M. P. 1971. Role of regression analysis in plant ecology. *Proceedings of the Ecological Society of Australia*, 6, 63–75.

Austin, M. P. 1985. Continuum concept, ordination methods, and niche theory. *Annual Review of Ecology and Systematics*, 16, 39–61.

Austin, M. P. 1992. Modeling the environmental niche of plants – implications for plant community response to elevated CO_2 levels. *Australian Journal of Botany*, 40, 615–630.

Austin, M. P. 2002. Spatial prediction of species distribution: an interface between ecological theory and statistical modelling. *Ecological Modelling*, 157, 101–118.

Austin, M. P. 2007. Species distribution models and ecological theory: a critical assessment and some possible new approaches. *Ecological Modelling*, 200, 1–19.

Austin, M. P. and Van Niel, K. P. 2011. Improving species distribution models for climate change studies: variable selection and scale. *Journal of Biogeography*, 38, 1–8.

Austin, M. P., Nicholls, A. O. and Margules, C. R. 1990. Measurement of the realized qualitative niche: environmental niches of 5 *Eucalyptus* species. *Ecological Monographs*, 60, 161–177.

Ba, J., Hou, Z., Platvoet, D., Zhu, L. and Li, S. 2010. Is *Gammarus tigrinus* (Crustacea, Amphipoda) becoming cosmopolitan through shipping? Predicting its potential invasive range using ecological niche modeling. *Hydrobiologia*, 649, 183–194.

Bahn, V. and McGill, B. J. 2007. Can niche-based distribution models outperform spatial interpolation? *Global Ecology and Biogeography*, 16, 733–742.

Bahn, V., J O'Connor, R. and B Krohn, W. 2006. Importance of spatial autocorrelation in modeling bird distributions at a continental scale. *Ecography*, 29, 835–844.

Barbet-Massin, M., Thuiller, W. and Jiguet, F. 2010. How much do we overestimate future local extinction rates when restricting the range of occurrence data in climate suitability models? *Ecography* 33, 878–886.

Barbet-Massin, M., Jiguet, F., Albert, C. H. and Thuiller, W. 2012. Selecting pseudo-absences for species distribution models: how, where and how many? *Methods in Ecology and Evolution*, 3, 327–338.

Barry, S. and Elith, J. 2006. Error and uncertainty in habitat models. *Journal of Applied Ecology*, 43, 413–423.

Bartholome, E. and Belward, A. S. 2005. GLC2000: a new approach to global land cover mapping from Earth observation data. *International Journal of Remote Sensing*, 26, 1959–1977.

Barve, N., Barve, V., Jiménez-Valverde, A., et al. 2011. The crucial role of the accessible area in ecological niche modeling and species distribution modeling. *Ecological Modelling*, 222, 1810–1819.

Bascompte, J., Jordano, P., Melian, C. J. and Olesen, J. M. 2003. The nested assembly of plant–animal mutualistic networks. *Proceedings of the National Academy of Sciences of the United States of America*, 100, 9383–9387.

Baselga, A. and Araújo, M. B. 2009. Individualistic vs community modelling of species distributions under climate change. *Ecography*, 32, 55–65.

Beale, C. M., Lennon, J. J., Yearsley, J. M., Brewer, M. J. and Elston, D. A. 2010. Regression analysis of spatial data. *Ecology Letters*, 13, 246–264.

Beale, C. M., Baker, N. E., Brewer, M. J. and Lennon, J. J. 2013. Protected area networks and savannah bird biodiversity in the face of climate change and land degradation. *Ecology Letters*, 16, 1061–1068.

Beaumont, L. J. and Hughes, L. 2002. Potential changes in the distributions of latitudinally restricted Australian butterfly species in response to climate change. *Global Change Biology*, 8, 954–971.

Beaumont, L. J., Gallagher, R. V., Thuiller, W., et al. 2009. Different climatic envelopes among invasive populations may lead to underestimations of current and future biological invasions. *Diversity and Distributions*, 15, 409–420.

Beerling, D. J., Huntley, B. and Bailey, J. P. 1995. Climate and the distribution of *Fallopia japonica*: use of an introduced species to test the predictive capacity of response surfaces. *Journal of Vegetation Science*, 6, 269–282.

Bell, D. M., Bradford, J. B. and Lauenroth, W. K. 2014. Early indicators of change: divergent climate envelopes between tree life stages imply range shifts in the western United States. *Global Ecology and Biogeography*, 23, 168–180.

Bellard, C., Bertelsmeier, C., Leadley, P., Thuiller, W. and Courchamp, F. 2012. Impacts of climate change on the future of biodiversity. *Ecology Letters*, 15, 365–377.

Belward, A. S., Estes, J. E. and Kline, K. D. 1999. The IGBP-DIS global 1-km landcover data set DISCover: a project overview. *Photogrammetric Engineering and Remote Sensing*, 65, 1013–1020.

Bendall, M. L., Stevens, S. L., Chan, L.-K., et al. 2016. Genome-wide selective sweeps and gene-specific sweeps in natural bacterial populations. *The ISME Journal*, 10, 1589–1601.

Berrocal, V. J., Craigmile, P. F. and Guttorp, P. 2012. Regional climate model assessment using statistical upscaling and downscaling techniques. *Environmetrics*, 23, 482–492.

Berteaux, D., Humphries, M. M., Krebs, C. J., et al. 2006. Constraints to projecting the effects of climate change on mammals. *Climate Research*, 32, 151–158.

Bertrand, R., Lenoir, J., Piedallu, C., et al. 2011. Changes in plant community composition lag behind climate warming in lowland forests. *Nature*, 479, 517–520.

Bertrand, R., Perez, V. and Gégout, J. C. 2012. Disregarding the edaphic dimension in species distribution models leads to the omission of crucial spatial information under climate change: the case of *Quercus pubescens* in France. *Global Change Biology*, 18, 2648–2660.

Binzenhofer, B., Schroder, B., Strauss, B., Biedermann, R. and Settele, J. 2005. Habitat models and habitat connectivity analysis for butterflies and burnet moths: the example of *Zygaena carniolica* and *Coenonympha arcania*. *Biological Conservation*, 126, 247–259.

Blondel, J. and Aronson, J. 1995. Biodiversity and ecosystem function in the Mediterranean basin: human and non-human determinants. In Davis, G.W. and Richardson, D. M. (eds), *Ecological Studies*. Berlin: Springer-Verlag, pp. 43–119.

Bocedi, G., Zurell, D., Reineking, B. and Travis, J. M. J. 2014. Mechanistic modelling of animal dispersal offers new insights into range expansion dynamics across fragmented landscapes. *Ecography*, 37, 1240–1253.

Bombi, P. and D'Amen, M. 2012. Scaling down distribution maps from atlas data: a test of different approaches with virtual species. *Journal of Biogeography*, 39, 640–651.

Bombi, P., Salvi, D., Vignoli, L. and Bologna, M. A. 2009. Modelling Bedriaga's rock lizard distribution in Sardinia: an ensemble approach. *Amphibia–Reptilia*, 30, 413–424.

Booth, T. H., Nix, H. A., Busby, J. R. and Hutchinson, M. F. 2014. BIOCLIM: the first species distribution modelling package, its early applications and relevance to most current MAXENT studies. *Diversity and Distributions*, 20, 1–9.

Botkin, D. B., Saxe, H., Araujo, M. B., et al. 2007. Forecasting the effects of global warming on biodiversity. *Bioscience*, 57, 227–236.

Boucher, D. H., James, S. and Keeler, K. H. 1982. The ecology of mutualism. *Annual Review of Ecology and Systematics*, 13, 315–347.

Boucher, F. C., Thuiller, W., Roquet, C., et al. 2012. Reconstructing the origins of high-alpine niches and cushion life form in the genus *Androsace* sl (Primulaceae). *Evolution*, 66, 1255–1268.

Boulangeat, I., Gravel, D. and Thuiller, W. 2012a. Accounting for dispersal and biotic interactions to disentangle the drivers of species distributions and their abundances. *Ecology Letters*, 15, 584–593.

Boulangeat, I., Philippe, P., Abdulhak, S., et al. 2012b. Improving plant functional groups for dynamic models of biodiversity: at the crossroads between functional and community ecology. *Global Change Biology*, 18, 3464–3475.

Boulangeat, I., Georges, D. and Thuiller, W. 2014. FATE-HD: a spatially and temporally explicit integrated model for predicting vegetation structure and diversity at regional scale. *Global Change Biology*, 20, 2368–2378.

Box, E. O. 1981. *Macroclimate and Plant Forms: An Introduction to Predictive Modeling in Phytogeography*, The Hague: Junk.

Boyce, M. S., Vernier, P. R., Nielsen, S. E. and Schmiegelow, F. K. A. 2002. Evaluating resource selection functions. *Ecological Modelling*, 157, 281–300.

Braconnot, P., Otto-Bliesner, B., Harrison, S., et al. 2007. Results of PMIP2 coupled simulations of the mid-holocene and last glacial maximum. Part I: experiments and large-scale features. *Climate of the Past*, 3, 261–277.

Breiman, L. 1996. Bagging predictors. *Machine Learning*, 24, 123–140.

Breiman, L. 2001. Random forests. *Machine Learning*, 45, 5–32.

Breiman, L., Friedman, J. H., Olshen, R. A. and Stone, C. J. 1984. *Classification and Regression Trees*, New York: Chapman and Hall.

Breiner, F. T., Guisan, A., Bergamini, A. and Nobis, M. P. 2015. Overcoming limitations of modelling rare species by using ensembles of small models. *Methods in Ecology and Evolution*, 6(10), 1210–1218.

Broennimann, O. and Guisan, A. 2008. Predicting current and future biological invasions: both native and invaded ranges matter. *Biology Letters*, 4, 585–589.

Broennimann, O., Thuiller, W., Hughes, G., et al. 2006. Do geographic distribution, niche property and life form explain plants' vulnerability to global change? *Global Change Biology*, 12, 1079–1093.

Broennimann, O., Treier, U. A., Müller-Schärer, H., et al. 2007. Evidence of climatic niche shift during biological invasion. *Ecology Letters*, 10, 701–709.

Broennimann, O., Fitzpatrick, M. C., Pearman, P. B., et al. 2012. Measuring ecological niche overlap from occurrence and spatial environmental data. *Global Ecology and Biogeography*, 21, 481–497.

Broennimann, O., Mráz, P., Petitpierre, B., Guisan, A. and Müller-Schärer, H. 2014a. Contrasting spatio-temporal climatic niche dynamics during the eastern and western invasions of spotted knapweed in North America. *Journal of Biogeography*, 41, 1126–1136.

Broennimann, O., Di Cola, V. and Guisan, A. (2016). *ecospat: Spatial Ecology Miscellaneous Methods. R package version 2.1.1.* http://CRAN.R-project.org/package=ecospat.

Broennimann, O., Ursenbacher, S., Meyer, A., et al. 2014c. Influence of climate on the presence of colour polymorphism in two montane reptile species. *Biology Letters*, 10, 20140638.

Broitman, B., Szathmary, P., Mislan, K., Blanchette, C. and Helmuth, B. 2009. Predator–prey interactions under climate change: the importance of habitat vs body temperature. *Oikos*, 118, 219–224.

Brook, B. W., Akçakaya, H. R., Keith, D. A., et al. 2009. Integrating bioclimate with population models to improve forecasts of species extinctions under climate change. *Biology Letters*, 5, 723–725.

Brotons, L., Thuiller, W., Araújo, M. B. and Hirzel, A. H. 2004. Presence–absence vs presence-only modelling methods for predicting bird habitat suitability. *Ecography*, 27, 165–172.

Brown, A. M., Warton, D. I., Andrew, N. R., et al. 2014. The fourth-corner solution: using predictive models to understand how species traits interact with the environment. *Methods in Ecology and Evolution*, 5, 344–352.

Brown, J. H. 1971. Mechanisms of competitive exclusion between two species of chipmunks. *Ecology*, 52, 305–311.

Brown, J. H., Stevens, G. C. and Kaufman, D. M. 1996. The geographic range: size, shape, boundaries, and internal structure. *Annual Review of Ecology and Systematics*, 597–623.

Brown, J. H., Gillooly, J. F., Allen, A. P., Savage, V. M. and West, G. B. 2004. Toward a metabolic theory of ecology. *Ecology*, 85, 1771–1789.

Bruno, J. F., Stachowicz, J. J. and Bertness, M. D. 2003. Inclusion of facilitation into ecological theory. *Trends in Ecology and Evolution*, 18, 119–125.

Brzeziecki, B., Kienast, F. and Wildi, O. 1995. Modelling potential impacts of climate change on the spatial distribution of zonal forest communities in Switzerland. *Journal of Vegetation Science*, 6, 257–268.

Buckley, L. B. 2008. Linking traits to energetics and population dynamics to predict lizard ranges in changing environments. *The American Naturalist*, 171, E1-E19.

Buckley, L. B., Davies, T. J., Ackerly, D. D., et al. 2010. Phylogeny, niche conservatism and the latitudinal diversity gradient in mammals. *Proceedings of the Royal Society of London B: Biological Sciences*, 277, 2131–2138.

Buisson, L., Thuiller, W., Lek, S., Lim, P. and Grenouillet, G. 2008. Climate change hastens the turnover of stream fish assemblages. *Global Change Biology*, 14, 2232–2248.

Buisson, L., Thuiller, W., Casajus, N., Lek, S. and Grenouillet, G. 2010. Uncertainty in ensemble forecasting of species distribution. *Global Change Biology*, 16, 1145–1157.

Buisson, L., Grenouillet, G., Villéger, S., Canal, J. and Laffaille, P. 2013. Toward a loss of functional diversity in stream fish assemblages under climate change. *Global Change Biology*, 19, 387–400.

Burgman, M. A. and Fox, J. C. 2003. Bias in species range estimates from minimum convex polygons: implications for conservation and options for improved planning. *Animal Conservation*, 6, 19–28.

Burnham, K. P. and Anderson, D. R. 2002. *Model Selection and Multimodel Inference: A Practical Information-Theoretic Approach*. Berlin: Springer.

Busby, J. R. 1991. BIOCLIM: a bioclimate analysis and prediction system. In Margules, C. R. and Austin, M. P. (eds.), *Nature Conservation: Cost Effective Biological Surveys and Data Analysis*. Canberra, Australia: CSIRO, pp. 64–68.

Calenge, C. and Basille, M. 2008. A general framework for the statistical exploration of the ecological niche. *Journal of Theoretical Biology*, 252, 674–685.

Caley, M. J. and Schluter, D. 1997. The relationship between local and regional diversity. *Ecology*, 78, 70–80.

Callaway, R. M. 1995. Positive interactions among plants (interpreting botanical progress). *The Botanical Review*, 61, 306–349.

Calvete, C., Estrada, R., Miranda, M. A., et al. 2008. Modelling the distributions and spatial coincidence of bluetongue vectors *Culicoides imicola* and the *Culicoides obsoletus* group throughout the Iberian peninsula. *Medical and Veterinary Entomology*, 22, 124–134.

Carl, G. and Kühn, I. 2007. Analyzing spatial autocorrelation in species distributions using Gaussian and logit models. *Ecological Modelling*, 207, 159–170.

Carlson, B. Z., Georges, D., Rabatel, A., et al. 2014. Accounting for tree line shift, glacier retreat and primary succession in mountain plant distribution models. *Diversity and Distributions*, 20, 1379–1391.

Carnaval, A. C., Hickerson, M. J., Haddad, C. F., Rodrigues, M. T. and Moritz, C. 2009. Stability predicts genetic diversity in the Brazilian Atlantic forest hotspot. *Science*, 323, 785–789.

Carpenter, G., Gillison, A. N. and Winter J. 1993. DOMAIN: a flexible modelling procedure for mapping potential distributions of plants and animals. *Biodiversity and Conservation*, 2, 667–680.

Carroll, C., Johnson, D. S., Dunk, J. R. and Zielinski, W. J. 2010. Hierarchical Bayesian spatial models for multispecies conservation planning and monitoring. *Conservation Biology*, 24, 1538–1548.

Carvalho, S. B., Brito, J. C., Crespo, E. G., Watts, M. E. and Possingham, H. P. 2011. Conservation planning under climate change: toward accounting for uncertainty in predicted species distributions to increase confidence in conservation investments in space and time. *Biological Conservation*, 144, 2020–2030.

Chase, J. M. and Leibold, M. A. 2003. *Ecological Niches*. Chicago, IL: The University of Chicago Press.

Chave, J. 2013. The problem of pattern and scale in ecology: what have we learned in 20 years? *Ecology Letters*, 16, 4–16.

Chefaoui, R. M. and Lobo, J. M. 2008. Assessing the effects of pseudo-absences on predictive distribution model performance. *Ecological Modelling*, 210, 478–486.

Chen, I. C., Hill, J. K., Ohlemuller, R., Roy, D. B. and Thomas, C. D. 2011. Rapid range shifts of species associated with high levels of climate warming. *Science*, 333, 1024–1026.

Christie, K. S., Gilbert, S. L., Brown, C. L., Hatfield, M. and Hanson, L. 2016. Unmanned aircraft systems in wildlife research: current and future applications of a transformative technology. *Frontiers in Ecology and the Environment*, 14, 241–251.

Civantos, E., Thuiller, W., Maiorano, L., Guisan, A. and Araújo, M. B. 2012. Potential impacts of climate change on ecosystem services in Europe: the case of pest control by vertebrates. *Bioscience*, 62, 658–666.

Claeskens, G. and Hjort, N. L. 2008. *Model Selection and Model Averaging*. Cambridge, UK: University Press Cambridge.

Clark, J. S., Dietze, M., Chakraborty, S., et al. 2007. Resolving the biodiversity paradox. *Ecology Letters*, 10, 647–659.

Clark, J. S., Gelfand, A. E., Woodall, C. W. and Zhu, K. 2014. More than the sum of the parts: forest climate response from joint species distribution models. *Ecological Applications*, 24, 990–999.

Clausen, J., Keck, D. and Hiesey, W. M. 1948. Experimental studies on the nature of species. III. Environmental responses of climatic races of *Achillea*. *Carnegie Inst. Wash.*, 581.

Coetzee, B. W. T., Robertson, M. P., Erasmus, B. F. N., van Rensburg, B. J. and Thuiller, W. 2009. Ensemble models predict important bird areas in southern Africa will become less effective for conserving endemic birds under climate change. *Global Ecology and Biogeography*, 18, 701–710.

Cohen, J. 1960. A coefficient of agreement for nominal scales. *Educational and Psychological Measurement*, 20, 37–46.

Cohen, J. 1968. Weighted kappa: nominal scale agreement provision for scaled disagreement or partial credit. *Psychological Bulletin*, 70, 213.

Colwell, R. K. and Futuyma, D. J. 1971. Measurement of niche breadth and overlap. *Ecology*, 52, 567–576.

Colwell, R. K. and Rangel, T. F. 2009. Hutchinson's duality: the once and future niche. *Proceedings of the National Academy of Sciences*, 106, 19651–19658.

Connell, J. H. 1961. The influence of interspecific competition and other factors on the distribution of the barnacle *Chthamalus stellatus*. *Ecology*, 42, 710–723.

Corsi, F., de Leeuw, J. and Skidmore, A. 2000. Modeling Species Distribution with GIS. In Boitani, L. and Fuller, T. K. (eds), *Research Techniques in Animal Ecology: Controversies and Consequences*. New York: Columbia University Press, pp. 389–434.

Côté, I. M. and Reynolds, J. D. 2002. Predictive ecology to the rescue? *Science*, 298, 1181–1182.

Cox, B. 2001. The biogeographic regions reconsidered. *Journal of Biogeography*, 28, 511–523.

Crase, B., Liedloff, A. C. and Wintle, B. A. 2012. A new method for dealing with residual spatial autocorrelation in species distribution models. *Ecography*, 35, 879–888.

Crase, B., Liedloff, A., Vesk, P. A., Fukuda, Y. and Wintle, B. A. 2014. Incorporating spatial autocorrelation into species distribution models alters forecasts of climate-mediated range shifts. *Global Change Biology*, 20, 2566–2579.

Cressie, N. 1993. Geostatistics: a tool for environmental models. In Goodchild, M. F., Parks, B. O. and Steyaert, L. T. (eds.), *Environmental Modeling with GIS*. Oxford, UK: Oxford University Press, pp. 414–421.

Crimmins, S. M., Dobrowski, S. Z., Greenberg, J. A., Abatzoglou, J. T. and Mynsberge, A. R. 2011. Changes in climatic water balance drive downhill shifts in plant species' optimum elevations. *Science*, 331, 324–327.

Cutler, D. R., Edwards, T. C., Beard, K. H., Cutler, A. and Hess, K. T. 2007. Random forests for classification in ecology. *Ecology*, 88, 2783–2792.

D'Amen, M., Zimmermann, N. E. and Pearman, P. B. 2013. Conservation of phylogeographic lineages under climate change. *Global Ecology and Biogeography*, 22, 93–104.

D'Amen, M., Dubuis, A., Fernandes, R. F., et al. 2015a. Using species richness and functional traits predictions to constrain assemblage predictions from stacked species distribution models. *Journal of Biogeography*, 42, 1255–1266.

D'Amen, M., Zimmermann, N. E., Rahbek, C. and Guisan, A. 2015b. Spatial prediction at the community level: state of the art and future perspectives. *Biological Reviews*, 92, 169–187.

Daly, C., Neilson, R. P. and Phillips, D. L. 1994. A statistical topographic model for mapping climatological precipitation over mountainous terrain. *Journal of Applied Meteorology*, 33, 140–158.

Davis, E. B., McGuire, J. L. and Orcutt, J. D. 2014. Ecological niche models of mammalian glacial refugia show consistent bias. *Ecography*, 37, 1133–1138.

Davis, M. B. 1989. Lags in vegetation response to greenhouse warming. *Climatic Change*, 15, 75–82.

Dawson, T. P., Curran, P. J. and Plummer, S. E. 1998. The biochemical decomposition of slash pine needles from reflectance spectra using neural networks. *International Journal of Remote Sensing*, 19, 1433–1438.

De'Ath, G. 2007. Boosted trees for ecological modeling and prediction. *Ecology*, 88, 243–251.

De'Ath, G. and Fabricius, K. E. 2000. Classification and regression trees: a powerful yet simple technique for ecological data analysis. *Ecology*, 81, 3178–3192.

de Oliveira, S. V., Escobar, L. E., Peterson, A. T. and Gurgel-Gonçalves, R. 2013. Potential geographic distribution of hantavirus reservoirs in Brazil. *Plos One*, 8, e85137.

de Witte, L. C. and Stöcklin, J. 2010. Longevity of clonal plants: why it matters and how to measure it. *Annals of botany*, 1–12.

Defossez, E., Courbaud, B., Marcais, B., et al. 2011. Do interactions between plant and soil biota change with elevation? A study on *Fagus sylvatica*. *Biology Letters*, 7, 699–701.

DeFries, R., Hansen, M., Steininger, M., et al. 1997. Subpixel forest cover in central Africa from multisensor, multitemporal data. *Remote Sensing of Environment*, 60, 228–246.

DeFries, R. S., Townshend, J. R. G. and Hansen, M. C. 1999. Continuous fields of vegetation characteristics at the global scale at 1-km resolution. *Journal of Geophysical Research: Atmospheres*, 104, 16911–16923.

Delisle, F., Lavoie, C., Jean, M. and Lachance, D. 2003. Reconstructing the spread of invasive plants: taking into account biases associated with herbarium specimens. *Journal of Biogeography*, 30, 1033–1042.

Del Monte-Luna, P., Brook, B. W., Zetina-Rejón, M. J. and Cruz-Escalona, V. H. 2004. The carrying capacity of ecosystems. *Global Ecology and Biogeography*, 13, 485–495.

Delzon, S., Urli, M., Samalens, J.-C., et al. 2013. Field evidence of colonisation by Holm oak, at the northern margin of its distribution range, during the Anthropocene period. *Plos One*, 8 e80443.

Dennis, R. L., Sparks, T. H. and Hardy, P. B. 1999. Bias in butterfly distribution maps: the effects of sampling effort. *Journal of Insect Conservation*, 3, 33–42.

DeVaney, S. C., McNyset, K. M., Williams, J. B., Peterson, A. T. and Wiley, E. O. 2009. A tale of four "carp": invasion potential and ecological niche modeling. *Plos One*, 4.

Di Cola, V., Broennimann, O., Petitpierre, B., et al. 2017. ecospat: an R package for the support of spatial analyses and modelling of species niches and distributions. *Ecography*, 40, 774–787.

Dietz, H. and Edwards, P. J. 2006. Recognition that causal processes change during plant invasion helps explain conflicts in evidence. *Ecology*, 87, 1359–1367.

Diniz, J. A. F., Bini, L. M., Rangel, T. F., et al. 2009. Partitioning and mapping uncertainties in ensembles of forecasts of species turnover under climate change. *Ecography*, 32, 897–906.

Diniz-Filho, J. A. F., Bini, L. M. and Hawkins, B. A. 2003. Spatial autocorrelation and red herrings in geographical ecology. *Global Ecology and Biogeography*, 12, 53–64.

Dirnbock, T., Greimler, J., Lopez, P. and Stuessy, T. F. 2003. Predicting future threats to the native vegetation of Robinson Crusoe Island, Juan Fernandez Archipelago, Chile. *Conservation Biology*, 17, 1650–1659.

Dobrowski, S. Z., Thorne, J. H., Greenberg, J. A., et al. 2011. Modeling plant ranges over 75 years of climate change in California, USA: temporal transferability and species traits. *Ecological Monographs*, 81, 241–257.

Dobrowski, S. Z., Abatzoglou, J., Swanson, A. K., et al. 2013. The climate velocity of the contiguous United States during the 20th century. *Global Change Biology*, 19, 241–251.

Doebeli, M. and Dieckmann, U. 2003. Speciation along environmental gradients. *Nature*, 421, 259–264.

Dolédec, S., Chessel, D. and Gimaret-Carpentier, C. 2000. Niche separation in community analysis: a new method. *Ecology*, 81, 2914–2927.

Dorazio, R. M. 2014. Accounting for imperfect detection and survey bias in statistical analysis of presence-only data. *Global Ecology and Biogeography*, 23, 1472–1484.

Dormann, C. F. 2007a. Assessing the validity of autologistic regression. *Ecological Modelling*, 207, 234–242.

Dormann, C. F. 2007b. Effects of incorporating spatial autocorrelation into the analysis of species distribution data. *Global Ecology and Biogeography*, 30, 609–628.

Dormann, C. F., McPherson, J. M., Araujo, M. B., et al. 2007. Methods to account for spatial autocorrelation in the analysis of species distributional data: a review. *Ecography*, 30, 609–628.

Dormann, C. F., Purschke, O., Marquez, J. R. G., Lautenbach, S. and Schroder, B. 2008. Components of uncertainty in species distribution analysis: a case study of the great grey shrike. *Ecology*, 89, 3371–3386.

Dormann, C. F., Schymanski, S. J., Cabral, J., et al. 2012. Correlation and process in species distribution models: bridging a dichotomy. *Journal of Biogeography*, 39, 2119–2131.

Dormann, C. F., Elith, J., Bacher, S., et al. 2013. Collinearity: a review of methods to deal with it and a simulation study evaluating their performance. *Ecography*, 36, 27–46.

Drake, J. M., Randin, C. and Guisan, A. 2006. Modelling ecological niches with support vector machines. *Journal of Applied Ecology*, 43, 424–432.

Dray, S., Chessel, D. and Thioulouse, J. 2003. Co-inertia analysis and the linking of ecological data tables. *Ecology*, 84, 3078–3089.

Dubuis, A., Pottier, J., Rion, V., et al. 2011. Predicting spatial patterns of plant species richness: a comparison of direct macroecological and species stacking modelling approaches. *Diversity and Distributions*, 17, 1122–1131.

Dubuis, A., Giovanettina, S., Pellissier, L., et al. 2013. Improving the prediction of plant species distribution and community composition by adding edaphic to topo-climatic variables. *Journal of Vegetation Science*, 24, 593–606.

Duckworth, J., Bunce, R. and Malloch, A. 2000. Vegetation–environment relationships in Atlantic European calcareous grasslands. *Journal of Vegetation Science*, 11, 15–22.

Dullinger, S., Gattringer, A., Thuiller, W., et al. 2012. Extinction debt of high-mountain plants under twenty-first-century climate change. *Nature Climate Change*, 2, 619–622.

Edwards, T. C., Cutler, D. R., Zimmermann, N. E., Geiser, L. and Alegria, J. 2005. Model-based stratifications for enhancing the detection of rare ecological events. *Ecology*, 86, 1081–1090.

Edwards, T. C., Cutler, D. R., Zimmermann, N. E., Geiser, L. and Moisen, G. G. 2006. Effects of sample survey design on the accuracy of classification tree models in species distribution models. *Ecological Modelling*, 199, 132–141.

Efron, B. and Gong, G. 1983. A leisurely look at the bootstrap, the jackknife, and cross- validation. *American Statistician*, 37, 36–48.

Efron, B. and Tibshirani, R. 1993. *An Introduction to the Bootstrap*, New York: Chapman and Hall.

Efron, B. and Tibshirani, R. 1997. Improvements on cross-validation: The .632+ bootstrap method. *Journal of the American Statistical Association*, 92, 548–560.

Elith, J. 2002. *Predicting the distribution of plants.* PhD Thesis. Melbourne, Australia: The University of Melbourne.

Elith, J. and Leathwick, J. R. 2009. Species distribution models: ecological explanation and prediction across space and time. *Annual Review of Ecology Evolution and Systematics*, 40, 677–697.

Elith, J., Burgman, M. A. and Regan, H. M. 2002. Mapping epistemic uncertainties and vague concepts in predictions of species distribution. *Ecological Modelling*, 157, 313–329.

Elith, J., Ferrier, S., Huettmann, F. and Leathwick, J. 2005. The evaluation strip: a new and robust method for plotting predicted responses from species distribution models. *Ecological Modelling*, 186, 280–289.

Elith, J., Graham, C. H., Anderson, R. P., et al. 2006. Novel methods improve prediction of species' distributions from occurrence data. *Ecography*, 29, 129–151.

Elith, J., Leathwick, J. R. and Hastie, T. 2008. A working guide to boosted regression trees. *Journal of Animal Ecology*, 77, 802–813.

Elith, J., Kearney, M. and Phillips, S. 2010. The art of modelling range-shifting species. *Methods in Ecology and Evolution*, 1, 330–342.

Elith, J., Phillips, S. J., Hastie, T., et al. 2011. A statistical explanation of MaxEnt for ecologists. *Diversity and Distributions*, 17, 43–57.

Ellenberg, H. 1953. Physiologisches und ökologisches Verhalten derselben Pflanzenarten. *Berichte der Deutschen Botanischen Gesellschaft*, 65, 351–362.

Ellenberg, H. 1954. Ueber einige fortschritte der kausalen Vegetationskunde. *Vegetatio*, 5–6, 199–211.

Ellenberg, H. 1988. *Vegetation Ecology of Central Europe*, 4th edn. Cambridge, UK: Cambridge University Press.

Elton, C. 1927. *Animal Ecology*, London: Sidgwick and Jackson.

Emery, N. C., Ewanchuck, P. J. and Bertness, M. D. 2001. Competition and salt marsh plant zonation: stress tolerators my be dominant competitors. *Ecology*, 82, 2471–2485.

Engler, R. and Guisan, A. 2009. MigClim: predicting plant distribution and dispersal in a changing climate. *Diversity and Distributions*, 15, 590–601.

Engler, R., Guisan, A. and Rechsteiner, L. 2004. An improved approach for predicting the distribution of rare and endangered species from occurrence and pseudo-absence data. *Journal of Applied Ecology*, 41, 263–274.

Engler, R., Randin, C. F., Vittoz, P., et al. 2009. Predicting future distributions of mountain plants under climate change: does dispersal capacity matter? *Ecography*, 32, 34–45.

Engler, R., Randin, C. F., Thuiller, W., et al. 2011a. 21st century climate change threatens mountain flora unequally across Europe. *Global Change Biology*, 17, 2330–2341.

Engler, R., Randin, C. R., Thuiller, W., et al. 2011b. Climate change impacts on European mountain plant diversity. *Global Change Biology*, 17, 2330–2341.

Engler, R., Hordijk, W. and Guisan, A. 2012. The MIGCLIM R package–seamless integration of dispersal constraints into projections of species distribution models. *Ecography*, 35, 872–878.

Espíndola, A., Pellissier, L., Maiorano, L., et al. 2012. Predicting present and future intra-specific genetic structure through niche hindcasting across 24 millennia. *Ecology Letters*, 15, 649–657.

Estrada-Peña, A. and Thuiller, W. 2008. An assessment of the effect of data partitioning on the performance of modelling algorithms for habitat suitability for ticks. *Medical and Veterinary Entomology*, 22, 248–257.

Estrada-Peña, A. and Venzal, J. M. 2007. Climate niches of tick species in the Mediterranean region: modeling of occurrence data, distributional constraints, and impact of climate change. *Journal of Medical Entomology*, 44, 1130–1138.

Evans, M. E., Smith, S. A., Flynn, R. S. and Donoghue, M. J. 2009. Climate, niche evolution, and diversification of the "bird-cage" evening primroses (*Oenothera*, Sections Anogra and Kleinia). *The American Naturalist*, 173, 225–240.

Farber, O. and Kadmon, R. 2003. Assessment of alternative approaches for bioclimatic modeling with special emphasis on the Mahalanobis distance. *Ecological Modelling*, 160, 115–130.

Fernandes, R. F., Vicente, J. R., Georges, D., et al. 2014. A novel downscaling approach to predict plant invasions and improve local conservation actions. *Biological Invasions*, 16, 2577–2590.

Ferreira, M. P., Zortea, M., Zanotta, D. C., Shimabukuro, Y. E. and de Souza Filho, C. R. 2016. Mapping tree species in tropical seasonal semi-deciduous forests with hyperspectral and multispectral data. *Remote Sensing of Environment*, 179, 66–78.

Ferrier, S. 1984. *The status of the Rufous Scrub-Bird Atrichornis rufescens: habitat, geographical variation and abundance*. PhD Thesis. Armidale, Australia: University of New England.

Ferrier, S. and Guisan, A. 2006. Spatial modelling of biodiversity at the community level. *Journal of Applied Ecology*, 43, 393–404.

Ferrier, S. and Watson, G. 1997. *An Evaluation of the Effectiveness of Environmental Surrogates and Modelling Techniques in Predicting the Distribution of Biological Diversity*. Canberra, Australia: NSW National Parks and Wildlife Service.

Ferrier, S., Drielsma, M., Manion, G. and Watson, G. 2002. Extended statistical approaches to modelling spatial pattern in biodiversity in north-east New South Wales. II. Community-level modelling. *Biodiversity and Conservation*, 11, 2309–2338.

Ferrier, S., Manion, G., Elith, J. and Richardson, K. 2007. Using generalized dissimilarity modelling to analyse and predict patterns of beta diversity in regional biodiversity assessment. *Diversity and Distributions*, 13, 252–264.

Fielding, A. H. 2002. What are the appropriate characteristics of an accuracy measure? In Scott, J. M., Heglund, P. J., Morrison, M. L., et al. (eds), *Predicting Species Occurrences: Issues of Accuracy and Scale*. Covelo, California: Island Press.

Fielding, A. H. and Bell, J. F. 1997. A review of methods for the assessment of prediction errors in conservation presence–absence models. *Environmental Conservation*, 24, 38–49.

Filchak, K. E., Roethele, J. B. and Feder, J. L. 2000. Natural selection and sympatric divergence in the apple maggot *Rhagoletis pomonella*. *Nature*, 407, 739–742.

Fithian, W., Elith, J., Hastie, T. and Keith, D. A. 2015. Bias correction in species distribution models: pooling survey and collection data for multiple species. *Methods in Ecology and Evolution*, 6, 424–438.

Fitzpatrick, M. C. and Hargrove, W. W. 2009. The projection of species distribution models and the problem of non-analog climate. *Biodiversity and Conservation*, 18, 2255–2261.

Fitzpatrick, M. C., Weltzin, J. F., Sanders, N. J. and Dunn, R. R. 2007. The biogeography of prediction error: why does the introduced range of the fire ant over-predict its native range? *Global Ecology and Biogeography*, 16, 24–33.

Foden, W., Midgley, G. F., Hughes, G., et al. 2007. A changing climate is eroding the geographical range of the Namib Desert tree aloe through population declines and dispersal lags. *Diversity and Distributions*, 13, 645–653.

Fondi, M., Karkman, A., Tamminen, M. V., et al. 2016. "Every gene is everywhere but the environment selects": global geolocalization of gene sharing in environmental samples through network analysis. *Genome Biology and Evolution*, 8, 1388–1400.

Forbes, A. D. 1995. Classification-algorithm evaluation: five performance measures based on confusion matrices. *Journal of Clinical Monitoring*, 11, 189–206.

Fordham, D. A., Resit Akçakaya, H., Araújo, M. B., et al. 2012. Plant extinction risk under climate change: are forecast range shifts alone a good indicator of species vulnerability to global warming? *Global Change Biology*, 18, 1357–1371.

Fraker, M. E. and Peacor, S. D. 2008. Statistical tests for biological interactions: a comparison of permutation tests and analysis of variance. *Acta Oecologica-International Journal of Ecology*, 33, 66–72.

Franklin, J. 1995. Predictive vegetation mapping: geographic modelling of biospatial patterns in relation to environmental gradients. *Progress in Physical Geography*, 19, 474–499.

Franklin, J. 2010a. *Mapping Species Distribution: Spatial Inference and Prediction.* Cambridge, UK: Cambridge University Press.

Franklin, J. 2010b. Moving beyond static species distribution models in support of conservation biogeography. *Diversity and Distributions*, 16, 321–330.

Franklin, J., Davis, F. W., Ikegami, M., et al. 2013. Modeling plant species distributions under future climates: how fine scale do climate projections need to be? *Global Change Biology*, 19, 473–483.

Fraser, L. H. and Keddy, P. A. 2005. Can competitive ability predict structure in experimental plant communities? *Journal of Vegetation Science*, 16, 571–578.

Freeman, E. A. and Moisen, G. G. 2008. A comparison of the performance of thresh-old criteria for binary classification in terms of predicted prevalence and kappa. *Ecological Modelling*, 217, 48–58.

Friedl, M. A., Sulla-Menashe, D., Tan, B., et al. 2010. MODIS Collection 5 global land cover: algorithm refinements and characterization of new datasets. *Remote Sensing of Environment*, 114, 168–182.

Friedman, J. H. 2001. Greedy function approximation: a gradient boosting machine. *Annals of Statistics*, 29, 1189–1232.

Friedman, J. H. 2002. Stochastic gradient boosting. *Computational Statistics and Data Analysis*, 38, 367–378.

Friedman, J. H. and Meulman, J. J. 2003. Multiple additive regression trees with application in epidemiology. *Statistics in Medicine*, 22, 1365–1381.

Friedman, J. H., Hastie, T. J. and Tibshirani, R. 2000. Additive logistic regression: a statistical view of boosting. *Annals of Statistics*, 28, 337–374.

Gallagher, R. V., Beaumont, L. J., Hughes, L. and Leishman, M. R. 2010. Evidence for climatic niche and biome shifts between native and novel ranges in plant species introduced to Australia. *Journal of Ecology*, 98, 790–799.

Gallant, D., Slough, B. G., Reid, D. G. and Berteaux, D. 2012. Arctic fox versus red fox in the warming Arctic: four decades of den surveys in north Yukon. *Polar Biology*, 35, 1421–1431.

Gallien, L., Münkemüller, T., Albert, C. H., Boulangeat, I. and Thuiller, W. 2010. Predicting potential distributions of invasive species: where to go from here? *Diversity and Distributions*, 16, 331–342.

Gallien, L., Douzet, R., Pratte, S., Zimmermann, N. E. and Thuiller, W. 2012. Invasive species distribution models: how violating the equilibrium assumption can create new insights? *Global Ecology and Biogeography*, 21, 1126–1136.

Gallien, L., Mazel, F., Lavergne, S., et al. 2015. Contrasting the effects of environment, dispersal and biotic interactions to explain the distribution of invasive plants in alpine communities. *Biological Invasions*, 17, 1407–1423.

Gao, B. C. 1996. NDWI: A normalized difference water index for remote sensing of vegetation liquid water from space. *Remote Sensing of Environment,* 58, 257–266.

Gause, G. F. 1936. *The Struggle for Existence*. Baltimore, MD: Williams and Wilkins.

Gehrig-Fasel, J., Guisan, A. and Zimmermann, N. E. 2007. Tree line shifts in the Swiss Alps: climate change or land abandonment? *Journal of Vegetation Science*, 18, 571–582.

Gelfand, A. E., Silander, J. A., Wu, S., et al. 2006. Explaining species distribution patterns through hierarchical modeling. *Bayesian Analysis*, 1, 41–92.

Gellrich, M. and Zimmermann, N. E. 2007. Investigating the regional-scale pattern of agricultural land abandonment in the Swiss mountains: a spatial statistical modelling approach. *Landscape and Urban Planning*, 79, 65–76.

Goldewijk, K. K. 2001. Estimating global land use change over the past 300 years: the HYDE Database. *Global Biogeochemical Cycles*, 15, 417–433.

Golding, N. and Purse, B. V. 2016. Fast and flexible Bayesian species distribution modelling using Gaussian processes. *Methods in Ecology and Evolution*.

Gotelli, N. J., Graves, G. R. and Rahbek, C. 2010. Macroecological signals of species interactions in the Danish avifauna. *Proceedings of the National Academy of Sciences*, 107, 5030–5035.

Grace, J. B. and Wetzel, R. G. 1981. Habitat partitioning and competitive displacement in Cattails (*Typha*): experimental field studies. *American Naturalist*, 118, 463–474.

Graham, C. H., Ferrier, S., Huettman, F., Moritz, C. and Peterson, A. T. 2004a. New developments in museum-based informatics and applications in biodiversity analysis. *Trends in Ecology and Evolution*, 19, 497–503.

Graham, C. H., Ron, S. R., Santos, J. C., Schneider, C. J. and Moritz, C. 2004b. Integrating phylogenetics and environmental niche models to explore speciation mechanisms in dendrobatid frogs. *Evolution*, 58, 1781–1793.

Grant, P. R. and Grant, B. R. 2009. The secondary contact phase of allopatric speciation in Darwin's finches. *Proceedings of the National Academy of Sciences of the United States of America*, 106, 20141–20148.

Gravel, D., Massol, F., Canard, E., Mouillot, D. and Mouquet, N. 2011. Trophic theory of island biogeography. *Ecology Letters*, 14, 1010–1016.

Graves, G. R. and Rahbek, C. 2005. Source pool geometry and the assembly of continental avifaunas. *Proceedings of the National Academy of Sciences of the United States of America*, 102, 7871–7876.

Green, R. H. 1971. Multivariate statistical approach to Hutchinsonian niche: bivalve molluscs of central Canada. *Ecology*, 52, 543–556.

Green, R. H. 1979. *Sampling Design and Statistical Methods for Environmental Biologists.* New York: John Wiley.

Grenouillet, G., Buisson, L., Casajus, N. and Lek, S. 2011. Ensemble modelling of species distribution: the effects of geographical and environmental ranges. *Ecography*, 34, 9–17.

Grinnell, J. 1917. The niche-relationships of the California Trasher. *Auk*, 34, 131–135.

Guillera-Arroita, G., Lahoz-Monfort, J. J. and Elith, J. 2014. Maxent is not a presence–absence method: a comment on Thibaud et al. *Methods in Ecology and Evolution*, 5, 1192–1197.

Guisan, A. and Harrell, F. E. 2000. Ordinal response regression models in ecology. *Journal of Vegetation Science*, 11, 617–626.

Guisan, A. and Hofer, U. 2003. Predicting reptile distributions at the mesoscale: relation to climate and topography. *Journal of Biogeography*, 30, 1233–1243.

Guisan, A. and Rahbek, C. 2011. SESAM: a new framework integrating macroecological and species distribution models for predicting spatio-temporal patterns of species assemblages. *Journal of Biogeography*, 38, 1433–1444.

Guisan, A. and Theurillat, J. P. 2000. Equilibrium modeling of alpine plant distribution: how far can we go? *Phytocoenologia*, 30, 353–384.

Guisan, A. and Thuiller, W. 2005. Predicting species distribution: offering more than simple habitat models. *Ecology Letters*, 8, 993–1009.

Guisan, A. and Zimmermann, N. E. 2000. Predictive habitat distribution models in ecology. *Ecological Modelling*, 135, 147–186.

Guisan, A., Weiss, S. B. and Weiss, A. D. 1999. GLM versus CCA spatial modeling of plant species distribution. *Plant Ecology*, 143, 107–122.

Guisan, A., Edwards, J., Thomas, C. and Hastie, T. 2002. Generalized linear and generalized additive models in studies of species distributions: setting the scene. *Ecological Modelling*, 157, 89–100.

Guisan, A., Broennimann, O., Engler, R., et al. 2006a. Using niche-based models to improve the sampling of rare species. *Conservation Biology*, 20, 501–511.

Guisan, A., Lehmann, A., Ferrier, S., et al. 2006b. Making better biogeographical predictions of species' distributions. *Journal of Applied Ecology*, 43, 386–392.

Guisan, A., Graham, C. H., Elith, J., Huettmann, F. and NCEAS Species Distribution Modelling Group 2007a. Sensitivity of predictive species distribution models to change in grain size. *Diversity and Distributions*, 13, 332–340.

Guisan, A., Zimmermann, N. E., Elith, J., et al. 2007b. What matters for predicting the occurrences of trees: techniques, data, or species' characteristics? *Ecological Monographs*, 77, 615–630.

Guisan, A., Petitpierre, B., Broennimann, O., et al. 2012. Response to comment on "Climatic niche shifts are rare among terrestrial plant invaders". *Science*, 338, 193.

Guisan, A., Tingley, R., Baumgartner, J. B., et al. 2013. Predicting species distributions for conservation decisions. *Ecology Letters*, 16, 1424–1435.

Guisan, A., Petitpierre, B., Broennimann, O., Daehler, C. and Kueffer, C. 2014. Unifying niche shift studies: insights from biological invasions. *Trends in Ecology and Evolution*, 29, 260–269.

Hair, J. F., Black, W. C., Babin, B. J., Anderson, R. E. and Tatham, R. L. 2006. *Multivariate Data Analysis*. Upper Saddle River, NJ: Pearson Prentice Hall.

Hakkarainen, H., Mykra, S., Kurki, S., Tornberg, R. and Jungell, S. 2004. Competitive interactions among raptors in boreal forests. *Oecologia*, 141, 420–424.

Halvorsen, R., Mazzoni, S., Bryn, A. and Bakkestuen, V. 2015. Opportunities for improved distribution modelling practice via a strict maximum likelihood interpretation of MaxEnt. *Ecography*, 38, 172–183.

Hanberry, B. B., He, H. S. and Palik, B. J. 2012. Pseudoabsence generation strategies for species distribution models. *Plos One*, 7.

Hansen, M. C., Defries, R. S., Townshend, J. R. G. and Sohlberg, R. 2000. Global land cover classification at 1 km spatial resolution using a classification tree approach. *International Journal of Remote Sensing*, 21, 1331–1364.

Hansen, M. C., DeFries, R. S., Townshend, J. R. G., et al. 2002. Towards an operational MODIS continuous field of percent tree cover algorithm: examples using AVHRR and MODIS data. *Remote Sensing of Environment*, 83, 303–319.

Hanski, I. and Gilpin, M. E. 1997. *Metapopulation Biology*, San Diego, CA: Academic Press.

Hanspach, J., Kühn, I., Pompe, S. and Klotz, S. 2010. Predictive performance of plant species distribution models depends on species traits. *Perspectives in Plant Ecology, Evolution and Systematics*, 12, 219–225.

Hanssen, A. J. and Kuipers, W. J. 1965. On the relationship between the frequency of rain and various meteorological parameters. *Meded Verhand*, 81, 2–15.

Harrell, F. E. 2001. *Regression Modeling Strategies: With Applications to Linear Models, Logistic Regression, and Survival Analysis*. Berlin: Springer.

Harrell, F. E., Lee, K. L. and Mark, D. B. 1996. Multivariable prognostic models: Issues in developing models, evaluating assumptions and adequacy, and measuring and reducing errors. *Statistics in Medicine*, 15, 361–387.

Harris, D. J. 2015. Generating realistic assemblages with a joint species distribution model. *Methods in Ecology and Evolution*, 6, 465–473.

Harte, J. 2011. *Maximum Entropy and Ecology: A Theory of Abundance, Distribution, and Energetics*. London: Oxford University Press.

Hastie, T. and Tibishirani, R. 1986. Generalized additive models. *Statistical Science*, 1, 297–318.

Hastie, T. J. and Tibishirani, R. 1990. *Generalized Additive Models*, London: Chapman and Hall.

Hastie, T., Tibishirani, R. and Buja, A. 1994. Flexible discriminant analysis by optimal scoring. *Journal of the American Statistical Association*, 89(428), 1255–1270.

Hastie, T., Tibishirani, R. and Friedman, J. 2009. *The Elements of Statistical Learning: Data Mining, Inference, and Prediction*. Berlin Springer.

Hastings, D. A., Dunbar, P. K., Elphingstone, G. M., et al. 1999. *The Global Land One-kilometer Base Elevation (GLOBE) Digital Elevation Model*, Version 1.0. Boulder, CO: National Oceanic and Atmospheric Administration, National Geophysical Data Center.

Hausser, J. (ed.) 1995. *Säugetiere der Schweitz. Atlas des Mammifères de Suisse. Mammiferi della Svizzera*. Basel: Birkhäuser Verlag.

Hautier, Y., Randin, C. F., Stocklin, J. and Guisan, A. 2009. Changes in reproductive investment with altitude in an alpine plant. *Journal of Plant Ecology*, 2, 125–134.

Hawkins, B. A. 2012a. Are multiple regression models of spatially structured data to be trusted? *Journal of Biogeography*, 39, 998.

Hawkins, B. A. 2012b. Eight (and a half) deadly sins of spatial analysis. *Journal of Biogeography*, 39, 1–9.

Hector, A., von Felten, S., Hautier, Y., Weilenmann, M. and Bruelheide, H. 2012. Effects of dominance and diversity on productivity along Ellenberg's experimental water table gradients. *Plos One*, 7, e43358.

Heikkinen, R. K., Luoto, M., Araujo, M. B., et al. 2006. Methods and uncertainties in bioclimatic envelope modelling under climate change. *Progress in Physical Geography*, 30, 751–777.

Heikkinen, R. K., Luoto, M., Virkkala, R., Pearson, R. G. and Korber, J. H. 2007. Biotic interactions improve prediction of boreal bird distributions at macroscales. *Global Ecology and Biogeography*, 16, 754–763.

Heinanen, S. and von Numers, M. 2009. Modelling species distribution in complex environments: an evaluation of predictive ability and reliability in five shorebird species. *Diversity and Distributions*, 15, 266–279.

Heinrichs, J. A., Bender, D. J., Gummer, D. L. and Schumaker, N. H. 2010. Assessing critical habitat: evaluating the relative contribution of habitats to population persistence. *Biological Conservation*, 143, 2229–2237.

Heller, H. C. and Gates, D. 1971. Altitudinal zonation of chipmunks (*Eutamias*): energy budgets. *Ecology*, 52, 424–443.

Hernandez, P. A., Graham, C. H., Master, L. L. and Albert, D. L. 2006. The effect of sample size and species characteristics on performance of different species distribution modeling methods. *Ecography*, 29, 773–785.

Hijmans, R., Phillips, S., Leathwick, J. and Elith, J. 2013. dismo: Species distribution modeling. *R package version 0.9–..* http://CRAN.R-project.org/package=dismo.

Hijmans, R. J. 2012. Cross-validation of species distribution models: removing spatial sorting bias and calibration with a null model. *Ecology*, 93, 679–688.

Hijmans, R. J., Cameron, S. E., Parra, J. L., Jones, P. G. and Jarvis, A. 2005. Very high resolution interpolated climate surfaces for global land areas. *International Journal of Climatology*, 25, 1965–1978.

Hintikka, V. 1963. Über das Großklima einiger Pflanzenareale in zwei Klimakoordinatensystemen dargestellt. *Annales Botanici Societatis Zoologicae Botanicae Fennicae 'Vanamo'.* 34, 1–64.

Hirzel, A. and Guisan, A. 2002. Which is the optimal sampling strategy for habitat suitability modelling. *Ecological Modelling*, 157, 331–341.

Hirzel, A., Helfer, V. and Metral, F. 2001. Assessing habitat-suitability models with a virtual species. *Ecological Modelling*, 145, 111–121.

Hirzel, A. H. and Arlettaz, R. 2003. Modeling habitat suitability for complex species distributions by environmental-distance geometric mean. *Environmental Management*, 32, 614–623.

Hirzel, A. H., Hausser, J., Chessel, D. and Perrin, N. 2002a. Ecological niche factor analysis: how to compute habitat suitability maps without absence data? *Ecology*, 83, 2027–2036.

Hirzel, A. H., Hausser, J. and N., P. 2002b. *Biomapper 2.0*, Bern: Division of Conservation Biology.

Hirzel, A. H., Posse, B., Oggier, P. A., et al. 2004. Ecological requirements of reintroduced species and the implications for release policy: the case of the bearded vulture. *Journal of Applied Ecology*, 41, 1103–1116.

Hirzel, A. H., Le Lay, G., Helfer, V., Randin, C. and Guisan, A. 2006. Evaluating the ability of habitat suitability models to predict species presences. *Ecological Modelling*, 199, 142–152.

Hoerl, A. E. and Kennard, R. W. 1970. Ridge regression: biased estimation for nonorthogonal problems. *Technometrics*, 12, 55–67.

Holdridge, L. R. 1967. *Life Zone Ecology*, San Jose, Costa Rica: Tropical Science Center.

Homer, C. H., Fry, J. A. and Barnes, C. A. 2012. The national land cover database. *US Geological Survey Fact Sheet 2012–3020*. Reston, VA: USGS. Available at: http://pubs. usgs. gov/fs/2012/3020/fs2012-3020.pdf.

Hooten, M. B., Larsen, D. R. and Wikle, C. K. 2003. Predicting the spatial distribution of ground flora on large domains using a hierarchical Bayesian model. *Landscape Ecology*, 18, 487–502.

Hortal, J., De Marco Jr, P., Santos, A. and Diniz-Filho, J. A. F. 2012. Integrating biogeographical processes and local community assembly. *Journal of Biogeography*, 39, 627–628.

Hoskin, C. J., Higgie, M., McDonald, K. R. and Moritz, C. 2005. Reinforcement drives rapid allopatric speciation. *Nature*, 437, 1353–1356.

Howard, C., Stephens, P. A., Pearce-Higgins, J. W., Gregory, R. D. and Willis, S. G. 2014. Improving species distribution models: the value of data on abundance. *Methods in Ecology and Evolution*, 5, 506–513.

Howe, A. and Chain, P. S. 2015. Challenges and opportunities in understanding microbial communities with metagenome assembly (accompanied by IPython Notebook tutorial). *Frontiers in Microbiology*, 6, 678.

Hubbell, S. P. 2001. *The Unified Neutral Theory of Biodiversity and Biogeography*. Princeton, NJ: Princeton University Press.

Huete, A. R. 1988. A soil-adjusted vegetation index (SAVI). *Remote Sensing of Environment*, 25, 295–309.

Hugall, A., Moritz, C., Moussalli, A. and Stanisic, J. 2002. Reconciling paleodistribution models and comparative phylogeography in the Wet Tropics rainforest land snail *Gnarosophia bellendenkerensis* (Brazier 1875). *Proceedings of the National Academy of Sciences*, 99, 6112–6117.

Hui, F. K., Warton, D. I. and Foster, S. D. 2015. Multi-species distribution modeling using penalized mixture of regressions. *The Annals of Applied Statistics*, 9, 866–882.

Huntley, B., Green, R. E., Collingham, Y. C., et al. 2004. The performance of models relating species geographical distributions to climate is independent of trophic level. *Ecology Letters*, 7, 417–426.

Hurtt, G. C., Chini, L. P., Frolking, S., et al. 2011. Harmonization of land-use scenarios for the period 1500–2100: 600 years of global gridded annual land-use transitions, wood harvest, and resulting secondary lands. *Climatic Change*, 109, 117–161.

Huston, M. A. 2002. Introductory essay: critical issues for improving predictions. In Scott, J. M., Heglund, P. J., Morrison, M. L., et al. (eds.) *Predicting Species Occurrences: Issues of Accuracy and Scale*. Covelo, California: Island Press, pp. 7–21.

Hutchinson, G. E. 1957. Population studies. Animal ecology and demography: concluding remarks. *Cold Spring Harbor Symposia on Quantitative Biology*, 22, 415–427.

Hutchinson, M. F. 1995. Interpolating mean rainfall using thin-plate smoothing splines. *International Journal of Geographical Information Systems*, 9, 385–403.

Ings, T. C., Montoya, J. M., Bascompte, J., et al. 2009. Ecological networks: beyond food webs. *Journal of Animal Ecology*, 78, 253–269.

IPCC 2007. *Climate Change 2007: The Physical Science Basis. Contribution of Working Group I to the Fourth Assessment Report of the Intergovernmental Panel on Climate Change.* Cambridge, UK and New York, NY: Cambridge University Press.

IPCC 2013. *Climate Change 2013: The Physical Science Basis. Contribution of Working Group I to the Fifth Assessment Report of the Intergovernmental Panel on Climate Change.* Cambridge, UK and New York, NY: Cambridge University Press.

IUCN 2001. *IUCN Red List Categories and Criteria: Version 3.1. IUCN Species Survival Commissio.n* Gland, Switzerland and Cambridge, UK: IUCN.

Iverson, L. R. and Prasad, A. M. 1998. Predicting abundance of 80 tree species following climate change in the eastern United States. *Ecological Monographs*, 68, 465–485.

Iverson, L. R. and Prasad, A. M. 2001. Potential changes in tree species richness and forest community types following climate change. *Ecosystems*, 4, 186–199.

Iverson, L. R., Prasad, A. M., Matthews, S. N. and Peters, M. 2008. Estimating potential habitat for 134 eastern US tree species under six climate scenarios. *Forest Ecology and Management*, 254, 390–406.

Ives, A. R. and Helmus, M. R. 2011. Generalized linear mixed models for phylogenetic analyses of community structure. *Ecological Monographs*, 81, 511–525.

Jaberg, C. and Guisan, A. 2001. Modelling the distribution of bats in relation to landscape structure in a temperate mountain environment. *Journal of Applied Ecology*, 38, 1169–1181.

Jackson, S. T. and Overpeck, J. T. 2000. Responses of plant populations and communities to environmental changes of the late Quaternary. *Paleobiology*, 26, 194–220.

Jackson, S. T. and Williams, J. W. 2004. Modern analogs in Quaternary paleoecology: here today, gone yesterday, gone tomorrow? *Annual Review of Earth and Planetary Sciences*, 32, 495–537.

Jalas, J. and Suominen, J. 1972–1996. *The Atlas Florae Europaeae.* Helsinki, Committee for Mapping the Flora of Europe and Societas Biologica Fennica Vanamo.

James, G., Witten, D., Hastie, T. and Tibshirani, R. 2013. *An Introduction to Statistical Learning.* Berlin: Springer.

Jamil, T., Ozinga, W. A., Kleyer, M. and ter Braak, C. J. 2013. Selecting traits that explain species–environment relationships: a generalized linear mixed model approach. *Journal of Vegetation Science*, 24, 988–1000.

Jardine, N. 1972. Computational methods in the study of plant distributions. In Valentine D. H. (ed.), *Taxonomy, Phytogeography and Evolution.* London: Academic Press, pp. 381–393.

Jeltsch, F., Moloney, K. A., Schurr, F. M., Kochy, M. and Schwager, M. 2008. The state of plant population modelling in light of environmental change. *Perspectives in Plant Ecology Evolution and Systematics*, 9, 171–189.

Jensen, O. P., Seppelt, R., Miller, T. J. and Bauer, L. J. 2005. Winter distribution of blue crab Callinectes sapidus in Chesapeake Bay: application and cross-validation of a two-stage generalized additive model. *Marine Ecology-Progress Series*, 299, 239–255.

Jimenez-Valverde, A. and Lobo, J. M. 2007. Threshold criteria for conversion of probability of species presence to either-or presence–absence. *Acta Oecologica – International Journal of Ecology*, 31, 361–369.

Jimenez-Valverde, A., Lobo, J. M. and Hortal, J. 2009. The effect of prevalence and its interaction with sample size on the reliability of species distribution models. *Community Ecology*, 10, 196–205.

Jimenez-Alfaro, B., Draper, D. and Nogues-Bravo, D. 2012. Modeling the potential area of occupancy at fine resolution may reduce uncertainty in species range estimates. *Biological Conservation*, 147, 190–196.

Johnson, C. J. and Gillingham, M. P. 2005. An evaluation of mapped species distribution models used for conservation planning. *Environmental Conservation*, 32, 117–128.

Johnson, C. J. and Gillingham, M. P. 2008. Sensitivity of species-distribution models to error, bias, and model design: an application to resource selection functions for woodland caribou. *Ecological Modelling*, 213, 143–155.

Johnson, J. B. and Omland, K. S. 2004. Model selection in ecology and evolution. *Trends in Ecology and Evolution*, 19, 101–108.

Johnston, T. H. 1924. The relation of climate to the spread of prickly pear. *Transactions of the Royal Society of South Australia*, 48, 269–295.

Jones, H. G. 1992. *Plants and Microclimate. A Quantitative Approach to Environmental Plant Physiology*, Cambridge, UK: Cambridge University Press.

Ju, J. C., Kolaczyk, E. D. and Gopal, S. 2003. Gaussian mixture discriminant analysis and sub-pixel land cover characterization in remote sensing. *Remote Sensing of Environment*, 84, 550–560.

Kadmon, R., Farber, O. and Danin, A. 2004. Effect of roadside bias on the accuracy of predictive maps produced by bioclimatic models. *Ecological Applications*, 14, 401–413.

Kaplan, J. O., Krumhardt, K. M. and Zimmermann, N. E. 2009. The prehistoric and preindustrial deforestation of Europe. *Quaternary Science Reviews*, 28, 3016–3034.

Kauth, R. J. and Thomas, G. S. 1976. The tasselled cap: a graphic description of the spectral-temporal development of agricultural crops as seen by Landsat. *LARS Symposia*, 159.

Kearney, M. 2006. Habitat, environment and niche: what are we modelling? *Oikos*, 115, 186–191.

Kearney, M. and Porter, W. P. 2004. Mapping the fundamental niche: physiology, climate, and the distribution of a nocturnal lizard. *Ecology*, 85, 3119–3131.

Kearney, M. and Porter, W. 2009. Mechanistic niche modelling: combining physiological and spatial data to predict species' ranges. *Ecology Letters*, 12, 334–350.

Keil, P., Belmaker, J., Wilson, A. M., Unitt, P. and Jetz, W. 2013. Downscaling of species distribution models: a hierarchical approach. *Methods in Ecology and Evolution*, 4, 82–94.

Keil, P., Wilson, A. M. and Jetz, W. 2014. Uncertainty, priors, autocorrelation and disparate data in downscaling of species distributions. *Diversity and Distributions*, 20, 797–812.

Keith, D. A., Akcakaya, H. R., Thuiller, W., et al. 2008. Predicting extinction risks under climate change: coupling stochastic population models with dynamic bioclimatic habitat models. *Biology Letters*, 4, 560–563.

Kerr, J. T. and Ostrovsky, M. 2003. From space to species: ecological applications for remote sensing. *Trends in Ecology and Evolution*, 18, 299–305.

Kharouba, H. M., McCune, J. L., Thuiller, W. and Huntley, B. 2013. Do ecological differences between taxonomic groups influence the relationship between species' distributions and climate? A global meta-analysis using species distribution models. *Ecography*, 36, 657–664.

King, A. 1991. Translating models across scales in a landscape. In Turner, M. G. and Gardner, R. H. (eds.), *Quantitative Methods in Landscape Ecology*. New York, NY: Springer, pp. 479–517.

Kissling, W. D., Dormann, C. F., Groeneveld, J., et al. 2012. Towards novel approaches to modelling biotic interactions in multispecies assemblages at large spatial extents. *Journal of Biogeography*, 39, 2163–2178.

Knutti, R. 2010. The end of model democracy? *Climatic Change*, 102, 395–404.

Knutti, R., Furrer, R., Tebaldi, C., Cermak, J. and Meehl, G. A. 2010. Challenges in combining projections from multiple climate models. *Journal of Climate*, 23, 2739–2758.

Körner, C. 2003. *Alpine Plant Life: Functional Plant Ecology of High Mountain Ecosystems; With 47 Tables*. Berlin: Springer.

Körner, C. 2007. The use of "altitude" in ecological research. *Trends in Ecology and Evolution*, 22, 569–574.

Körner, C. and Paulsen, J. 2004. A world-wide study of high altitude treeline temperatures. *Journal of Biogeography*, 31, 713–732.

Kremen, C., Cameron, A., Moilanen, A., et al. 2008. Aligning conservation priorities across taxa in Madagascar with high-resolution planning tools. *Science*, 320, 222–226.

Kühn, I. 2007. Incorporating spatial autocorrelation may invert observed patterns. *Diversity and Distributions*, 13, 66–69.

Kühn, I. and Dormann, C. F. 2012. Less than eight (and a half) misconceptions of spatial analysis. *Journal of Biogeography*, 39, 995–998.

Kühn, I., Nobis, M. P. and Durka, W. 2009. Combining spatial and phylogenetic eigenvector filtering in trait analysis. *Global Ecology and Biogeography*, 18, 745–758.

Lahoz-Monfort, J. J., Guillera-Arroita, G. and Wintle, B. A. 2014. Imperfect detection impacts the performance of species distribution models. *Global Ecology and Biogeography*, 23, 504–515.

Landis, J. R. and Koch, G. G. 1977. The measurement of observer agreement for categorical data. *Biometrics*, 159–174.

Larsen, P. E., Scott, N., Post, A. F., et al. 2015. Satellite remote sensing data can be used to model marine microbial metabolite turnover. *The ISME Journal*, 9, 166–179.

Larson, M. A., Thompson, F. R., Millspaugh, J. J., Dijak, W. D. and Shifley, S. R. 2004. Linking population viability, habitat suitability, and landscape simulation models for conservation planning. *Ecological Modelling*, 180, 103–118.

Lassueur, T., Joost, S. and Randin, C. F. 2006. Very high resolution digital elevation models: Do they improve models of plant species distribution? *Ecological Modelling*, 198, 139–153.

Lavergne, S. and Molofsky, J. 2007. Increased genetic variation and evolutionary potential drive the success of an invasive grass. *Proceedings of the National Academy of Sciences of the United States of America*, 104, 3883–3888.

Lawler, J. J., White, D., Neilson, R. P. and Blaustein, A. R. 2006. Predicting climate-induced range shifts: model differences and model reliability. *Global Change Biology*, 12, 1568–1584.

Lawler, J. J., Shafer, S. L., White, D., et al. 2009. Projected climate-induced faunal change in the Western Hemisphere. *Ecology*, 90, 588–597.

Le Cessie, S. and van Houwelingen, J. C. 1992. Ridge estimators in logistic regression. *Applied Statistics*, 41, 191–201.

Leathwick, J., Elith, J. and Hastie, T. 2006. Comparative performance of generalized additive models and multivariate adaptive regression splines for statistical modelling of species distributions. *Ecological Modelling*, 199, 188–196.

Leathwick, J., Moilanen, A., Francis, M., et al. 2008. Novel methods for the design and evaluation of marine protected areas in offshore waters. *Conservation Letters*, 1, 91–102.

Leathwick, J. R. 1998. Are New Zealand's *Nothofagus* species in equilibrium with their environment? *Journal of Vegetation Science*, 9, 719–732.

Leathwick, J. R. 2002. Intra-generic competition among *Nothofagus* in New Zealand's primary indigenous forests. *Biodiversity and Conservation*, 11, 2177–2187.

Leathwick, J. R. and Austin, M. P. 2001. Competitive interactions between tree species in New Zealand's old-growth indigenous forests. *Ecology*, 82, 2560–2573.

Leathwick, J. R., Rowe, D., Richardson, J., Elith, J. and Hastie, T. 2005. Using multivariate adaptive regression splines to predict the distributions of New Zealand's freshwater diadromous fish. *Freshwater Biology*, 50, 2034–2052.

Legendre, P. 1993. Spatial autocorrelation: trouble or new paradigm? *Ecology*, 74, 1659–1673.

Legendre, P., Dale, M. R., Fortin, M. J., et al. 2002. The consequences of spatial structure for the design and analysis of ecological field surveys. *Ecography*, 25, 601–615.

Legendre, P., Borcard, D. and Peres-Neto, P. R. 2005. Analyzing beta diversity: partitioning the spatial variation of community composition data. *Ecological Monographs*, 75, 435–450.

Lehmann, A., Overton, J. M. and Leathwick, J. R. 2002. GRASP: generalized regression analysis and spatial prediction. *Ecological Modelling*, 157, 189–207.

Leibold, M. A. 1995. The niche concept revisited: mechanistic models and community context. *Ecology*, 76, 1371–1382.

Leibold, M. A. and McPeek, M. A. 2006. Coexistence of the niche and neutral perspectives in community ecology. *Ecology*, 87, 1399–1410.

Lek, S., Delacoste, M., Baran, P., et al. 1996. Application of neural networks to modelling nonlinear relationships in ecology. *Ecological Modelling*, 90, 39–52.

Lenoir, J. and Svenning, J. C. 2015. Climate-related range shifts–a global multidimensional synthesis and new research directions. *Ecography*, 38, 15–28.

Lenoir, J., Gégout, J.-C., Marquet, P. A., de Ruffray, P. and Brisse, H. 2008. A significant upward shift in plant species optimum elevation during the 20th century.
Science, 320, 1768–1771.

Lenoir, J., Gégout, J.-C., Guisan, A., et al. 2010. Going against the flow: potential
mechanisms for unexpected downslope range shifts in a warming climate.
Ecography, 33, 295–303.

Lima-Ribeiro, M. S., Nogues-Bravo, D., Marske, K. A., et al. 2012. Human arrival
scenarios have a strong influence on interpretations of the late Quaternary
extinctions. *Proceedings of the National Academy of Sciences of the United States of
America*, 109, E2409–E2410.

Linacre, J. 2008. The expected value of a point-biserial (or similar) correlation. *Rasch
Measurement Transactions*, 22, 1154–1157.

Linder, H. P., Bykova, O., Dyke, J., et al. 2012. Biotic modifiers, environmental modu-
lation and species distribution models. *Journal of Biogeography*, 39, 2179–2190.

Link, W. A. and Barker, R. J. 2006. Model weights and the foundations of multimodel
inference. *Ecology*, 87, 2626–2635.

Lira-Noriega, A. and Peterson, A. T. 2014. Range-wide ecological niche comparisons
of parasite, hosts and dispersers in a vector-borne plant parasite system. *Journal
of Biogeography*, 41, 1664–1673.

Lischke, H., Zimmermann, N. E., Bolliger, J., Rickebusch, S. and Loffler, T. J. 2006.
TreeMig: A forest-landscape model for simulating spatio-temporal patterns
from stand to landscape scale. *Ecological Modelling*, 199, 409–420.

Lischke, H., Löffler, T. J., Thornton, P. E. and Zimmermann, N. E. 2007. Model
up-scaling in landscape research. In Kienast, F., Wildi, O. and Ghosh, S.
(eds), *A Changing World: Challenges for Landscape Research*. Dordrecht, The
Netherlands: Springer.

Liu, C., White, M. and Newell, G. 2013. Selecting thresholds for the prediction of species occurrence with presence-only data. *Journal of Biogeography*, 40, 778–789.

Liu, C. R., Berry, P. M., Dawson, T. P. and Pearson, R. G. 2005. Selecting thresholds
of occurrence in the prediction of species distributions. *Ecography*, 28, 385–393.

Liu, C. R., White, M. and Newell, G. 2011. Measuring and comparing the accuracy of species distribution models with presence–absence data. *Ecography*, 34,
232–243.

Loarie, S. R., Duffy, P. B., Hamilton, H., et al. 2009. The velocity of climate change.
Nature, 462, 1052-U111.

Lobo, J. M., Jimenez-Valverde, A. and Real, R. 2008. AUC: a misleading measure of the performance of predictive distribution models. *Global Ecology and
Biogeography*, 17, 145–151.

Lobo, J. M. and Tognelli, M. F. 2011. Exploring the effects of quantity and location of
pseudo-absences and sampling biases on the performance of distribution models with limited point occurrence data. *Journal for Nature Conservation*, 19, 1–7.

Loiselle, B. A., Howell, C. A., Graham, C. H., et al. 2003. Avoiding pitfalls of using
species distribution models in conservation planning. *Conservation Biology*, 17,
1591–1600.

Loiselle, B. A., Jorgensen, P. M., Consiglio, T., et al. 2008. Predicting species distributions from herbarium collections: does climate bias in collection sampling
influence model outcomes? *Journal of Biogeography*, 35, 105–116.

Lomba, A., Pellissier, L., Randin, C., et al. 2010. Overcoming the rare species modelling paradox: a novel hierarchical framework applied to an Iberian endemic plant. *Biological Conservation*, 143, 2647–2657.

Lomolino, M. V., Riddle, B. R., Whittaker, R. J. and Brown, J. H. 2010. *Biogeography*. Sunderland, MA: Sinauer Associates.

Lorenzen, E. D., Nogues-Bravo, D., Orlando, L., et al. 2011. Species-specific responses of Late Quaternary megafauna to climate and humans. *Nature*, 479, 359–U195.

Lortie, C. J., Brooker, R. W., Choler, P., et al. 2004. Rethinking plant community theory. *Oikos*, 107, 433–438.

Lowry, J., Ramsey, R. D., Thomas, K., et al. 2007. Mapping moderate-scale landcover over very large geographic areas within a collaborative framework: a case study of the Southwest Regional Gap Analysis Project (SWReGAP). *Remote Sensing of Environment*, 108, 59–73.

Luoto, M., Heikkinen, R. K., Poyry, J. and Saarinen, K. 2006. Determinants of the biogeographical distribution of butterflies in boreal regions. *Journal of Biogeography*, 33, 1764–1778.

Luoto, M., Virkkala, R. and Heikkinen, R. K. 2007. The role of land cover in bioclimatic models depends on spatial resolution. *Global Ecology and Biogeography*, 16, 34–42.

Lyet, A., Thuiller, W., Cheylan, M. and Besnard, A. 2013. Fine-scale regional distribution modelling of rare and threatened species: bridging GIS Tools and conservation in practice. *Diversity and Distributions*, 19, 651–663.

Lyons, K. G. and Schwartz, M. W. 2001. Rare species loss alters ecosystem function–invasion resistance. *Ecology Letters*, 4, 358–365.

Lyons, K. G., Brigham, C., Traut, B. and Schwartz, M. W. 2005. Rare species and ecosystem functioning. *Conservation Biology*, 19, 1019–1024.

Mac Nally, R. 2000. Regression and model-building in conservation biology, biogeography and ecology: the distinction between – and reconciliation of – "predictive" and "explanatory" models. *Biodiversity and Conservation*, 9, 655–671.

MacArthur, R. H. 1968. The theory of the niche. In Lewontin, R. C. (ed.), *Population Biology and Evolution*. Syracuse, NY: Syracuse University Press.

MacArthur, R. H. 1972. *Geographical Ecology*. New-York: Harper & Row.

Maggini, R., Lehmann, A., Zimmermann, N. E. and Guisan, A. 2006. Improving generalized regression analysis for the spatial prediction of forest communities. *Journal of Biogeography*, 33, 1729–1749.

Maher, S. P., Randin, C. F., Guisan, A. and Drake, J. M. 2014. Pattern-recognition ecological niche models fit to presence-only and presence–absence data. *Methods in Ecology and Evolution*, 5, 761–770.

Maiorano, L., Cheddadi, R., Zimmermar., N. E., et al. 2013. Building the niche through time: using 13,000 years of data to predict the effects of climate change on tree species in Europe. *Global Ecology and Biogeography*, 22, 302–317.

Malanson, G. P., Westman, W. E. and Yan, Y.-L. 1992. Realized versus fundamental niche functions in a model of chaparral response to climatic change. *Ecological Modelling*, 64, 261–277.

Manel, D., Dias, J. M., Buckton, S. T. and Ormerod, S. J. 1999a. Alternative methods for predicting species distribution: an illustration with Himalayan river birds. *Journal of Applied Ecology*, 36, 734–747.

Manel, S., Dias, J.-M. and Ormerod, S. J. 1999b. Comparing discriminant analysis, neural networks and logistic regression for predicting species distributions: a case study with a Himalayan river bird. *Ecological Modelling*, 120, 337–347.

Manel, S., Williams, H. C. and Ormerod, S. J. 2001. Evaluating presence–absence models in ecology: the need to account for prevalence. *Journal of Applied Ecology*, 38, 921–931.

Manly, B. F. J. (ed.) 2006. *Randomization, Bootstrap and Monte Carlo Methods in Biology. 3rd edition.* London: Chapman & Hall/CRC.

Marcelli, M., Polednik, L., Polednikova, K. and Fusillo, R. 2012. Land use drivers of species re-expansion: inferring colonization dynamics in Eurasian otters. *Diversity and Distributions.*

Margules, C. R. and Austin, M. P. 1991. *Nature Conservation: Cost Effective Biological Survey and Data Analysis.* Canberra, Australia: CSIRO.

Marion, G., McInerny, G. J., Pagel, J., et al. 2012. Parameter and uncertainty estimation for process-oriented population and distribution models: data, statistics and the niche. *Journal of Biogeography*, 39, 2225–2239.

Marmion, M., Parviainen, M., Luoto, M., Heikkinen, R. K. and Thuiller, W. 2009. Evaluation of consensus methods in predictive species distribution modelling. *Diversity and Distributions*, 15, 59–69.

Martínez-Meyer, E., Peterson, A. T. and Hargrove, W. W. 2004. Ecological niches as stable distributional constraints on mammal species, with implications for Pleistocene extinctions and climate change projections for biodiversity. *Global Ecology and Biogeography*, 13, 305–314.

Mateo, R. G., Felicísimo, Á. M., Pottier, J., Guisan, A. and Muñoz, J. 2012. Do stacked species distribution models reflect altitudinal diversity patterns? *Plos One*, 7, e32586.

Mateo, R. G., Broennimann, O., Petitpierre, B., et al. 2015. What is the potential of spread in invasive bryophytes? *Ecography*, 38(5), 480–487.

Mathys, L., Guisan, A., Kellenberger, T. W. and Zimmermann, N. E. 2009. Evaluating effects of spectral training data distribution on continuous field mapping performance. *ISPRS Journal of Photogrammetry and Remote Sensing*, 64, 665–673.

Mauri, A., Davis, B. A. S. and Kaplan, J. O. 2014. The influence of atmospheric circulation on the mid-holocene climate of Europe: a data–model comparison. *Climate of the Past*, 10, 1925–1938.

McCullagh, P. and Nelder, J. A. 1989a. *Generalized Linear Models.* London: Chapman & Hall.

McCullagh, P. and Nelder, J. A. 1989b. *Generalized Linear Models. 2nd edition.* London: Chapman & Hall.

McGill, B. J. 2010. Matters of scale. *Science*, 328, 575–576.

McInerny, G. J. and Purves, D. W. 2011. Fine-scale environmental variation in species distribution modelling: regression dilution, latent variables and neighbourly advice. *Methods in Ecology and Evolution*, 2, 248–257.

McPherson, J. M., Jetz, W. and Rogers, D. J. 2004. The effects of species' range sizes on the accuracy of distribution models: ecological phenomenon or statistical artefact? *Journal of Applied Ecology*, 41, 811–823.

McPherson, J. M., Jetz, W. and Rogers, D. J. 2006. Using coarse-grained occurrence data to predict species distributions at finer spatial resolutions-possibilities and limitations. *Ecological Modelling*, 192, 499–522.

Meehl, G. A., Goddard, L., Murphy, J., et al. 2009. Decadal prediction: can it be skilful? *Bulletin of the American Meteorological Society*, 90, 1467–1485.

Meentemeyer, R. K., Anacker, B. L., Mark, W. and Rizzo, D. M. 2008. Early detection of emerging forest disease using dispersal estimation and ecological niche modeling. *Ecological Applications*, 18, 377–390.

Meier, E. S., Kienast, F., Pearman, P. B., et al. 2010. Biotic and abiotic variables show little redundancy in explaining tree species distributions. *Ecography*, 33, 1038–1048.

Meier, E. S., Edwards, T. C., Kienast, F., Dobbertin, M. and Zimmermann, N. E. 2011. Co-occurrence patterns of trees along macro-climatic gradients and their potential influence on the present and future distribution of *Fagus sylvatica* L. *Journal of Biogeography*, 38, 371–382.

Meier, E. S., Lischke, H., Schmatz, D. R. and Zimmermann, N. E. 2012. Climate, competition and connectivity affect future migration and ranges of European trees. *Global Ecology and Biogeography*, 21, 164–178.

Meller, L., Cabeza, M., Pironon, S., et al. 2014. Ensemble distribution models in conservation prioritization: from consensus predictions to consensus reserve networks. *Diversity and Distributions*, 20, 309–321.

Menard, S. W. 2002. *Applied Logistic Regression Analysis.* Thousand Oaks, CA: Sage.

Merow, C., Smith, M. J., Edwards, T. C., et al. 2014. What do we gain from simplicity versus complexity in species distribution models? *Ecography*, 37, 1267–1281.

Mesgaran, M. B., Cousens, R. D. and Webber, B. L. 2014. Here be dragons: a tool for quantifying novelty due to covariate range and correlation change when projecting species distribution models. *Diversity and Distributions*, 20, 1147–1159.

Meyer, C., Kreft, H., Guralnick, R. and Jetz, W. 2015. Global priorities for an effective information basis of biodiversity distributions. *Nature Communications*, 6.

Meynard, C. N. and Kaplan, D. M. 2013. Using virtual species to study species distributions and model performance. *Journal of Biogeography*, 40, 1–8.

Meynard, C. N. and Quinn, J. F. 2007. Predicting species distributions: a critical comparison of the most common statistical models using artificial species. *Journal of Biogeography*, 34, 1455–1469.

Meyneeke, J. O. 2004. Effects of global climate change on geographic distributions of vertebrates in North Queensland. *Ecological Modelling*, 174, 347–357.

Mitchell, M. S., Lancia, R. A. and Gerwin, J. A. 2001. Using landscape-level data to predict the distribution of birds on a managed forest: effects of scale. *Ecological Applications*, 11, 1692–1708.

Mod, H. K., Scherrer, D., Luoto, M. and Guisan, A. 2016. What we use is not what we know: environmental predictors in plant distribution models. *Journal of Vegetation Science*, 27(6), 1308–1322.

Moisen, G. G. and Frescino, T. S. 2002. Comparing five modelling techniques for predicting forest characteristics. *Ecological Modelling*, 157, 209–225.

Mokany, K., Harwood, T. D., Williams, K. J. and Ferrier, S. 2012. Dynamic macroecology and the future for biodiversity. *Global Change Biology*, 18, 3149–3159.

Moloney, K. A. and Jeltsch, F. 2008. Space matters: novel developments in plant ecology through spatial modelling. *Perspectives in Plant Ecology Evolution and Systematics*, 9, 119–120.

Monserud, R. A. and Leemans, R. 1992. Comparing global vegetation maps with the Kappa statistic. *Ecological Modelling*, 62, 275–293.

Montgomery, D. C. and Peck, E. A. 1982. *Introduction to Linear Regression Analysis.* New York, Wiley.

Morales, J. M., Moorcroft, P. R., Matthiopoulos, J., et al. 2010. Building the bridge between animal movement and population dynamics. *Philosophical Transactions of the Royal Society B: Biological Sciences*, 365, 2289–2301.

Moretti, M., Conedera, M., Moresi, R. and Guisan, A. 2006. Modelling the influence of change in fire regime on the local distribution of a Mediterranean pyrophytic plant species (*Cistus salviifolius*) at its northern range limit. *Journal of Biogeography*, 33, 1492–1502.

Morgan, J. N. and Sonquist, J. A. 1963. Problems in the analysis of survey data, and a proposal. *Journal of the American Statistical Association*, 58, 415–434.

Moritz, C., Patton, J. L., Conroy, C. J., et al. 2008. Impact of a century of climate change on small-mammal communities in Yosemite National Park, USA. *Science*, 322, 261–264.

Mouillot, D., Bellwood, D. R., Baraloto, C., et al. 2013. Rare species support vulnerable functions in high-diversity ecosystems. *PLoS Biology*, 11, e1001569.

Münkemüller, T., De Bello, F., Meynard, C., et al. 2012. From diversity indices to community assembly processes: a test with simulated data. *Ecography*, 35, 468–480.

Münkemüller, T., Boucher, F. C., Thuiller, W. and Lavergne, S. 2015. Phylogenetic niche conservatism: common pitfalls and ways forward. *Functional Ecology*, 29(5), 627–639.

Murphy, A. H. and Winkler, R. L. 1987. A general framework for forecast verification. *Monthly Weather Review*, 115, 1330–1338.

Naujokaitis-Lewis, I. R., Curtis, J. M., Tischendorf, L., et al. 2013. Uncertainties in coupled species distribution–metapopulation dynamics models for risk assessments under climate change. *Diversity and Distributions*, 19, 541–554.

Nenzén, H. K. and Araújo, M. 2011. Choice of threshold alters projections of species range shifts under climate change. *Ecological Modelling*, 222, 3346–3354.

Nenzén, H. K., Swab, R. M., Keith, D. A. and Araújo, M. B. 2012. demoniche: an R-package for simulating spatially-explicit population dynamics. *Ecography*, 35, 577–580.

Nesme, J., Achouak, W., Agathos, S. N., et al. 2016. Back to the future of soil metagenomics. *Frontiers in Microbiology*, 7, 73.

Newbold, T., Reader, T., El-Gabbas, A., et al. 2010. Testing the accuracy of species distribution models using species records from a new field survey. *Oikos*, 119, 1326–1334.

Nieto-Lugilde, D., Lenoir, J., Abdulhak, S., et al. 2015. Tree cover at fine and coarse spatial grains interacts with shade tolerance to shape plant species distributions across the Alps. *Ecography*, 38, 578–589.

Nix, H., McMahon, J. and Mackenzie, D. 1977. Potential areas of production and the future of pigeon pea and other grain legumes in Australia. In Wallis, E. Amp, S., and Whiteman, P. C. (eds), *The Potential for Pigeon Pea in Australia. Proceedings of Pigeon Pea (Cajanus cajan (L.) Millsp.) Field Day.* Queensland, Australia: University of Queensland, pp. 5/1–5/12.

Nogues-Bravo, D. 2009. Predicting the past distribution of species climatic niches. *Global Ecology and Biogeography*, 18, 521–531.

Nogues-Bravo, D., Rodiguez, J., Hortal, J., Batra, P. and Araujo, M. B. 2008. Climate change, humans, and the extinction of the woolly mammoth. *Plos Biology*, 6, 685–692.

Normand, S., Treier, U. A., Randin, C., Vittoz, P., Guisan, A. and Svenning, J. C. 2009. Importance of abiotic stress as a range-limit determinant for European plants: insights from species responses to climatic gradients. *Global Ecology and Biogeography*, 18, 437–449.

Normand, S., Ricklefs, R. E., Skov, F., Bladt, J., Tackenberg, O. and Svenning, J.-C. 2011. Postglacial migration supplements climate in determining plant species ranges in Europe. *Proceedings of the Royal Society of London B: Biological Sciences*, 278, 3644–3653.

Normand, S., Randin, C., Ohlemueller, R., et al. 2013. A greener Greenland? Climatic potential and long-term constraints on future expansions of trees and shrubs. *Philosophical Transactions of the Royal Society B-Biological Sciences*, 368.

O'Neill, R. V., DeAngelis, D. L., Waide, J. B. and Allen, T. F. H. 1986. *A hierarchical concept of ecosystems*, Princeton, NJ, USA, Princeton University Press.

Olden, J. D. 2003. A species-specific approach to modeling biological communities and its potential for conservation. *Conservation Biology*, 17, 854–863.

Olwoch, J. M., Rautenbach, C. J. D., Erasmus, B. F. N., Engelbrecht, F. A. and van Jaarsveld, A. S. 2003. Simulating tick distributions over sub-Saharan Africa: the use of observed and simulated climate surfaces. *Journal of Biogeography*, 30, 1221–1232.

Osborne, P. E. and Suarez-Seoane, S. 2002. Should data be partitioned spatially before building large-scale distribution models? *Ecological Modelling*, 157, 249–259.

Osborne, P. E., Foody, G. M. and Suárez-Seoane, S. 2007. Non-stationarity and local approaches to modelling the distributions of wildlife. *Diversity and Distributions*, 13, 313–323.

Ottaviani, D., Lasinio, G. J. and Boitani, L. 2004. Two statistical methods to validate habitat suitability models using presence-only data. *Ecological Modelling*, 179, 417–443.

Oulas, A., Pavloudi, C., Polymenakou, F., et al. 2015. Metagenomics: tools and insights for analyzing next-generation sequencing data derived from biodiversity studies. *Bioinformatics and Biology Insights*, 9, 75.

Ovaskainen, O. and Meerson, B. 2010. Stochastic models of population extinction. *Trends in Ecology and Evolution*, 25, 643–652.

Ovaskainen, O. and Soininen, J. 2011. Making more out of sparse data: hierarchical modeling of species communities. *Ecology*, 92, 289–295.

Pagel, J. and Schurr, F. M. 2012. Forecasting species ranges by statistical estimation of ecological niches and spatial population dynamics. *Global Ecology and Biogeography*, 21, 293–304.

Parviainen, M., Marmion, M., Luoto, M., Thuiller, W. and Heikkinen, R. K. 2009. Using summed individual species models and state-of-the-art modelling techniques to identify threatened plant species hotspots. *Biological Conservation*, 142, 2501–2509.

Patsiou, T. S., Conti, E., Zimmermann, N. E., Theodoridis, S. and Randin, C. F. 2014. Topo-climatic microrefugia explain the persistence of a rare endemic plant in the Alps during the last 21 millennia. *Global Change Biology*, 20, 2286–2300.

Pearce, J. and Ferrier, S. 2000. Evaluating the predictive performance of habitat models developed using logistic regression. *Ecological Modelling*, 133, 225–245.

Pearce, J. and Ferrier, S. 2001. The practical value of modelling relative abundance of species for regional conservation planning: a case study. *Biological Conservation*, 98, 33–43.

Pearce, J. and Lindenmayer, D. 1998. Bioclimatic analysis to enhance reintroduction biology of the endangered helmeted honeyeater (*Lichenostomus melanops cassidix*) in southeastern Australia. *Restoration Ecology*, 6, 238–243.

Pearce, J. L. and Boyce, M. S. 2006. Modelling distribution and abundance with presence-only data. *Journal of Applied Ecology*, 43, 405–412.

Pearlstine, L. G., Smith, S. E., Brandt, L. A., et al. 2002. Assessing state-wide biodiversity in the Florida Gap analysis project. *Journal of Environmental Management*, 66, 127–144.

Pearman, P. B., Guisan, A., Broennimann, O. and Randin, C. F. 2008a. Niche dynamics in space and time. *Trends in Ecology and Evolution*, 23, 149–158.

Pearman, P. B., Randin, C. F., Broennimann, O., et al. 2008b. Prediction of plant species distributions across six millennia. *Ecology Letters*, 11, 357–369.

Pearman, P. B., D'Amen, M., Graham, C. H., Thuiller, W. and Zimmermann, N. E. 2010. Within-taxon niche structure: niche conservatism, divergence and predicted effects of climate change. *Ecography*, 33, 990–1003.

Pearson, R. G. and Dawson, T. P. 2003. Predicting the impacts of climate change on the distribution of species: are bioclimate envelope models useful? *Global Ecology and Biogeography*, 12, 361–372.

Pearson, R. G., Dawson, T. P., Berry, P. M. and Harrison, P. A. 2002. SPECIES: a spatial evaluation of climate impact on the envelope of species. *Ecological Modelling*, 154, 289–300.

Pearson, R. G., Dawson, T. P. and Liu, C. 2004. Modelling species distributions in Britain: a hierarchical integration of climate and land-cover data. *Ecography*, 27, 285–298.

Pearson, R. G., Thuiller, W., Araujo, M. B., et al. 2006. Model-based uncertainty in species range prediction. *Journal of Biogeography*, 33, 1704–1711.

Pearson, R. G., Raxworthy, C. J., Nakamura, M. and Peterson, A. T. 2007. Predicting species distributions from small numbers of occurrence records: a test case using cryptic geckos in Madagascar. *Journal of Biogeography*, 34, 102–117.

Pellet, J., Guisan, A. and Perrin, N. 2004. A concentric analysis of the impact of urbanization on the threatened European tree frog in an agricultural landscape. *Conservation Biology*, 18, 1599–1606.

Pellissier, L., Brathen, K. A., Pottier, J., et al. 2010. Species distribution models reveal apparent competitive and facilitative effects of a dominant species on the distribution of tundra plants. *Ecography*, 33, 1004–1014.

Pellissier, L., Alvarez, N. and Guisan, A. 2012a. Pollinators as drivers of plant distribution and assemblage into communities. In Patiny, S. (ed.), *Evolution of Plant–Pollinator Interactions*. Cambridge: Cambridge University Press.

Pellissier, L., Fiedler, K., Ndribe, C., et al. 2012b. Shifts in species richness, herbivore specialization, and plant resistance along elevation gradients. *Ecology and Evolution*, 2, 1818–1825.

Pellissier, L., Bråthen, K. A.,Vittoz, P. A., et al. 2013a.Thermal niches are more conserved at cold than warm limits in arctic-alpine plant species. *Global Ecology and Biogeography*, 22, 933–941.

Pellissier, L., Meltofte, H., Hansen, J., et al. 2013b. Suitability, success and sinks: how do predictions of nesting distributions relate to fitness parameters in high arctic waders? *Diversity and Distributions*, 19, 1496–1505.

Pellissier, L., Pinto-Figueroa, E., Niculita-Hirzel, H., et al. 2013c. Plant species distributions along environmental gradients: do belowground interactions with fungi matter? *Frontiers in plant science*, 4, 1–9.

Pellissier, L., Rohr, R. P., Ndiribe, C., et al. 2013d. Combining food web and species distribution models for improved community projections. *Ecology and Evolution*, 3, 4572–4583.

Peltier, W. R. 2004. Global glacial isostasy and the surface of the ice-age earth: The ice-5G (VM2) model and grace. *Annual Review of Earth and Planetary Sciences*, 32, 111–149.

Peppler-Lisbach, C. and Schroder, B. 2004. Predicting the species composition of Nardus stricta communities by logistic regression modelling. *Journal of Vegetation Science*, 15, 623–634.

Peters, J., De Baets, B.,Verhoest, N. E. C., et al. 2007. Random forests as a tool for ecohydrological distribution modelling. *Ecological Modelling*, 207, 304–318.

Peters, R. H. 1991. *A Critique for Ecology*, Cambridge, UK: Cambridge University Press.

Peterson, A. T. 2003. Predicting the geography of species' invasions via ecological niche modeling. *Quarterly Review of Biology*, 78, 419–433.

Peterson, A. T. 2006. Ecologic niche modeling and spatial patterns of disease transmission. *Emerging Infectious Diseases*, 12, 1822–1826.

Peterson, A. T. 2011. Ecological niche conservatism: a time-structured review of evidence. *Journal of Biogeography*, 38, 817–827.

Peterson, A. T., Ortega-Huerta, M. A., Bartley, J., et al. 2002a. Future projections for Mexican faunas under global climatic change scenarios. *Nature*, 416, 626–629.

Peterson, A.T., Sanchez-Cordero,V., Ben Beard, C. and Ramsey, J. M. 2002b. Ecologic niche modeling and potential reservoirs for Chagas disease, Mexico. *Emerging Infectious Diseases*, 8, 662–667.

Peterson, A. T., Papes, M. and Soberon, J. 2008a. Rethinking receiver operating characteristic analysis applications in ecological niche modeling. *Ecological Modelling*, 213, 63–72.

Peterson, A. T., Stewart, A., Mohamed, K. I. and Araújo, M. B. 2008b. Shifting global invasive potential of European plants with climate change. *PLoS One*, 3, e2441.

Peterson, A.T., Soberon, J., Pearson, R. G., et al. 2011. *Ecological Niches and Geographic Distributions*, Princeton, NJ: Princeton University Press.

Petitpierre, B., Kueffer, C., Broennimann, O., et al. 2012. Climatic niche shifts are rare among terrestrial plant invaders. *Science*, 335, 1344–1348.

Petitpierre, B., McDougall, K., Seipel, T., et al. 2016. Will climate change increase the risk of plant invasions into mountains? *Ecological Applications*, 26, 530–544.

Petitpierre, B., Broennimann, O., Kueffer, C., Daehler, C. and Guisan, A. 2017. Selecting predictors to maximize the transferability of species distribution models: lessons from cross-continental plants invasions. *Global Ecology and Biogeography*, 26(3), 275–287.

Phillips, S. J. and Elith, J 2010. POC plots: calibrating species distribution models with presence-only data. *Ecology*, 91, 2476–2484.

Phillips, S. J. and Dudík, M. 2008. Modeling of species distributions with Maxent: new extensions and a comprehensive evaluation. *Ecography*, 31, 161–175.

Phillips, S. J., Dudík, M. and Schapire, R. E. 2004. A maximum entropy approach to species distribution modeling. In *Proceedings of the 21st International Conference on Machine Learning*. Association for Computing Machinery: Banff, Canada, p. 83.

Phillips, S. J., Anderson, R. P. and Schapire, R. E. 2006. Maximum entropy modeling of species geographic distributions. *Ecological Modelling*, 190, 231–259.

Phillips, S. J., Dudik, M., Elith, J., et al. 2009. Sample selection bias and presence-only distribution models: implications for background and pseudo-absence data. *Ecological Applications*, 19, 181–197.

Pickett, S. T. A. and Bazzaz, F. A. 1978. Organization of an assemblage of early successional species on a soil-moisture gradient. *Ecology*, 59, 1248–1255.

Pio, D. V., Engler, R., Linder, H. P., et al. 2014. Climate change effects on animal and plant phylogenetic diversity in southern Africa. *Global Change Biology*, 20, 1538–1549.

Polis, G. A., Anderson, W. B. and Holt, R. D. 1997. Toward an integration of landscape and food web ecology: the dynamics of spatially subsidized food webs. *Annual Review of Ecology and Systematics*, 28, 289–316.

Pollock, L. J., Morris, W. K. and Vesk, P. A. 2012. The role of functional traits in species distributions revealed through a hierarchical model. *Ecography*, 35, 716–725.

Pollock, L. J., Tingley, R., Morris, W. K., et al. 2014. Understanding co-occurrence by modelling species simultaneously with a joint species distribution model (JSDM). *Methods in Ecology and Evolution*, 5, 397–406.

Pongratz, J., Reick, C., Raddatz, T. and Claussen, M. 2008. A reconstruction of global agricultural areas and land cover for the last millennium. *Global Biogeochemical Cycles*, 22, GB3018, 1–16.

Pottier, J., Dubuis, A., Pellissier, L., et al. 2013. The accuracy of plant assemblage prediction from species distribution models varies along environmental gradients. *Global Ecology and Biogeography*, 22, 52–63.

Pottier, J., Malenovský, Z., Psomas, A., et al. 2014. Modelling plant species distribution in alpine grasslands using airborne imaging spectroscopy. *Biology Letters*, 10, 20140347.

Pradervand, J.-N., Dubuis, A., Pellissier, L., Guisan, A. and Randin, C. 2014. Very high resolution environmental predictors in species distribution models: moving beyond topography? *Progress in Physical Geography*, 38, 79–96.

Prasad, A. M., Iverson, L. R. and Liaw, A. 2006. Newer classification and regression tree techniques: bagging and random forests for ecological prediction. *Ecosystems*, 9, 181–199.

Pulliam, H. R. 2000. On the relationship between niche and distribution. *Ecology Letters*, 3, 349–361.

Quetier, F., Rivoal, F., Marty, P., et al. 2010. Social representations of an alpine grass-land landscape and socio-political discourses on rural development. *Regional Environmental Change*, 10, 119–130.

Quinlan, J. R. 1986. Induction of decision trees. *Machine Learning*, 1, 81–106.

Ramankutty, N. and Foley, J. A. 1999. Estimating historical changes in global land cover: croplands from 1700 to 1992. *Global Biogeochemical Cycles*, 13, 997–1027.

Randin, C. F., Dirnbock, T., Dullinger, S., et al. 2006. Are niche-based species distribution models transferable in space? *Journal of Biogeography*, 33, 1689–1703.

Randin, C. F., Engler, R., Normand, S., et al. 2009. Climate change and plant distribution: local models predict high-elevation persistence. *Global Change Biology*, 15, 1557–1569.

Randin, C. F., Paulsen, J., Vitasse, Y., et al. 2013. Do the elevational limits of deciduous tree species match their thermal latitudinal limits? *Global Ecology and Biogeography*, 22, 913–923.

Raxworthy, C. J., Martinez-Meyer, E., Horning, N., et al. 2003. Predicting distributions of known and unknown reptile species in Madagascar. *Nature*, 426, 837–841.

Regan, H. M., Hierl, L. A., Franklin, J., et al. 2008. Species prioritization for monitoring and management in regional multiple species conservation plans. *Diversity and Distributions*, 14, 462–471.

Reid, P. C., Lancelot, C., Gieskes, W. W. C., Hagmeier, E. and Weichart, G. 1990. Phytoplankton of the North-Sea and its dynamics: a review. *Netherlands Journal of Sea Research*, 26, 295–331.

Reineking, B. and Schroder, B. 2006. Constrain to perform: regularization of habitat models. *Ecological Modelling*, 193, 675–690.

Renner, I. W. and Warton, D. I. 2013. Equivalence of MAXENT and Poisson point process models for species distribution modeling in ecology. *Biometrics*, 69, 274–281.

Renner, I. W., Elith, J., Baddeley, A., et al. 2015. Point process models for presence-only analysis. *Methods in Ecology and Evolution*, 6, 366–379.

Richardson, D. M., Pysek, P., Rejmanek, M., et al. 2000. Naturalization and invasion of alien plants: concepts and definitions. *Diversity and Distributions*, 6, 93–107.

Ricklefs, R. E. 1987. Community diversity: relative roles of local and regional processes. *Science*, 235, 167–171.

Ricklefs, R. E. 2008. Disintegration of the ecological community. *The American Naturalist*, 172, 741–750.

Ridgeway, G. 1999. The state of boosting. *Computing Science and Statistics*, 31, 172–181.

Ripley, B. D. 1996. *Pattern Recognition and Neural Networks*. Cambridge: Cambridge University Press.

Robertson, M. P., Caithness, N. and Villet, M. H. 2001. A PCA-based modelling technique for predicting environmental suitability for organisms from presence records. *Diversity and Distributions*, 7, 15–27.

Robertson, M. P., Villet, M. H. and Palmer, A. R. 2004. A fuzzy classification technique for predicting species' distributions: applications using invasive alien plants and indigenous insects. *Diversity and Distributions*, 10.

Robinson, A. P., Lane, S. E. and Thérien G. 2011. Fitting forestry models using generalized additive models: a taper model example. *Canadian Journal of Forest Research*, 41, 1909–1916.

Rödder, D., Nekum, S., Cord, A. F. and Engler, J. O. 2016. Coupling satellite data with species distribution and connectivity models as a tool for environmental management and planning in matrix-sensitive species. *Environmental Management*, 1–14.

Rodríguez, J. P., Brotons, L., Bustamante, J. and Seoane, J. 2007. The application of predictive modelling of species distribution to biodiversity conservation. *Diversity and Distributions*, 13, 243–251.

Root, T. L., Price, J. T., Hall, K. R., et al. 2003. Fingerprints of global warming on wild animals and plants. *Nature*, 421, 57–60.

Roura-Pascual, N., Suarez, A. V., Gomez, C., et al. 2004. Geographical potential of Argentine ants (*Linepithema humile* Mayr) in the face of global climate change. *Proceedings of the Royal Society B: Biological Sciences*, 271, 2527–2534.

Roura-Pascual, N., Suarez, A. V., McNyset, K. M., et al. 2006. Niche differentiation and fine-scale projections for Argentine ants based on remotely sensed data. *Ecological Applications*, 16, 1832–1841.

Rouse J. W., Haas R. H., Schell J. A. and Deering D. W. 1974. Monitoring vegetation systems in the Great Plains with ERTS. In Fraden S. C., Marcanti E. P. and Becker M. A. (eds), *Third ERTS-1 Symposium, 10–14 Dec. 1973, NASA SP-351*, Washington DC: NASA, pp. 309–317.

Rykiel, E. J. J. 1996. Testing ecological models: the meaning of validation. *Ecological Modelling*, 90, 229–244.

Sala, O. E., Chapin, F. S., Armesto, J. J., et al. 2000. Global biodiversity scenarios for the year 2100. *Science*, 287, 1770–1774.

Salisbury, E. J. 1929. The biological equipment of species in relation to competition. *Journal of Ecology*, 17, 197–222.

Saltre, F., Saint-Amant, R., Gritti, E. S., et al. 2013. Climate or migration: what limited European beech post-glacial colonization? *Global Ecology and Biogeography*, 22, 1217–1227.

Sandel, B., Arge, L., Dalsgaard, B., et al. 2011. The Influence of Late Quaternary climate-change velocity on species endemism. *Science*, 334, 660–664.

Sargent, R. D. and Ackerly, D. D. 2008. Plant–pollinator interactions and the assembly of plant communities. *Trends in Ecology and Evolution*, 23, 123–130.

Scherrer, D., Massy, S., Meier, S., Vittoz, P. and Guisan, A. 2017. Assessing and predicting shifts in mountain forest composition across 25 years of climate change. *Diversity and Distributions*, 23, 517–528.

Schlossberg, S. and King, D. I. 2009. Modeling animal habitats based on cover types: a critical review. *Environmental Management*, 43, 609–618.

Schneider, D. C. 2001. The rise of the concept of scale in ecology. *Bioscience*, 51, 545–553.

Schoener, T. W. 1989. Food webs from the small to the large. *Ecology*, 70, 1559–1589.

Schorr, G., Holstein, N., Pearman, P. B., Guisan, A. and Kadereit, J. W. 2012. Integrating species distribution models (SDMs) and phylogeography for two species of Alpine primula. *Ecology and Evolution*, 2, 1260–1277.

Schorr, G., Pearman, P. B., Guisan, A. and Kadereit, J. W. 2013. Combining palaeodistribution modelling and phylogeographical approaches for identifying glacial refugia in Alpine primula. *Journal of Biogeography*, 40, 1947–1960.

Schurr, F. M., Pagel, J., Cabral, J. S., et al. 2012. How to understand species' niches and range dynamics: a demographic research agenda for biogeography. *Journal of Biogeography*, 39, 2146–2162.

Schwartz, M. W. 2012. Using niche models with climate projections to inform conservation management decisions. *Biological Conservation*, 155, 149–156.

Schwarz, M. and Zimmermann, N. E. 2005. A new GLM-based method for mapping tree cover continuous fields using regional MODIS reflectance data. *Remote Sensing of Environment*, 95, 428–443.

Scott, J. M., Davis, F., Csuti, B., et al. 1993. Gap analysis: a geographic approach to protection of biological diversity. *Wildlife Monographs*, 123, 1–41.

Scott, J. M., Davis, F. W., McGhie, R. G., et al. 2001. Nature reserves: do they capture the full range of America's biological diversity? *Ecological Applications*, 11, 999–1007.

Scott, J. M., Heglund, P. J., Haufler, J. B., et al. (eds) 2002. *Predicting Species Occurrences: Issues of Accuracy and Scale*, Covelo, CA: Island Press.

Segurado, P. and Araujo, M. B. 2004. An evaluation of methods for modelling species distributions. *Journal of Biogeography*, 31, 1555–1568.

Segurado, P., Araujo, M. B. and Kunin, W. E. 2006. Consequences of spatial autocorrelation for niche-based models. *Journal of Applied Ecology*, 43, 433–444.

Seo, C., Thorne, J. H., Hannah, L. and Thuiller, W. 2009. Scale effects in species distribution models: implications for conservation planning under climate change. *Biology Letters*, 5, 39–43.

Serra-Diaz, J. M., Franklin, J., Ninyerola, M., et al. 2014. Bioclimatic velocity: the pace of species exposure to climate change. *Diversity and Distributions*, 20, 169–180.

Serra-Varela, M., Grivet, D., Vincenot, L., Broennimann, O., et al. 2015. Does phylogeographical structure relate to climatic niche divergence? A test using maritime pine (*Pinus pinaster* Ait.). *Global Ecology and Biogeography*, 24, 1302–1313.

Sexton, J. O., Song, X.-P., Feng, M., et al. 2013. Global, 30-m resolution continuous fields of tree cover: Landsat-based rescaling of MODIS vegetation continuous fields with lidar-based estimates of error. *International Journal of Digital Earth*, 6, 427–448.

Shipley, B., Vile, D. and Garnier, E. 2006. From plant traits to plant communities: a statistical mechanistic approach to biodiversity. *Science*, 314, 812–814.

Shipley, B., Laughlin, D. C., Sonnier, G. and Otfinowski, R. 2011. A strong test of a maximum entropy model of trait-based community assembly. *Ecology*, 92, 507–517.

Silvertown, J. 2004. Plant coexistence and the niche. *Trends in Ecology and Evolution*, 19, 605–611.

Silvertown, J., Dodd, M., Gowing, D., Lawson, C. and McConway, K. 2006. Phylogeny and the hierarchical organization of plant diversity. *Ecology*, 87, S39–S49.

Simmons, R. E., Barnard, P., Dean, W. R. J., et al. 2004. Climate change and birds: perspectives and prospects from southern Africa. *Ostrich*, 75, 295–308.

Smith, T. M. and Smith, R. L. 2015. *Elements of Ecology*, 9th edn. San Francisco, CA: Pearson Education Ltd.

Snell, R. S., Huth, A., Nabel, J. E. M. S., et al. 2014. Using dynamic vegetation models to simulate plant range shifts. *Ecography*, 37, 1184–1197.

Soberón, J. 2007. Grinnellian and Eltonian niches and geographic distributions of species. *Ecology Letters*, 10, 1115–1123.

Soberón, J. and Nakamura, M. 2009. Niches and distributional areas: concepts, methods, and assumptions. *Proceedings of the National Academy of Sciences*, 106, 19644–19650.

Sokol, E. R., Benfield, E., Belden, L. K. and Valett, H. M. 2011. The assembly of ecological communities inferred from taxonomic and functional composition. *The American Naturalist*, 177, 630–644.

Stachowicz, J. J. 2001. Mutualism, facilitation, and the structure of ecological communities. *Bioscience*, 51, 235–246.

Stockwell, D. 1999. The GARP modelling system: problems and solutions to automated spatial prediction. *International Journal of Geographical Information Science*, 13, 143–158.

Stockwell, D. R. B. and Peterson, A. T. 2002a. Controlling bias in biodiversity data. In Scott, J. M., Heglund, P. J., Morrison, M. L., Haufler, J. B., Raphael, M. G., Wall, W. A. and Samson, F. B. (eds), *Predicting Species Occurrences: Issues of Accuracy and Scale*. Covelo, CA: Island Press, pp. 537–546.

Stockwell, D. R. B. and Peterson, A. T. 2002b. Effects of sample size on accuracy of species distribution models. *Ecological Modelling*, 148, 1–13.

Strobl, C., Boulesteix, A. L., Kneib, T., Augustin, T. and Zeileis, A. 2008. Conditional variable importance for random forests. *BMC Bioinformatics*, 9(1), 307.

Strobl, C., Malley, J. and Tutz, G. 2009. An introduction to recursive partitioning: rationale, application, and characteristics of classification and regression trees, bagging, and random forests. *Psychological Methods*, 14, 323.

Suarez-Seoane, S., Virgos, E., Terroba, O., Pardavila, X. and Barea-Azcon, J. M. 2014. Scaling of species distribution models across spatial resolutions and extents along a biogeographic gradient. The case of the Iberian mole *Talpa occidentalis*. *Ecography*, 37, 279–292.

Sutherst, R. W., Maywald, G. F. and Bourne, A. S. 2007. Including species interactions in risk assessments for global change. *Global Change Biology*, 13, 1843–1859.

Svenning, J. C. and Skov, F. 2004. Limited filling of the potential range in European tree species. *Ecology Letters*, 7, 565–573.

Svenning, J. C. and Skov, F. 2007. Could the tree diversity pattern in Europe be generated by postglacial dispersal limitation? *Ecology Letters*, 10, 453–460.

Svenning, J. C., Normand, S. and Skov, F. 2006. Range filling in European trees. *Journal of Biogeography*, 33, 2018–2021.

Svenning, J. C., Normand, S. and Kageyama, M. 2008. Glacial refugia of temperate trees in Europe: insights from species distribution modelling. *Journal of Ecology*, 96, 1117–1127.

Svenning, J. C., Flojgaard, C., Marske, K. A., Nogues-Bravo, D. and Normand, S. 2011. Applications of species distribution modeling to paleobiology. *Quaternary Science Reviews*, 30, 2930–2947.

Swets, J. A. 1988. Measuring the accuracy of diagnostic systems. *Science*, 240, 1285–1293.

Syfert, M. M., Smith, M. J. and Coomes, D. A. 2013. The effects of sampling bias and model complexity on the predictive performance of MaxEnt species distribution models. *Plos One*, 8, e55158.

Talluto, M., Boulangeat, I., Ameztegui, A., et al. 2016. Cross-scale integration of knowledge for predicting species ranges: a metamodeling framework. *Global Ecology and Biogeography*, 25, 238–249.

Tang, L. and Shao, G. 2015. Drone remote sensing for forestry research and practices. *Journal of Forestry Research*, 26, 791–797.

Tautz, D. 2003. Evolutionary biology: splitting in space. *Nature*, 421, 225–226.

Team, R. C. 2014. R: A language and environment for statistical computing. In *R Foundation For Statistical Computing, V.* www.R-project.org/

Tebaldi, C. and Knutti, R. 2007. The use of the multi-model ensemble in probabilistic climate projections. *Philosophical Transactions of the Royal Society A: Mathematical Physical and Engineering Sciences*, 365, 2053–2075.

ter Braak, C. J. F. 1986. Canonical correspondence analysis: a new eigenvector technique for multivariate direct gradient analysis. *Ecology*, 67, 1167–1179.

Thibaud, E., Petitpierre, B., Broennimann, O., Davison, A. C. and Guisan, A. 2014. Measuring the relative effect of factors affecting species distribution model predictions. *Methods in Ecology and Evolution*, 5(9), 947–955.

Thomas, C. D., Cameron, A., Green, R. E., et al. 2004. Extinction risk from climate change. *Nature*, 427, 145–148.

Thornton, P. E. and Running, S. W. 1999. An improved algorithm for estimating incident daily solar radiation from measurements of temperature, humidity, and precipitation. *Agricultural and Forest Meteorology*, 93, 211–228.

Thornton, P. E., Running, S. W. and White, M. A. 1997. Generating surfaces of daily meteorological variables over large regions of complex terrain. *Journal of Hydrology*, 190, 214–251.

Thornton, P. E., Thornton, M. M., Mayer, B. W., et al. 2017. Daymet: daily surface weather data on a 1-km grid for North America. Version 3. Oak Ridge, TN: ORNL DAAC. Available at: https://doi.org/10.3334/ORNLDAAC/1328.

Thuiller, W. 2003. BIOMOD: optimizing predictions of species distributions and projecting potential future shifts under global change. *Global Change Biology*, 9, 1353–1362.

Thuiller, W. 2004. Patterns and uncertainties of species' range shifts under climate change. *Global Change Biology*, 10, 2020–2027.

Thuiller, W. 2007. Biodiversity: climate change and the ecologist. *Nature*, 448, 550–552.

Thuiller, W., Araujo, M. B. and Lavorel, S. 2003a. Generalized models vs. classification tree analysis: predicting spatial distributions of plant species at different scales. *Journal of Vegetation Science*, 14, 669–680.

Thuiller, W., Vayreda, J., Pino, J., et al. 2003b. Large-scale environmental correlates of forest tree distributions in Catalonia (NE Spain). *Global Ecology and Biogeography*, 12, 313–325.

Thuiller, W., Araujo, M. B., Pearson, R. G., et al. 2004a. Uncertainty in predictions of extinction risk. *Nature*, 430, 34.

Thuiller, W., Brotons, L., Araújo, M. B. and Lavorel, S. 2004b. Effects of restricting environmental range of data to project current and future species distributions. *Ecography*, 27, 165–172.

Thuiller, W., Lavorel, S., Midgley, G., Lavergne, S. and Rebelo, T. 2004c. Relating plant traits and species distributions along bioclimatic gradients for 88 Leucadendron taxa. *Ecology*, 85, 1688–1699.

Thuiller, W., Lavorel, S., Araujo, M. B., Sykes, M. T. and Prentice, I. C. 2005a. Climate change threats to plant diversity in Europe. *Proceedings of the National Academy of Sciences of the United States of America*, 102, 8245–8250.

Thuiller, W., Richardson, D. M., Pysek, P., et al. 2005b. Niche-based modelling as a tool for predicting the risk of alien plant invasions at a global scale. *Global Change Biology*, 11, 2234–2250.

Thuiller, W., Midgley, G. F., Rouget, M. and Cowling, R. M. 2006. Predicting patterns of plant species richness in megadiverse South Africa. *Ecography*, 29, 733–744.

Thuiller, W., Albert, C., Araujo, M. B., et al. 2008. Predicting global change impacts on plant species' distributions: future challenges. *Perspectives in Plant Ecology Evolution and Systematics*, 9, 137–152.

Thuiller, W., Lafourcade, B., Engler, R. and Araujo, M. B. 2009. BIOMOD: a platform for ensemble forecasting of species distributions. *Ecography*, 32, 369–373.

Thuiller, W., Albert, C. H., Dubuis, A., Randin, C. and Guisan, A. 2010. Variation in habitat suitability does not always relate to variation in species' plant functional traits. *Biology Letters*, rsbl20090669.

Thuiller, W., Lavergne, S., Roquet, C., et al. 2011. Consequences of climate change on the tree of life in Europe. *Nature*, 470, 531–534.

Thuiller, W., Münkemüller, T., Lavergne, S., et al. 2013. A road map for integrating eco-evolutionary processes into biodiversity models. *Ecology Letters*, 16, 94–105.

Thuiller, W., Guéguen, M., Georges, D., et al. 2014a. Are different facets of plant diversity well protected against climate and land cover changes? A test study in the French Alps. *Ecography*, 37, 1254–1266.

Thuiller, W., Münkemüller, T., Schiffers, K. H., et al. 2014b. Does probability of occurrence relate to population dynamics? *Ecography*, 37, 1155–1166.

Thuiller, W., Pironon, S., Psomas, A., et al. 2014c. The European functional tree of bird life in the face of global change. *Nature Communications*, 5.

Thuiller, W., Pollock, L. J., Gueguen, M. and Münkemüller, T. 2015. From species distributions to meta-communities. *Ecology Letters* 18(12), 1321–1328.

Tibshirani, R. 1996. Regression shrinkage and selection via the LASSO. *Journal of the Royal Statistical Society Series B: Methodological*, 58, 267–288.

Tibshirani, R. 1997. The LASSO method for variable selection in the Cox model. *Statistics in Medicine*, 16, 385–395.

Tingley, M. W., Monahan, W. B., Beissinger, S. R. and Moritz, C. 2009. Birds track their Grinnellian niche through a century of climate change. *Proceedings of the National Academy of Sciences of the United States of America*, 106, 19637–19643.

Tsoar, A., Allouche, O., Steinitz, O., Rotem, D. and Kadmon, R. 2007. A comparative evaluation of presence-only methods for modelling species distribution. *Diversity and Distributions*, 13, 397–405.

Tuanmu, M. N., Vina, A., Roloff, G. J., et al. 2011. Temporal transferability of wildlife habitat models: implications for habitat monitoring. *Journal of Biogeography*, 38, 1510–1523.

Tulloch, A. I., Sutcliffe, P., Naujokaitis-Lewis, I., et al. 2016. Conservation planners tend to ignore improved accuracy of modelled species distributions to focus on multiple threats and ecological processes. *Biological Conservation*, 199, 157–171.

Vale, C. G., Tarroso, P. and Brito, J. C. 2014. Predicting species distribution at range margins: testing the effects of study area extent, resolution and threshold selection in the Sahara-Sahel transition zone. *Diversity and Distributions*, 20, 20–33.

Van Horne, B. 2002. Approaches to Habitat Modeling: The Tensions between Pattern and Process and between Specificity and Generality. In Scott, J. M., Heglund, P. J., Haufler, J. B., et al. (eds), *Predicting Species Occurrences: Issues of Accuracy and Scale*. Covelo, CA: Island Press.

Van Niel, K. P. and Austin, M. P. 2007. Predictive vegetation modeling for conservation: Impact of error propagation from digital elevation data. *Ecological Applications*, 17, 266–280.

Vandermeer, J. H. 1972. Niche theory. *Annual Review of Ecology and Systematics*, 3, 107–132.

VanDerWal, J., Shoo, L. P., Graham, C. and William, S. E. 2009. Selecting pseudo-absence data for presence-only distribution modeling: how far should you stray from what you know? *Ecological Modelling*, 220, 589–594.

Vaughan, I. P. and Ormerod, S. J. 2005. The continuing challenges of testing species distribution models. *Journal of Applied Ecology*, 42, 720–730.

Veloz, S. D. 2009. Spatially autocorrelated sampling falsely inflates measures of accuracy for presence-only niche models. *Journal of Biogeography*, 36, 2290–2299.

Venables, W. N. and Ripley, B. D. 2002. *Modern Applied Statistisc with S*. Dordrecht, The Netherlands: Springer.

Verbyla, D. L. and Litvaitis, J. A. 1989. Resampling methods for evaluating classification accuracy of wildlife habitat models. *Environmental Management*, 13, 783–787.

Verner, J., Morrison, M. L. and Ralph, C. J. 1986. *Wildlife 2000: Modelling Habitat Relationships of Terrestrial Vertebrates*, Madison, WI: University of Wisconsin Press.

Vesk, P. A. 2013. How traits determine species responses to environmental gradients. *Journal of Vegetation Science*, 24, 977–978.

Vetaas, O. R. 2002. Realized and potential climate niches: a comparison of four Rhododendron tree species. *Journal of Biogeography*, 29, 545–554.

Vicente, J., Fernandes, R., Randin, C., et al. 2013. Will climate change drive alien invasive plants into areas of high protection value? An improved model-based regional assessment to prioritise the management of invasions. *Journal of Environmental Management*, 131, 185–195.

Vincent, P. J. and Haworth, J. M. 1983. Poisson regression models of species abundance. *Journal of Biogeography*, 10, 153–160.

Vittoz, P. and Engler, R. 2007. Seed dispersal distances: a typology based on dispersal modes and plant traits. *Botanica Helvetica*, 117, 109–124.

Walther, G. R., Berger, S. and Sykes, M. T. 2005. An ecological 'footprint' of climate change. *Proceedings of the Royal Society B: Biological Sciences*, 272, 1427–1432.

Ward, G., Hastie, T., Barry, S., Elith, J. and Leathwick, J. R. 2009. Presence-only data and the EM algorithm. *Biometrics*, 65, 554–563.

Warren, D. L., Glor, R. E. and Turelli, M. 2008. Environmental niche equivalency versus conservatism: quantitative approaches to niche evolution. *Evolution*, 62, 2868–2883.

Warton, D. I., Blanchet, F. G., O'Hara, et al. 2015. So many variables: joint modeling in community ecology. *Trends in Ecology and Evolution*, 30, 766–779.

Weiher, E. and Keddy, P. 2001. *Ecological Assembly Rules: Perspectives, Advances, Retreats*, Cambridge, UK: Cambridge University Press.

Weisberg, S. 1980. *Applied Linear Regression*, New York, NY: Wiley.

Wenger, S. J. and Olden, J. D. 2012. Assessing transferability of ecological models: an underappreciated aspect of statistical validation. *Methods in Ecology and Evolution*, 3, 260–267.

Wenger, S. J., Som, N. A., Dauwalter, D. C., et al. 2013. Probabilistic accounting of uncertainty in forecasts of species distributions under climate change. *Global Change Biology*, 19, 3343–3354.

Westman, W. E. 1991. Measuring realized niche spaces: climatic response of chaparral and coastal sage scrub. *Ecology*, 72, 1678–1684.

Wharton, T. N. and Kriticos, D. J. 2004. The fundamental and realized niche of the Monterey pine aphid, *Essigella californica* (Essig) (Hemiptera: Aphididae): implications for managing softwood plantations in Australia. *Diversity and Distributions*, 10, 253–262.

Whittaker, R. H. 1967. Gradient analysis of vegetation. *Biological Reviews*, 42, 207–264.

Whittaker, R. H., Levin, S. A. and Root, R. B. 1973. Niche, habitat, and ecotope. *American Naturalist*, 107, 321–338.

Whittingham, M. J., Stephens, P. A., Bradbury, R. B. and Freckleton, R. P. 2006. Why do we still use stepwise modelling in ecology and behaviour? *Journal of Animal Ecology*, 75, 1182–1189.

Wiens, J. A. 1989. Spatial scaling in ecology. *Functional Ecology*, 3, 385–397.

Wiens, J. J. and Graham, C. H. 2005. Niche conservatism: Integrating evolution, ecology, and conservation biology. *Annual Review of Ecology Evolution and Systematics*, 36, 519–539.

Wiens, T. S., Dale, B. C., Boyce, M. S. and Kershaw, G. P. 2008. Three way *k*-fold cross-validation of resource selection functions. *Ecological Modelling*, 212, 244–255.

Williams, J. W. and Jackson, S. T. 2007. Novel climates, no-analog communities, and ecological surprises. *Frontiers in Ecology and the Environment*, 5, 475–482.

Williams, J. W., Jackson, S. T. and Kutzbacht, J. E. 2007. Projected distributions of novel and disappearing climates by 2100 AD. *Proceedings of the National Academy of Sciences of the United States of America*, 104, 5738–5742.

Williams, R. A. J., Fasina, F. O. and Peterson, A. T. 2008. Predictable ecology and geography of avian influenza (H5N1) transmission in Nigeria and West Africa. *Transactions of the Royal Society of Tropical Medicine and Hygiene*, 102, 471–479.

Willis, K. J. and McElwain, J. C. 2002. *The Evolution of Plants*. Oxford, UK: Oxford University Press.

Willis, K. J. and Whittaker, R. J. 2002. Ecology. Species diversity, Scale matters. *Science*, 295, 1245–1248.

Wintle, B. A., McCarthy, M. A., Volinsky, C. T. and Kavanagh, R. P. 2003. The use of Bayesian model averaging to better represent uncertainty in ecological models. *Conservation Biology*, 17, 1579–1590.

Wintle, B. A., Elith, J. and Potts, J. M. 2005. Fauna habitat modelling and mapping: a review and case study in the Lower Hunter Central Coast region of NSW. *Austral Ecology*, 30, 719–738.

Wisz, M. S. and Guisan, A. 2009. Do pseudo-absence selection strategies influence species distribution models and their predictions? An information-theoretic approach based on simulated data. *BMC Ecology*, 9, 8.

Wisz, M. S., Hijmans, R. J., Li, J., et al. 2008. Effects of sample size on the performance of species distribution models. *Diversity and Distributions*, 14, 763–773.

Wisz, M. S., Pottier, J., Kissling, W. D., et al. 2013. The role of biotic interactions in shaping distributions and realised assemblages of species: implications for species distribution modelling. *Biologica. Reviews*, 88, 15–30.

Wisz, M. S., Broennimann, O., Grønkjær, P., et al. 2015. Arctic warming will promote Atlantic–Pacific fish interchange. *Nature Climate Change*, 5, 261–265.

Wolmarans, R., Robertson, M. P. and van Rensburg, B. J. 2010. Predicting invasive alien plant distributions: how geographical bias in occurrence records influences model performance. *Journal of Biogeography*, 37, 1797–1810.

Wood, S. 2006. *Generalized Additive Models: An Introduction with R.* London: CRC Press.

Wood, S. N., Goude, Y. and Shaw, S. 2015. Generalized additive models for large data sets. *Journal of the Royal Statistical Society: Series C (Applied Statistics)*, 64, 139–155.

Woodward, F. I. 1987. *Climate and Plant Distribution*, Cambridge, UK: Cambridge University Press.

Woodward, F. I. and Kelly, C. K. 2003. Why are species not more widely distributed? Physiological and environmental limits. In Blackburn, T. M. and Gaston, K. J. (eds), *Macroecology*. Oxford, UK: Blackwell.

Wu, J. 1999. Hierarchy and scaling: extrapolating information along a scaling ladder. *Canadian Journal of Remote Sensing*, 25, 367–380.

Wu, J. G. 2004. Effects of changing scale on landscape pattern analysis: scaling relations. *Landscape Ecology*, 19, 125–138

Yañez-Arenas, C., Peterson, A. T., Mokondoko, P., Rojas-Soto, O. and Martínez-Meyer, E. 2014. The use of ecological niche modeling to infer potential risk areas of snakebite in the Mexican state of Veracruz. *PLoS One*, 9, e100957.

Yates, C. J., McNeill, A., Elith, J. and Midgley, G. F. 2010. Assessing the impacts of climate change and land transformation on *Banksia* in the South West Australian Floristic Region. *Diversity and Distributions*, 16, 187–201.

Zanini, F., Pellet, J. and Schmidt, B. R. 2009. The transferability of distribution models across regions: an amphibian case study. *Diversity and Distributions*, 15, 469–480.

Zhang, J., Hu, J., Lian, J., et al. 2016. Seeing the forest from drones: testing the potential of lightweight drones as a tool for long-term forest monitoring. *Biological Conservation*, 198, 60–69.

Zhu, G. and Peterson, A. T. 2014. Potential geographic distribution of the novel avian-origin influenza A (H7N9) Virus. *PLoS One*, 9, e93390.

Zimmermann, N. E., Yoccoz, N. G., Edwards, T. C., et al. 2009. Climatic extremes improve predictions of spatial patterns of tree species. *Proceedings of the National Academy of Sciences of the United States of America*, 106, 19723–19728.

Zimmermann, N. E., Edwards, T. C., Graham, C. H., Pearman, P. B. and Svenning, J. C. 2010. New trends in species distribution modelling. *Ecography*, 33, 985–989.

Zobel, M. 1997. The relative role of species pools in determining plant species richness: an alternative explanation of species coexistence. *Trends in Ecology and Evolution*, 12, 266–269.

Zurell, D., Berger, U., Cabral, J. S., et al. 2010. The virtual ecologist approach: simulating data and observers. *Oikos*, 119, 622–635.

Zurell, D., Elith, J. and Schröder, B. 2012. Predicting to new environments: tools for visualizing model behaviour and impacts on mapped distributions. *Diversity and Distributions*, 18, 628–634.

Index